Lecture Notes in Physics

The Lecture Notes in Physics

The series Lecture Notes in Physics (LNP), founded in 1969, reports new developments in physics research and teaching – quickly and informally, but with a high quality and the explicit aim to summarize and communicate current knowledge in an accessible way. Books published in this series are conceived as bridging material between advanced graduate textbooks and the forefront of research to serve the following purposes:

• to be a compact and modern up-to-date source of reference on a well-defined topic;

• to serve as an accessible introduction to the field to postgraduate students and nonspecialist researchers from related areas;

• to be a source of advanced teaching material for specialized seminars, courses and schools.

Both monographs and multi-author volumes will be considered for publication. Edited volumes should, however, consist of a very limited number of contributions only. Proceedings will not be considered for LNP.

Volumes published in LNP are disseminated both in print and in electronic formats, the electronic archive is available at springerlink.com. The series content is indexed, abstracted and referenced by many abstracting and information services, bibliographic networks, subscription agencies, library networks, and consortia.

Proposals should be sent to a member of the Editorial Board, or directly to the managing editor at Springer:

Dr. Christian Caron
Springer Heidelberg
Physics Editorial Department I
Tiergartenstrasse 17
69121 Heidelberg/Germany
christian.caron@springer.com

Eric Beaurepaire   Hervé Bulou
Fabrice Scheurer   Jean-Paul Kappler   (Eds.)

Magnetism: A Synchrotron Radiation Approach

 Springer

Editors

Eric Beaurepaire
Hervé Bulou
Fabrice Scheurer
Jean-Paul Kappler
CNRS
IPCMS
23, rue du Loess BP 43
67034 Strasbourg Cedex 2, France
Email: Eric.Beaurepaire@ipcms.u-strasbg.fr
        Fabrice.Scheurer@ipcms.u-strasbg.fr
        Herve.Bulou@ipcms.u-strasbg.fr
        Jean-Paul.Kappler@ipcms.u-strasbg.fr

E. Beaurepaire et al., *Magnetism: A Synchrotron Radiation Approach*, Lect. Notes Phys.
697 (Springer, Berlin Heidelberg 2006), DOI 10.1007/b11594864

Library of Congress Control Number: 2006923355

ISSN 0075-8450
ISBN-10 3-540-33241-3 Springer Berlin Heidelberg New York
ISBN-13 978-3-540-33241-1 Springer Berlin Heidelberg New York

Springer is a part of Springer Science+Business Media
springer.com
© Springer-Verlag Berlin Heidelberg 2006
Printed in The Netherlands

Typesetting: by the authors and techbooks using a Springer LATEX macro package
Cover design: *design & production* GmbH, Heidelberg

Printed on acid-free paper   SPIN: 11594864   54/techbooks   5 4 3 2 1 0

Preface

This volume contains the lecture notes of the fourth school on Magnetism and Synchrotron Radiation,[1] held in Mittelwihr, France, from 10 to 15 October 2004. The school was meant to introduce basic knowledge on magnetism and synchrotron radiation, and to present the state-of-the-art in the important domains using synchrotron radiation-related techniques for the analysis of magnetic properties of new materials. This edition introduces the magnetism community towards the european synchrotron radiation centers.

In this book the reader will find:

(i) three introductory chapters on the basics of magnetism, on electron spectroscopies in a large sense, and on synchrotron radiation;
(ii) four chapters on spectroscopic, diffraction, and imaging techniques for magnetic studies;
(iii) three chapters on the theory and recent developments of magnetic dichroism;
(iv) five chapters on specialized topics (spin-dynamics, magnetism under pressure, and magnetism in mineralogy) and new magnetic materials (molecular magnets, and material for spin-electronics).

We would like to express our thanks to the teachers who agreed to animate this school, and took on the difficult job of writing the lectures. We also acknowledge the kind hospitality of the Centre de Mittelwihr and the Communauté des Communes de Ribeauvillé. This school would not have been possible without the support of many people, in particular the support from Institut de Physique et Chimie des Matériaux de Strasbourg, from Laboratoire de Spectroscopie Electronique, Mulhouse, from the local organization and scientific committees, and from the office of the *Formation Permanente de la Délégation Alsace du CNRS*. It would not have been possible without financial support from Formation Permanente du CNRS; Université de Haute Alsace; Université Louis Pasteur; Institut de Physique et Chimie des Matériaux de Strasbourg; Conseil Général du Haut-Rhin; Collège Doctoral

---

[1] The edition of the lecture notes of the previous Mittelwihr schools (1996, 2000) on *Magnetism and Synchrotron Radiation* are available at Les Editions de Physique (1997) and Springer Verlag (2001), respectively.

Européen des Universités de Strasbourg; Crédit Mutuel; France Telecom; MECA 2000; M.K.S. Instruments; Sairem and Varian.

This edition of the book benefited from the expert help of R. Allenspach, M. Alouani, Ch. Brouder, J.-L. Gallani, F. Gautier, C. Hague, J.-M. Mariot, V. Pierron-Bohnes, W. Weber, and D. Weinmann.

Strasbourg   *E. Beaurepaire*
January 2006   *H. Bulou*
                                                                              F. Scheurer
                                                                            J.-P. Kappler

Contents

Magnetism under Pressure with Synchrotron Radiation

Matteo d'Astuto, Alessandro Barla, Nolwell Kernavanois, Jean-Pascal Rueff, Francois Baudelet, Rudolf Rüffer, Luigi Paolasini, and Bernard Couzinet

X-ray Spectroscopy and Magnetism in Mineralogy

Philippe Sainctavit, Sandrine Brice-Profeta, Emilie Gaudry, Isabelle Letard, and Marie-Anne Arrio

Materials for Spintronics

Agnès Barthélémy and Richard Mattana

List of Contributors

Massimo Altarelli
European XFEL Project Team
DESY
Notkestr. 85
22607 Hamburg, Germany
and
Abdus Salam International Centre
for Theoretical Physics
Strada Costiera 11
34100 Trieste, Italy
altarell@ictp.it

Christian. H. Back
Institut für Experimentelle
und Angewandte Physik
Universität Regensburg
Universitätsstr. 31
93040 Regensburg, Germany
back@physik.uni-regensburg.de

Agnès Barthélémy
Unité Mixte de
Physique CNRS-Thales
Domaine de Corbeville
91404 Orsay, France
and
Université Paris Paris-Sud
91405 Orsay Cedex, France
barthelemy@thalesgroup.com

Stephen J. Blundell
Oxford University
Department of Physics
Clarendon Laboratory, Parks Road
Oxford OX1 3PU
United Kingdom
s.blundell@physics.ox.ac.uk

Matteo d'Astuto
Institut de Minéralogie et de
Physique des Milieux Condensés –
CNRS URM7590 Université
Pierre et Marie Curie, B 77
4 Place Jussieu
75252 Paris, France
dastuto@ccr.jussieu.fr

Catherine Dufour
Laboratoire Physique des
Matériaux
Université de Nancy 1
BP 239, 54506 Vandoeuvre
France
dufour@lpm.u-nancy.fr

Mikael Eriksson
MAXlab, Lund University
Box 118, 221 00 Lund, Sweden
Mikael.Eriksson@maxlab.lu.se

Wolfgang Kuch
Freie Universität
Berlin Institut für Experimental-
physik
Arnimallee 14
D-14195 Berlin, Germany
kuch@physik.fu-berlin.de

Gerrit van der Laan
Daresbury Laboratory
Warrington WA4 4AD, UK
g.vanderlaan@dl.ac.uk

Claudine Lacroix
Laboratoire Louis Néel
CNRS BP166
38042 Grenoble, France
lacroix@grenoble.cnrs.fr

Jürg Osterwalder
Physik-Institut
Universität Zürich
Winterthurerstr. 190
CH-8057 Zürich, Switzerland
osterwal@physik.unizh.ch

Andrei Rogalev
ESRF
6 rue Jules Horowitz BP220
38043 GRENOBLE Cedex, France
rogalev@esrf.fr

Philippe Sainctavit
Institut de Minénalogie et de
Physique des Milieux Condensés
Université Pierre et Marie Curie
4 place Jussieu
75252 Paris, France
Sainctavit@impmc.jussieu.fr

Didier Sébilleau
Équipe de Physique des
Surfaces et des Interfaces
Laboratoire de Physique des Atomes
Lasers, Molécules et Surfaces
Université de Rennes-1
35042 Rennes, France
sebilleau@univ-rennes1.fr

Alexander Yaresko
Max Planck Institute for the
Physics of Complex Systems
01187 Dresden, Germany
yaresko@mpipks-dresden.mpg.de

Introduction to Magnetism

Claudine Lacroix

Laboratoire Louis Néel, CNRS, BP 166, 38042, Grenoble Cedex 9, France
lacroix@grenoble.cnrs.fr

Abstract. In this article some aspects of the microscopic basis of magnetism will be reviewed. The main interactions responsible for the stability of magnetic moments will be described, as well as the various types of exchange interactions which induce cooperative magnetic ordering; some temperature effects are also discussed; this is not an exhaustive review, but only a discussion of some important points. No details of calculation are given, but more details can be found in textbooks. A short list is given at the end of this chapter.

1 Magnetism of Isolated Atoms and Ions

1.1 Magnetism is due to Unfilled Atomic Shells

Magnetic moments of atoms or ions are essentially due to the electrons of the unfilled shells; nuclei often have a magnetic moment, but its value is much smaller (by a factor 10^{-3} to 10^{-4}). In isolated atoms or ions, the electrons are moving around the nucleus in a potential due to the nucleus and the other electrons; this potential has very important consequences on the eigenstates of electrons: each state is characterized by 3 quantum numbers:

(i) $n = 1, 2, 3 \ldots$ is the principal quantum number; it determines the spatial variation of the wave function of each electronic shell. The larger n is, the more extended the wave function is (Fig. 1).
(ii) l is the orbital quantum number. l can take all integer values between 0 and $n - 1$. It determines the value of kinetic orbital momentum: $\langle l^2 \rangle = l(l+1)$.
(iii) m is the projection of kinetic orbital momentum on axis z; it is also quantized: $m = -l, -l+1, \ldots l-1, l$.

The orbital magnetic moment is proportional to the kinetic orbital moment: $\boldsymbol{m}_L = -\mu_B \boldsymbol{l}$, μ_B being the Bohr magneton.

Besides the orbital magnetic moment, electrons have a spin magnetic moment $\boldsymbol{m}_s = -2\mu_B \boldsymbol{s}$, where the spin $\boldsymbol{s}$ is an intrinsic kinetic moment of electron, which is also quantized ($s_z = \pm 1/2$). The total magnetic moment of each electron is the sum of both contributions: $\boldsymbol{m}_T = -\mu_B(\boldsymbol{l} + 2\boldsymbol{s})$. It should be pointed out that $\boldsymbol{m}_T$ is not collinear with the total kinetic moment

C. Lacroix: *Introduction to Magnetism*, Lect. Notes Phys. **697**, 1–13 (2006)
www.springerlink.com   © Springer-Verlag Berlin Heidelberg 2006

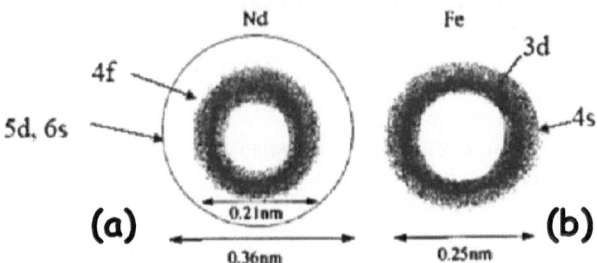

Fig. 1. Spatial extension of the wave functions for rare earth (**a**) and transition metal ions (**b**)

$l_t = l + s$. For a given atom or ion, the total magnetic moment is the sum of the magnetic moments of each electron; again there are 2 contributions: an orbital one $M_L = -\mu_B \sum_i l_i$, and a spin one: $M_S = -2\mu_B \sum_i s_i$. For filled shells both sums vanish so that only unfilled shells contribute to the magnetic moment. It is important to take into account the spin-orbit coupling (whose origin is relativistic), specially for heavy atoms: this coupling is written $H_{s-o} = \lambda L.S$, where λ, the spin orbit coupling parameter, increases strongly with the atomic number Z. In presence of this coupling, L and S are no longer good quantum numbers then a good quantum number is $J = L + S$. These rules apply well to transition and rare earth ions on which we will concentrate in the following. However, for heavier atoms (i.e. $5f$), because λ is larger, the ground state should be determined differently, taking spin-orbit coupling in a non-perturbative way (jj coupling). For more details see the van der Laan lecture.

1.2 The Ground State of Atoms and Ions

The ground state is obtained by taking into account several rules.

The first one is the Pauli principle which forbids 2 electrons in the same quantum state. Thus $l = 0$ shells contain 2 electrons (spin $\pm\frac{1}{2}$), p shells are filled with 6 electrons ($m = -1$, 0 or +1 and spin $\frac{1}{2}$)... In the ground state, electrons fill the lowest energy states. However Coulomb interactions between electrons are large and lead to some constraints, called Hund's rules, on the ground state configuration of the unfilled shells:

(i) the first rule says that total spin S should be as large as possible,
(ii) the second rule says that, within the maximum spin states, the ground state is obtained for the largest possible value of orbital moment, L.

Finally L and S are coupled by spin-orbit interaction; λ is positive for less than half-filled shells, and negative for more than half-filled shells; depending on the sign of λ, the ground state is obtained for $J = L + S$ or $J = |L - S|$.

It should be pointed out that magnetism is a consequence of the Pauli principle and intra-atomic Coulomb interactions between electrons which give rise to Hund's rules. In fact it is possible to give a qualitative explanation of the first rule: due to Pauli principle, two electrons on the same atom prefer to have the same spins, because, they cannot in this case occupy the same orbital state and thus cannot come too close from each other, thus their electrostatic interaction energy is reduced. This effect favors large spin values.

1.3 Magnetic Field Effect: The Landé Factor

Magnetic field is coupled through the Zeeman interaction to the total magnetic moment $\boldsymbol{M}_T = \mu_B(\boldsymbol{L} + 2\boldsymbol{S})$ and not to $\boldsymbol{J}$. However it is possible to define an effective magnetic moment as $\boldsymbol{M}_{eff} = -g_J\mu_B\boldsymbol{J}$, where g_J is the Landé factor:

$$g_J = \frac{3}{2} + \frac{S(S+1) - L(L+1)}{2J(J+1)} , \tag{1}$$

which is equal to 2 only if orbital moment vanishes. The saturation moment is then given by $g_J\mu_B J$ and the high temperature susceptibility obeys a Curie law with an effective moment $g_J\mu_B\sqrt{J(J+1)}$.

This is well verified for rare earth ions, but for $3d$ transition metal ions, as we will see in Sect. 2.1, the orbital moment is partially or totally quenched and effective moment often depends on environment.

All these rules determine the ground state values of L, S and J for isolated atoms or ions. In fact almost all atoms or ions have a magnetic moment when they are isolated. Absence of magnetic moment is obtained either in ions with only filled shells ($L = S = 0$ in this case) or if $J = L - S$ vanishes, as for example in Eu^{3+} ($4f^6$) or Cr^{2+}($3d^4$).

However, in pure solids only 15 elements are found to be magnetically ordered. These are materials with unfilled d and f shells (mainly $3d$ and $4f$). There are several reasons for that which will be explained in the following; magnetism can be unstable either because the atomic rules leading to determination of the ground state are not valid or because intersite interactions are not sufficient to induce long range ordering; this often occurs in metals where the band kinetic energy competes with local Coulomb interactions (Sect. 3.3).

2 Atoms and Ions in Solids

2.1 Difference Between 3d and 4f Magnetic Ions

In solids, electrons are not in a spherical potential but, due to neighboring ions, there is a crystal potential which has the symmetry of the environment. This crystal potential is an electrostatic potential due the charge distribution around the magnetic ion. Because this potential is not spherical, the orbital moment is no longer a good quantum number. In fact, the effect of this "crystal field" is different for d and f ions because the order of magnitude of the energies involved is different:

(i) In $3d$ ions, the crystal field is much larger than spin-orbit coupling. Thus Hund's rules determine the ground state values of L and S and the largest perturbation is the crystal field.
(ii) In $4f$ ions, the spin-orbit coupling is much larger than the crystal field. J is determined by the sign of the spin-orbit coupling, and crystal field is a smaller perturbation.

The reasons for this difference are: (i) spin orbit coupling is much larger in $4f$ than in $3d$ since it increases rapidly with Z; (ii) $4f$ wave functions are more localized than the $3d$ ones and thus they are less sensitive to the crystal potential due to neighboring ions.

2.2 Quenching of Orbital Moment in 3d Ions

For $3d$ ions a cubic crystal field removes the degeneracy of the five d orbitals (Fig. 2). These are split in e_g and t_{2g} states. In these new states, the orbital moment vanishes to first order: all diagonal matrix elements of the orbital moment $\langle \alpha | L_i | \alpha \rangle$ vanish. Thus the main contribution to the magnetic moment comes from the spins; orbital moment can often be neglected. For example the ground state of Fe^{2+} has an effective moment of $5.9\,\mu_B$, corresponding to $L = S = 2, J = 4$, while in FeO, the effective moment is reduced to $5.3\,\mu_B$, indicating a reduction of the orbital moment (not a total quenching). Another consequence of the large crystal field is the possibility of violating the Hund's rules: if the crystal field is large, the electrons can fill the low energy crystal field states, without taking into account Hund's rules. In particular, the first

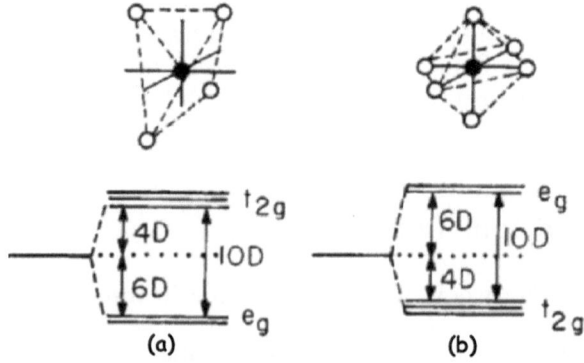

Fig. 2. Crystal field splitting of $3d$ orbitals for tetrahedral (**a**) and tetragonal (**b**) environment

rule may be violated, leading to a "low spin" state (eventually this low spin state can be non-magnetic). For example if one consider again Fe^{2+} with 6 electrons, the ground state given by the first Hund's rule is $S = 2$, but if the crystal field is large these 6 electrons can fill the 3 t_{2g} states, leading to $S = 0$.

Thus in $3d$ ions crystal field effects are an important and can modify both orbital and spin magnetic moments. Spin-orbit coupling is a perturbation which leads to anisotropy energies which are usually small.

2.3 Crystal Field for $4f$ Ions

In this case, the ground state is determined in first approximation by quantum numbers J, L and S. Degeneracy of the ground state is $2J + 1$; the crystal field lifts this degeneracy and leads to a large single ion anisotropy energy.

In some cases the resulting ground state is a doublet corresponding to $J_z = \pm J$ and it has then a strong Ising-like behavior. It can also occur, if J is integer, that the ground state has $J_z = 0$ and the ground state is then non-magnetic (no Zeeman effect).

2.4 Magneto-Crystalline Anisotropy

Magneto-crystalline anisotropy energy reflects the fact that the moments prefer to stay along some particular directions in the crystal: this is due to the non spherical charge distribution in the presence of a crystal field; the spin is coupled to the orbital moment trough spin-orbit coupling, and it is sensitive to the charge distribution. In cubic crystals for example this energy is written:

$$E = K_1(m_x^2 m_y^2 + m_y^2 m_z^2 + m_z^2 m_x^2) + K_2(m_x^2 m_y^2 m_z^2) , \qquad (2)$$

the constants K_i strongly depend on the system.

2.5 Itinerant Magnetism in Metals

In metals, magnetic moments of "ions" are not determined by the rules given previously in Sect. 1. The hybridization with conduction electrons of the metal strongly influences the magnetic properties. It leads in some cases to the Kondo effect, i.e. to the formation of a singlet non-magnetic state at low temperature. If the hybridization is sufficiently large magnetic moments disappear, the electrons of the magnetic ions being strongly delocalized.

3 Exchange Interactions

Magnetic long range order occurs because the local magnetic moments interact with each other. The simplest interaction energy is the Heisenberg exchange interaction between localized spins: $J_{ij}\boldsymbol{S}_i.\boldsymbol{S}_j$. There are several mechanisms responsible for this exchange. Besides this bilinear interaction other interaction terms are possible: biquadratic exchange, multi-site interactions, Dzyaloshinskii-Moriya interaction which will not be discussed here.

3.1 Origin of Exchange: Pauli Principle and Electrostatic Interactions

In solids, magnetic moments interact and this interaction leads to long range magnetic ordering. There are several mechanisms which are responsible for this interaction. However the dipolar interaction, which is always present, is usually too small to explain the observed values of ordering temperatures. Again, the interaction is due to Coulomb interactions and Pauli principle. The simple case of the H_2 molecule is studied in many textbooks, the main results are summarized below: H_2 molecule has 2 electrons and the wave function is a product of a spatial part $\Psi(r_1, r_2)$ and a spin part $\Phi(\sigma_1, \sigma_2)$. Due to Pauli principle, the wave function should be antisymmetric with respect to interchange of the 2 electrons. Thus there are two possibilities:

(i) $\Psi(r_1, r_2)$ is symmetric and $\Phi(\sigma_1, \sigma_2)$ is antisymmetric; this corresponds to a singlet arrangement of the spins: $\frac{1}{\sqrt{2}}(|+\frac{1}{2}, -\frac{1}{2}\rangle - |-\frac{1}{2}, +\frac{1}{2}\rangle)$ which corresponds to total spin $S = 0$.

(ii) $\Psi(r_1, r_2)$ is antisymmetric and $\Phi(\sigma_1, \sigma_2)$ is symmetric; this corresponds to the (triply degenerate) triplet arrangement of the spins: $|+\frac{1}{2}, +\frac{1}{2}\rangle, \frac{1}{\sqrt{2}}(|+\frac{1}{2}, -\frac{1}{2}\rangle + |+\frac{1}{2}, +\frac{1}{2}\rangle), |-\frac{1}{2}, -\frac{1}{2}\rangle)$.

These 3 states have total spin $S = 1$, and $S_z = 1, 0$ or -1 respectively.

Electrons in these 2 states do not have the same electrostatic energy because the spatial wave functions are different. This difference can be written as: $E_1 - E_2 = J_{12}\boldsymbol{S}_1.\boldsymbol{S}_2$, J_{12} being the interatomic exchange interaction whose sign depends on distance between the two H nucleus.

This mechanism is known as direct exchange. It was the first proposed mechanism for exchange interaction between spins. However in most situations other mechanisms are more important than direct exchange. Some of them are described in the following; it should be noticed that magnetic exchange is always due to Coulomb repulsion between electrons.

3.2 Superexchange

Superexchange arises frequently in transition metal oxides where the $3d$ magnetic ions are separated by non-magnetic oxygen ions. Thus there is no direct overlap between d orbitals of magnetic ions; the interaction is indirect and occurs because d orbitals hybridize with p orbitals of oxygen ions; if we consider 2 magnetic ions M_1 and M_2 separated by an oxygen ion O, d electrons

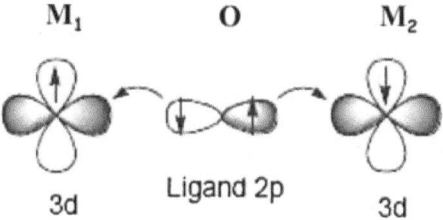

Fig. 3. Superexchange mechanism

are partially delocalized on oxygen and the delocalization energy is different if the two magnetic moments are parallel or antiparallel (Fig. 3).

The sign and value of this superexchange interaction depends on the number of electrons, of the type of d orbitals involved, and on the angle M_1-O-M_2. It is most often negative, leading to antiferromagnetic interaction as in MnO, but it can be positive as in EuO. Finally it is always a short range interaction because it involves hybridization between neighboring ions, M and O.

3.3 Double Exchange

This interaction is important in $3d$ systems where the $3d$ ions are in a mixed valence state: for example in the manganites $(La,Sr)MnO_3$, the Mn ions are either Mn^{3+} with 4 electrons in the $3d$ shell ($S = 2$), or Mn^{4+} with 3 d-electrons ($S = 3/2$). On both ions the t_{2g} orbitals of a given spin direction are filled, forming a spin 3/2; on Mn^{3+}, there is one electron more in an e_g state coupled ferromagnetically to the t_{2g} spin $S - 3/2$, because of first Hund's rule. This electron can propagate on a neighboring Mn^{4+} ion, only if this neighboring spin is parallel, in order to satisfy Hund's rule (Fig. 4).

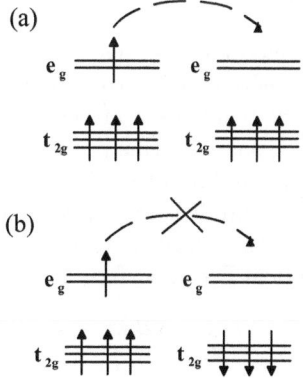

Fig. 4. Double exchange mechanism. In case (**a**) hopping is possible, while in (**b**) Hund's rule forbids hopping of e_g electron

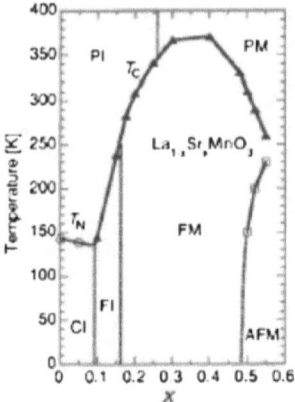

Fig. 5. Phase diagram of the system La$_{1-x}$Sr$_x$MnO$_3$ showing the competition between ferromagnetism (FM) and antiferromagnetism (AFM, CI) when the number of e$_g$ electrons changes

This can be described by introducing an effective transfer integral between neighboring sites, which depends on the angle θ between the t$_{2g}$ spins: $t_{eff} = t\sin(\frac{\theta}{2})$, vanishing if $\theta = \pi$, i.e. if the 2 spins are antiparallel. Thus, there is a competition between kinetic energy of the e$_g$ conduction electrons and intra-atomic exchange, responsible for Hund's rule. Ferromagnetism is then stabilized if intra-atomic exchange is large enough compared to the e$_g$ bandwidth. In these systems, there is a strong connection between magnetism and transport, and this leads to large variations of electric transport under applied field (colossal magnetoresistance). Ferromagnetism induced by such mechanism is observed in manganites (Fig. 5).

3.4 RKKY Interaction

This interaction arises in metallic systems when magnetic moments remain localized, as in rare earth metals or intermetallics. This is again an indirect interaction, mediated by the band electrons: each localized moment polarizes locally the conduction electrons, and this polarization propagates to another magnetic ion, leading to a coupling of the two magnetic moments. This interaction is long ranged, because it is mediated by conduction electrons; it is an oscillating interaction, with a wavelength related to the Fermi wave vector k_F; calculation for a spherical Fermi surface gives:

$$J(r) \propto \frac{2k_F r \cos 2k_F r - \sin 2k_F r}{(2k_F r)^4} . \tag{3}$$

Thus it strongly depends on distance and on the band filling through the Fermi wave vector. This oscillating interaction may lead to a large variety of

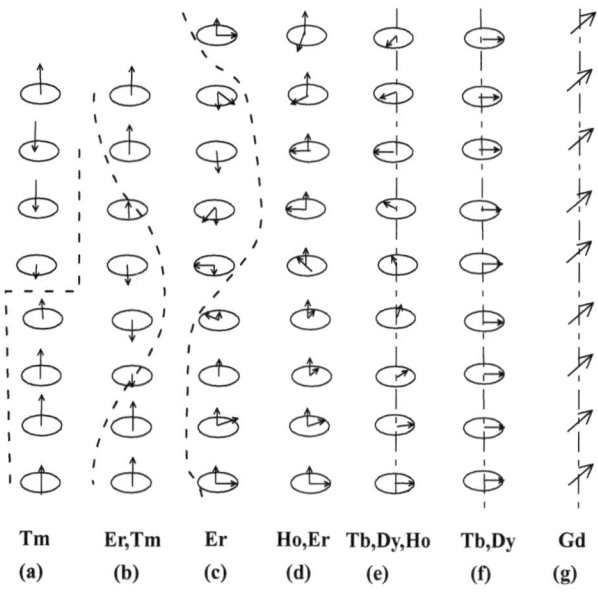

Tm	Er,Tm	Er	Ho,Er	Tb,Dy,Ho	Tb,Dy	Gd
(a)	(b)	(c)	(d)	(e)	(f)	(g)

Fig. 6. In rare earth a large number magnetic orderings are observed because of the oscillating nature of RKKY interactions

magnetic orderings as observed in rare earth metals: ferromagnetic, antiferromagnetic, sinusoidal, helicoidal (Fig. 6).

This interaction plays an important role in multilayers: the exchange coupling between two ferromagnetic layers separated by a non-magnetic layer is of the same type and the dependence of $J(r)$ with r can be observed by varying the non-magnetic layer thickness.

3.5 Itinerant Magnetism

In metallic $3d$ magnetic systems, like Fe, Co or Ni it is not possible to describe the magnetic properties in terms of atomic or ionic moments coupled by intersite exchange interaction. The same electrons are responsible for both magnetic moments and exchange interactions through local Coulomb repulsion and kinetic energy.

The simplest model describing interacting band electrons is the Hubbard model, in this model the Hamiltonian is written:

$$H = \sum_{k,\sigma} E(k)n_{k\sigma} + \sum_{i} U n_{i\uparrow}n_{i\downarrow} \, , \qquad (4)$$

where $E(k)$ is the band energy and U the on-site Coulomb repulsion between electrons of opposite spin. Degeneracy of the d orbitals is here neglected

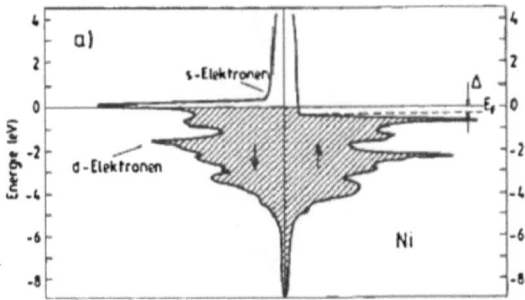

Fig. 7. d band of Ni schema

but it can be taken into account. If the interaction U is small, the metal is paramagnetic. Moreover, ferromagnetism occurs if U exceeds a critical value U_c given by the Stoner criterion:

$$U_c N(E_F) = 1 \qquad (5)$$

where $N(E_F)$ is the density of paramagnetic states per spin at the Fermi level. Above this critical value, band ferromagnetism is stable, with a spin magnetic moment which is not fixed by atomic rules, but by the position of the Fermi level in the conduction band: the ordered magnetic moment is $2.2\,\mu_B$ per atom in Fe, $0.6\,\mu_B$ per atom in Ni. Ferromagnetism is thus favored by a high density of states at the Fermi level. In the ferromagnetic state the spin subbands $\sigma(\uparrow, \downarrow)$ are split by the exchange field Δ. This simple scheme is qualitatively valid for Ni, Co and Fe, as obtained by LSDA calculation (Fig. 7). In the presence of disorder, the Stoner criterion can be qualitatively applied locally, using the local density of states. This local density of states strongly depends on the environment; thus magnetism in metals is very sensitive to environment: a $3d$ atom can be magnetic or not, depending on many different factors such as crystal structure, nature of surrounding atoms.... In thin films and multilayers, this effect plays a significant role: magnetic moments are often different from their bulk value near a surface; a non-magnetic metal like Pd shows interface magnetism near a ferromagnetic layer (Fig. 8).

Finally this Stoner criterion can be generalized to describe other types of magnetic instability (antiferromagnetism...): in this case $N(E_F)$ has to be replaced by the q-dependent susceptibility $\chi(q)$ where q is the wave number of the spin density wave.

For a half-filled band and at very large values of U, this Hubbard model describes localized electrons interacting through antiferromagnetic super-exchange interaction. With increasing value of U, a Hubbard model is thus able to describe the crossover from paramagnetic metal, to itinerant ferromagnet and finally localized spin system. A metal-insulator occurs at half-filling for a value of U of the order of the bandwidth.

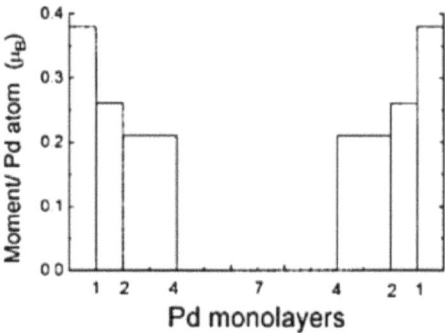

Fig. 8. Magnetic moments of Pd in Fe/Pd multilayers: only the Pd near the interface are magnetic

4 Magnetism at Finite Temperature

4.1 Mean Field Description

The mean field approximation predicts the temperature dependence of magnetization of a magnetic system and allows to calculate the Curie temperature T_C: this Curie temperature is proportional to exchange interaction and to the number of neighbors: this explains qualitatively the reduction of T_C often observed in thin films. At low T, in this approximation the magnetization decreases exponentially, while near T_C, it varies as $\sqrt{T_C - T}$; Above T_C, the magnetic susceptibility is found to follow a Curie-Weiss law: $\chi(T) = \frac{C}{T - T_C}$. All these behaviour coincide only qualitatively with experimental observations and improvements of mean field approximation are necessary both at low and high temperatures.

4.2 Low Temperature: Effect of Spin Waves

There are low energy excitations, not included in mean field approximation, which dominate the low temperature behavior of magnetization. These spin waves are collective excitations of the spin system; their dispersion law at low energy can be written: $\omega(k) = Dk^2$ for a ferromagnet, or $\omega(k) = Ck$ for an antiferromagnet (Fig. 9). Due to these collective excitations, the magnetization decreases much more rapidly than in mean field: it varies as $T^{3/2}$ for a ferromagnet and T^2 for an antiferromagnet. This applies to both localized and itinerant magnetic systems.

Dimensionality effects are absent in the mean field approximation where T_C is proportional to the number of neighbors. However it is known from the Mermin-Wagner theorem that magnetism cannot be stable at finite temperature in 1 and 2-dimensional isotropic spin systems. Spin wave theory is in agreement with this theorem: In 1- and 2-dimensional systems, spin

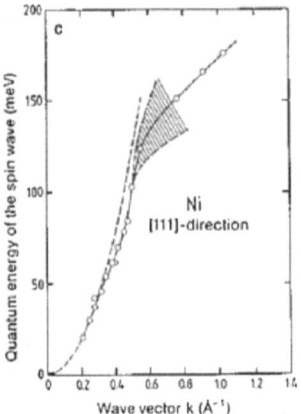

Fig. 9. Magnetic excitations measured in Ni

waves excitations at $T = 0$ destroy the long range magnetic order. Only if anisotropy is present, magnetism can remain stable in 2-D, as obtained in the Ising model for example.

The value of T_C calculated in mean field approximation is usually too high because low energy excitations are not well described; spin fluctuations strongly decrease this value and give values much closer from the experimental ones.

4.3 The Magnetic Phase Transition

Near a second-order phase transition at T_C, magnetization is small and the free energy can be expanded in powers of magnetization M; such expansion is called Landau expansion:

$$F = F_0 + a(T)M^2 + b(T)M^4 + c(T)(\nabla M)^2 - MH \ . \qquad (6)$$

The Curie temperature is the temperature where a(T) vanishes. Below T_C, if magnetization is uniform, on has $M(T) = \sqrt{T_C - T}$ as in mean field; the susceptibility above T_C is also found to obey the Curie-Weiss law. However near T_C the magnetization is strongly non-uniform and the term $(\nabla M)^2$ in the free energy cannot be neglected. This term varies with a characteristic length, the critical length ξ, which becomes infinite at T_C. Thus near T_C, the physical quantities are not given by mean field expressions; they obey some scaling laws with critical exponents which are universal numbers depending on the dimensionality of the system and the type of interaction. For example in 3-dimensional ferromagnets, for a Heisenberg interaction one has: $M(T) \propto (T_C - T)^\beta$ with $\beta = 0.36$, $\chi(T) = (T - T_C)^{-\gamma}$ with $\gamma = 1.39$, while for an Ising-type system $\beta = 0.32$ and $\gamma = 1.24$; these values are quite far from the mean field ones (Fig. 10).

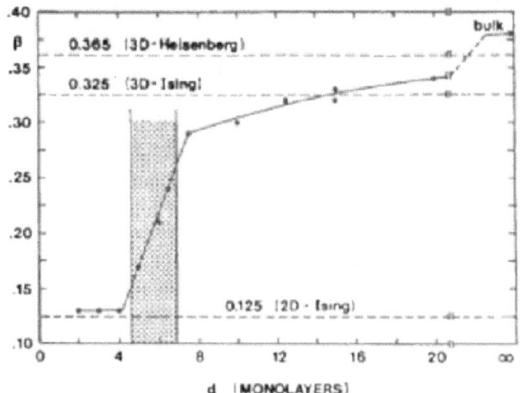

Fig. 10. Critical exponent β in Ni thin film as a function of thickness: there is a transition from 2- to 3-dimensional behavior around 6 monolayers

Bibliography

There are many books in which more details can be found on the various topics presented here:

- Magnétisme, edited by E. de Lacheisserie, EDP Sciences
- Magnetism in Condensed Matter, S. Blundell, Oxford University Press
- The theory of magnetism, D.C. Mattis (Springer series in Solid State Sciences vol 17 and 55)
- Spin fluctuations in itinerant electron magnetism, T. Moriya (Springer)
- J.B. Goodenough, Magnetism and the chemical bond (Wiley)
- H.J. Zeiger and G.W. Pratt, Magnetic interactions in solids (Clarendon press, Oxford)
- P.W. Anderson, Basic notions in condensed matter physics (Frontiers in Physics 55)
- K. Yoshida, Theory of magnetism (Spinger serie in solid state sciences 122)

X-ray and Electron Spectroscopies: An Introduction

Didier Sébilleau

Équipe de Physique des Surfaces et des Interfaces, Laboratoire de Physique des Atomes, Lasers, Molécules et Surfaces, UMR CNRS 6627, Université de Rennes-1, 35042 Rennes-Cédex, France
didier.sebilleau@univ-rennes1.fr

Abstract. We give a basic introduction to the physical phenomena underlying x-ray and electron spectroscopies. We review as well all the side effects, such as manybody ones, that can affect the spectra measured experimentally. Some elements of multiple-scattering theory are given to formulate a description of these techniques. Finally, in the particular case of the determination of structural parameters, a tentative methodology is outlined.

1 Introduction

Spectroscopies are among the most widely used techniques to study the physical and chemical properties of materials. They are now extensively present in many fields of science such as physics, chemistry or biology. The term "spectroscopy" which was still characterizing the study of light by means of a prism in the 19th century, has now, since the advent of quantum theory, a much wider meaning. Indeed, according to the 15th edition of *The New Encyclopaedia Britannica*, "spectroscopy is the study of the absorption and emission of light and other radiation, as related to the wavelength of the radiation". If we add scattering to the two previous physical processes, the quantum nature of particles makes this definition cover most types of experiment that can be performed nowadays. Such a typical experiment is sketched in Fig. 1. It should be noted that the incoming and outgoing particles are not necessarily the same. When they are of an identical nature, the corresponding technique can be performed either in the reflection or in the transmission mode.

The idea behind these spectroscopies is that information about the sample can be obtained from the analysis of the outgoing particles. Depending on the type of spectroscopy, this information can be related to the crystallographic structure, to the electronic structure or to the magnetic structure of the sample. Or it can be about a reaction that takes place within the sample. With this in mind, the type and energy of the detected particles will directly determine the region of the sample from which this information can be traced back, and therefore the sensitivity of the technique to the bulk or the surface. For instance, low-energy electrons or ions will not travel much more than ten

D. Sébilleau: *X-ray and Electron Spectroscopies: An Introduction*, Lect. Notes Phys. **697**, 15–57 (2006)
www.springerlink.com

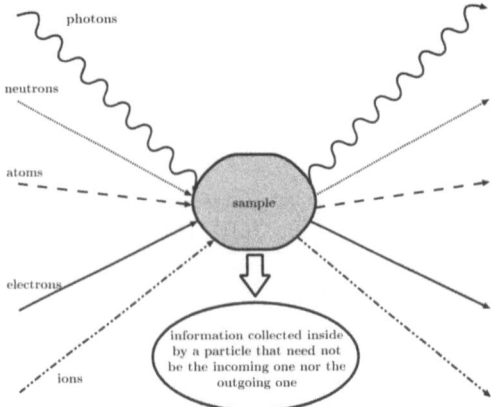

Fig. 1. Typical spectroscopic experiment

interatomic distances in a solid while electromagnetic radiations will be able to emerge from much deeper layers. In diffraction techniques using significantly penetrating particles, surface sensitivity can nevertheless be achieved with grazing incidence angles. Note also that the counterpart to detecting particles with a small mean free path is the necessity to work under ultra high vacuum conditions so that the outgoing particles can effectively reach the detector. Figure 2 gives the variation of the electron mean free path λ_e in solids as a function of the electron kinetic energy. We see clearly here that in the range 10–1000 eV, electrons coming from within a sample do not originate from much deeper than 5 to 20 Å. As a consequence, spectroscopies detecting

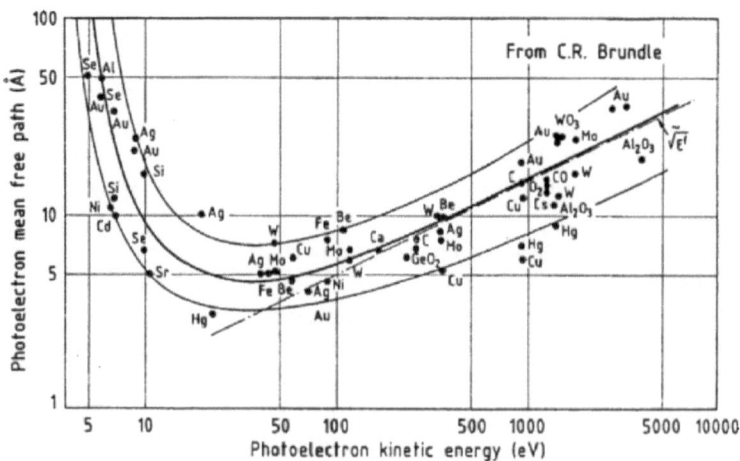

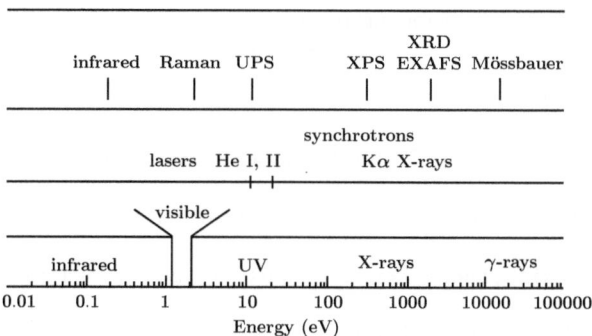

Fig. 3. The electromagnetic spectrum, along with the common photon sources and some spectroscopies based on photons [3]

electrons in this energy range will be essentially sensitive to the surface structure as the particle detected will carry information from the very topmost layers.

Although, as we have just seen, many techniques can provide information on a sample and its surface (if any), we will mainly restrict ourselves here to some of them specifically related to synchrotron radiation: x-ray spectroscopies. X-ray spectroscopies are characterized by the fact that the incoming beam is composed of photons in the range of about 100 eV to 10 keV. Higher energies, in the γ-ray region, will not be considered here as in this latter range photons will be scattered significantly not only by the electrons but also by the nuclei [2] giving rise to entirely different spectroscopies such as Mössbauer spectroscopy. Figure 3, taken from [3], gives a sketch of the electromagnetic spectrum with some common photon sources and the spectroscopies corresponding to the various energy ranges.

The acronyms of some common spectroscopies are given in Table 1.

This chapter is organized as follows. Section 2 recalls the basics of photon-electron interaction. Section 3 describes some other physical processes that can result from this interaction, especially many-body ones, and discusses the various ways these spectroscopies can be made sensitive to the magnetic properties. Section 4 presents four spectroscopies either well-established (XAS, PhD, DAFS) or very promising (APECS). X-ray magnetic circular dichroism (XMCD) being a major technique to study magnetic samples, the physical principles it relies on are reviewed in the paragraph devoted to XAS.

Table 1. The acronyms of the some standard spectroscopies

AED   : Auger Electron Diffraction

APECS : Auger PhotoElectron Coincidence Spectroscopy

BIS   : Bremsstrahlung Isochromat Spectroscopy

DANES : Diffraction Anomalous Near Edge Structure

DAFS   : Diffraction Anomalous Fine Structure

EELS   : Electron Energy Loss Spectroscopy

EXAFS : Extended X-ray Absorption Fine Structure

IS   : Ion Scattering

LEED   : Low Energy Electron Diffraction

PhD   : Photoelectron Diffraction (or PED)

STM   : Scanning Tunnelling Microscopy

XANES : X-ray Absorption Near Edge Structure

XAS   : X-ray Absorption Spectroscopy

XMCD : X-ray Magnetic Circular Dichroism

XRMS : X-ray Resonant Magnetic Scattering

XRD   : X-Ray Diffraction

XSW   : X-ray Standing Waves

---

2 The Photon-Atom Interaction

2.1 The Interaction Hamiltonian

Before considering the interaction of an incoming photon with an atom, let us consider first the case when the electromagnetic field encounters a particle of mass m, charge q and spin direction s. The electromagnetic field will be characterized by its wave vector q and its polarization directions $\hat{e}_q$. Let us further suppose that the particle is subject to an external potential V, which can be the potential of a nucleus for instance. Within the non-relativistic theory, the Hamiltonian of this system can be written as [4]

$$H = \frac{1}{2m}\left[\boldsymbol{p} - q\boldsymbol{A}\right]^2 + qV - g\frac{q}{2m}\boldsymbol{s}\cdot\boldsymbol{B} + H_R . \tag{1}$$

The first term is the kinetic energy of the particle in the presence of the electromagnetic field. $\boldsymbol{A}$ is the vector potential of this field. The third term represents the interaction between the spin of the particle and the magnetic field of the radiation. g is the Landé factor of the particle. The Hamiltonian H_R of the excitation field is given by

$$H_R = \sum_{q,\hat{e}_q} \hbar\omega_q \left[a^{\dagger}_{q,\hat{e}_q} a_{q,\hat{e}_q} + \frac{1}{2}\right] \tag{2}$$

and the quantized electromagnetic field by

$$\boldsymbol{A}(\boldsymbol{r}) = \sum_{q,\hat{e}_q} \sqrt{\frac{\hbar}{2\epsilon_0 \omega_q \mathcal{V}}} \left[a_{q,\hat{e}_q} e^{i\boldsymbol{q}\cdot\boldsymbol{r}}\hat{\boldsymbol{e}}_q + a^{\dagger}_{q,\hat{e}_q} e^{-i\boldsymbol{q}\cdot\boldsymbol{r}}\hat{\boldsymbol{e}}^*_q\right] . \tag{3}$$

Here $a_{q,\hat{e}_q}$ annihilates a photon while $a^{\dagger}_{q,\hat{e}_q}$ creates one. The sum is over the wave vector $\boldsymbol{q}$ and the two polarizations $\hat{\boldsymbol{e}}_q$ of the wave field. The time evolution is obtained by replacing $\boldsymbol{q}\cdot\boldsymbol{r}$ by $(\boldsymbol{q}\cdot\boldsymbol{r} - \omega_q t)$ in the previous expression. $\mathcal{V}$ is the normalization volume. For x-ray radiation, the inverse of q can be large compared to the extension of the core orbitals involved in the transition. In this case, $\boldsymbol{q}\cdot\boldsymbol{r}$ is small and the exponential can be replaced by 1 (dipole approximation) or by $1 + i\boldsymbol{q}\cdot\boldsymbol{r}$. For instance, for $1s$ electrons and photon energies varying between $250\,\text{eV}$ and $10\,\text{keV}$, the value of $\boldsymbol{q}\cdot\boldsymbol{r}$ is of the order of 0.04 [5].

Within the Coulomb gauge ($\boldsymbol{\nabla}\cdot\boldsymbol{A} = 0$), the Hamiltonian H of the system can be reorganized as

$$H = H_P + H_R + V_I \tag{4}$$

with the Hamiltonian of the particle alone given by

$$H_P = \frac{\boldsymbol{p}^2}{2m} + qV \tag{5}$$

and the interaction potential by

$$V_I = -\frac{q}{m}\boldsymbol{p}\cdot\boldsymbol{A} + \frac{q^2}{2m}\boldsymbol{A}^2 - g\frac{q}{2m}\boldsymbol{s}\cdot\boldsymbol{B} . \tag{6}$$

This result is valid within the framework of non-relativistic theory. If relativistic corrections are needed i.e., if the ratio v/c, where v is the speed of the particle and c that of the light, is not completely negligible, then it can be shown that a new important term of the order $(v/c)^2$ arises [6]. This term, called the spin-orbit term, gives two contributions [7]. One, the spin-orbit contribution of the electron alone can be included into H_P. The second one, $-q^2/(2m^2c^2)\,\boldsymbol{s}\cdot(\partial\boldsymbol{A}/\partial t\wedge\boldsymbol{A})$, gives rise to a spin-dependent scattering [8]. This term and the $\boldsymbol{s}\cdot\boldsymbol{B}$ one are responsible for magnetic scattering.

Before going further, it is interesting to look at the order of magnitude of the various terms which compose V_I. If we rewrite (6) as

$$V_I = V_I^{(1)} + V_I^{(2)} + V_I^{(3)} \tag{7}$$

it can be shown [9] that

$$\left\{\begin{array}{l} \dfrac{V_I^{(3)}}{V_I^{(1)}} \sim \dfrac{\hbar q}{p} \\[2em] \dfrac{V_I^{(2)}}{V_I^{(1)}} \sim \dfrac{V_I^{(1)}}{H_P} \end{array}\right\} \tag{8}$$

For a bound electron and low-energy photons, the ratio $\hbar q/p$ is generally much smaller than 1 and therefore $V_I^{(1)}$ will dominate over $V_I^{(3)}$. Likewise, for low intensity radiation beams, $V_I^{(1)}$ will be dominant with respect to $V_I^{(2)}$. Furthermore, it should be noted that magnetic scattering is a very small effect compared to charge scattering and hence needs high intensity synchrotron radiation sources to be detected. Indeed, the ratio of the cross-sections is about $(\hbar\omega_q/mc^2)^2$ [7]. High-energy photons are therefore needed and resonance phenomena can be used as well to improve the detection as they often lead to a huge enhancement of the cross-section [10].

We see from the expression (6) that the interaction Hamiltonian is proportional to the inverse of the mass of the particle. Consequently, the cross-section for the phenomena resulting from this interaction will behave like $1/m^2$. This means that in the x-ray regime ($\hbar\omega_q < 1\,\mathrm{MeV}$), the scattering of photons by electrons in the atomic shells will largely dominate as the nuclei are much heavier than electrons. Scattering from the nucleus will become sufficient to be detected only for γ-rays [11], or for hard x-rays in the resonant mode and making use of the fact that scattering by electrons is almost instantaneous compared to nuclear scattering, which allows to discriminate them [12].

Therefore, we will restrict here the Hamiltonian for the interaction of x-rays with an atom to a sum over all the electrons. In the atomic case, (1) will therefore have a sum over these electrons, plus a Coulomb term that describes the interaction between the electrons [4] which will be included in H_P as it does not depend upon the radiation. However, for the sake of simplicity, we will restrict the discussion here to the case of a single electron of charge $-e$. Let us drop the magnetic terms in 6 to focus on the leading contributions i.e.,

$$V_I = \frac{e}{m}\boldsymbol{p}\cdot\boldsymbol{A} + \frac{e^2}{2m}\boldsymbol{A}^2\,. \tag{9}$$

We can treat V_I as a perturbation to the Hamiltonian $H_P + H_R$. We know in this case from standard perturbation theory that we can express the solutions of the total Hamiltonian H in terms of the eigenfunctions $|\phi\rangle$ of the unperturbed Hamiltonian $H_0 = H_P + H_R$. These unperturbed solutions can be written as

$$\left\{ \begin{array}{l} |\phi_i\rangle = |\varphi_i\rangle|\boldsymbol{q}_i, \hat{\boldsymbol{e}}_{\boldsymbol{q}_i}\rangle \\[2mm] |\phi_f\rangle = |\varphi_f\rangle|\boldsymbol{q}_f, \hat{\boldsymbol{e}}_{\boldsymbol{q}_f}\rangle \ . \end{array} \right\} . \tag{10}$$

Here the labels i and f indicate respectively the initial and the final state. $|\varphi\rangle$ is a particle state (or an atomic state if we keep the sum over the electrons), eigenstate of $H_P = \boldsymbol{p}^2/2m + qV$ and $|\boldsymbol{q}, \hat{\boldsymbol{e}}_{\boldsymbol{q}}\rangle$ is a photon state, eigenstate of H_R. The eigenvalues of H_0 are then

$$\left\{ \begin{array}{l} H_0|\phi_i\rangle = E_i|\phi_i\rangle \ \text{with} \ E_i = \mathcal{E}_i + \hbar\omega_{\boldsymbol{q}_i} \\[2mm] H_0|\phi_f\rangle = E_f|\phi_f\rangle \ \text{with} \ E_f = \mathcal{E}_f + \hbar\omega_{\boldsymbol{q}_f} \end{array} \right\} . \tag{11}$$

$\mathcal{E}$ and $\hbar\omega_{\boldsymbol{q}}$ are the eigenvalues associated to $|\varphi\rangle$ and $|\boldsymbol{q}, \hat{\boldsymbol{e}}_{\boldsymbol{q}}\rangle$ for the corresponding Hamiltonian. We are interested here in transitions from $|\phi_i\rangle$ to $|\phi_f\rangle$ that can result from the action of the interaction term V_I. From time-dependent perturbation theory [13], we know that the transition probability per unit time from state $|\phi_i\rangle$ to state $|\phi_f\rangle$ under the effect of the perturbation V_I is given exactly by

$$W_{i \to f} = \frac{2\pi}{\hbar} |\langle\phi_f|T_I|\phi_i\rangle|^2 \rho_f \ . \tag{12}$$

Here ρ_f is the density of final states and T_I is the transition operator associated to the interaction potential V_I. T_I can be expressed by the relation [14]

$$T_I = V_I + V_I G(E_i) V_I \ , \tag{13}$$

where $G(E_i)$ is the resolvent of the total Hamiltonian H at the energy of the initial state. This resolvent is defined by

$$G(E_i) = \lim_{\epsilon \to 0^+} \frac{1}{E_i - H + i\epsilon} \ . \tag{14}$$

The infinitesimal imaginary part is introduced so that the denominator is always non zero. The action of $G(E_i)$ on any eigenstate $|\psi_n\rangle$ of H will give

$$G(E_i)|\psi_n\rangle = \frac{|\psi_n\rangle}{E_i - E_n} \ . \tag{15}$$

At the lowest order in V_I, the transition operator can be approximated by the perturbation V_I. When inserted in 12, this gives second Fermi's Golden Rule [13]. To second order in V_I, it can be replaced by

$$T_I \approx V_I + V_I G_0(E_i) V_I \ , \tag{16}$$

which gives the first Fermi's Golden Rule [15]. $G_0(E_i)$ is the unperturbed resolvent and is obtained from 14 by replacing H by H_0.

2.2 The Physical Processes Induced

If we replace V_I by its expression (9) in the previous equation, we obtain to the second order in e/m

$$T_I^{(2)} = \frac{e}{m} \boldsymbol{p} \cdot \boldsymbol{A} + \left(\frac{e}{m}\right)^2 \left[\frac{m}{2} \boldsymbol{A} \cdot \boldsymbol{A} + \boldsymbol{p} \cdot \boldsymbol{A} \, G_0(E_i) \, \boldsymbol{p} \cdot \boldsymbol{A}\right] + \mathcal{O}\left(\frac{e^3}{m^3}\right) . \quad (17)$$

The term $(e/m) \, \boldsymbol{p} \cdot \boldsymbol{A}$ corresponds to a first-order process while those in $(e/m)^2$ are second-order processes. Moreover, we see from this expression that only one photon is involved in the first order (only one $\boldsymbol{A}$ in the operator) while two photons are present in the second-order processes ($\boldsymbol{A}$ occurs twice in each operator). These second-order processes involving two photons are called scattering processes. Let us look in more details at the various physical processes involved. With the transition operator $T_I^{(2)}$ expressed as (17), they can be of three kind: emission, absorption or scattering of a photon. To describe them more accurately, we will suppose now that the electron involved in the interaction with the photon belongs to an atom that will evolve from state $|i\rangle$ to state $|f\rangle$. The case of emission is sketched as a Feynman diagram in Fig. 4 [16].

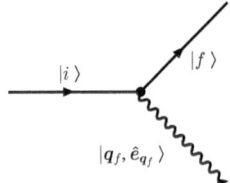

Fig. 4. Feynman diagram corresponding to an emission process

Due to energy conservation, the energy of state $|f\rangle$ will be lower than that of the initial state of the atom. The Hamiltonian describing the electrons in the atom has two kind of solutions: bound (discrete) states and continuum states (we will not consider here resonance states). We will suppose that all the bound states are below the lowest energy continuum state. In this case, there are three possibilities for such an emission process. If it takes place between two bound states, it will be called a radiative deexcitation. When it takes place from a continuum state to a discrete state, it means that the electron is trapped by the atom. In the last case, when two continuum states are involved, this is the *bremsstrahlung* process: the electron arriving on the atom emits a photon, but its kinetic energy is still too high to get trapped and it leaves at a lower speed.

The second first-order process is absorption, sketched in Fig. 5. In this case, a photon is absorbed by the atom and an electron in a bound state can be excited towards another discrete state or towards a continuum state. The

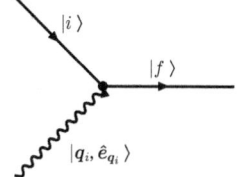

Fig. 5. Feynman diagram corresponding to an absorption process

latter process is called photoionization. But the absorption process can result as well in the acceleration of an electron approaching an atom. This is the inverse *bremsstrahlung* [17].

From the expression (3) of the quantized vector potential $\boldsymbol{A}$, we see that for emission and absorption, we will have the transition operators

$$
\left\{
\begin{aligned}
T_{I,\text{ab}} &= \frac{e}{m} \sum_{q,\hat{e}_q} \sqrt{\frac{\hbar}{2\epsilon_0 \omega_q \mathcal{V}}} \, (\hat{e}_q \cdot \boldsymbol{p}) \, a_{q,\hat{e}_q} \, e^{i\boldsymbol{q}\cdot\boldsymbol{r}} \\
T_{I,\text{em}} &= \frac{e}{m} \sum_{q,\hat{e}_q} \sqrt{\frac{\hbar}{2\epsilon_0 \omega_q \mathcal{V}}} \, (\hat{e}_q^* \cdot \boldsymbol{p}) \, a_{q,\hat{e}_q}^\dagger \, e^{-i\boldsymbol{q}\cdot\boldsymbol{r}}
\end{aligned}
\right\}
\tag{18}
$$

The absorption cross-section is obtained by dividing the transition probability $W_{i\rightarrow f}$ summed over all possible final states by the incoming flux of photons. Writing this flux as $c/\mathcal{V}$ [18], we obtain directly

$$
\sigma_{\text{abs}} = 4\pi^2 \alpha \frac{\hbar}{m^2 \omega_q} \sum_f \left| \langle \varphi_f | \hat{e}_q \cdot \boldsymbol{p} \, e^{i\boldsymbol{q}\cdot\boldsymbol{r}} | \varphi_i \rangle \right|^2 \rho(\mathcal{E}_f = \mathcal{E}_i + \hbar\omega_q)
\tag{19}
$$

where $\rho(\mathcal{E}_f = \mathcal{E}_i + \hbar\omega_q)$ is the density of states of the electrons in the final state. Note that the expression of this density of states depends on the normalization condition used for $|\varphi_f\rangle$ (see Friedrich [19] for different examples). Here, $\alpha = e^2/(4\pi\epsilon_0\hbar c)$ is the fine structure constant, a dimensionless coefficient whose numerical value is $\alpha \sim 1/137$.

This simplifies, under the dipole approximation (see Sect. 2.3), as

$$
\sigma_{\text{abs}} = 4\pi^2 \alpha \, \hbar\omega_q \sum_f \left| \langle \varphi_f | \hat{e}_q \cdot \boldsymbol{r} | \varphi_i \rangle \right|^2 \rho(\mathcal{E}_f = \mathcal{E}_i + \hbar\omega_q) \,.
\tag{20}
$$

The corresponding emission cross-section can be derived accordingly. In this case, there is no incoming flux of photons to divide by, but (12) shows that the density of photon final states must be taken into account.

The scattering second-order term in the expression (17) of $T_I^{(2)}$ involves two photons, one created and one absorbed. There are obviously three possibilities for such a time-dependent process. The first one, which corresponds

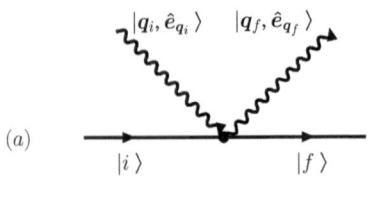

(a)

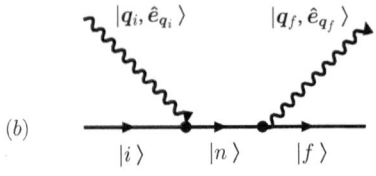

(b)

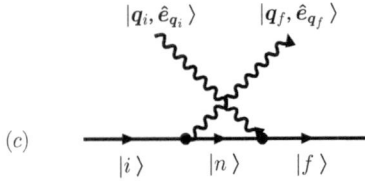

(c)

Fig. 6. Feynman diagrams corresponding to the scattering processes

to the term in $\boldsymbol{A} \cdot \boldsymbol{A}$, is that the incoming photon is absorbed at the same time as the outgoing photon is created. This process is sketched in Fig. 6(a). It leads to the so-called "seagull" diagram [20]. The second possibility, often termed the direct process, is that the outgoing photon is created after the absorption of the incoming one. It corresponds to (b) in Fig. 6. In this case, a virtual electron is excited in the time between the two processes. Finally, the exchange process in Fig. 6(c) is associated to the creation of the outgoing photon *before* the absorption of the incoming one. Both processes (b) and (c) differ only by the nature of the intermediate state. For (b), it corresponds to the atom in state $|n\rangle$ with an excited electron and no photon. In contrast, case (c) has the atom in state $|n\rangle$ plus two photons.

We can now calculate the transition operators for these processes from which the corresponding cross-sections can be deduced. Taking the matrix elements between the initial and final states, we find [20] for case (a)

$$\langle \phi_f | T_{I(a)}^{(2)} | \phi_i \rangle = \frac{e^2}{2m} \frac{\hbar}{2\epsilon_0 \mathcal{V}} \frac{1}{\sqrt{\omega_{q_f} \omega_{q_i}}} \langle \phi_f | (\hat{\boldsymbol{e}}_{\boldsymbol{q}_f}^* \cdot \hat{\boldsymbol{e}}_{\boldsymbol{q}_i}) e^{i(\boldsymbol{q}_i - \boldsymbol{q}_f) \cdot \boldsymbol{r}}$$
$$\left[a_{\boldsymbol{q}_i, \hat{\boldsymbol{e}}_{\boldsymbol{q}_i}} a_{\boldsymbol{q}_f, \hat{\boldsymbol{e}}_{\boldsymbol{q}_f}}^{\dagger} + a_{\boldsymbol{q}_f, \hat{\boldsymbol{e}}_{\boldsymbol{q}_f}}^{\dagger} a_{\boldsymbol{q}_i, \hat{\boldsymbol{e}}_{\boldsymbol{q}_i}} \right] | \phi_i \rangle . \tag{21}$$

In the dipole approximation, where the spatial variation of $\boldsymbol{A}$ is neglected, this becomes

$$\langle \phi_f | T_{I\,(a)}^{(2)} | \phi_i \rangle = \frac{e^2}{2m} \frac{\hbar}{\epsilon_0 \mathcal{V}} \frac{1}{\sqrt{\omega_{q_f} \omega_{q_i}}} (\hat{e}_{q_f}^* \cdot \hat{e}_{q_i}) \langle \varphi_f | \varphi_i \rangle . \tag{22}$$

Therefore, this term only gives a contribution if the initial state $|\varphi_i\rangle$ of the electron involved in the atom is the same as its final state $|\varphi_f\rangle$. In this case, the energy conservation imposes $\omega_{q_f} = \omega_{q_i}$.

When the momentum transfer vector $\boldsymbol{Q} = \boldsymbol{q}_f - \boldsymbol{q}_i$ is not negligible, the dipole approximation is no more valid and the overlap term $\langle \varphi_f | \varphi_i \rangle$ in 22 should be replaced by $\langle \varphi_f | e^{-i\boldsymbol{Q}\cdot\boldsymbol{r}} | \varphi_i \rangle$. For elastic scattering, the initial state and the final state of the atom are the same and this quantity becomes

$$f_T(\boldsymbol{Q}) = \int |\varphi(\boldsymbol{r})|^2 e^{-i\boldsymbol{Q}\cdot\boldsymbol{r}} d\boldsymbol{r} . \tag{23}$$

It is called the atomic form factor and it is the Fourier transform of the electronic density around the atom considered if one can make the approximation that each electron in the atom scatters the photons independently of the others.

The contributions of the two other terms (b) et (c) can be worked out similarly although it is more involved. Recalling that

$$G_0(E_i)|\phi_n\rangle = \lim_{\epsilon \to 0^+} (E_i - H_P - H_R + i\epsilon)^{-1}$$
$$\times |\varphi_n\rangle |\boldsymbol{q}_n, \hat{e}_{q_n}\rangle$$
$$= \lim_{\epsilon \to 0^+} \frac{|\varphi_n\rangle |\boldsymbol{q}_n, \hat{e}_{q_n}\rangle}{E_i - \mathcal{E}_n - \hbar\omega_{q_n} + i\epsilon} \tag{24}$$

with $E_i = \mathcal{E}_i + \hbar\omega_{q_i}$ and $\hbar\omega_{q_n}$ the energy of the photons present in the intermediate state $|\phi_n\rangle$ if any. In the two cases, this gives for the denominator

$$\left\{ \begin{array}{ll} (b) : 0 \text{ photon} & \mathcal{E}_i - \mathcal{E}_n + \hbar\omega_{q_i} + i\epsilon \\[2mm] (c) : \omega_{q_i} \text{ and } \omega_{q_f} \text{ photons} & \mathcal{E}_i - \mathcal{E}_n - \hbar\omega_{q_f} + i\epsilon \end{array} \right\} \tag{25}$$

Replacing into the expression (17) of $T_I^{(2)}$ gives [20–22]

$$\langle \phi_f | T_{I\,(b)}^{(2)} | \phi_i \rangle = \left(\frac{e}{m}\right)^2 \frac{\hbar}{2\epsilon_0 \mathcal{V}} \frac{1}{\sqrt{\omega_{q_f} \omega_{q_i}}}$$
$$\times \sum_n \frac{\langle \varphi_f | \hat{e}_{q_f}^* \cdot \boldsymbol{p}\, e^{-i\boldsymbol{q}_f\cdot\boldsymbol{r}} | \varphi_n \rangle \langle \varphi_n | \hat{e}_{q_i} \cdot \boldsymbol{p}\, e^{i\boldsymbol{q}_i\cdot\boldsymbol{r}} | \varphi_i \rangle}{\mathcal{E}_i - \mathcal{E}_n + \hbar\omega_{q_i} + i\epsilon} \tag{26}$$

$$\langle \phi_f | T_{I\,(c)}^{(2)} | \phi_i \rangle = \left(\frac{e}{m}\right)^2 \frac{\hbar}{2\epsilon_0 \mathcal{V}} \frac{1}{\sqrt{\omega_{q_f} \omega_{q_i}}}$$
$$\times \sum_n \frac{\langle \varphi_f | \hat{e}_{q_i} \cdot \boldsymbol{p}\, e^{i\boldsymbol{q}_i\cdot\boldsymbol{r}} | \varphi_n \rangle \langle \varphi_n | \hat{e}_{q_f}^* \cdot \boldsymbol{p}\, e^{-i\boldsymbol{q}_f\cdot\boldsymbol{r}} | \varphi_i \rangle}{\mathcal{E}_i - \mathcal{E}_n - \hbar\omega_{q_f} + i\epsilon} . \tag{27}$$

The sum here is over all the unoccupied states $|\varphi_n\rangle$ and the limit has been omitted to simplify the notation.

Adding the three terms together, dividing by the incoming flux $c/\mathcal{V}$ and multiplying by the density of photons in the final state [according to 12] given by [18]

$$\frac{\mathcal{V}}{(2\pi)^3} \frac{(\hbar\omega_{q_f})^2}{\hbar^3 c^3} .$$

(28)

the cross-section for the scattering case becomes

$$\frac{d\sigma}{d\Omega} = r_o^2 \frac{\omega_{q_f}}{\omega_{q_i}} \left| \hat{e}_{q_f}^* \cdot \hat{e}_{q_i} \langle\varphi_f| e^{-i\boldsymbol{Q}\cdot\boldsymbol{r}} |\varphi_i\rangle \right.$$

$$\frac{1}{m} \left[\sum_n \frac{\langle\varphi_f|\hat{e}_{q_f}^* \cdot \boldsymbol{p} \, e^{-i\boldsymbol{q}_f\cdot\boldsymbol{r}}|\varphi_n\rangle\langle\varphi_n|\hat{e}_{q_i} \cdot \boldsymbol{p} \, e^{i\boldsymbol{q}_i\cdot\boldsymbol{r}}|\varphi_i\rangle}{\mathcal{E}_i - \mathcal{E}_n + \hbar\omega_{q_i} + i\epsilon} \right.$$

$$\left. \left. + \frac{\langle\varphi_f|\hat{e}_{q_i} \cdot \boldsymbol{p} \, e^{i\boldsymbol{q}_i\cdot\boldsymbol{r}}|\varphi_n\rangle\langle\varphi_n|\hat{e}_{q_f}^* \cdot \boldsymbol{p} \, e^{-i\boldsymbol{q}_f\cdot\boldsymbol{r}}|\varphi_i\rangle}{\mathcal{E}_i - \mathcal{E}_n - \hbar\omega_{q_f} + i\epsilon} \right] \right|^2 .$$

(29)

Here $d\Omega$ is a solid angle about $\boldsymbol{q}_f$ and $r_o = e^2/(4\pi\epsilon_0 mc^2)$ is the classical radius of the electron. This result is known as the Kramers–Heisenberg cross-section for the scattering of light by atomic electrons.

Elastic scattering corresponds to $\omega_{q_f} = \omega_{q_i}$ while in the case of inelastic scattering, $\omega_{q_f} \neq \omega_{q_i}$ and therefore $|\varphi_f\rangle \neq |\varphi_i\rangle$. Low-energy (i.e., $\hbar\omega_{q_i} \ll \mathcal{E}_I$, the ionization energy of the atom) elastic scattering is called Rayleigh scattering while high-energy ($\hbar\omega_{q_i} \gg \mathcal{E}_I$) scattering is called Thomson scattering. Low-energy inelastic scattering is called Raman scattering. When $\hbar\omega_{q_f} < \hbar\omega_{q_i}$, it is called Raman Stokes while the opposite case is called Raman anti-Stokes. Finally, high-energy inelastic scattering is called Compton scattering.

The cross-section of Rayleigh scattering can be shown to be proportional to $\omega_{q_i}^4$ [22,23]. This low-energy behaviour is responsible for the color of the sky as blue wavelengths will be more scattered than red [23]. In the case of Thomson scattering, the denominator of the second and third term in (29) becomes very large and these contributions can be neglected. For this reason, the first term is usually termed the Thomson term. If we look again at the expression (29) of the scattering cross-section, we see that the first dispersive term, the direct contribution, can become singular when the energy of the incoming photon is equal to $(\mathcal{E}_n - \mathcal{E}_i)$ i.e., to the excitation energy of the electron. Actually, it does not become completely singular as the intermediate state $|\varphi_n\rangle$ is short-lived [1] and so it has a finite and small lifetime that should

---

[1] As discussed by Hague [65], the term "virtual" electron should be restricted to the case where the energy of the intermediate state is far away from an absorption

be added to the denominators in (29). Nevertheless, a strong enhancement of the cross-section will result when the energy of the incoming wavefield goes through an absorption edge. This is called resonance (or anomalous) scattering. For this reason, the direct term in the Kramers–Heisenberg cross-section is often called the resonant term while the third term, which cannot be singular, is called the non-resonant one.

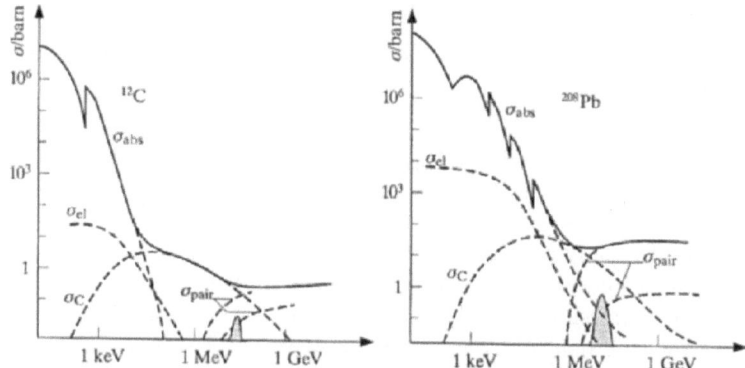

Fig. 7. Energy dependence of the cross-sections for the interaction of x-rays with a carbon and lead atom. The full line corresponds to the total cross-section σ_{tot}, σ_{abs} is the photoelectric absorption, σ_{el} corresponds to elastic scattering and σ_C to inelastic (Compton) scattering, σ_{pair} to the creation of electron-positron pairs. The shaded areas are the nuclear absorption cross-section [24]

Since x-ray scattering is a second-order process (in e/m), it is expected to have a much lower cross-section than first-order optical processes such as x-ray absorption or photoemission. This is indeed the case as can be seen from Fig. 7. Here, we can see that the cross-section for the various scattering processes is several orders of magnitude lower than that of photoabsorption. Resonance scattering is therefore more favorable as far as detection is concerned.

Under resonance conditions, the non-resonant term in expression (29) of the cross-section i.e., the second dispersive term, can usually be neglected. It is then customary to decompose the resonant term, the only complex term remaining as the Thomson term is real, into its real and imaginary part, noted respectively $f'(\hbar\omega_{q_i})$ and $f''(\hbar\omega_{q_i})$. Using the well-known relation [25]

$$\frac{1}{x - x_o + i\epsilon} = \mathcal{P}\left(\frac{1}{x - x_o}\right) - i\pi\,\delta(x - x_o)\,, \tag{30}$$

where the symbol $\mathcal{P}$ stands for Cauchy's principal value, defined as

---

edge i.e., for non resonant processes. Otherwise, relaxation can occur before the decay of this state which does not make it "virtual" anymore.

$$\mathcal{P}\left(\frac{1}{x-x_o}\right) = \frac{1}{2} \lim_{\epsilon \to 0^+} \left[\frac{1}{x-x_o+i\epsilon} + \frac{1}{x-x_o-i\epsilon}\right] \qquad (31)$$

we obtain for the imaginary part

$$f''(\hbar\omega_{q_i}) = -\frac{\pi}{m} \sum_n \langle\varphi_f|\hat{e}_{q_f}^* \cdot p\, e^{-iq_f \cdot r}|\varphi_n\rangle$$

$$\times \langle\varphi_n|\hat{e}_{q_i} \cdot p\, e^{iq_i \cdot r}|\varphi_i\rangle\, \delta(\mathcal{E}_i - \mathcal{E}_n + \hbar\omega_{q_i}) . \qquad (32)$$

In the dipole approximation, the factors $e^{-iq_f \cdot r}$ and $e^{iq_i \cdot r}$ disappear from the matrix elements.

2.3 The Dipole Approximation

We have mentioned several times in the previous subsections the dipole approximation. In many case, this approximation is sufficient to account for the experimental results so we will look at it in more details now.

From the expression of the Kramers–Heisenberg cross-section (29) or from the corresponding one for emission or absorption processes 20, we see that we need to calculate matrix elements of the form $\langle\varphi_n|\hat{e}_q \cdot p|\varphi_m\rangle$ where $|\varphi_m\rangle$ and $|\varphi_n\rangle$ are eigensolutions of the Hamiltonian H_P. The momentum vector p can be rewritten as [26]

$$p = i\frac{m}{\hbar} [H_P, r] . \qquad (33)$$

We deduce from this result that

$$\langle\varphi_n|\hat{e}_q \cdot p|\varphi_m\rangle = i\frac{m}{\hbar} (\mathcal{E}_n - \mathcal{E}_m) \langle\varphi_n|\hat{e}_q \cdot r|\varphi_m\rangle . \qquad (34)$$

As the interaction potentials $\hat{e}_q \cdot p$ and $\hat{e}_q^* \cdot p$ induce a transition of energy $\hbar\omega_q$ corresponding respectively to the absorption or the emission of a photon, we have finally

$$\begin{cases} \langle\varphi_n|\hat{e}_q \cdot p|\varphi_m\rangle = im\,\omega_q\, \langle\varphi_n|\hat{e}_q \cdot r|\varphi_m\rangle \\[2mm] \langle\varphi_n|\hat{e}_q^* \cdot p|\varphi_m\rangle = -im\,\omega_q\, \langle\varphi_n|\hat{e}_q^* \cdot r|\varphi_m\rangle . \end{cases} \qquad (35)$$

This latter matrix element can be calculated easily once the eigensolutions of the particle (atomic) Hamiltonian H_P are known. When H_P contains a potential that depends only on the radial distance r, the angular part of the eigensolutions $\langle r|\varphi_n\rangle$ is then conveniently described using spherical harmonics $Y_l^m(\hat{r})$.

The angular part of the matrix element $\langle\varphi_n|\hat{e}_q \cdot r|\varphi_m\rangle$ can be expressed easily in terms of the Wigner's $3j$ symbols. The definition of the latter implies the so-called dipole selection rule

$$\left\{ \begin{array}{l} l_n = l_m \pm 1 \\ \\ m_n = m_m, m_m \pm 1 \end{array} \right\} \tag{36}$$

Transitions from $|\varphi_m\rangle$ to $|\varphi_n\rangle$ that do not follow this rule will have a zero matrix element.

The interaction potential $\hat{e}_q \cdot p$ does not act on spin and therefore the transitions allowed will conserve the spin ($s_n = s_m$), where s_n is the spin quantum number of the state $|\varphi_n\rangle$. In this case, when the eigenstates should be described by the (j, m_j) quantum numbers instead of (l, m_l, s), the selection rules will give

$$\left\{ \begin{array}{l} j_n = j_m, j_m \pm 1 \\ \\ m_n = m_m, m_m \pm 1 \end{array} \right\} \tag{37}$$

Note that usually the transition $l_m \to l_m + 1$ dominates [27].

3 By-Products of the Photon-Atom Interaction

3.1 The Various Physical Processes

In the previous section, the atom was modelled as an electron plus a potential that binds it. This was sufficient for the discussion of the basics of electron-atom interaction but it is certainly not realistic enough to account for the experiments as many-body effects will strongly influence the experimental spectra. Reality differs at least in two significant ways from this simplified model. First, the atom is generally not alone in the sample, but it is surrounded by other atoms. This will have a dramatic effect on the resulting signal as the excited – or escaping – electron will feel the surrounding. Then the absorption of a photon resulting in a core level excitation leaves a core hole behind. The remaining electrons on the photoabsorber and neighbouring atoms will react by screening and then filling the core hole. This will either affect the outcome of the experiment or give rise to new phenomena that can be exploited to gain some more information on the sample. Fig. 8 shows the various physical processes that can occur during such a photon-sample interaction. It has been restricted to photon energies below the γ-ray region. Above this range, new physical processes, such as the production of electron-positron pairs or photonuclear absorption, can occur. The typical cross-sections for these processes were shown in Fig. 7.

3.2 The Deexcitation Processes

As sketched in Fig. 8, there are basically two decay channels by which the core hole will be filled by a higher energy electron. These deexcitation processes are represented schematically in Fig. 9.

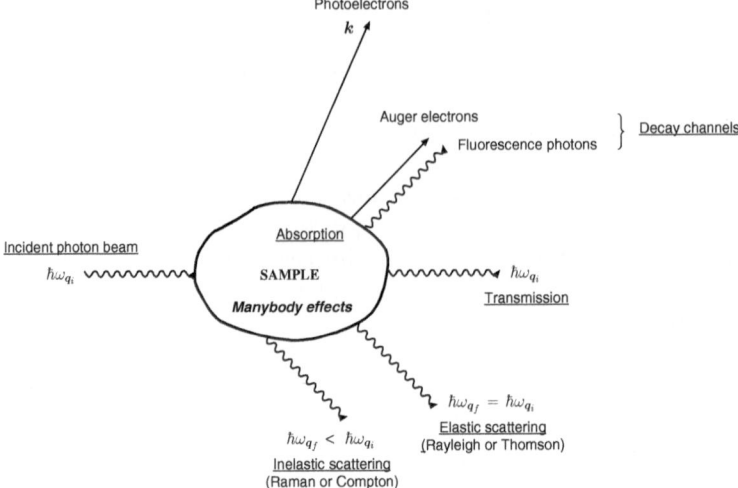

Fig. 8. The different physical processes resulting from the interaction of an x-ray photon beam with matter

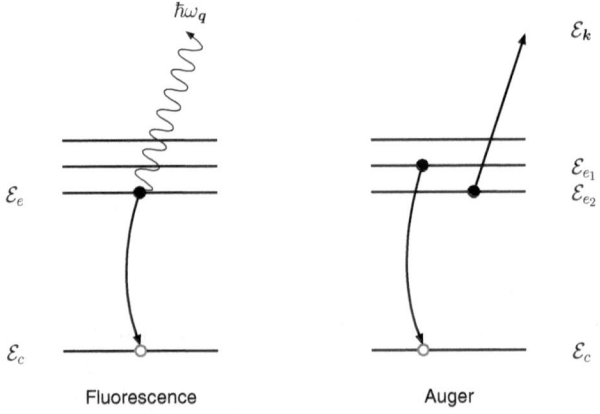

Fig. 9. Graphical representation of the fluorescence and Auger decay channels

In the case of fluorescence (radiative decay), the extra energy coming from the core hole filling is released in the form of a photon of energy $\hbar\omega_q = \mathcal{E}_e - \mathcal{E}_c$, $\mathcal{E}_e$ being the binding energy of the electron filling the core hole and $\mathcal{E}_c$ that of the latter. When resulting from the deexcitation of a valence electron, this photon carries information on the atom it originates from and its immediate surroundings. It can therefore be used to gain information on the sample.

The second decay channel is when the electron that fills the core hole communicates its energy to another electron. This latter electron, labelled e_2 in Fig. 9, can then escape from the sample if this energy is sufficient. This is the Auger effect. Neglecting relaxation effects due to the core hole, energy conservation implies now that $\mathcal{E}_k = \mathcal{E}_{e_1} - \mathcal{E}_c - \mathcal{E}_{e_2}$. Here again, the escaping electron will carry information on the atom. Moreover, as it will encounter the surrounding atoms on its way out, its wave function will also contain information on the neighboring atoms. When the electron that fills the core hole is on the same shell as the core hole, the corresponding Auger process is termed Coster–Kronig. Because of the strong overlap of the wave functions, this kind of transition will occur much faster than that for standard Auger processes. When the three electrons involved belong to the same shell, one speaks of a super Coster–Kronig process.

These decay channels will influence directly the energy resolution of the spectra. Indeed, if we call as before $W_{i \to f}$ the transition probability per unit time from state $|\phi_i\rangle$ of the atom to state $|\phi_f\rangle$ under one of these two deexcitation processes, then the total transition probability from state $|\phi_i\rangle$ is

$$P_i = \sum_f W_{i \to f} \,. \tag{38}$$

The lifetime of the core hole is then

$$t_i = \frac{1}{P_i} \,. \tag{39}$$

This finite lifetime will induce, according to the uncertainty relations, an energy broadening (Lorentzian as a first approximation) given by

$$\Delta \mathcal{E}_i = \frac{\hbar}{t_i} = \hbar P_i = 2\Gamma_i \,. \tag{40}$$

Here, Γ_i is the half-width at half-maximum. So, in principle, as soon as we know how to calculate the transition probability $W_{i \to f}$ of the various processes, according to the expression (12), we can compute the broadening Γ due to the core hole lifetime as $\Gamma = \sum_i \Gamma_i$, the sum being on all the decay channels. Figure 10 gives the contribution to the energy broadening of the two decay channels as a function of the atomic number Z for the K edge excitation. We see from this figure that the Auger process dominates up to the atomic number of gallium while fluorescence is the main decay channel for heavier elements. A similar trend is observed for higher excitation edges [27] with the Auger decay being dominant for even higher Z.

Strictly speaking, these two physical processes should be considered as second-order ones as they cannot be separated from the photoionization process they originate from.

3.3 Many-Body Effects

As mentioned before, the simple one-electron picture we gave of the photon-atom interaction does not account for many features that can appear in the

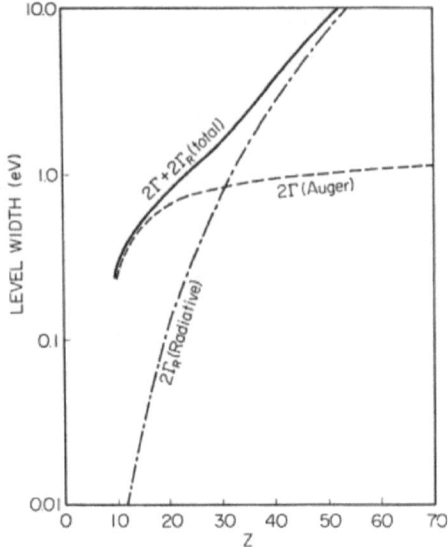

Fig. 10. K-level width due to the fluorescence and Auger decay channels as a function of the atomic number Z [28]

experimental spectra. These features, coming from many-body effects, are the fingerprint of the response of the constituents of the sample (electrons, nuclei, ...) to the creation of the excited electron-core hole pair. We will review now the most common ones and discuss their effect on the spectra.

Electron-Core Hole-Interaction

Close to the excitation edge, when the core hole is not sufficiently screened, it will interact strongly with the excited electron via a long-range Coulomb potential. This can affect dramatically the spectra. For instance, this inter-action is held responsible for the strong deviation observed in the branching ratio of the $L_{2,3}$ edge of early $3d$ metals i.e., the ratio of the jumps of the absorption coefficient at the L_3 and L_2 edges, from the ideal value of 2:1 [29]. Many approaches have been devised to include this effect as accurately as possible in the calculations. A simplified approach by Fonda [30] keeping the one-electron view describes this interaction as an extra coulombic potential which is then included into the multiple-scattering treatment of the photo-electron interaction with the sample. Other methods more involved, based on many-body approaches, have also been proposed to include this effect [31,32].

Plasmons and Core Hole Screening

The creation of the core hole and therefore that of the core hole-photoelectron dipole, disturbs the atom and leaves it in an unstable state. As seen before, the surrounding electrons will eventually react by filling the core hole so that the atom can lower its energy. But before the filling takes places, the core hole will generally be screened by the surrounding electrons. In other words, the removal of a core electron significantly alters the effective potential seen by the other electrons. As a consequence, they will relax into eigenstates of a new Hamiltonian. This is the screening process. The effects induced will essentially depend on the timescale of this change of effective potential [33]. When this change is slow i.e., for low kinetic energies of the excited electron, the other electrons will be always in the ground state corresponding to the instantaneous effective potential as they will have time to adjust their wavefunction. This is the so-called adiabatic approximation. The escaping electron being in this case slow to move away from the atom, it will be able to gain the relaxation energy, both from the other electrons of the atom (intraatomic relaxation) and from the conduction electrons in metals (extraatomic relaxation). On the opposite, when the kinetic energy of the photoelectron is relatively high (sudden approximation), the photoelectron and the core hole will be screened separately. This screening process will now appear in the lower energy side of the photoemission spectra as plasmon peaks both intrinsic (screening of the core hole) and extrinsic (screening of the photoelectron outside the atom). Note that only intrinsic plasmons peaks carry information about the core hole.

Electron-Hole Pairs

In the case of metals, where the Fermi level is within the conduction band, it is extremely easy to excite electrons to energies right above this Fermi level as it does not cost much energy. This excitation of electron-hole pairs can be done in two ways. One, which is usually neglected, is due to the interaction of the escaping electron with the Fermi sea electrons. The second process appears within the sudden approximation: the sudden appearance of the core hole-photoelectron dipole will create a strong local potential (in the adiabatic approximation this potential can be considered as screened) that will be able to excite such electron-hole pairs. This was first investigated theoretically by Nozières, De Dominicis [34] and Mahan [35]. In the case of the photoemission spectra for instance, the creation of electron-hole pairs induces an asymetric profile for the core level peak as extra losses will manifest on the lower energy side, following a $1/\mathcal{E}^{1-\alpha}$ rule, with α being the asymmetry parameter. Here, $\mathcal{E}$ is the energy relative to the edge. For x-ray absorption spectra, it leads to a complicated pattern resulting from two competing effects: a threshold singularity giving an enhancement of the intensity near the edge [35], and a reduction of this intensity due to the so-called Anderson orthogonality catastrophe [36]. This leads to a $1/\mathcal{E}^{\alpha_l}$ decrease law after the edge. The exponent $\alpha_l = 2\delta_l/\pi - \alpha$ reflects the two competing effects, the enhancement induced by the phaseshift δ_l of the core hole potential (l corresponds to the electron final state) and the reduction by the asymmetry parameter α.

Phonons

There are three basic processes for photoelectron-phonon interaction. The first one is the interaction of the escaping electron with phonons on its way outside the sample. This is usually described by including a specific damping in the photoelectron wave function. It has generally the effect of smoothing the spectra. A second mechanism comes from the recoil energy communicated by the photoelectron to the absorbing atom [37]. Except for very light elements or very energetic photoelectrons, this is usually negligible. The last mechanism involves phonons due to the relaxation of the nuclei that follow the sudden appearance of the core hole. This will affect the lineshape of the core level photoemission peak by adding a Gaussian broadening [38].

Electronic Correlations

The model developed so far does not take into account two-particle interactions such as the Coulomb repulsion. These effects can become important when the absorber has a predominantly atomic character so that there is a possibility of substantial overlap between the wavefunction of the electron involved in the spectroscopy and valence electrons on the absorber. This leads to the so-called multiplet effect. It is particularly important, for low-energy excitations (XANES for instance) of shallow core levels in materials with incomplete valence shell such as rare-earth, actinide or transition metal compounds (final state effect). A similar effect occurs when an electron is ejected from an incomplete shell (initial state effect). This effect can be understood in terms of configurations. Let us consider the excitation of a $3p$ electron in a material with an incomplete $3d$ shell. If we suppose all the lower shells to be full and the upper ones empty, we have for dipole selection rules ($\Delta l = \pm 1$)

$$\hbar\omega + 3p^6 3d^N \quad \longrightarrow \quad \left\{ \begin{array}{c} 3p^5 3d^{N+1} \\ \\ 3p^5 3d^N 4s^1 \end{array} \right\} \tag{41}$$

The transition towards $4s$ levels being usually less probable, we will not consider it any further. The $3p^5 3d^{N+1}$ state is highly degenerate. For instance, $3p^5 3d^2$ offers 270 possibilities and $3p^5 3d^5$ up to 1512 [5]. The various interactions (spin-orbit, Coulomb, exchange, crystal field, ...) will lift this degeneracy and so, the numerous different levels that can be reached following the excitation selection rules will appear with different weights and energies. The energy shift will depend on the magnitude of the interaction potentials. Due to the different broadenings affecting the spectrum (and hence the resolution of the experiment), only a few of them will be visible with a position depending on the interplay between the dominant interactions. Therefore, for systems or spectroscopies where these interactions are important, the spectrum will exhibit the dominant multiplet components in the form of a main line and satellite lines (that can appear as shoulders if not sufficiently resolved).

3.4 Characteristic Timescales

It is interesting to have an idea of the timescale on which the various processes we have just discussed occur. This helps to construct a suitable theory to incorporate them, and it is important as well for the interpretation of the results obtained. Figure 11, adapted from Gadzuk [33] gives a summary of these timescales.

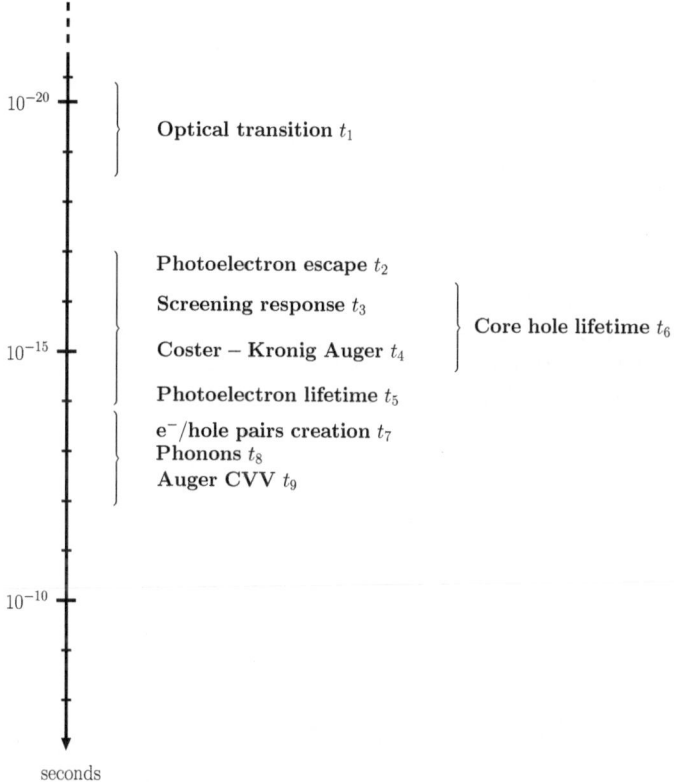

Fig. 11. Timescales in x-ray spectroscopies [33]

   The timescale t_1 of the optical transition is defined by $t_1 = 1/W_{i \to f}$. The escape time t_2 of the photoelectron outside the limits of the atom can be estimated as $t_2 \sim a_0 \sqrt{m/E_k}$ [39] where a_0 is the Bohr radius. The screening time t_3 can be approximated as $t_3 \sim \hbar/E_R$ where E_R is the relaxation energy. t_4, like t_1 and t_7, is the inverse of the corresponding transition probability per unit time. The lifetime of the photoelectron can be estimated as $t_5 \sim \lambda_e/\sqrt{2E_k/m}$ [40]. That of the core hole t_6 depends on the deexcitation processes involved to fill this hole. Typical lifetimes can be extracted from the width represented in Fig. 10.

It is important to note here that the characteristic times t_3 and t_2 can be of the same order. Hence, when the photoelectron escapes the absorbing atom faster than it takes time for the screening process (i.e., in the high kinetic energy range) to be completed, then the electron will feel the screening process such as plasmons oscillations and therefore plasmon peaks will appear in the photoelectron spectra. As we have already seen, this is called the sudden approximation regime because it can be mimicked by assuming that both the photoelectron and the core hole are created suddenly. On the opposite, when the photon energy is close to an absorption edge, the surrounding electrons will have time to completely screen the core hole before the photoelectron escapes. This is the adiabatic range. In this case, it is the screened core hole that the photoelectron will see on its way outside the limits of the atom. Hence, no collective excitations are expected there. Actually the fact that the lifetime of the electrons produced as a result of the creation of the core hole (photoelectron, Auger) overlaps that of the various relaxation processes indicates that spectroscopies based on them should be described within a one-step model.

We can sum up the various timescales in the following way [33]. The absorption of the photon is almost instantaneous. For high kinetic energies, the potential of the core hole appears very suddenly while a photoelectron excited just above the threshold will stay around the hole quite a long time with almost no net charge seen from the outside; this is the adiabatic limit. Once the potential of the hole begins to appear, the response of the solids sets in. Screening by plasmons starts ($\sim 10^{-16}$ s). Then ($\sim 10^{-15}$ s), various Auger processes can occur, involving essentially core electrons, or maybe one valence electron. This can result in the moving of the hole from one core state to another. However, this will not influence significantly the electron gas excitation as there will still be a hole localized on the atom. By now the hole potential is totally screened. In the case of metals, low-energy electron-hole pairs begin to respond to the presence of this screened potential. For molecules on surfaces this is also the time where intra-molecular vibrational modes are excited. Later ($\sim 10^{-13}$ s), phonons can be excited in the sample. Finally, the key decay process for low Z elements, CVV Auger, begins between 10^{-14} s and 10^{-12} s. Therefore, any relaxation process needing a longer time will be irrelevant because the core hole is now filled.

3.5 Sensitivity to Magnetic and Other Properties

The properties of the sample that can be investigated are numerous, and the particular technique(s) required to get this information will depend on both the nature and the scale of the structure of interest. As discussed by Van Hove [41], spectroscopies cover basically three different scales. The mesoscale will deal with morphology, islands, domains and boundaries, chemical composition and defects, while on the nanoscale one is generally interested by polyatomic assemblies. On the other hand, the atomic scale will give access to the local crystallographic structure, the bandstructure and the bonding or the spin arrangement. All these types of information will be accessible by spectroscopies, but the way to retrieve it from the experimental spectra will mainly depend on the scale of interest. Roughly speaking, the mesoscopic scale will only require qualitative structure determination and therefore no detailed theoretical analysis and computer modelling will be necessary. On the contrary, the atomic scale will heavily rely on sophisticated calculations based on a state of the art theoretical model to extract the information with the required accuracy. The nanoscale lying in between, it will usually involve a mix of qualitative and quantitative structure determination. A detailed comparison of the main atomic scale techniques for the determination of surfaces and interfaces (namely LEED, XRD, PhD, Surface EXAFS, IS and STM) has been carried out by Van Hove and we refer the reader to it for a more detailed account.

From the point of view of magnetic properties, we can see from Sect. 2 that the sensitivity of the x-ray spectroscopies can be achieved in three ways:

(i)   by the spin detection of the photoelectron when the spectroscopy involves the measure of an electron flux like in photoemission

(ii)   by making use of the dichroic effect, i.e., the difference between the signals obtained by reversing the photon polarization or the magnetization. This magnetic dichroic effect results from an interplay of the spin-orbit interaction and the magnetic exchange coupling. Sum rules can then relate the measured quantities to spin or orbital magnetic moments

(iii) by focusing, for second-order optical spectroscopies, on the two magnetic terms giving rise to spin-dependent scattering. Although their amplitude is smaller by $\hbar\omega/mc^2$ than that of the Thomson term [7], the use of circularly polarized light will give an important interference term where spin and orbital magnetization densities can be discriminated. Moreover, the resonant mode will lead to an enhancement of the magnetic scattering relative to the charge scattering

(i) and (ii) are based on first-order processes while (iii) correspond to second-order ones. These magnetic spectroscopies will be described in more details in separate chapters.

4 An Overview of Some Particular Spectroscopies

We will review here very briefly the basics of four particular spectroscopies. The first three ones, XAS, PhD and DAFS are widely used and originate from first-order and second-order optical processes. Moreover, they are intimately related to one another through the optical theorem. The fourth one, APECS, is more recent but seems very promising and quite original as it couples together two processes and makes use of the coincidence of these processes to gain more information on the sample.

4.1 X-ray Absorption Spectroscopy

Standard X-ray Absorption Spectroscopy

X-ray absorption spectroscopy is based on the measure of the absorption of photons as a function of the energy of the beam in the vicinity of an edge i.e., of an energy corresponding to the binding energy of a given core level. Typical x-ray absorption spectra are shown in Fig. 12 both for the case of an isolated atom and when the absorbing atom is embedded into a lattice. Three regions can be distinguished in the spectra. When the photon energy

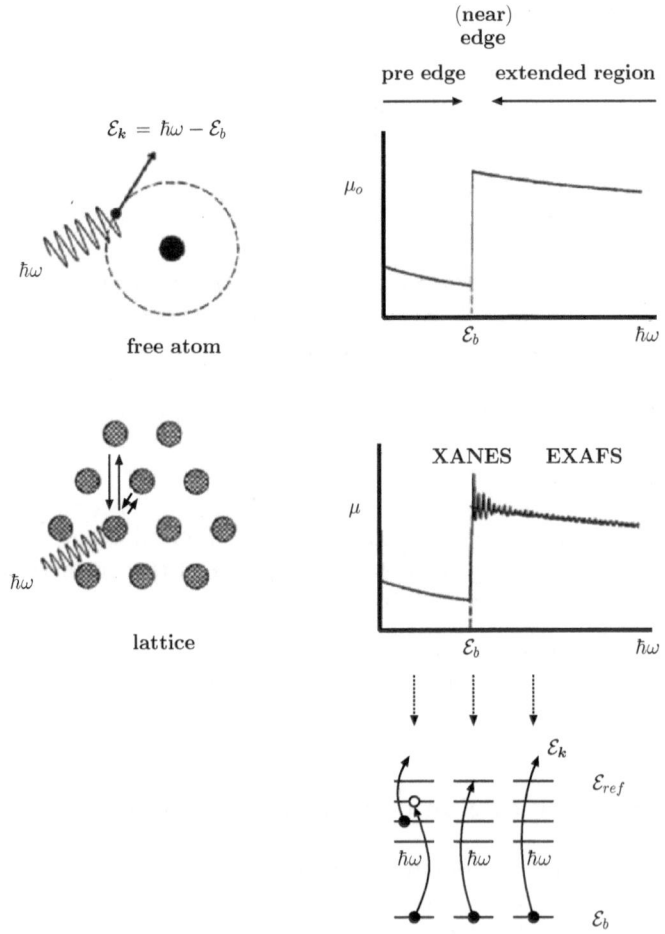

Fig. 12. Absorption of x-rays as a function of the photon energy $\hbar\omega$ by a free atom and by atoms in a lattice. The transitions contributing to the various regions are represented at the bottom (adapted from [3])

$\hbar\omega$ is lower than the binding energy $\mathcal{E}_b$ of the core level considered, the corresponding electron cannot be excited. This pre-edge region contains essentially contributions from the excitation of less energetic core levels (the tails of the corresponding absorption edges) if any or, for ions for instance, it corresponds to the excitation of the core electron to empty states of lower binding energy if any. Then the edge corresponds to $\hbar\omega = \mathcal{E}_b$. When $\hbar\omega$ is slightly higher than E_b, the excited electron will start to populate the empty states of the sample. This is the near-edge region and here the spectrum reflects the electronic structure. For non correlated materials, this is essentially the density of empty states while for correlated systems, the multiplet structure dominates the spectra. XAS is then called XANES (X-ray Absorption Near Edge Spectroscopy). Finally, for higher energies, the photoelectron is excited into the continuum and it leaves the atom or dies out through multiple energy losses. This is the extended region and XAS is called here EXAFS (Extended X-ray Absorption Fine Structure). It still reflects the density of states of the sample.

For isolated and non correlated atoms, as sketched in Fig. 12, the near edge and extended regions are essentially structureless. For metals, their behaviour is as $1/(\hbar\omega - \mathcal{E}_b)^{\alpha_l}$ [35] (see Sect. 3). But as soon as the absorbing atom is surrounded by other atoms, oscillation-like structures appear in both regions. Roughly speaking, these oscillations reflect the density of empty states in the XANES region and the local arrangement of atoms around the absorber in the EXAFS region. In both cases, these oscillations are caused by the interference between the photoelectron wave function and the parts of it backscattered by the neighboring atoms towards the absorber. Local structure around the absorber can therefore be extracted from the EXAFS spectrum, and especially the radial distribution around it. Actually, the EXAFS oscillations observed are essentially due to the nearest neighbors shell with the other neighbors contributing as a more or less weak perturbation, depending on the backscattering properties of the other neighbors and their distance to the absorber. This can be seen very clearly on Fig. 13. Here, the Cu K edge EXAFS spectra at 100 K of Cu and Cu-Ru clusters on SiO_2 are shown together with their Fourier transform (which gives the radial distribution) and the inverse transform of the contribution of the first coordination shell. The similarity between the true EXAFS spectrum and the one resulting from the first shell of neighbors is striking.

Historically, XAS spectra were recorded in transmission by comparing the initial intensity I_0 of the beam to the transmitted one I. The absorption coefficient μ, for a sample of thickness t is then defined as

$$\mu = \frac{1}{t} \ln\left[\frac{I_0}{I}\right]. \tag{42}$$

For surface studies, this detection mode is not possible as the bulk information will then dominate the signals. In this case, other detection modes

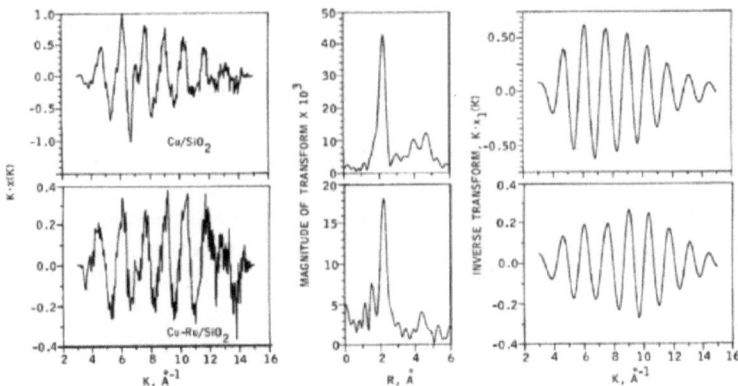

Fig. 13. Normalized EXAFS data at 100 K, with associated Fourier transforms and inverse transforms, for silica supported copper and ruthenium-copper clusters. The inverse transforms correspond to nearest neighbours only [44]

are used. They are based on the fact that the cross-section of the decay channels, Auger deexcitation and fluorescence, is generally proportional to the absorption coefficient μ. A more comprehensive discussion, including the cases where this is not completely true, can be found in [42, 43]. Due to the low escape depth of electrons, the Auger decay channel will be particularly suited to surface studies.

X-ray Magnetic Circular Dichroism

XMCD has been in the last decade one of the most successful technique to investigate magnetic properties in materials. Being derived from XAS, we give here a short account of the basic ideas underlying this spectroscopy.

Historically, "dichroism" was referring to the change of color of some material when absorbing light along two different directions, hence the name. Nowadays, in the case of absorption, this term is more generally used to characterize the dependence of the measured spectra on the polarization of the incident light.

The physical grounds on which dichroism can appear can be better understood within the simplified (one-electron) two-step model proposed by Stöhr and Wu [45]. As we discussed it in Sect. 2, the absorption process is strongly dominated by the dipole term $\hat{e}_q \cdot r$. This term imposes selection rules (see Sect. 2.3) to the electron excitation process. This means that the absorbed photon transfers its angular momentum to the electron. As the dipole operator does not affect spin variables, no spin polarization is expected from this process if the initial core state is a pure state. But this changes as soon as the orbital and spin angular momenta of the core electron are coupled. Indeed, if spin-orbit interaction is present, the photon angular momentum will be transferred to the total angular momentum $j = l + s$ of the electron and therefore part of it will be transferred to the spin s. Upon reversal of the helicity of the light, the sign of the photon angular momentum will change and as a consequence, the amount transferred to the spin will be opposite. Therefore, absorption of a circularly polarized photon by a spin-orbit-resolved core level creates an internal source of spin-polarized electrons. The degree of spin-polarization depends both on the helicity of the photon and on the edge considered. For instance, for a L_2 edge ($j = l - s$), left circularly polarized (LCP) light will excite 1/4 of spin up electrons and 3/4 of spin down electrons, with opposite values for right circularly polarized (RCP) light. Spin up and spin down refer to the direction of the light. The convention taken here is that of Born and Wolfe (see [46] for a discussion of the two conventions) so that LCP corresponds to a helicity of +1. In the case of the L_3 edge ($j = l + s$), LCP will excite 5/8 of spin up electrons and 3/8 of spin down electrons. The spin polarization is opposite at the two edges as expected from the expressions of their momentum j.

The second step of the model concerns the final state. As long as the number of empty spin up and empty spin down states in the valence shell is the same, the two types of photons will be absorbed identically. But for materials where the exchange interaction splits the final states, a desequilibrium appears between these numbers of empty states, giving rise to a net magnetic moment. Up and down now refer to the direction of magnetization. The dipole operator being spin-independent, no spin-flip is possible in the excitation process, which means that spin up core electrons will only populate the empty spin up valence levels and likewise for the opposite spin. As a consequence, the exchange-split valence states will act as a spin detector. The transition intensity will then be proportional to the number of empty states of a given spin and so absorption becomes sensitive to the spin. This dichroic effect scales as the cosine of the angle between the photon direction and the magnetization direction. It will be therefore maximum when these directions are parallel or antiparallel.

We have discussed so far the reversal of the helicity but the magnetization of the sample can also be reversed. As demonstrated formally by Brouder and Kappler [42], changing the sign of the magnetization is completely equivalent to reversing the helicity of the photon in the case of absorption. Note however that this is not always true for dichroism in photoemission.

This simple two-step model, although not valid for all materials, shows qualitatively how the use of circularly polarized light in absorption can allow access to information on the spin of the final state in magnetic materials, with the further advantage of elemental specificity which is common to all core electron spectroscopies. The quantitative analysis of the experimental data can be made in two ways: by comparison to simulations performed within a suitable theoretical model (monoelectronic such as multiple-scattering or band structure approaches, or using multiplet theory when electron correlations become important) or by directly extracting the information through sum rules. These sum rules (see Altarelli and Sainctavit [47] for a thorough discussion) relate the ground state expectation values $\langle L_z \rangle$ and $\langle S_z \rangle$ to integrated features in the XAS and XMCD spectra. They are not always valid (they derive from a one-electron picture) but have proved extremely useful in many cases, especially for magnetic transition metals and their compounds.

4.2 Photoelectron Diffraction

Photoelectron diffraction, and more generally photoemission, consists in measuring the intensity of photoelectrons that escape from the sample as a result of an excitation by photons of energy $\hbar\omega$. The spectra obtained look like the one in Fig. 14. An Auger and the $2s$ and $2p$ core level peaks are clearly visible, together with plasmons peaks on the low-energy side of the $2p$ peak.

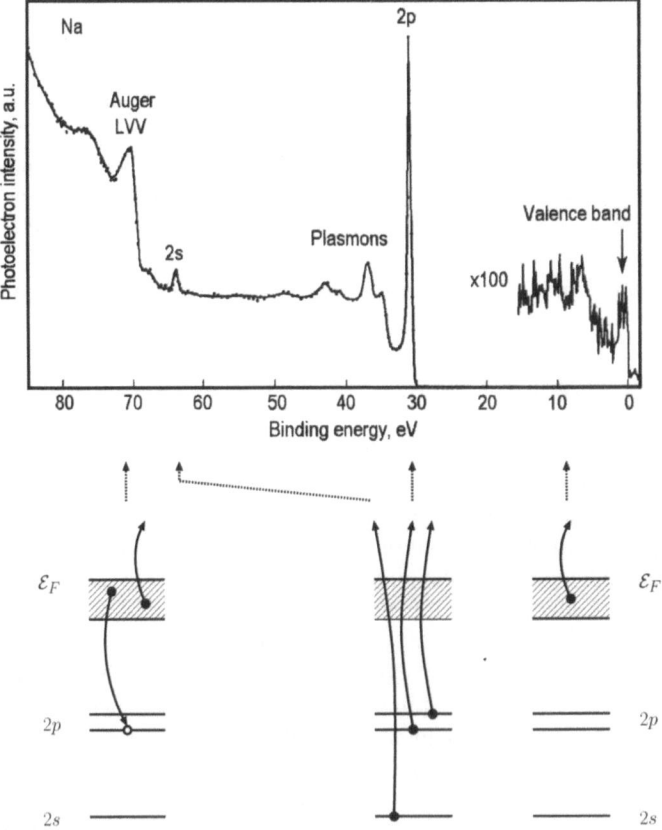

Fig. 14. X-ray photoemission spectrum of Na excited by synchrotron radiation at 100 eV photon energy. Auger emission and plamons peaks are also indicated. The various mechanisms responsible for these peaks are sketched at the bottom [51]

The high intensity in the first part of the spectrum, before the Auger *LVV* peak, is due to secondary electrons which have suffered inelastic losses. Note the low intensity of the valence band spectrum compared to that of core level peaks.

Photoelectron diffraction is obtained by monitoring one of these peaks as a function of either the energy or the outgoing direction of the photoelectron. Standard photelectron diffraction is obtained by monitoring a core level whereas Auger electron diffraction comes from an Auger peak. This results in a series of modulations that reflect the local structure around the absorbing atom. Valence photoelectron diffraction is also possible [48] but here the localization of the core hole is more questionable and this will affect the local nature of the information extracted from the spectra. Note as well that similar modulations are obtained when monitoring other structures such as plasmon peaks [49], the branching ratio of a spin-orbit split core level [50] or even the background of secondary electrons.

Apart from the appearance of loss structures (secondary electrons, plasmons) or deexcitation processes (Auger peaks) which reflect the response of the sample to the creation of a hole, the photoemission spectrum is globally a fingerprint of the density of the occupied states of the material. For this reason, angular-resolved photoemission is an invaluable tool for the study of the band structure and the Fermi surface of materials [52].

Like EXAFS, core level PhD reflects the local environment around the absorbing atom. The main difference with the absorption technique here is that due to the detection of the escaping electron, the information it bears will be more sensitive to bond directions. Technically speaking, this comes directly from the fact that in EXAFS only closed paths of the excited photoelectron contribute to the absorption signal (paths starting and ending on the photoabsorber), while in PhD also open paths contribute (from the photoabsorber to the detector. See Fig. 15).

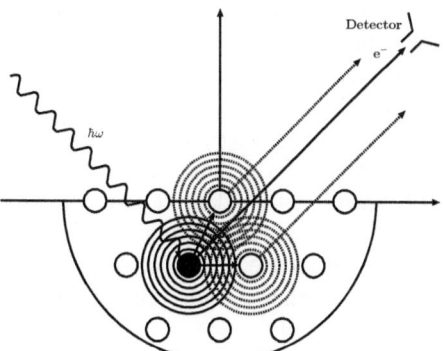

Fig. 15. Pictorial representation of the photoelectron diffraction process. The essential mechanism comes from the interference of open paths

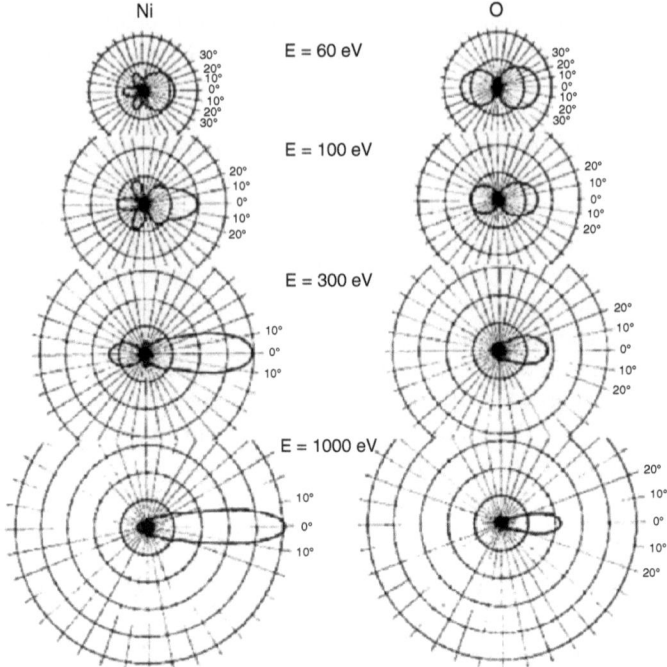

Fig. 16. Modulus of the scattering factor $f(\theta)$ at several energies for Ni and O atoms [53]

Here as well, two energy regimes can be distinguished, based on the shape of the scattering factor. Figure 16 shows typical shapes of the scattering factor $f(\theta)$ (which is the amplitude of the scattered electronic wave off a given atom in the direction θ) for nickel and oxygen atoms in the 60–1000 eV range. We see from this figure that at low energy (the meaning of "low" being element dependent, but roughly below 100 eV) the scattered amplitude can be important for many directions while at high energies, $f(\theta)$ always exhibit a strong peak in the forward direction ($\theta = 0$). The straightforward consequence of this energy behaviour is that at high energy, the spectra should consist essentially in a series of peaks corresponding to the main directions of neighbors. Hence, qualitative information can be obtained directly by looking at the position of the prominent peaks. At low energy, more scattering directions will contribute to the signal, making any direct interpretation of the spectra impossible; but in that case, the electron will probe much more thoroughly its immediate neighborhood and consequently the resulting spectra will contain more information.

When the energy resolution of the experimental set-up is sufficiently high, PhD and angular-resolved photoemission in general can further enhance their sensitivity to the surface (which is already high due to the small mean free

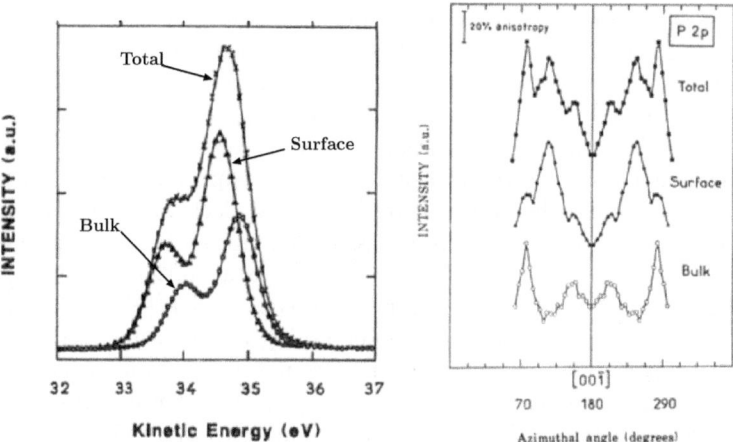

Fig. 17. The *left hand* side represents a typical P $2p$ core level peak with the surface and bulk contributions separated out. In the *right hand* side, the azimuthal modulations of the total, surface and bulk contributions are plotted [55]

path of the electron and can be reinforced by playing with the polarization of the incoming beam of photons) by separating out the surface contribution to the core level peak from that of the bulk. Indeed, due to different local environments, photoelectrons originating from surface atoms will appear in the spectrum at a slightly different energy from those coming from bulk atoms. This is the surface core level shift [54]. More generally, such a shift appears when two absorbers of the same chemical species have different local environments or different chemical form (it is then called chemical shift). As the PhD signal reflects the local structure around the absorbing atom, different modulations are expected for the surface and the bulk component of a core level. This can readily be seen on Fig. 17 which represents the azimuthal modulations of the components of the P $2p$ core level peak of InP(110). Here, the kinetic energy of the photoelectrons detected is about 35 eV and therefore the structures in the modulations do not correspond to scattering directions. Actually, at such a low energy, a nearest neighbor direction can even lead to a dip in the spectrum. But comparison of the modulations of the surface contribution to this $2p$ peak to a multiple-scattering calculation allowed to determine with a high accuracy the parameters of the reconstruction of the InP(110) surface [55].

4.3 Diffraction Anomalous Fine Structure

Diffraction anomalous fine structure is a resonant spectroscopy based on the scattering of x-rays. It consists in monitoring the intensity of a Bragg diffraction peak when the photon energy is varied through an absorption edge. As we have seen from (29), a strong enhancement of the second term in the cross-section is expected when the energy of the incoming beam of light goes through an absorption edge. Like in XAS, three energy regions can be distinguished: the pre-resonance one, the resonance region (where the spectroscopy is named DANES) and the extended region (this is the DAFS area itself). Likewise, since the excitation of a core electron is involved (through a second-order process), the measured signal is expected to reflect the local environment around the atom where the virtual electron has been excited.

The Thomson term in the cross-section depends on the momentum transfer vector $Q = q_f - q_i$. It is therefore proportional to the Fourier transform of the atomic electronic charge distribution (i.e., $\sim \int e^2 |\varphi(r)|^2 e^{-iQ \cdot r} dr$).

The two other terms normally depend on q_i and q_f separately. However, in practice, it is found that this dependence is very weak, in other words the dipole approximation can be applied there [56]. As a consequence, these two terms are considered as if $Q = 0$ (forward scattering approximation). Within this scheme, the cross-section can be written as

$$\frac{d\sigma}{d\Omega} = |f_T(Q) + \Delta f(\hbar\omega)|^2 \ , \tag{43}$$

where $f_T(Q)$ is the real Thomson scattering amplitude and $\Delta f(\hbar\omega)$ the anomalous – or resonant – scattering amplitude. The latter amplitude is customarily decomposed into its real part $f'(\hbar\omega)$ and its imaginary part $f''(\hbar\omega)$ which are related by the Kramers–Kronig relation. $f'(\hbar\omega)$ is a dispersion correction to the Thomson amplitude $f_T(Q)$ while $f''(\hbar\omega)$ is an absorption correction. The important point here is that $f''(\hbar\omega)$ is proportional to $\hbar\omega\mu$, where μ is the absorption coefficient measured in XAS, when the transfer vector $Q = 0$ (i.e., for forward scattering and $\hat{e}_{q_i} = \hat{e}_{q_f}$). This can be readily seen from the expression (32) of $f''(\hbar\omega)$. DAFS being from a measure of the elastic Bragg peak, the final state of the atom must be the same as its initial state. Therefore, according to 32, one must have

$$f''(\hbar\omega) = -\frac{\pi}{m} \sum_n \langle \varphi_i | \hat{e}_{q_f}^* \cdot p | \varphi_n \rangle \langle \varphi_n | \hat{e}_{q_i} \cdot p | \varphi_i \rangle \, \delta(\mathcal{E}_i - \mathcal{E}_n + \hbar\omega) \ . \tag{44}$$

When both incoming and outgoing beams have the same polarization, this reduces to

$$f''(\hbar\omega, \hat{e}_{q_f} = \hat{e}_{q_i}) = -\frac{\pi}{m} \sum_n |\langle \varphi_n | \hat{e}_{q_i} \cdot p | \varphi_i \rangle|^2 \, \delta(\mathcal{E}_i - \mathcal{E}_n + \hbar\omega) \ , \tag{45}$$

which is, to a factor, the XAS cross-section (see (20)).

As a consequence the DAFS signal should exhibit oscillations like those of EXAFS and can be modelled within the same framework.

Typical DAFS (and DANES) spectra are shown on Fig. 18. Here, the intensity of the Cu(111) and Cu(222) Bragg diffraction peaks have been measured in the range of the K edge of copper and compared to the corresponding

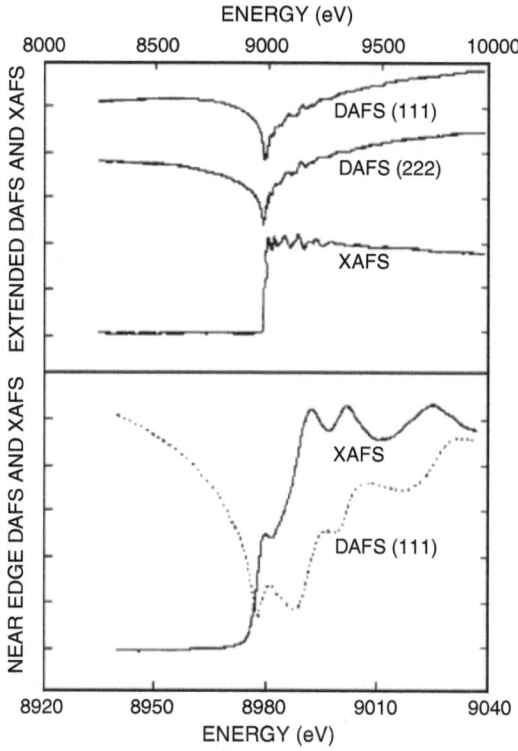

Fig. 18. Comparison of the raw DAFS and XAS signals [56]

EXAFS and XANES signal. It is clear from these raw data that strong similarities connects the DAFS(111) and (222) on the one hand, and the XAS on the other hand for both the near edge and the extended region. Comparison of the modulation functions χ is given in Fig. 19 together with the corresponding Fourier transform (which gives the radial distribution function). Now the similarity is striking between the DAFS and the EXAFS modulations. The small discrepancies observed are not sufficient to change the radial distribution as it it clear from the Fourier transform. This demonstrates that DAFS and EXAFS contain the same local structural information.

Obviously, being a second-order technique, DAFS needs a higher intensity beam to be detected. But it has the invaluable advantage of combining the information contained in XRD (ordered long-range atomic structure through the Bragg peaks) with that contained in XAS (local structure around specifically excited atoms – the resonant atoms). Thanks to the diffraction condition (the choice of the Bragg peak), the DAFS technique can select a subset of atoms in the sample and probe the structure around these atoms. Likewise, atoms of the same atomic number but on different crystallographic sites in the unit cell can be distinguished provided that the long-range ordered is known

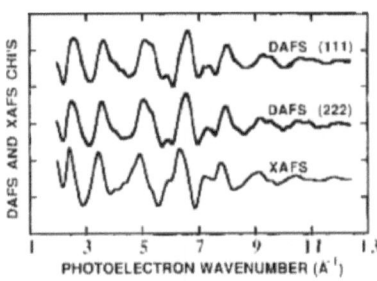 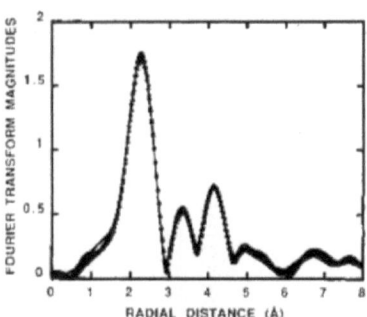

Fig. 19. The *left hand* side represents the background-subtracted and normalized DAFS and EXAFS signals. In the *right hand* side, the Fourier transform of these signals are plotted [56]. The symbols are as follows: circles, Cu(111) DAFS; triangles, Cu(222) DAFS; and lines, Cu EXAFS

precisely. Indeed, the contribution of these atoms is mixed in the DAFS signal according to the crystallographic structure factor which will generally differ between various Bragg reflections. Therefore, it becomes possible to disentangle the different contributions once the structure factor is known. This is the great difference with XAS where inequivalent absorbers will be mixed according to their number density [57]. But like in XAS, the chemical species is selected through the choice of the absorption edge.

Finally, let us mention that DAFS, like XAS and PhD, can be used as well to probe site-specific local magnetic information [58].

4.4 Auger Photoelectron Coincidence Spectroscopy

Although proposed for the first time in 1978 [59], Auger photoelectron coincidence spectroscopy and related techniques have only become popular in the recent years [60], probably because of the experimental difficulties involved. The basics of the technique relies in the simultaneous detection of a photoelectron and of the Auger electron resulting from the same photoionization process. Here "simultaneous" should be understood in view of the characteristic times discussed in Sect. 3.4.

The experimental set-up involves two electron energy analyzers, one being tuned to the energy of a core level and the other to the appropriate Auger decay energy. When two electrons are detected "simultaneously" in each analyzer, the event is recorded. Typical values for this "simultaneous" detection are about 10 to 20 ns [59, 61]. Such a value should ensure that both electrons are associated to the same photoionization process. An important consequence of this modus operandi is that secondary electrons, which are inelastically scattered, will not appear in the coincidence spectra as they are not correlated to Auger electrons [62]. Another consequence which was pointed out in the pioneering work by Haak et al. [59] is an enhancement

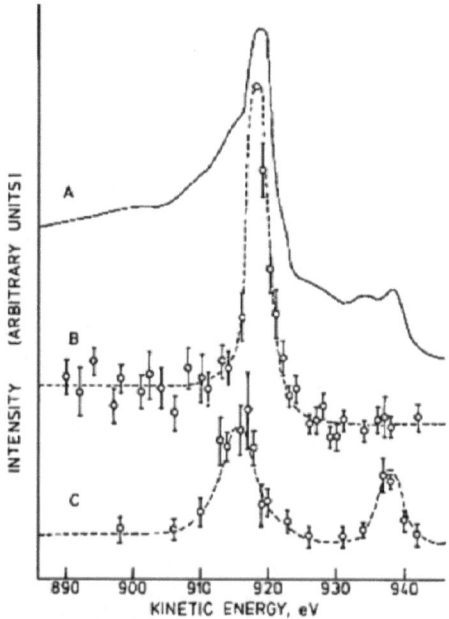

Fig. 20. Auger and Auger-photoelectron coincidence of copper. The curve marked A is the normal Auger spectrum, B is the Auger spectrum in coincidence with the $2p_{3/2}$ photoelectrons, and C is the spectrum in coincidence with the $2p_{1/2}$ photoelectrons [59]

of the sensitivity to the surface with respect to the photoelectron or Auger spectroscopy because both electrons must escape without loss to be detected coincidentally. More precisely, the mean free path of the combined coincidence electrons will be given by $1/\lambda = 1/\lambda_P + 1/\lambda_A$ where the indices P and A represent respectively the photoelectron and the Auger electron [63]. Furthermore, it has a unique capability of disentangling overlapping components of an Auger or photoelectron spectrum. This can be seen in Fig. 20 where the normal $L_{2,3}M_{4,5}M_{4,5}$ Auger spectrum of copper, curve A, is compared to the corresponding Auger spectrum recorded in coincidence with the $2p_{3/2}$ photoelectrons (curve B) and the $2p_{1/2}$ photoelectrons (curve C). The comparison between A and B shows strong coincidences between the $L_3M_{4,5}M_{4,5}$ and the $2p_{3/2}$ and none between these photoelectrons and the $L_2M_{4,5}M_{4,5}$. The comparison with C is even more informative: two peaks appear here, one corresponding to a coincidence between the $2p_{1/2}$ and the $L_2M_{4,5}M_{4,5}$, and another, below the $L_3M_{4,5}M_{4,5}$ peak, which comes from a Coster–Kronig transition (which is always fast) followed by an Auger one [59]. We can see as well from this figure that the low energy side of the Auger peaks usually attributed to inelastically scattered electrons is essentially removed from the coincidence spectra as anticipated.

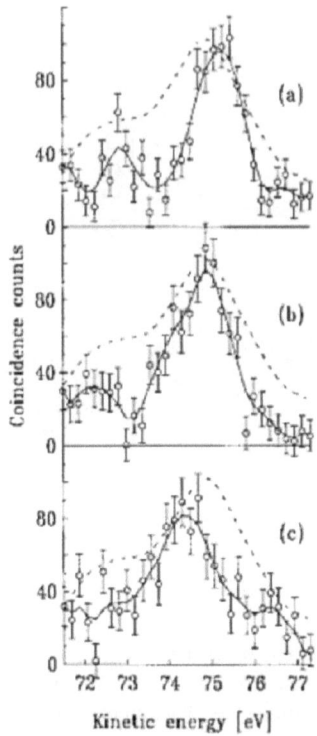

Fig. 21. Photoemission spectra of the Cu(100) $3p$ lines in coincidence with the M_3VV Auger line taken at 150 eV photon energy. The data are open circles with error bars, the *solid lines* are digital smooths to the data, and the *dashed lines* are the background substracted singles spectra. **(a)** An Auger analyzer setting of 60.6 eV, **(b)** an Auger energy of 61.1 eV, and **(c)** an Auger energy of 61.6 eV [62]

The simultaneous detection of the photoelectron and Auger electron has another extremely interesting consequence: the possibility to remove the core hole lifetime from the photoelectron spectrum. This was demonstrated by Jepsen et al. [62] and their main results are shown in Fig. 21. In these plots, the Cu(100) $3p$ photemission peak is figured as a dashed line with the $3p_{1/2}$ being weaker than the $3p_{3/2}$. This peak is background substracted. Comparison to the raw coincidence data (circles with error bars) indicates that the inelastic background on the low kinetic energy side has been removed by the coincidence. The three figures correspond to different settings of the Auger analyzer in the Auger M_3VV transition region: 60.6 eV (a), 61.1 eV (b) and 61.6 eV (c). Three settings were chosen to cover this broad Auger peak. We see on each of them that the $3p_{1/2}$ component is almost removed. The small remainder on the two first figures is due in fact to the Coster–Kronig decay of the $3p_{1/2}$ core hole into the $3p_{3/2}$ core hole followed by a M_3VV decay. But more importantly, the coincidence core lines are significantly narrower than the photoemission one. More precisely, the width of the $3p$ can be estimated to 2.2 eV for the photoemission peak. This large value is due to the strong overlap between the $3p$ and the $3d$ levels which results in a very rapid Auger decay and therefore induces a wide broadening. The width of the coincidence spectra is 1.0 eV.

The argument given by Jepsen et al. [62] to explain this narrowing is the following. When a system has a single long-lived Auger final state, resulting from the decay of a short-lived core hole, the energy conservation imposes that the sum of the kinetic energies of the Auger and of the photoelectron is fixed and equal to the photon energy minus the energy of the Auger final state. Actually, it is fixed whatever the uncertainty on the kinetic energies introduced by the core lifetime broadening. In other words, if a photoelectron is produced with a kinetic energy higher than the average value, the corresponding Auger electron will be detected with a lower than average kinetic energy. Therefore, the width of the coincidence peak will correspond to that of the final state, the double ionized state, which is much narrower than the core hole state (the core hole has a very short lifetime as we have just seen it)[2]. Let us call E^N the ground state energy of the atom containing N electrons and $\mathcal{E}_P$ and $\mathcal{E}_A$ respectively the energies of the photoelectron and the Auger electron. In the initial state, the energy of the system is $E_i = E^N + \hbar\omega$. In the final state, it becomes $E_f = E^{N-2} + \mathcal{E}_P + \mathcal{E}_A$. The system being conservative we have $E_f = E_i$ and therefore

$$(\mathcal{E}_P + \mathcal{E}_A) = (E^N - E^{N-2}) + \hbar\omega . \tag{46}$$

This result demonstrates that the sum of the photoelectron and the Auger electron energies has a width which is determined only by the two-hole final state (see Sawatzky [63] for a more detailed discussion).

This interpretation of the narrowing of the peak implies that the photoelectron and the Auger electron cannot be described independently in APECS. In other words, the transition rate of the Auger decay depends on the photoionization process that created the core hole.

An important point here is that the photoemission one core hole state is here only an *intermediate* state. But in a coincidence experiment, only events coming from the same atom and with the *same intermediate* state are measured [63]. This is why with a badly resolved multi-component photoelectron spectrum and a corresponding well resolved multi-component Auger spectrum, the coincidence experiment will separate the various contributions to the total photoelectron spectrum and vice versa (see Fig. 20 for an example).

Finally, it should be noted that APECS can be performed in an angular-resolved mode. This leads to modulations like in photoelectron diffraction

---

[2] Actually, the possibility to remove the core hole lifetime from the experimental spectrum is a common feature of all the second-order techniques involving the creation of this hole as the first step. Indeed, in all these cases, the core hole lifetime will only contribute to the intermediate state [64, 65]

but that can vary from them [60]. Differences in the modulations between the coincidence and the PhD spectra can originate from the difference in sensitivity to the surface of the two techniques, or to correlations between the two electrons. Therefore, a proper description of APECS should include these correlations within a one step model. Furthermore, by choosing carefully the emission angle of both electrons, angular-resolved APECS can allow to select the specific magnetic sublevel from which the Auger electron originates [66].

To summarize, APECS allows to discriminate more easily different events such as [66]:

(i)   isolate individual sites in a solid and probe their local atomic structure
(ii)   separate overlapping multiplet structure
(iii) eliminate uncorrelated secondary electron background
(iv)   eliminate core level lifetime broadening of spectral features
(v)   distinguish between intrinsic and extrinsic secondary electron emission
(vi)   extract information about the orbital occupancy

We refer the reader to the review article by Sawatzky [63] for a more detailed description of APECS, both theoretical and experimental.

Standard APECS experiments rely on the fact that both electrons can be discriminated by their energy. When the kinetic energies of the two electrons are equal, then the distinction between them is lost, despite the time delay, and the electron exchange induces strong interference effects in the angular and energy distributions of these electrons, especially if they are ejected at small mutual angles [67]. This is particularly true for electrons ejected from isolated atoms. Likewise, the Coulomb interaction between the three particles (the two electrons plus the ionized atom) can also distort the spectrum, especially when their kinetic energy is small. This is the so-called post-collision interaction [68,69]. Note that this case is always possible due to the different nature of photoelectrons and Auger electrons: the kinetic energy of the photoelectron follows that of the photons while the kinetic energy of the Auger electron is fixed, independent of the photon energy.

4.5 Links Between these Techniques

The spectroscopies presented here are four among many others. They were chosen as examples of first-order and second-order processes but also because, despite their differences, they are intimately linked one to another. XAS and PhD are first-order techniques while DAFS and APECS are second-order ones. The link between PhD and APECS is obvious. Let us explore more thoroughly now the link between PhD, XAS and DAFS. It is actually XAS that is the common denominator and the connection is made through the optical theorem. This theorem, which was known anciently as the Bohr–Peierls–Placzek relation, is an old but essential tool of scattering theory [70] that was originally discovered by Rayleigh for classical physics and extended to quantum theory by Feenberg. Let us consider a wavefield incoming upon a sample. Part of these waves will be absorbed by the target and the rest transmitted and scattered. The optical theorem is just another formulation of the conservation of the number of particles (unitarity principle): it states that the sum of the outgoing and of the absorbed waves must equal the incoming wave. In the case of elastic scattering, there is no absorption of the incoming particles and therefore the optical theorem implies that the intensity of the incident wave after scattering i.e., the intensity measured in the forward direction, must be reduced by the exact amount of the waves scattered in all the other directions. This is often referred to as the "shadow effect". If $f(\boldsymbol{k}_{in}, \boldsymbol{k}_{sc})$ is the scattering amplitude of the scattered wave, with an incident plane wave $e^{i\boldsymbol{k}_{in}\cdot\boldsymbol{r}}$ directed upon the sample, the wavefield after scattering, far away from the scattering area, can be written as

$$\psi(\boldsymbol{r}) = e^{i\boldsymbol{k}_{in}\cdot\boldsymbol{r}} + f(\boldsymbol{k}_{in}, \boldsymbol{k}_{sc})\frac{e^{ikr}}{r}, \tag{47}$$

with $k = ||\boldsymbol{k}_{in}|| = ||\boldsymbol{k}_{sc}||$ for elastic scattering. $f(\boldsymbol{k}_{in}, \boldsymbol{k}_{sc})$ is also called the scattering factor. The optical theorem will now write as

$$\sigma_{\text{tot}} = \frac{4\pi}{k} \Im [f(\boldsymbol{k}_{sc} = \boldsymbol{k}_{in})], \tag{48}$$

where $\Im[\ ]$ designates the imaginary part.

DAFS consists in monitoring an elastic scattering peak of photons through an absorption edge. Therefore, the optical theorem will connect the absorption signal (XAS) to the sum of the DAFS signals over all directions in space.

In a XAS experiment, the absorption process results in the excitation of an electron. To simplify the discussion, let us suppose that we are sufficiently far away from the edge so that the electron is ejected of the atom. PhD measures the resulting photocurrent in a given direction. So, if we neglect inelastic losses, the sum of the PhD signal over all directions of space will give the XAS modulations. This was recognized more than thirty years ago by Lee [71] in the early days of PhD and EXAFS and is demonstrated numerically in Fig. 22. Here, the PhD and EXAFS signals have been calculated using a multiple-scattering series expansion approach on a 9-atom MgO cluster. No damping of any sort was included. The integration of the PhD signal over the 4π steradian was performed using a Gaussian quadrature formula with 1202 weighted points, as derived by Lebedev and Laikov [73]. The series expansion was carried out up to the fourth scattering order for EXAFS and to the fifth one for PhD (because of the strong forward peaking of the scattering factor when the energy increases, PhD does not converge as fast as EXAFS). The agreement between the two is clearly excellent providing thereby a numerical demonstration of the optical theorem. Small discrepancies arise in the low energies (between 100 and 140 eV). They are due to an insufficient convergence of the multiple-scattering series in this energy range (below 50 eV, the multiple-scattering series can even diverge [74]). And it

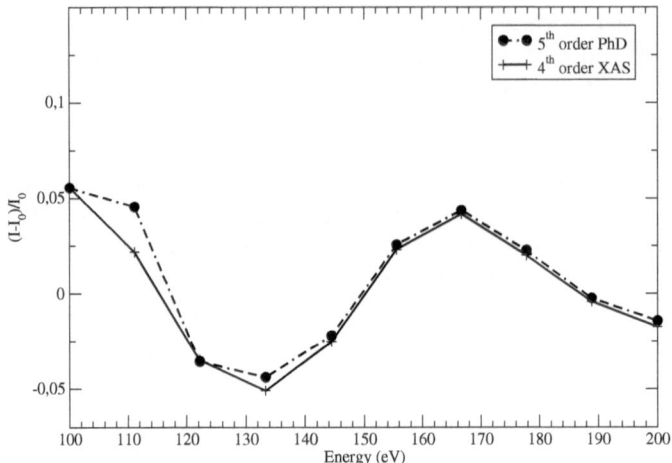

Fig. 22. Test of EXAFS *vs* PhD summed over all directions in space. A 9-atom cluster of MgO was used and no damping was included [72]

is a consequence of the fact that the optical theorem is only valid on fully converged signals: integrated single and double scattering PhD signals never matched low-order scattering EXAFS signals in this test case.

5 Conclusion

We have presented in this chapter a general overview of a large class of spectroscopies. All the experiments detailed here have in common to involve photons as the exciting particles and electrons as the probe particles, although these electrons are not necessarily detected outside the sample. But in all cases, the information that can be extracted from the experimental spectra is the result of the interaction of these probe electrons with their surrounding. This overview includes the basic physical processes (up to the second order) these spectroscopies rely on. The influence of many-body effects on the spectra has also been discussed in detail. Finally, four particular techniques, x-ray absorption spectroscopy (including its spin-sensitive version x-ray magnetic circular dichroism), photoelectron diffraction, diffraction anomalous fine structure and Auger photoelectron coincidence spectroscopy have been documented as an illustration.

References

1. C.R. Brundle: Surf. Sci. **48**, 99 (1975)
2. R. Karazija: *Introduction to the Theory of X-Ray and Electronic Spectra of Free Atoms* (Plenum Press, New York, 1996)

3. J.W. Niemantsverdriet: *Spectroscopy in Catalysis* (VCH, Weinheim, 1993)
4. C. Cohen-Tannoudji, J. Dupont-Roc and G. Grynberg: *Processus d'interaction entre photons et atomes* (InterEditions/Editions du CNRS, Paris, 1988), p 601
5. F.M.F. de Groot: Coordination Chemistry Reviews **249**, 31 (2005)
6. L.I. Schiff: *Quantum Mechanics* (McGraw-Hill/Kogakusha, Tokyo, 1968), p 479
7. M. Blume: J. Appl. Phys. **57**, 3615 (1985)
8. P. Rennert: Phys. Rev. B **48**, 13559 (1993)
9. ibid. reference [4], p 603
10. J.M. Tonnerre, L. Sève, D. Raoux, G. Soullié, B. Rodmacq and P. Wolfers: Phys. Rev. Lett. **75**, 740 (1995)
11. ibid. reference [2], p 261
12. J.B. Hastings, D.P. Siddons, U. Van Bürck, R. Hollatz and U. Bergmann: Phys. Rev. Lett. **66**, 770 (1991)
13. ibid. reference [6], p 314
14. ibid. reference [4], p 220
15. ibid. reference [6], p 336
16. ibid. reference [4], p 67
17. ibid. reference [4], p 80
18. ibid. reference [4], p 36
19. H. Friedrich: *Theoretical Atomic Physics* (Springer, Berlin Heidelberg New York, 1990), p 96
20. B.R. Holstein: *Topics in Advanced Quantum Mechanics* (Addison-Wesley, Redwood City, 1992), p 157
21. W.H. Louisell: *Quantum Statistical Properties of Radiation* (Wiley, New York, 1973), p 296
22. M.D. Scadron: *Advanced Quantum Theory* (Springer, Berlin Heidelberg New York, 1979), p 168
23. ibid. reference [4], p 500
24. V. V. Balashov: *Interaction of Particles and Radiation with Matter*, (Springer, Berlin Heidelberg New York, 1997), p 136
25. R.G. Newton: *Scattering Theory of Waves and Particles*, 2nd edn, (Springer, Berlin Heidelberg New York, 1982), p 177
26. L. Landau and E. Lifschitz: *Mécanique quantique*, (édition MIR, Moscow, 1980), p 72
27. J.-M. Mariot and C. Brouder: Spectroscopy and Magnetism: An introduction. In *Magnetism and Synchrotron Radiation*, ed by E. Beaurepaire, F. Scheurer, G. Krill and J.-P. Kappler, (Springer, Berlin Heidelberg New York, 2001), p 40
28. A. Kotani and Y. Toyozawa: in *Synchrotron Radiation. Techniques and Applications*, ed by C. Kunz, (Springer, Berlin Heidelberg New York, 1979), p 195
29. J. Fink, Th. Müller-Heinzerling, B. Scheerer, W. Speier, F.U. Hillebrecht, J.C. Fuggle, J. Zaanen and G.A. Sawatzky: Phys. Rev. B **32**, 4899 (1985)
30. L. Fonda: J. Phys. B: At. Mol. Opt. Phys. **30**, 3173 (1997)
31. E.L. Shirley: Phys. Rev. Lett. **80**, 794 (1998)
32. J. Schwitalla and H. Ebert: Phys. Rev. Lett. **80**, 4586 (1986)
33. J.W. Gadzuk: Many-body Effects in Photoemission. In *Photemission and the Electronic Properties of Surfaces*, ed by B. Feuerbacher, B. Fitton and R.F. Willis, (Wiley, New York, 1978), p 111
34. P. Nozières and C.T. De Dominicis: Phys. Rev. **178**, 1097 (1969)
35. G.D. Mahan: Phys. Rev. B **11**, 4814 (1975)

36. P.W. Anderson: Phys. Rev. Lett. **18**, 1049 (1967)
37. C.P. Flynn: Phys. Rev. Lett. **37**, 1445 (1976)
38. P.H. Citrin and D.R. Hamann: Phys. Rev. B **15**, 2923 (1977)
    P.H. Citrin, G.K. Wertheim and Y. Baer: Phys. Rev. B **16**, 4256 (1977)
39. ibid. reference [2], p 165
40. J.B. Pendry: Electron Emission from Solids. In *Photemission and the Electronic Properties of Surfaces*, ed by B. Feuerbacher, B. Fitton and R.F. Willis, (Wiley, New York, 1978), p 87
41. M.A. Van Hove: Surf. Interface Anal. **28**, 36 (1999)
42. C. Brouder and J.-P. Kappler: Prolegomena to Magnetic Circular Dichroism in X-Ray Absorption Spectroscopy. In *Magnetism and Synchrotron Radiation – Mittelwhir 1996*, ed by E. Beaurepaire, B. Carrière and J.-P. Kappler, (Les éditions de physique, Les Ulis, 1997), p 19
43. C. Noguera: Scan. Elect. Micr. **II**, 521 (1985)
44. J.H. Sinfelt, G.H. Via and F.W. Lytle: J. Chem. Phys. **72**, 4832 (1980)
45. J. Stöhr and Y. Wu: Circular dichroism: basic concepts and theory for $3d$ transition metal atoms. In *New Directions in Research with Third-Generation Soft X-Ray Synchrotron Radiation Sources*, ed by A.S. Schlachter and F.J. Wuilleumier, (Kluwer Academic Publishers, Netherlands, 1994) p 221
46. T. Funk, A. Deb, S.J. George, H. Wang and S.P. Cramer: Coordination Chemistry Reviews **249**, 3 (2005)
47. M. Altarelli and P. Sainctavit: Sum Rules for XMCD. In *Magnetism and Synchrotron Radiation*, ed by E. Beaurepaire, F. Scheurer, G. Krill and J.-P. Kappler, (Springer, Berlin Heidelberg New York, 2001), p 65
48. J. Osterwalder, T. Greber, S. Hüfner and L. Schlapbach: Phys. Rev. Lett. **64**, 2683 (1990)
49. J. Osterwalder, T. Greber, S. Hüfner and L. Schlapbach: Phys. Rev. B **41**, 12495 (1990)
50. E.L. Bullock, R. Gunnella, C.R. Natoli, H.W. Yeom, S. Kono, R. Uhrberg and L.S.O. Johansson: Surf. Sci. **352/354**, 352 (1996)
51. K. Oura, V.G. Lifshits, A.A. Saranin, A.V. Zotov and M. Katayama: *Surface Science. An Introduction*, (Springer, Berlin Heidelberg New York, 2003), p 102
52. S. Hüfner: *Photoelectron Spectroscopy*, (Springer, Berlin Heidelberg New York, 1995), p 297
53. M.L. Xu, J.J. Barton and M.A. Van Hove: Phys. Rev. B **39**, 8275 (1989)
54. Tran Minh Duc, C. Guillot, Y. Lassailly, J. Lecante, Y. Jugnet and J.C. Vedrine: Phys. Rev. Lett. **43**, 789 (1979)
55. S. Gota, R. Gunnella, Z.-Y. Wu, G. Jézéquel, C.R. Natoli, D. Sébilleau, E.L. Bullock, F. Proix, C. Guillot and A. Quémerais: Phys. Rev. Lett. **71**, 3387 (1993)
56. H. Stragier, J.O. Cross, J.J. Rehr, L.B. Sorensen, C.E. Bouldin and J.C. Woicik: Phys. Rev. Lett. **69**, 3064 (1992)
57. J.O. Cross: Analysis of Diffraction Anomalous Fine Structure. PhD thesis, University of Washington, Seattle (1996)
58. L.B. Sorensen, J.O. Cross, M. Newville, B. Ravel, J.J. Rehr, H. Stragier, C.E. Bouldin and J.C. Woicik: Diffraction anomalous fine structure: unifying x-ray diffraction and x-ray absorption with DAFS. In *Resonant Anomalous X-ray Scattering - Theory and Applications*, ed by G. Materlik, C.J. Sparks and K. Fischer, (North-Holland, Amsterdam, 1994)

59. H.W. Haak, G.A. Sawatzky and T.D. Thomas: Phys. Rev. Lett. **41**, 1825 (1978)
60. G. Stefani, S. Iacobucci, A. Ruocco and R. Gotter: J. Elec. Spec. Relat. Phenom. **127**, 1 (2002)
61. R.A. Bartynski, S. Yang, S.L. Hubert, C.-C. Kao, M. Weinert and D.M. Zehner: Phys. Rev. Lett. **68**, 2247 (1992)
62. E. Jensen, R.A. Bartynski, S.L. Hulbert, E.D. Johnson and R. Garrett: Phys. Rev. Lett. **62**, 71 (1989)
63. G.A. Sawatzky: Auger Photoelectron Coincidence Spectroscopy. In *Auger Electron Spectroscopy*, ed by C.L. Briant and R.P. Messmer, (Academic Press, Boston, 1988)
64. J.-M. Mariot: Recent Progress in High-Resolution X-Ray Fluorescence. In *Magnetism and Synchrotron Radiation - Mittelwhir 1996*, ed by E. Beaurepaire, B. Carrière and J.-P. Kappler, (Les éditions de physique, Les Ulis, 1997), p 307
65. C.F. Hague: Resonant Inelastic X-ray Scattering. In *Magnetism and Synchrotron Radiation*, ed by E. Beaurepaire, F. Scheurer, G. Krill and J.-P. Kappler, (Springer, Berlin Heidelberg New York, 2001), p 273
66. R. Gotter, A. Ruocco, M. T. Butterfield, S. Iacobucci, G. Stefani and R.A. Bartynski: Phys. Rev. B **67**, 033303 (2003)
67. L. Végh and J.H. Macek: Phys. Rev. A **50**, 4031 (1994)
68. S.A. Sheinerman and V. Schmidt: J. Phys. B: At. Mol. Phys. **30**, 1677 (1997)
69. S. Rioual, B. Rouvellou, L. Avaldi, G. Battera, R. Camilloni, G. Stefani and G. Turri: Phys. Rev. Lett. **86**, 1470 (2001)
70. R.G. Newton: Am. J. Phys. **44**, 639 (1976)
71. P.A. Lee: Phys. Rev. B **13**, 5261 (1971)
72. D. Sébilleau: unpublished results
73. V.I. Lebedev, and D.N. Laikov: Doklady Mathematics **59**, 477 (1999)
74. C.R. Natoli and M. Benfatto: J. de Phys. C **8**, 11 (1986)

Synchrotron Radiation

Mikael Eriksson

MAXlab, Lund University, Box 118, 221 00 Lund, Sweden
Mikael.Eriksson@maxlab.lu.se

Abstract. This lecture focuses on the properties of synchrotron radiation. This subject is extensively treated in many excellent articles and textbooks. (See for instance [1,2]). We have also excellent and user-friendly computer codes, as SPECTRA (Spring-8) and SRW (ESRF) which you can get on the net, so few of us are now actually using the analytic formulas to calculate radiation spectra. This lecture is an effort to transmit an intuitive picture of these properties which might be used for a brief understanding of the results you get from the computer codes.

1 Dipole Radiation from Accelerating Charges

Let us first look at a radio antenna emitting radio waves. A charge is moving up and down in this antenna. Let us assume that the charge is at the dark position at time $t = 0$ and is suddenly moved a moment later to the grey position at $t = \tau$. The electric field lines for the two positions are indicated in Fig. 1.

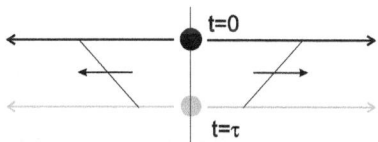

Fig. 1. Electric field lines for an accelerating charge

Just after acceleration, the field pattern close to the charge must be the one in grey. But the change of the field lines cannot go faster than the speed of light. For distances larger than $l = \tau c$ (c is the speed of light), the field lines will be given by the dark lines. We can thus see how an electromagnetic wave (the transition between the dark and grey cases) is travelling outwards from the accelerating charge. The amplitude of this field will be at maximum horizontally and zero in the vertical direction where the two cases overlap.

The strength of the radiated field will thus be proportional to $\cos(\theta)$ and the intensity, proportional to the field squared, will be proportional to

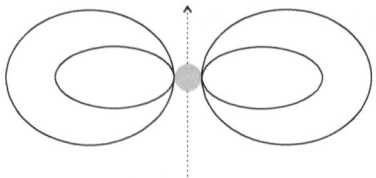

Fig. 2. Intensity distribution of the dipole radiation

$(\cos(\theta))^2$. The intensity distribution is illustrated in Fig. 2, it is rotationally symmetric around the direction of the acceleration.

We can now already notice some characteristics of the radiation:

1. The electric field vector of the radiation is parallel to the acceleration vector.
2. The intensity at half-maximum is at an angle of $\pi/4$.

We are now going to the case of a moving electron being bent in a constant magnet field B. The electron orbit is curved and the electron is thus being accelerated inwards. Following the radio transmitter case described above, we should then see a radiation distribution as indicated in Fig. 3.

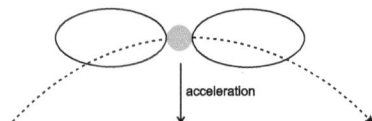

Fig. 3. Classical radiation of an electron in a magnet field

In Fig. 4 we assume a relativistic electron, that is an electron moving with a speed close to that of light. We have now to apply the relativistic Lorentz transformation of angles for emission. In the forward direction, these angles are decreased by a factor of γ, γ being the ratio between the total energy of the particle and the rest mass energy.

It should be noted that γ is generally quite large. The electron rest-mass is $0.5\,\mathrm{MeV}/c^2$ while the electron energy in a storage ring is $1\text{--}5\,\mathrm{GeV}$, so γ is

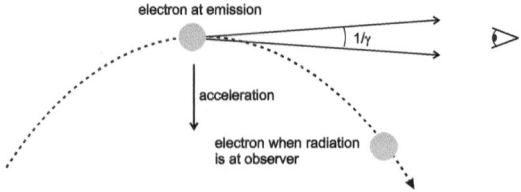

Fig. 4. Relativistic electron in a magnetic field

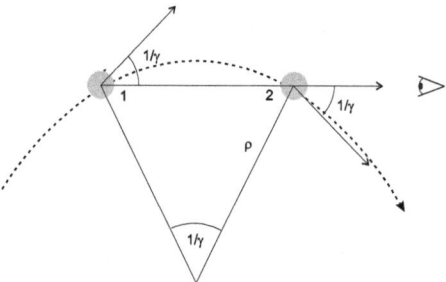

Fig. 5. Observer's view of light emission

somewhere between 2000 and 10000. The angle of emission has thus decreased to less than a mrad.

Let us now consider a relativistic electron moving in a constant magnetic field viewed from above. An observer will start to see the light emmited by the electron when it is at position 1 and the observer will stop seeing it when the electron is at position 2 (Fig. 5). The electron is travelling close to the speed of light, βc, where

$$\beta = \sqrt{1 - \frac{1}{\gamma^2}} \,,$$

and thus very close to 1. The length of the light pulse is then given by the segment length times the relative speed difference between the light and the moving electron:

$$l = \frac{\rho}{\gamma} \left(1 - \beta\right),$$

where ρ is the curvature radius of the electron trajectory. If $\gamma \gg 1$

$$l = \frac{\rho}{2\gamma^3} \,.$$

This light pulse can be expanded into plane waves. The amplitude of these plane waves will be constant down to wave-lengths comparable to 2π times the pulse length, where the amplitude starts to drop off quickly for shorter wavelengths.

This cut-off wavelength is thus given by

$$\lambda_{cut-off} = \pi \frac{\rho}{\gamma^3} \,.$$

A more rigorous treatment will define a critical wavelength

$$\lambda_c = \frac{4\pi}{3} \frac{\rho}{\gamma^3} \,,$$

where half of the radiated power is above the critical wave-length. In engineering units, the corresponding critical energy is given by

$$\varepsilon\left(keV\right) = 0.665E^{2}\left(GeV^{2}\right)B\left(T\right) .$$

To get the power emitted, we need to integrate over all photon energies. The critical energy is proportional to γ^{3}/ρ, so integration brings us to a power proportional to γ^{4}/ρ^{2}. The total energy lost by an electron for one turn in a storage ring is given by (engineering units)

$$\varDelta E_{\gamma}\left(keV\right) = 88.575\frac{E^{4}\left(GeV^{4}\right)}{\rho\left(m\right)} .$$

This is the very minimum voltage the radio-frequency (RF) system must deliver to the electron beam in a storage ring. The total power delivered to the beam is

$$P\left(kW\right) = \varDelta E_{\gamma}I_{circ} ,$$

where I_{circ} is the circulating current. Let us now look at an instant picture of the light emitted by an orbiting relativistic electron as in Fig. 6.

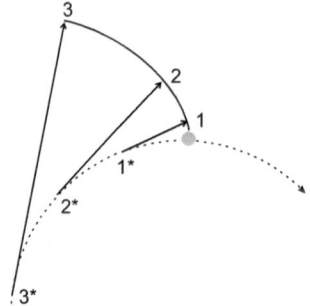

Fig. 6. Emission of wave front

The electron is moving along a circle (in a magnet field) and is just now at the indicated position. When it was at position 1*, it emitted light which now is at position 1 and so forth. The dashed line indicates the light wave front at the time when the electron is in its actual position. This picture might seem to contradict the picture of synchrotron light emitted as the light from the headlights of a car moving in a curve. The reader is encouraged to find out the difference.

The frozen picture of the light front emitted by an electron during little more than during one turn in a storage ring is seen in Fig. 7. The distance between the wave-fronts spiralling outwards is given by the circumference of the ring (o).

2 Synchrotron Radiation Damping in a Storage Ring

Let us first consider the vertical motion of a particle in a storage ring. The beam is focussed by the quadrupole lenses in the ring and we approximate the electron orbit to an harmonic oscillation (Fig. 8).

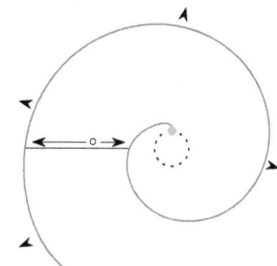

Fig. 7. Light spiral of synchrotron radiation

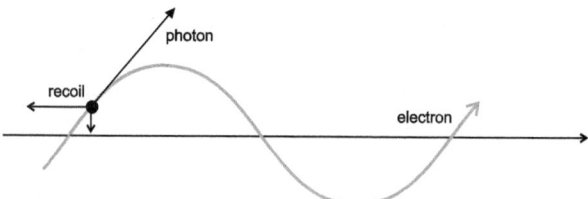

Fig. 8. Vertical damping

The recoil of a photon emitted of an electron passing a dipole magnet is directed against the velocity of the electron. This velocity has two components, one vertical and one longitudinal. Both components will be decreased by the recoil. The longitudinal one, will be restored by the acceleration of the RF cavity, but the vertical component will be damped. The damping time will be lattice dependent but we can expect a damping time of

$$\tau_{damp} = \frac{\Delta E_\gamma}{E} \frac{2\pi R}{c} \, ,$$

where E is the electron energy and R is the mean ring radius. In most storage rings, we find a damping time of 1–10 ms.

The situation in the horizontal direction is a little more complicated. The equilibrium orbit position in the dipole magnets is energy-dependent, that is the dispersion is non-zero. A particle suffering an energy loss will find a new equilibrium orbit as seen in Fig. 9. The electron will now start oscillating around the new equilibrium orbit, this is a heating process counteracting the damping one. An equilibrium beam size or emittance will be attained when heating and damping equals.

It is now evident that this equilibrium emittance, ε_H, is lattice-dependent. The smaller the dispersion, the smaller the equilibrium emittance. As seen from Fig. 9 the dispersion grows quadratically with magnet bending angle and the integrated dispersion is thus proportional to θ^3. It turns out that

$$\varepsilon_H \propto \frac{1}{\theta^3} \, .$$

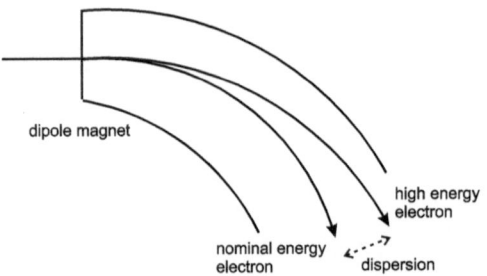

Fig. 9. Horizontal damping

This is the reason why low emittance storage rings have a large number of magnet cells.

The vertical emittance should go to very small values in a perfect storage ring. However, small quadrupole twists coupling to the horizontal motion and spurious vertical dispersion will give a vertical dispersion in the order of 1% of the horizontal one.

3 Insertion Devices

Third generation rings are characterized by the extensive usage of insertion devices (ID) used as light sources. We will here only treat the simplest IDs, namely the planar ones.

An electron is entering a magnet array as indicated in Fig. 10. We assume a sine-shaped magnet field in the median plane

$$B_y = B_0 \sin(ks)$$

with $k = \frac{2\pi}{\lambda_u}$.

The electron will now execute a harmonic oscillation with amplitude

$$x(s) = \frac{K}{\gamma k} \sin(ks),$$

where $K = \frac{eB_0}{km_e c^2}$.

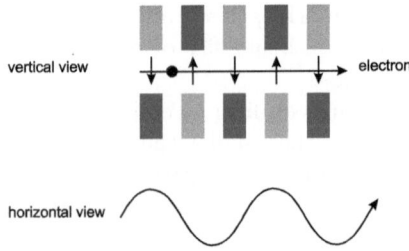

Fig. 10. Undulator magnet array and electron orbit

The maximum angle of deflection is $\frac{K}{\gamma}$. K is typically in the order of unity, so the angle of defection counts some tenths of a mrad and the orbit amplitude is just around some micrometers for a 3 GeV ring.

Positive interference will take place for wave-lengths

$$\lambda_i = \frac{\lambda_u}{2\gamma^2 i} \left(1 + \frac{K^2}{2} + \theta^2\gamma^2\right),$$

where θ is the angle of observation.

This is the undulator equation. It could be understood in the following way: The wave fronts, one being emitted one period length λ_u after the previous one, will be separated with a distance

$$\Delta = \lambda_u \left(1 - \beta^*\right),$$

in analogy with the previous discussion on wave-front widths. β^* is the mean electron velocity along the reference orbit. This velocity is thus lower then the absolute electron velocity due to this elongation of the electron orbit.

This can be expanded as earlier to

$$\Delta = \frac{\lambda_u}{2\gamma^{*2}},$$

where $\gamma^* = \frac{\gamma}{\sqrt{1+K^2/2}}$.

The angle dependence is purely geometrical. We will now get positive interference for wavelengths $\lambda_i = \frac{\Delta}{i}$.

It is also instructive to consider the length of the pulse emitted by one electron in an undulator. As seen in Fig. 11, a light train is emitted in from of the electron as this leaves the undulator. The length of this light pulse is

$$L = N_u \lambda_1,$$

where N_u is the number of undulator periods and λ_1 is the fundamental wave-length. Expanding this light train in plane waves gives us a decreasing

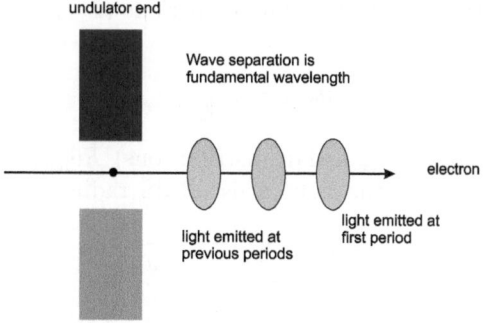

Fig. 11. Light emitted from electron in undulator

line-width as the number of undulator periods increases. The emitted energy of the light train increases as well with increasing number of periods. We can thus suspect that the energy density with respect to photon energy increases quadratically with N_u.

Since we can use some 100 periods in an undulator, we see that the energy density can be increased some 4 orders of magnitude compared to a dipole magnet source.

In the forward direction, we will see a series of peaks centered at the fundamental photon energy and at the odd harmonics. The width of the peaks is mainly due to the finite number of periods, the finite emittance and energy spread of the electron beam. The natural width, induced by the finite number of undulator periods is given by

$$\frac{\Delta\lambda_i}{\lambda_i} = \frac{1}{Ni} ,$$

where N is the number of undulator periods.

The two other broadening effects are given by the wave-length formula above

$$\frac{\Delta\lambda}{\lambda} = 2\frac{\Delta\gamma}{\gamma} ,$$

and

$$\frac{\Delta\lambda}{\lambda} = \frac{\theta^2\gamma^2}{1 + \frac{K}{2}} .$$

We can now see that we have an optimum number of undulator periods at $N_{opt} = \frac{500}{i}$ assuming an electron energy spread of 10^{-3}.

A large K-value will increase the number of harmonics. In case spacing between the harmonics becomes comparable to the peak broadening, we get a continuous spectrum resembling the dipole spectrum. We often then call the ID a wiggler. It should be noted though, that wigglers might very well show undulator characteristics at the low harmonics.

The relations allow to determine, provided we have a perfect electron beam without energy spread, the maximum angle of emission for a broadening given by the finite number of undulator periods (L_{und} is the total length of the undulator)

$$\theta_{\max} = \sqrt{\frac{\lambda_i}{L_{und}}} .$$

The diffraction relation (gaussian distributions) $R\theta_{\max} = \frac{\lambda}{4\pi}$ gives us the apparent source size or the light waist RMS radius in the middle of the undulator

$$R = \frac{1}{4\pi}\sqrt{\lambda_i L_{und}} .$$

4 Flux and Brilliance

The parameter of interest is the photon flux on the target. The flux and brilliance, defined below, are however often handy numbers to have as a start for calculating the performance of a beam-line.

The flux is integrated vertically and given per mrad horizontally and per 0.1% energy width of the spectrum. (The units are alas not exactly SI units)

$$F = \frac{Photons}{s, mrad, 0.1\%} \ .$$

The brilliance is the peak flux density in phase space

$$B = \frac{Photons}{s, mrad^2, mm^2, 0.1\%} \ .$$

Now we need to define the phase space area needed to calculate the brilliance above. Looking at one direction, for instance the horizontal one, we can draw the electron beam distribution and the light distribution as given by diffraction (Fig. 12).

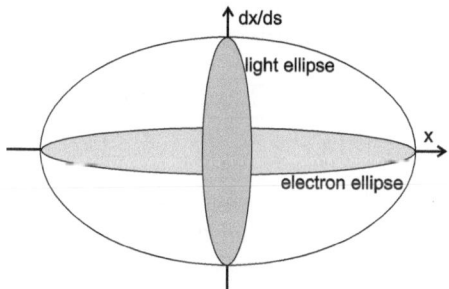

Fig. 12. Emittance ellipses

If we assume gaussian distributions, we have for instance for the electron beam:

$$Int\,(x) = \frac{1}{\sqrt{2\pi}\sigma_x} e^{-\frac{x^2}{2\sigma_x^2}} \ .$$

Even if the two phase-space areas are equally large, they do not match so we get a dilution and have to convolute the two distributions.

$$A_{phase-space} = 4\pi^2 \Sigma_x \Sigma_x' \Sigma_y \Sigma_y' \ ,$$

where for instance $\Sigma_x = \sqrt{\sigma_{x,e}^2 + R^2}$. As an example, the brilliance curves for MAX IV are in Fig. 13.

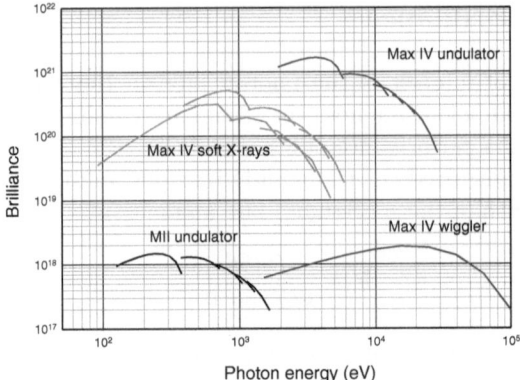

Fig. 13. Brilliance curves of MAX IV

5 The Computer Way

We have so far tried to give an intuitive picture of synchrotron radiation. Modern high-language computer codes offer a direct evaluation of the performance of synchrotron radiation, but some of the basic understanding might be lost. We give one example here. We start with retarded field expression for an accelerating charge (Liénard-Wiechert [3, 4])

$$\overline{E}\left(x,t\right)=\frac{e}{4\pi\varepsilon_0}\left[\frac{\overline{n}-\overline{\beta}}{\gamma^2\left(1-\overline{\beta}^*\overline{n}\right)^3 R^2}+\frac{1}{c}\frac{\overline{n}\times\left(\overline{n}-\overline{\beta}\right)\times\overline{\beta'}}{\left(1-\overline{\beta}^*\overline{n}\right)^3 R}\right]_{ret}.$$

$\overline{n}$ is the unit vector from the accelerating charge to the observer. The suffix *ret* means that if we go back in time to find the position of the accelerating charge to the place it was when it radiated. R is the distance between the radiating charge at this moment to the observer. The first term within brackets is the near-field contribution. This can often be disregarded at long distances and at high electron energies. The second term describes the far-field synchrotron radiation.

A code can now be written describing the electron motion. Fig. 14 shows an example from a 5-period undulator with 15 mm period length and an electron energy of 3 GeV and $K = 2$.

The transverse velocity is not quite sine-shaped but rather bell-shaped due to the high K-value. The electric field as a function of time seen in Fig. 15 is pretty spiky, again due to the relatively high K-value. The spikier the field pattern, the richer it is in harmonics.

The electric field seen above can be expanded in plane waves and this gives the amplitude of the fundamental and harmonics.

Finally, the analytic solution of the peak distribution is shown as a dotted line in Fig. 16.

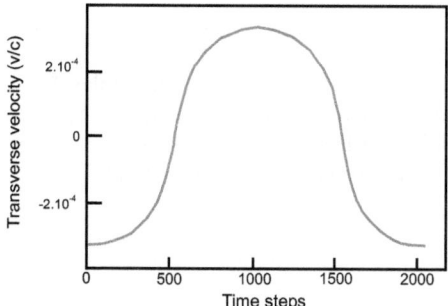

Fig. 14. Transverse velocity of an electron in an undulator

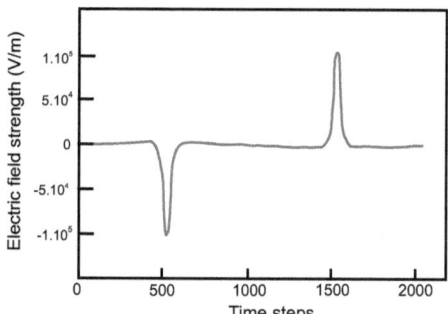

Fig. 15. Electric field from an electron in an undulator

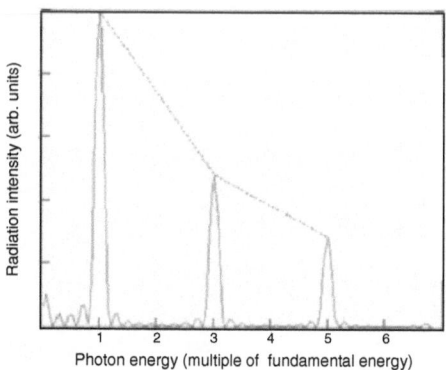

Fig. 16. Undulator spectrum

References

1. J.D. Jackson: *Classical Electrodynamics* (John Wiley & Sons, 1999)
2. Helmut Wiedemann: *Particle Accelerator Physics* (Springer-Verlag, 1993)
3. A. Lienard, L'Eclairage Electrique **16**, 5 (1898)
4. E. Wiechert, Archives Néerlandaises, **546**, (1907)

X-ray Magnetic Circular Dichroism: Historical Perspective and Recent Highlights

Andrei Rogalev[1], Fabrice Wilhelm[1], Nicolas Jaouen[1], José Goulon[1], and Jean-Paul Kappler[2]

[1] European Synchrotron Radiation Facility (ESRF) BP 220, 38043 Grenoble Cedex, France
`rogalev@esrf.fr`
[2] Institut de Physique et Chimie des Matériaux de Strasbourg, CNRS UMR7504, 23 rue du Loess BP 43, 67034 Strasbourg Cedex 2, France

Abstract. This chapter starts with a brief historical overview which shows how x-ray Magnetic Circular Dichroism (XMCD) has evolved from the very early days of x-ray physics to become a powerful spectroscopic technique with the derivation of the x-ray magneto-optical sum rules. We analyze the physical content of XMCD and its sum rules. It is the unique capability of XMCD to probe with elemental selectivity the magnetic properties of an electronic state of given symmetry which makes this method an outstanding tool to study magnetism. We decided to highlight two recent advances in XMCD: the first one deals with measurements of tiny magnetic moments induced either by a magnetic field in a Pauli paramagnet (Pd metal) or by hybridization with a transition metal; the second example concerns induced magnetism in ferromagnetic $3d/5d$ multilayers.

1 History of X-ray Magnetic Circular Dichroism

1.1 Search for Polarization of X-rays

On November 8th, 1895, Röntgen discovered, more or less accidentally, what he called later X rays. He spent the next eight weeks in his laboratory repeating his experiments and trying to determine the nature of these new rays. His first report, prepared in hurry, was entitled: *On a new kind of rays*, and got published very fast in the Proceedings of the Physical Medical Society of Würzburg on December 28th, 1895 [1]. Already on January 1rst, 1896, he distributed copies of his manuscript to several renowned physicists. Within *days*, his discovery became a worldwide sensation, unprecedented in the history of physics. Experiments on x-rays were run at almost every major laboratory in the world. A considerable amount of work was aimed to study the physical properties of these new rays and their interaction with matter. The absence of refraction of x-rays was demonstrated by J. Perrin [2] who rose a question whether these waves are transverse or longitudinal. To answer this question, a number of experiments were proposed to detect eventual polarization properties of x-rays. Already in the February issue of *Nature* in 1896, J.J. Thomson published a letter in which he wrote ... *of the three methods of producing polarisation in light - reflection, refraction and absorption -, only the latter is available for these rays*[1]. ... *The result of these experiments were entirely negative, for, although the tourmaline produced very considerable absorption of the rays, no difference was detected between the absorption when the axes were crossed and when they were parallel.* In March 1896 Galitzine and Karnojitzky [4] reported new results obtained with a plate of tourmaline and claimed that they had observed a tiny difference in absorption. This experiment led them to conclude that x-rays could be polarized. The week after, at the next meeting of the French Academy of Science, Sagnac reported on his search for x-ray dichroism on crystalline plates of quartz, calcite, tourmaline, mica and potassium ferrocyanide [5] but none of these crystals exhibited detectable dichroism with x-rays. Further attempts to polarize x-rays by transmission through an anisotropic crystal also proved to be unsuccessful and no clear evidence of the polarization of x-rays could yet be produced. In 1903, Blondlot [6] revisited the hypothesis that x-rays generated with a tube were polarized[2] and he tried to measure the rotary power of x-rays with a quartz single crystal and with a piece of sugar. He claimed that he had measured huge rotation angles: $40°$ with primary x-rays and up to $18°$ with secondary x-rays. This was unfortunately an *artefact*, and, tragically for him, not the last one. He nevertheless concluded his paper with a prophetic statement: ... *Il est extrêmement probable que la rotation magnétique existe aussi, tant pour les rayons X, que pour les rayons δ*[3]. He was the first to predict that magnetism could influence the interactions of x-rays with matter. The experiments by Blondlot stimulated von Lieben to study the polarization properties of scattered x-rays [8,9]. He may have observed an effect of polarization, but he admitted in a sunsequent paper [9] that his results were ambiguous. Finally, this is C.G. Barkla who demonstrated definitively that x-rays could be polarized by scattering at $90°$. In his famous paper on polarised x-rays [10], proved that the primary x-rays emitted by an x-ray tube were partially polarized with a rate of $\approx20\%$. In the same paper he wrote: *Shortly after I arrived at the conclusion as to the origin of secondary radiation from gases, Professor Wilberforce suggested to me the idea of producing a plane-polarised beam by means of a secondary radiator and of testing the polarisation by a tertiary radiator.* Barkla reported on this experiment in a second paper on the polarization properties of x-rays [11], in which he showed that it was possible to produce a linearly polarized x-ray beam with a polarization rate close to 100%. Bassler repeated Barkla's experiments with an improved apparatus and confirmed that x-rays could be polarized by scattering. The experimental details were described in Bassler's

---

[1] This may be the first allusion to X-ray dichroism.

[2] This idea had already been expressed earlier by Gifford [7], but had not been supported experimentally.

[3] It is extremely probable that magnetic rotation exists also, as well for x-rays as for δ rays (δ rays mean secondary x-rays emitted from a material)

dissertation thesis entitled: *Polarization of x-rays evidenced with secondary radiation* [12] which he defended in November 1908 at the Munich University. These results proved that x-rays were transverse electromagnetic waves just as for visible light.

1.2 Early Works of the Influence of Magnetism on X-ray Absorption

Starting with the assumption that x-ray radiation was kind of light of very short wavelength, Barkla suggested to Chapman to investigate a possible rotation of the plane of polarization of x-rays [13] using an experimental setup schematically reproduced in Fig. 1. x-rays emitted by the anode (A) of an x-ray tube, passed through the slits S_1 and S_2 in lead screens, and were scattered by a carbon plate (R_1). The x-rays scattered at 90° passed next through the slits S_3, S_4 and S_5 in further lead screens and were scattered in turn on a second carbon plate. The x-rays were detected by two similar electroscopes (E_1 horizontal and E_2 vertical). The sample was inserted between the slits S_3 and S_4, i.e. between two carbon polarizers.

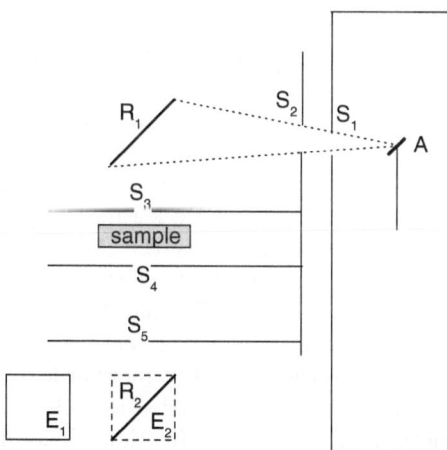

Fig. 1. Experimental setup used by Chapman to detect a rotation of the plane of polarization of x-rays. E_2 electroscope (*dotted lines*) was placed vertically above the second scattering plate; for other notations, see text [13]

Among various experiments performed with this setup, Chapman tried to measure the Faraday rotation of x-rays with a thin iron foil magnetized either parallel or perpendicular to the foil. Unfortunately, he could not see any effect. He also tried to detect a difference in the absorption of x-rays by an iron foil in magnetized and unmagnetized states, but, again, his experiments were negative. Chapman failed as well to measure any rotation of the polarization plane of x-rays in a quartz crystal and in an optically-active solution of sugar.

Not being discouraged by the negative results reported by Chapman, nor by his own inconclusive attempts to detect any influence of the magnetization on the absorption of x-rays by iron, Forman decided to build a new instrument which should allow him to detect changes in x-ray transmission of the order of 10^{-4} [14]. Tests performed with an iron foil magnetized in a direction perpendicular to the x-rays showed that the amplitude of the postulated effect would be well below the detection limit of his instrument. Two years later Forman reported [15] that he had built another instrument featuring the same sensitivity but with which he could set the direction of the magnetic field parallel to the x-rays. These very careful experiments performed by Forman may have revealed that there were some tiny changes ($\approx 5 \times 10^{-4}$) in the so-called *opacity* of an iron foil to x-rays when the foil was magnetized in a direction parallel to the incident x-rays. This result stimulated A.H. Compton [16] to look for a change in the γ-rays absorption of magnetized and unmagnetized iron. However, his results were negative either with the magnetic field parallel or perpendicular to the incident γ-rays beam. Alluding Forman's result he nevertheless stated ... *that the effect observed by Forman is of the order of magnitude to be expected if the electrons are rings of electricity which are oriented by the magnetic field.* Becker also repeated once again the experiments of Forman and Chapman and tried to investigate other materials than only iron [17,18]. Since the effect was undoubtly expected to be very small, Becker adopted a differential method that should have allowed him to reach a sensitivity which he claimed to be better than 10^{-4}. Experiments were made with a Coolidge x-ray tube (mean wavelength of about 0.3 Å) and with a magnetic field of *ca.* 1.8 T set either perpendicular or parallel to the x-ray propagation direction. Becker measured *aluminium, carbon, copper, iron, nickel, platinum, zinc and silver* [17]; he even tried *wood and lithium* [18]. In all cases, he found that the effect was smaller than the limit of accuracy of his experimental setup. His final, highly depressing conclusion was: ... *if the effect exists, then, it is very small.*

Kartschagin and Tschetwerikova went to more or less the same conclusion in a paper published in 1926 [19]. They tried as well to detect the rotation induced by a magnetic field on the polarization plane of x-rays in paraffin and, again, in an iron foil. They used partially crossed, large aperture 90° polarimeters and observed a very weak effect in iron but strictly no effect in paraffin. From a careful analysis of their results, they deduced that, far from the absorption edge, there was no hope to detect any rotation due the weakness of the effect. In contrast, they suspected that there could be a very weak rotation in the vicinity of an absorption edge for ferromagnetic materials although accurate measurements were reported to be most difficult. Five years later, Froman [20] confirmed the observation made by Kartschagin and Tschetwerikova, even though the rotation angle he measured was twice smaller.

Kurylenko [21] defended in 1939 a Ph.D. Thesis at the Sorbonne University in which he reported that that he had tried to measure the influence of a 1 T magnetic field on the absorption spectrum of an iron foil at the Fe K-edge. He claimed that he had found a shift of the absorption edge by *ca.* 2.5 eV when the magnetic field was applied in a direction *perpendicular* to the direction of x-rays. Reviewing the 104 pages written by Kurylenko [22] and which were published in 1946, Coster [23] disproved the results and wrote: *... Already years ago I myself looked in vain for this effect and I think it improbable that it exists. At any rate, Kurylenko's dispersion was certainly too small to measure such a shift.*

1.3 Discovery of XMCD and Derivation of the Sum Rules

The credit of the first theoretical investigation of the problem should be given to E.A. Stern et al. [24,25] who performed a band structure calculation and predicted a magnetic circular dichroism at the $M_{2,3}$ absorption edges of ferromagnetic nickel. Unfortunately, very few people paid attention to this paper and no serious attempt was made to check this theory experimentally, at least over the next 10 years.

In 1979, Hrdý et al. tried again to assess the upper limits of the x-ray Faraday rotation in nickel, cobalt and magnesium ferrite [26] and they went to the pessimistic conclusion that these limits were even much smaller than those predicted either by Froman [20] or by Kartschagin and Tschetwerikova [19] nearly 50 years earlier. Hart and Rodriguez also failed to measure a Faraday effect with x-rays [27]. Their upper limits determined for cobalt, nickel and iron were, however, in quantitative agreement with the results of Hrdý et al. Interestingly, it was concluded [27] that the detection of a Faraday rotation with x-rays could be possible only with the use of synchrotron radiation and only near absorption edges. In 1983, G. Schütz and her colleagues [28] concentrated their efforts on the use of circularly polarized x-rays in order to elucidate the influence of the magnetic state of iron on x-ray absorption spectra. In the first experiment, they exploited the circularly polarized L emission lines in oriented Ir isotopes after internal conversion decay resulting in the emission of photons with discrete energies ranging from 8 to 13 keV. Again, the sensitivity of the experimental setup was clearly not high enough to detect in this energy range any spin-dependent contribution to x-ray absorption.

In 1984, E.A. Stern et al. [29] were the first to try seriously to detect x-ray magnetic circular dichroism using synchrotron radiation at the L_3 edge of Gd in $Gd_{18}Fe_{82}$ alloy. Unfortunately, in this experiment, the circular polarization rate of the incident x-rays was only *ca.* 5% and the authors again failed to detect any reliable effect. Independently, using an atomic multiplet approach, Thole et al. [30] reported in 1985 new calculations which led them to conclude the existence of a quite significant x-ray magnetic *linear* dichroism at the $M_{4,5}$ edges of rare-earths. They pointed out that linearly polarized synchrotron radiation could be used most efficiently to determine accurately the magnitude and orientation of the local rare earth magnetic moments. One year later, they joined their efforts with other colleagues [31] and produced the very first experimental evidence of a strong x-ray magnetic linear dichroism at the Tb $M_{4,5}$ absorption edges in the ferrimagnetic garnet $Tb_3Fe_5O_{12}$. For that key experiment, they exploited the linearly polarized synchrotron radiation emitted by the storage ring ACO at LURE in Orsay.

Shortly later, the existence of X-ray Magnetic Circular Dichroism was proved experimentally by G. Schütz et al. [32] who reported in 1987 a weak, but significant effect at the Fe K edge in an iron foil. For this well known experiment, she exploited the elliptically polarized synchrotron radiation from a bending magnet at DESY in Hamburg. It was rapidly anticipated that magnetic circular dichroism should not be restricted to only the X-ray Absorption Near Edge Structure (XANES) region but should also manifest itself in the Extended X-ray Absorption Fine Structures (EXAFS). This was confirmed by nice experiments also performed by Schütz et al. [33] at the L edges of Gd in Gd metal and in the ferrimagnetic garnet $Gd_3Fe_5O_{12}$. The interest in this subject then suddenly exploded: an important step forward was made by C.T. Chen et al. [34] who were the first to extend x-ray magnetic *circular* dichroism experiments into the soft x-ray range: their contribution was not marginal since they confirmed the existence of huge dichroic signals at the $L_{2,3}$ edges of a ferromagnetic Ni film.

The beginning of modern days for XMCD started when B.T. Thole and P. Carra [35,54] derived, using a single ion model, a set of magneto-optical sum rules which made it possible to relate the integrated intensities of XMCD spectra to the orbital [35] and spin [54] moments carried by the absorbing atom. With these sum rules, the experimentalists were given a powerful tool to analyze XMCD spectra recorded at spin-orbit split absorption edges (e.g. $L_{2,3}$ edges) and to extract magnetic moments, both in magnitude and direction, with the full benefit of the element and orbital selectivity of x-ray absorption spectroscopy. This was a real breakthrough in the field of fundamental and applied magnetism [37]. Finally, the Agilent Technologies Europhysics Prize for outstanding achievement in condensed-matter physics has been awarded in the year 2000 to G. Schütz-Gmeineder, G. van der Laan and P.Carra for pioneering work in establishing the field of magnetic x-ray dichroism.

For the sake of completeness, one should add that the x-ray Faraday rotation was detected as well in 1990 in the x-ray range by Siddons et al. [38] who measured at the Co K edge very small rotation angles in a Co-Fe alloy and in a Co metallic glass. In addition to magnetic circular and linear dichroisms and to Faraday rotation, other interesting magneto-optical spectroscopies like the transverse and the longitudinal magneto-optical Kerr effects [39, 40], or the Voigt effect [41] could be finally observed with x-rays. Optical Activity in the x-ray range may deserve here also a special mention: it was unambiguously detected only in 1998 using X-ray Natural Circular Dichroism (XNCD) in a non-centrosymmetrical crystal of $LiIO_3$ [42]. More recently, time-reversal odd optical activity effects, also referred to as *non-reciprocal*, magnetochiral dichroism and non-reciprocal magnetic linear dichroism, were observed in the x-ray range [43,44]. New edge-selective sum rules for optical activity were derived by Carra et al. [45, 46] and revealed how closely related were optical activity and orbital magnetism.

2 Physical Content of XMCD and Sum Rules

The physical origin of XMCD can be explained most easily with a so-called *two-step* model originally proposed by G. Schütz et al. [32] and revisited later by J. Stöhr and R. Nakajima [47]. The first step describes the excitation of a core electron by a circularly polarized x-ray photon that carries an angular momentum ($+\hbar$ for right handed photon and $-\hbar$ for left handed photon), the corresponding helicity vector being parallel (right) or antiparallel (left) to the direction of propagation. As a consequence of the conservation of angular momentum in the absorption process, the angular momentum carried by the photon is entirely transferred to the excited photoelectron. Depending on the initial core state and on the nature of interactions, both the orbital moment and the spin of the photoelectron are affected by this momentum transfer. Let us consider a photoelectron excited from spin-orbit-split core levels, e.g. $2p_{1/2}$ and $2p_{3/2}$ involved in the L_2 and L_3 absorption edges: then part of the angular momentum carried by the photon will be converted into spin *via* spin-orbit coupling. This is known as the Fano effect [48]. The spin moment of the photoelectron is always parallel to the direction of propagation of the photon but its sign depends on the helicity of the incident x-ray photon and on the spin-orbit coupling ($l + s$ at the L_3 edge; $l - s$ at the L_2 edge). In the absence of spin-orbit coupling (as in the case of the excitation of atomic s core electrons), the angular momentum of the photon is entirely converted into $+\hbar$ ($-\hbar$) orbital moments and there cannot be any polarization of the photoelectron.

The magnetic properties of the sample are driving the second step. In the XANES region, x-ray absorption spectra primarily reflect the density of empty states for an angular momentum l given by the symmetry of the initial core state and selection rules of the transition. XMCD spectra simply reflect the difference in the density of states with different spin or orbital moments. In the case of spin-orbit split initial states, the excited photoelectron carries both a spin and an orbital momentum and any imbalance in either spin or orbital momentum in the final states will immediately give rise to a dichroic effect. However, if we *sum* the contributions of all electrons that can be excited from a split core level (taking properly into account their degeneracy), the result can only reflect a difference in the *orbital moments* of the final states because there is always a pair of photoelectron with opposite spin polarization excited from a spin-orbit split core level. In contrast, the *difference* in the dichroism intensity measured at two spin-orbit split edges will reflect a *spin imbalance* in the empty states because the orbital momentum transferred to the photoelectron has the same sign at the both edges. If there is no orbital moment in the final states, then the ratio of the dichroic signals at the two spin-orbit split edges should be negative and inversely proportional to the ratio of the corresponding degeneracies. This is precisely the content of the magneto-optical sum rules. We wish to draw attention here onto the fact that a summation over two spin-orbit split edges is more or less equivalent to what can be measured for a core level *without* spin-orbit interaction. This implies that a dichroism effect at K edges can only be due to the orbital polarization in the valence shell.

Even though the two-step model can help us to understand qualitatively the underlying physics, the discrimination between the two steps remains highly artificial. This does not apply to the magneto-optical sum rules that were first derived by Thole and Carra [35, 54] for spin-orbit split absorption edges and which were rederived later by other authors [49–51]. Such sum rules supply a firm basis to estimate directly from the experimental XMCD spectra the respective contributions of the orbital moment $(m_L = -\frac{\mu_B}{\hbar}\langle L_z\rangle)$ and of the spin moment $(m_S = -2\frac{\mu_B}{\hbar}\langle S_z\rangle)$ to the *total* magnetic moment associated with a specific state of given symmetry. Considering dipolar transitions from p core states towards empty states in the valence band, and assuming that the $p \rightarrow d$ transition is a dominant absorption channel, then the *orbital* and *spin* sum rules can be written:

$$\int (\Delta\mu_{j+} + \Delta\mu_{j-})\, dE = \frac{3}{2}\Omega_l\langle L_z\rangle, \tag{1}$$

$$\int (\Delta\mu_{j+})\, dE - 2 \times \int (\Delta\mu_{j-})\, dE = \frac{2}{3}\Omega_l\left[\langle S_z\rangle + \frac{7}{2}\langle T_z\rangle\right], \tag{2}$$

where $\Delta\mu_{j+(-)}$ are magnetic circular dichroism spectra at the absorption edges corresponding to the core states with $j^+ = 3/2$ and $j^- = 1/2$, $\langle T_z\rangle$ denotes here the expectation value of the magnetic dipole operator

$$\mathbf{T} = \sum_i (\mathbf{s}_i - 3\mathbf{r}_i(\mathbf{r}_i \cdot \mathbf{s}_i)/r_i^2),$$

which reflects some *asphericity* of the spin moment due either to an anisotropic charge distribution around the absorbing atom, or spin-orbit interaction. Fortunately, in a wide majority of practical cases, the expectation value of the magnetic dipole operator $\langle T_z\rangle$ is much smaller than the spin moment and, therefore, to a reasonable level of approximation, its contribution can be neglected. Note that the proportionality coefficient Ω_l in (1) and (2) is simply the absorption cross-section per hole of symmetry l in the valence band (in our case d hole) and is defined as

$$\Omega_l = \frac{1}{n_h}\int (\mu^0_{j^+ \rightarrow l} + \mu^0_{j^- \rightarrow l})\, dE, \tag{3}$$

where $\mu^0_{j+(-) \to l}$ denotes isotropic absorption cross section corresponding to given dipolar transitions and n_h is the number of holes. For a given absorbing atom and well specified transitions, Ω_l is a constant quantity [52], so that the sum rules establish a simple linear relationship between the magnitude of the XMCD signal and the magnetic moment.

To derive the magneto-optical sum rules, the authors adopted the single ion model combined with a scalar relativistic approach. For that reason it is often thought that their validity might be restricted to only systems with localized magnetic moments. However, another derivation of the sum rules was given by Ankudinov and Rehr [51] using the density matrix formalism: this clearly demonstrates that the aforementionned restriction does not hold true. Nevertheless, it can still be argued that the sum rules may not be directly applicable to systems where relativistic effects are very strong. As proved by Ebert,again to a fairly reasonable level of approximation, on the basis of first-principles spin-polarized relativistic multiple-scattering calculations, one may still apply the sum rules to such systems. The remarkable potentiality of the magneto-optical sum rules should not let us forget that a great number of intermediary approximations had to be made in order to derive them. Concerning $L_{2,3}$-edges, the most important ones are summarized below:

- the derivation is restricted to dipole allowed transitions[4];
- it ignores the exchange splitting for the core levels;
- it neglects the differences between $d_{3/2}$ and $d_{5/2}$ wavefunctions;
- it ignores any energy dependence of the wavefunctions.

Despite such limiting approximations, the validity of the sum rules appears to be now rather well established, at least in the cases of the $L_{2,3}$ absorption edges of $3d$ [55,56], $4d$ [57] and $5d$ [58] transition metals or for the $M_{4,5}$-edges of actinides [59].

3 Practice with Sum Rules: XMCD at Au $L_{2,3}$ Edges in Au$_4$Mn

3.1 Experimental Details

The quantity of interest in an XMCD experiment is the difference in the x-ray absorption cross-section measured with left and right circularly polarized photons while the sample magnetization is kept parallel or antiparallel to the wavevector of the incident x-ray beam. The spectra, which we reproduce below, were recorded at the ESRF beamline ID12 [60] using circularly polarized x-rays produced by a helical undulator of the Apple-II type. The fixed-exit double crystal monochromator was equipped with a pair of Si⟨111⟩ crystals

---

[4] Magneto-optical sum rules have been also derived for quadrupolar transitions [54], but practical applications are questionable.

80 A. Rogalev et al.

cooled down to $-140°$ C. Due to the ultra-low emittance of the source, the energy resolution was found to be very close to the theoretical limit, i.e. 2.3 eV and 2.9 eV at the L_3 (11.918 keV) and the L_2 (13.733 keV) absorption edges of gold, respectively. In all cases, the instrumental resolution was smaller than the core hole life-time broadening. At such rather high photon energies, the second harmonic of the undulator emission spectrum was used while the fundamental harmonic was attenuated by inserting a 25 μm thick aluminium foil upstream with respect to the monochromator: this concurred to reduce the heat load on the first crystal and improved the long term stability of the whole spectrometer. We did not use any mirror because the very weak 4th harmonic of the undulator has anyhow a very low transmission for a Si$\langle 111 \rangle$ double crystal monochromator, whereas the intensity of the 6th harmonic of an helical undulator can be taken as negligible. Polarimetry studies were performed using a quarter wave plate [61] allowed us to check that, under the described experimental conditions, the circular polarization rate of the *monochromatic* beam was of the order of *ca.* 90%. The sample was an optically polished piece of polycrystalline Au$_4$Mn alloy that had been prepared by arc-melting in Ar atmosphere using stoichiometric amounts of the pure constituents with subsequent annealing for 20 hours at 1070 K and 48 hours at 620 K. X-ray diffraction measurements show a single phase corresponding to the ordered NiMo$_4$-type crystal structure [62]. From magnetic measurements, we confirmed that the system is ferromagnetic below $T_C = 360$ K with a total magnetic moment of 4.15 $\mu_B/f.u.$ at T = 4.2 K and under saturating magnetic field of 3 T. For the XMCD experiment, the sample was attached to the cold finger of a liquid helium cryostat and was inserted in the bore of a superconducting electromagnet. XMCD spectra were recorded at T = 10 K and with a magnetic field of 6 T set either parallel or antiparallel to the direction of the x-ray beam propagation. All x-ray absorption spectra shown below were measured in the total fluorescence yield detection mode using a Si photodiode mounted directly inside the cryogenic shielding of the electromagnet.

3.2 Data Analysis

The raw experimental spectra had to be corrected for the imperfect circular polarization of the monochromatic beam and also for the self-absorption effects using an homographic transform described elsewhere [63]. The corrected and normalized Au $L_{2,3}$ isotropic x-ray absorption spectra of Au$_4$Mn are reproduced in Fig. 2a. In the particular case of this polycrystalline Au$_4$Mn sample, *quasi-isotropic* x-ray absorption spectra were obtained by averaging two spectra recorded consecutively with right and left circularly polarized x-ray photons. However, in the more general case of anisotropic samples, e.g. single crystals or epitaxially grown thin films, one had to average spectra recorded with x-ray photons linearly polarized successively along all three axes of the crystal. As one can notice from Fig. 2a, the L_3 absorption spectrum was normalized to 2.2 while the L_2 spectrum was normalized to unity: this is because the branching ratio of the L_3 and L_2 edge jumps is equal to 2.2. At first glance, it is quite surprising that this branching ratio is larger than 2, i.e. the value expected from the degeneracy of the core hole states. However, for the $2p$ core states with large spin-orbit splitting ($\approx 1.2\,\mathrm{keV}$ for Au), there is a difference in the $\langle 5d|r|2p_{3/2}\rangle$ and $\langle 5d|r|2p_{1/2}\rangle$ radial matrix elements which produces a deviation from the expected statistical ratio [64,65]. We would like to insist here that such a normalization is particularly important whenever XMCD data are concerned because in the formal derivation of the sum rules, the radial matrix elements were supposed to be the same at both absorption edges. This assumption is most probably acceptable at the L edges of $3d$ transition metals where spin-orbit splitting is of the order of $10\,\mathrm{eV}$, but this is not anymore the case in the hard x-ray range. One has therefore take this effect into account if we wish to deduce correct values of the spin and orbital moments using the magneto-optical sum rules. Figure 2b displays the Au $L_{2,3}$ XMCD spectra obtained, following a now well established convention [66,67]: what is plotted is the direct difference between the normalized x-ray absorption spectrum recorded with right circularly polarized beam *minus* the spectrum recorded with left polarized x-rays; the two spectra were recorded with a magnetic field of 6 T applied along the direction of propagation of the x-rays. We checked that strictly identical dichroism spectra - but of opposite sign - were obtained with a magnetic field set antiparallel to the x-ray propagation direction. The displayed XMCD signals are relatively small (about 3% of the edge jump) but the quality of the data is fair enough to show unambiguously the presence of a magnetic moment on the Au $5d$ electrons. This is the direct consequence of a strong hybridization of the Mn $3d$ and Au $5d$ states, but, clearly, this magnetic moment is expected to remain rather small because Au has a nearly filled $5d$ band. Moreover, the sign of the dichroism signals at the $L_{2,3}$ edges would tell us quite clearly that the Au $5d$ moment should be aligned parallel to the applied magnetic field and, therefore, should be coupled ferromagnetically to the Mn $3d$ magnetic moment.

In order to extract the values of the orbital and spin magnetic moments with the help of the magneto-optical sum rules one has to estimate as accurately as possible the absorption cross-section per hole in the valence band as defined in (3). For magnetic compounds involving $3d$ transition metals, Wu et al. [68] have proposed a procedure that ensures a quantitative determination of Ω_l: since, for a given metal and transition, Ω_l is a constant quantity [52], we expect a strict linear relation between the magnitude of the XMCD signal and the magnetic moment. Therefore, the magnetic moment of A in a compound AB could be determined by measuring the value of Ω_l in A metal. This procedure should hold true as far as one can find a reference metal that is magnetic in its pure metallic state. In all other cases, the

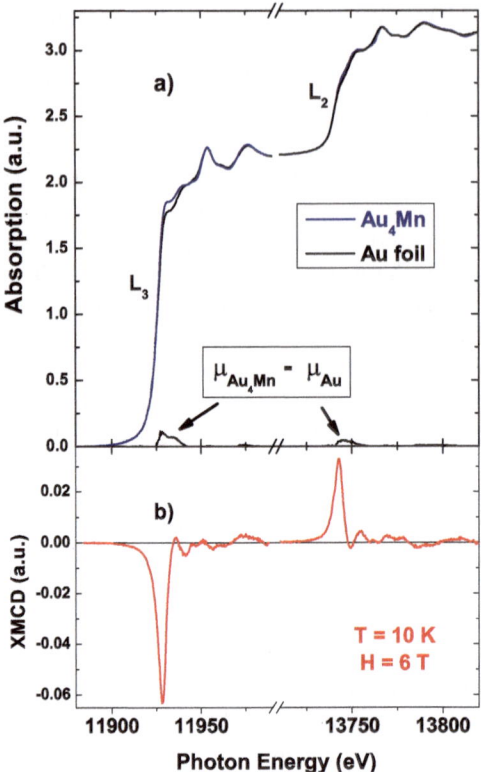

Fig. 2. (a) Au $L_{2,3}$ isotropic absorption spectra of Au$_4$Mn (blue line) and metallic Au foil (black line) and their difference (green line). (b) XMCD spectra of Au$_4$Mn recorded at $T = 10\,\mathrm{K}$ under an applied magnetic field of $6\,\mathrm{T}$

situation is getting far more complicated because $\mu^0_{j+(-) \rightarrow l}$ is only a part of the whole absorption spectrum: we are however interested in only particular transitions, e.g. $p \rightarrow d$, and n_h is the number of holes of the given symmetry. Indeed, the use of arctangent curves to account for the transitions towards the continuum states can only be taken as a very crude approximation whereas, unfortunately, the number of holes estimated from band structure calculations remains very sensitive to the selected method of calculation.

In some cases, it is nevertheless possible to determine directly the isotropic cross section per hole from experimental spectra. According to [52], the value of Ω_l can be obtained from:

$$\Omega_l = \int_{j^+ + j^-} \left[\frac{\mu^0_{sample}(E) - \mu^0_{ref}(E)}{n_h^{sample} - n_h^{ref}} \right] dE , \qquad (4)$$

where μ^0_{sample} et μ^0_{ref} denote the total isotropic absorption cross section measured either in the sample under study and in a reference sample. In the case of Au$_4$Mn, a metallic foil of gold appears as a suitable reference, and the corresponding normalized isotropic $L_{2,3}$ absorption spectra are also shown on Fig. 2a. Therefore, in our example, the (4) would become:

$$\Omega_l = \int_{j^+ + j^-} \left[\frac{\mu^0_{Au_4 Mn}(E) - \mu^0_{Au\ foil}(E)}{n_h^{Au_4 Mn} - n_h^{Au\ foil}} \right] dE . \tag{5}$$

Since the difference spectrum: $\mu^0_{Au_4 Mn}(E) - \mu^0_{Au\ foil}(E)$ shown on Fig. 2a is rather weak, this mat reflect a small change in the $5d$ hole counts between these two systems. To determine the change in $5d$ states occupation between Au$_4$Mn and Au metal, we follow a prescription given by by Mattheiss and Dietz [65] and refined more recently by Kuhn and Sham [69]. We would like to emphasize that this procedure is restricted to the case of strong spin-orbit coupling in the valence band. In other words, the crystal field splitting is supposed to be smaller than the spin-orbit interaction so that two sub-bands, associated with two different j values, can be defined. We shall assume next that: (i) there is no significant change in the contribution of the $2p \rightarrow 6s$ transitions to the $L_{2,3}$ spectra of Au$_4$Mn and of pure Au; (ii) the difference in the $\langle 5d_{3/2}|r|2p \rangle$ and $\langle 5d_{5/2}|r|2p \rangle$ radial matrix elements is negligible and they are energy independent over the $5d$ band. Then, we obtain:

$$\left(n_h^{5d_{5/2}, Au_4 Mn} - n_h^{5d_{5/2}, Au} \right) = \frac{1}{2C} \left[5 \frac{E_2}{E_3} A_3 - A_2 \right], \tag{6}$$

$$\left(n_h^{5d_{3/2}, Au_4 Mn} - n_h^{5d_{3/2}, Au} \right) = \left[\frac{3A_2}{C} \right], \tag{7}$$

where $A_{2(3)} = \int_{L_{2(3)}} \left[\mu^0_{Au_4 Mn}(E) - \mu^0_{Au}(E) \right]\ dE$, C is a constant proportional to $\langle 5d|r|2p \rangle$ and that is equal to 7.484×10^4 eV.cm^{-1} [69]; $E_{2(3)}$ is the electron binding energy of the $2p_{1/2}(2p_{3/2})$ states. Combining (5) with (6) and (7), we obtain that the difference in $5d$ number of holes in Au$_4$Mn and metallic Au is equal to 0.054. This value, which was determined experimentally, is not fairly different from the theoretical value of 0.07 that was obtained from a fully relativistic band structure calculations [70]. Now, we have all the necessary ingredients to apply the magneto-optical sum rules to our XMCD spectra. Assuming that the contribution of the magnetic dipole term is negligible in the case of Au$_4$Mn, the Au $5d$ spin and orbital magnetic moments can be estimated to be $0.035\,\mu_B$ and $0.005\,\mu_B$, respectively. These results when compared to $4.15\,\mu_B/f.u.$ as determined from macroscopic magnetization measurements, nicely illustrate the high sensitivity of the XMCD technique and its unique capability to access for a given element to the magnetic properties of a selected electronic state of well defined symmetry.

4 Highlights of Recent Advances in Hard X-ray Magnetic Circular Dichroism

4.1 XMCD Studies of Pauli Paramagnets

It is well known [71] that the *paramagnetic* susceptibility of a non-magnetic metal (e.g. Pd) results from the sum of two contributions:

- the Pauli susceptibility which describes the response of the spins of the system and is proportional to the density of states at the Fermi level;
- the orbital susceptibility which was shown by Kubo and Obata [72] to have the same origin as the van Vleck susceptibility in paramagnetic insulators.

The macroscopic magnetic susceptibility measured experimentally includes many more contributions: the diamagnetic susceptibility, the susceptibilities from ferromagnetic impurities as well as the susceptibility due to induced moments associated with the various conduction electrons (s, p, d). The wish of the experimentalis would be to disentangle each individual contribution. The current strategy was to analyze a variety of different phenomena [73]: specific heat, magnetization, NMR etc... but those methods are most often indirect and their interpretatations are relying on a number of assumptions. In contrast, XMCD which is inherently element and orbital selective, appears as a powerful technique which may allow us to disentangle the various contributions to the magnetic susceptibility, especially if XMCD is combined with the magneto-optical sum-rules.

In 2001, H. Ebert et al. [74] have calculated the XMCD spectrum of Pd metal in a magnetic field and pointed out that the signal should be large enough to be measured experimentally.

We performed this challenging experiment at the ESRF beamline ID12 [60] although the circular polarization rate of the monochromatic x-rays at Pd L edges is rather poor: 22% at the L_2 edge but only 12% at the L_3 edge. The experiment was performed on a Pd single crystal at temperatures ranging from 4 and 300 K and under a magnetic field of 7 T. The order of magnitude of the XMCD signals normalized to the edge-jumps were $\approx 3 \times 10^{-3}$ at the Pd L_3 edge and $\approx 2 \times 10^{-3}$ at the L_2 edge. When proper corrections were made to account for self-absorption [63] and low circular polarization rates, then the normalized XMCD signal was found to be *ca.* 4% at the L_3 edge and 1% at the L_2 edge. The corrected spectra are reproduced in Fig. 3. In order to make sure that the measured signal was not contaminated by eventual paramagnetic impurities in the sample, we have recorded the magnetization curves by monitoring the Pd L_3 XMCD signal as a function of the field at three different temperatures: 4, 100 and 300 K. These curves which are reproduced in the insert in Fig. 3 confirm unambiguously that the measured XMCD signal is truly due to the paramagnetism of Pd $4d$ electrons. Using the magneto-optical sum rules, we were able to determine the spin ($\approx 0.012\,\mu_B$) and orbital ($\approx 0.004\,\mu_B$) moments induced in the Pd $4d$ shell when an external

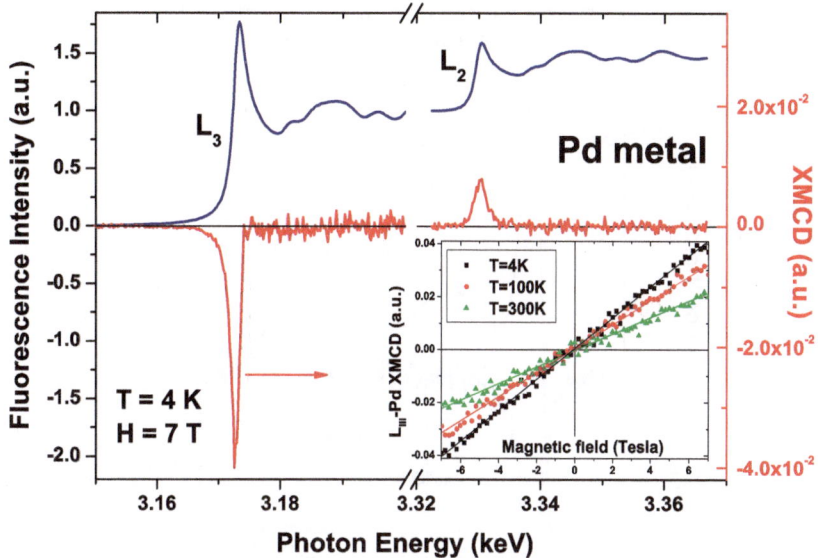

Fig. 3. Pd $L_{2,3}$ XANES and XMCD spectra of a Pd single crystal recorded at $T = 4\,\mathrm{K}$ with a magnetic field of $7\,\mathrm{T}$. The spectra were corrected for self-absorption and uncomplete circular polarization rate. Insert: Magnetization curves of Pd metal recorded at 4, 100 and 300 K by monitoring Pd L_3 XMCD signal

magnetic field of 7 Tesla is applied to the sample. These magnetic moments would correspond to a total paramagnetic susceptibility of the $4d$ electrons in Pd of $\approx 11.5 \times 10^{-6}\,\mathrm{cm^3/g}$. This value looks somewhat higher than the total susceptibility of Pd ($\approx 7 \times 10^{-6}\,\mathrm{cm^3/g}$) measured with another piece of the same single crystal. The latter value, however, includes other contributions such as the diamagnetic susceptibility and its experimental and theoretical evaluation remains still rather problematic.

The induced orbital magnetic moment of the $4d$ electrons is itself the sum of two contributions: the first one arises from the orbital susceptibility while the second one is due to the spin-orbit interactions. To disentangle these two parts, it may help to compare the ratios μ_L/μ_S in pure Pd (≈ 0.295) and in a very thin Pd layer (only 0.25 atomic layers) sandwiched between 30 atomic layers of Fe. In the latter case, the magnetism of Pd is obviously induced by the hybridization with the neighboring Fe atoms. We have reproduced in Fig. 4 the *as-measured* XANES and XMCD spectra recorded at the Pd $L_{2,3}$ absorption edges using a 35-element silicon drift diode detector that was installed recently at the ID12 beamline [75]. This experiment was particularly difficult, not only because of the poor circular polarization rate of the x-rays, but also because of the very intense background emission originating from the sample substrate. As can be seen from the insert of Fig. 4, the emission spectrum is strongly dominated by the Mg K_α emission line, while unresolved

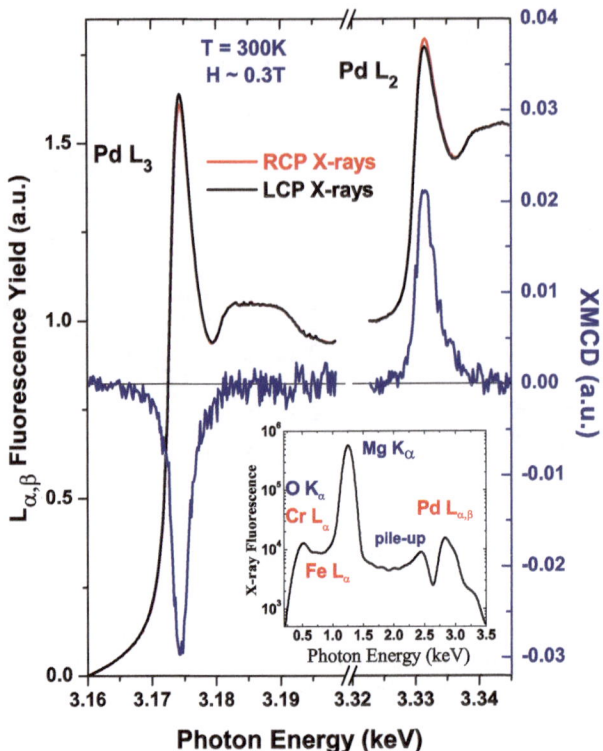

Fig. 4. Raw XANES and XMCD spectra of $Fe_{30ML}/Pd_{0.25ML}/Fe_{30ML}$ trilayer at the Pd $L_{2,3}$ edges. Insert: X-ray emission spectrum of the sample with excitation at the Pd L_2-edge

Cr L_{α}, Fe L_{α} and O K_{α} lines are of the same intensity as the Pd $L_{\alpha,\beta}$ lines. A rather intense XMCD signal at spin-orbit split Ledges of Pd was nevertheless measured at room temperature and with an applied magnetic field of 0.3 T. After correction for the low circular polarization rate, we tried to apply the magneto-optical sum rules: this led us to the conclusion that the ratio of the orbital-to-spin moments induced on the $4d$ states in Pd was ≈ 0.12. Let us recall that a similar value had already been measured for Pd in Pd/Fe multilayers [57].

There is a large difference between the ratios μ_L/μ_S measured in the field-induced experiment on pure Pd and in the Pd/Fe trilayer where the moments are induced by hybridization. This difference could be most probably assigned to large field-induced orbital moments because one may presume that the spin-orbit interactions in the $4d$ shell of Pd should be nearly the same in Pd metal or in the Fe/Pd multilayers. Thus, one may conclude that the orbital susceptibility of the $4d$ electrons in pure Pd due to the Kubo-Obata contribution would be of the order of 17% of the Pauli spin susceptibility.

This is a typical illustration of the potentiality of XMCD to disentangle the relative contributions to the total paramagnetic susceptibility of the $4d$ states of Pd: (i) the Pauli spin susceptibility being $\approx 8.9 \times 10^{-6}$ cm^3/g; (ii) the orbital susceptibility due to spin-orbit interaction being $\approx 1.1 \times 10^{-6}$ cm^3/g and (iii) the Kubo-Obata orbital susceptibility being finally $\approx 1.5 \times 10^{-6}$ cm^3/g.

4.2 Induced Magnetism in $3d/5d$ Multilayers

Multilayers combining ferromagnetic $3d$ metals and (normally) non-magnetic $5d$ elements are considered today as the most promising candidates for the next generation of magneto-optic recording media. It was realized rather recently that the magnetic polarization of the $5d$ atoms at interfaces is at the origin of the exceptional properties of these multilayers such as a large magnetic anisotropy, an enhanced magneto-optical Kerr effect and a giant magnetoresistance. The element specificity and the monolayer sensitivity of XMCD, in combination with theory, offer a unique opportunity to investigate the interfacial magnetism of both the ferromagnetic and the non-magnetic elements.

Due to their very sharp interfaces at the monolayer limit, Ni/Pt multilayers were excellent candidates to study the interface magnetism [77]. Moreover, in these multilayers the magnetic anisotropy can be either perpendicular or in-plane, depending on the thickness of the Pt layers. For a whole series of Ni$_n$/Pt$_m$ multilayers (with $2 \leq [n, m] \leq 13$), XMCD measurements were performed at 10 K with an applied magnetic field of up to 5 T [78]. The Pt $L_{2,3}$ XMCD spectra of a family of Ni$_2$/Pt$_m$ multilayers are reproduced in Fig. 5a. The existence of a XMCD signal proved the induced spin polarization of the Pt atoms and its sign revealed a ferromagnetic coupling between the Pt $5d$ and Ni $3d$ electrons. Since the measured XMCD signal decreases with the thickness of the Pt layers, the Pt magnetic polarization is expected to be much larger at the interfaces. A systematic XMCD study combining spectra recorded at the Pt $L_{2,3}$ edges using the ESRF beamline ID12-A (now ID12) [79] together with spectra recorded at the Ni $L_{2,3}$ edges using the former ESRF beamline ID12-B (now ID8) [80] has been performed for a whole series of Ni$_n$/Pt$_m$ multilayers: the results allowed us to reconstruct a monolayer-resolved magnetization profile in these systems. Such a profile for the Ni$_6$/Pt$_5$ multilayer is reproduced in Fig. 5b. The induced magnetic moments of the Pt atoms were found to be rather large ($\approx 0.29 \mu_B$/atom) at the interfaces but decrease quite sharply away from the interface. On the other hand, the Ni magnetic moments at the interfaces ($\approx 0.32 \mu_B$/atom) were found to be dramatically lower than in bulk Ni but the second Ni layer (away from the interface) already recovered nearly completely the bulk moment. These experimental data were compared with the results of ab initio calculations performed by the group of H. Ebert (LMTU-München, Germany) using the TB-LMTO method. The calculated magnetization profile (shown on Fig. 5c for the same Ni$_6$/Pt$_5$ multilayer) agreed remarkably well with the

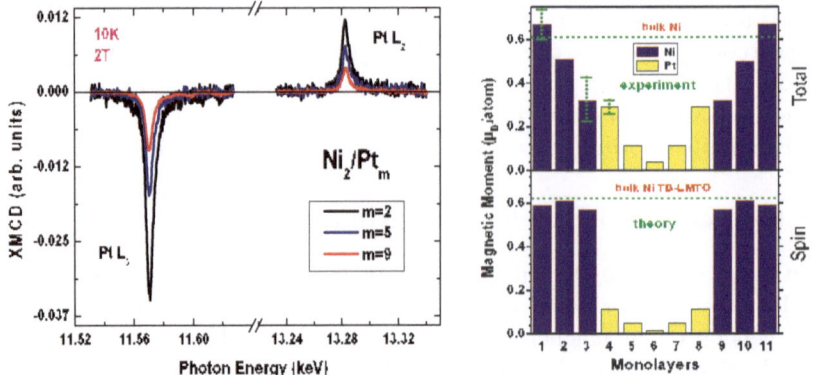

Fig. 5. *Left*: Experimental Pt $L_{2,3}$ XMCD spectra in a series of Ni_2/Pt_m multilayers. *Right*: Magnetization profile in Ni_6/Pt_5 multilayer deduced from the XMCD experiments and calculated magnetization profile in Ni_6/Pt_5 multilayer. Only the spin moments of Ni and Pt are shown

experimental one and confirmed that the Pt induced magnetic moments were due to the 3d-5d hybridization at the interfaces.

So far, mainly Pt induced magnetic moments in $3d/5d$ multilayers have been investigated [76], but the magnetic properties of multilayers with other interesting $5d$ metals (e.g. W, Ir or Au) remain nearly unexplored. Recently, we have shown [81] that the $5d$ electrons of W in W/Fe multilayers could also acquire an induced magnetic moment which is coupled antiferromagnetically to the moments carried by Fe. Surprisingly, we have found that the spin and the orbital magnetic moments of W were coupled parallel to each other whereas they were coupled antiparallel in the case of W impurity atoms in ferromagnetic Fe [82]. As a matter of fact, the antiparallel alignment of the $5d$ spin and orbital moments in W is in contradiction with the third Hund's rule [83]. In the case of Ir/Fe multilayers [81, 84], we have shown that the induced Ir spin and orbital magnetic moments were parallel to each other as expected from the third Hund's rule and, also, parallel to the Fe magnetic moment. Such an apparent *breakdown* of the atomic third Hund's rule for W could be understood if one considers not only intra-atomic spin-orbit interactions, but also inter-atomic $3d(spin)$-$5d(orbit)$ interactions in addition to the usual $3d(spin)$-$5d(spin)$ exchange coupling at the interfaces. In order to be consistent with the experimental results, the $3d(spin)$-$5d(orbit)$ interaction had to be stronger than the intra-atomic $5d$ spin-orbit coupling. In the past, the role of such interatomic spin-orbit interactions has already been considered in order to give a more accurate description of the properties of many-body systems [85].

Acknowledgement

The authors are indebted to G. Schmerber for the preparation and the characterization of Au_4Mn sample and to W. Grange for his assistance during the XMCD experiments on Au_4Mn. Special thanks are due to A. Derory for SQUID measurements on the Pd single crystal and on the Au_4Mn sample. We are grateful to J. Arabski for the fabrication of Fe/Pd/Fe trilayer. Very warm acknowledgements are finally due to M. Angelakeris, P. Poulopoulos and K. Baberschke for a long standing collaboration in the field of hard x-ray magnetic circular dichroism.

References

1. W.C. Röntgen, *Über eine neue Art von Strahlen. Vorlaufige Mitteilung*: Sitzber. Physik. Med. Ges. **137**, 132 (1895)(in German)
2. J. Perrin, *Quelques propriétés des rayons de Röntgen*: C.R. Acad. Sci. Paris **122**, 186 (1896)(in French)
3. J.J. Thomson, *The Röntgen Rays*: Nature **53**, 391 (1896)
4. B. Galitzine and A. Karnojitzky, *Recherches concernant les propriétés des rayons X*: C.R. Acad. Sci. Paris **122**, 717 (1896)(in French)
5. G. Sagnac, *Sur la diffraction et la polarisation des rayons de M. Röntgen*: C.R. Acad. Sci. Paris **122**, 783 (1896)(in French)
6. R. Blondlot, *Sur la polarisation des rayons X*: J. de Physique (4) **2**, 169; C.R. Acad. Sci. Paris **136**, 284 (1903)(in French)
7. J.W. Gifford, *Are Röntgen Rays Polarised?*: Nature **54**, 172 (1896)
8. R. von Lieben, Physik. Zeitschr. **4**, 471 (1903) (in German)
9. R. von Lieben, *Bemerkungen zur "Polarisation der Röntgenstrahlung"*: Physik. Zeitschr. **5**, 72 (1904)(in German)
10. C.G. Barkla, *Polarised Röntgen Radiation*: Phil. Trans. **A204**, 467 (1905)
11. C.G. Barkla, *Polarisation in Secondary Röntgen Radiation*: Proc. Roy. Soc. London A **77**, 247 (1906)
12. E. Bassler, *Polarisation der X-Strahlen, nachgewiesen mittels Sekundärstrahlung*: Annalen der Physik, **28**, 808 (1909)(in German)
13. J.C. Chapman, *Some experiments on polarized Röntgen radiation*: Phil. Mag. **25**, 792 (1913)
14. A.H. Forman, *The effect of magnetization on the opacity of iron to Röntgen rays*: Phys. Rev. **3**, 306 (1914)
15. A.H. Forman, *The effect of magnetization on the opacity of iron to Röntgen rays*: Phys. Rev. **7**, 119 (1916)
16. A.H. Compton, *The absorption of gamma rays by magnetized iron*: Phys. Rev. **17**, 38 (1921)
17. J.A. Becker, *The effect of the magnetic field on the absorption of X-rays*: Phys. Rev. **20**, 134 (1922)
18. J.A. Becker, *The effect of a magnetic field on the absorption of X-rays*: Phys. Rev. **22**, 320 (1923)
19. W. Kartschagin and E. Tschetwerikova, *Zur Frage nach der magnetischen Drehung der Polarisationsebene primärer Röntgenstrahlen*: Zeitsch. für Physik, **39**, 886 (1926)(in German)

20. D.K. Froman, *The Faraday Effect with X-rays*: Phys. Rev. **41**, 693 (1932)
21. C. Kurylenko, *Franges au voisinage de la discontinuité K des rayons X*: J. Phys. Rad. **1**, 133 (1940) (in French)
22. C. Kurylenko, *Oú en est la question de la discontinuité d'absorption K des rayons X et des franges qui l'accompagnent?*: Bull. Soc. R. Sci. Liège **15**, 236 (1946) (in French)
23. D. Coster, *Book Reviews*: Acta Cryst. **3**, 160 (1948)
24. H.S. Bennet and E.A. Stern, *Faraday effect in solids*: Phys. Rev. **137**, A448 (1965)
25. J.L. Erskine and E.A. Stern, *Calculation of the $M_{2,3}$ magneto-optical absorption spectrum of ferromagnetic nickel*: Phys. Rev. B **12**, 5016 (1975)
26. J. Hrdý, E. Krouský, and O. Renner, *A search for Faraday effect in X-ray region*: Phy. Stat. Sol. (a). **53**, 143 (1979)
27. M. Hart and A.R.D. Rodriguez, *Optical activity and the Faraday effect at X-ray frequencies*, Phil. Mag. B **43**, 321 (1981)
28. G. Schütz, E. Zech, E. Hagn, and P. Kienle, *Anisotropy of X-rays and spin dependence of the photoabsorption of circularly polarized soft X-rays in Magnetized Fe*: Hyperfine Interaction **16**, 1039 (1983)
29. E. Keller and E.A. Stern: Magnetic XANES. In Proc. of the EXAFS and Near Edge Structure III Conference, Ed.: K.O. Hodgson, B. Hedman and J.E. Penner-Hahn, p. 507 (1984)
30. B.T. Thole, G. van der Laan, and G.A. Sawatzky, *Strong magnetic dichroism predicted in the $M_{4,5}$ X-ray absorption spectra of magnetic rare-earth materials*: Phys. Rev. Lett. **55**, 2086 (1985)
31. G. van der Laan, B.T. Thole, G.A. Sawatzky, J.B. Goekoop, J.C. Fuggle, J.-M. Esteva, R. Karnatak, J.P. Remeika, and H.A. Dabkowska, *Experimental proof of magnetic x-ray dichroism*: Phys. Rev. B **34**, 6529 (1986)
32. G. Schütz, W. Wagner, W. Wilhelm, P. Kienle, R. Zeller, R. Frahm, G. Materlik, *Absorption of Circularly Polarized X-rays in Iron*: Phys. Rev. Lett. **58**, 737 (1987)
33. G. Schütz, R. Frahm, P. Mautner, R. Wienke, W. Wagner, W. Wilhelm and P. Kienle, *Spin dependent Extended X-ray Absorption Fine Structure: Probing Magnetic short-range order*: Phys. Rev. Lett. **62**, 2620 (1989)
34. C.T. Chen, F. Sette, Y. Ma, and S. Modesti, *Soft X-ray magnetic circular dichroism at the $L_{2,3}$ edges of nickel*: Phys. Rev. B **42**, 7262 (1990)
35. B.T. Thole, P. Carra, F. Sette, and G. van der Laan, *X-ray circular dichroism as a probe of orbital magnetization*: Phys. Rev. Lett. **68**, 1943 (1992)
36. P. Carra, B.T. Thole, M. Altarelli, and X. Wang, *X-ray circular dichroism and local magnetic fields*: Phys. Rev. Lett. **70**, 694 (1993)
37. J. Stöhr, *Exploring the microscopic origin of magnetic anisotropies with X-ray magnetic circular dichroism (XMCD) spectroscopy*: J. Magn. Magn. Mat. **200**, 470 (1999)
38. D.P. Siddons, M. Hart, Y. Amemiya, and J.B. Hastings, *X-ray Optical activity and the Faraday effect in Cobalt and its compounds*: Phys. Rev. Lett. **64**, 1967 (1990)
39. M. Sacchi and A. Mirone, *Resonant reflectivity from a Ni(110) crystal: Magnetic effects at the Ni 2p edges using linearly and circularly polarized photons*: Phys. Rev. B **57**, 8408 (1998)

40. O. Hellwig, J.B. Kortright, K. Takano, E.E. Fullerton, *Switching behavior of Fe-Pt/Ni-Fe exchange-spring films studied by resonant soft-x-ray magneto-optical Kerr effect*: Phys. Rev. B **62**, 11694 (2000)
41. H.-Ch. Mertins, P.M. Oppeneer, J. Kuneš, A. Gaupp, D. Abramsohn, and F. Schäfers, *Observation of the X-ray magneto-optical Voigt effect*: Phys. Rev. Lett. **87**, 047401 (2001)
42. J. Goulon, C. Goulon-Ginet, A. Rogalev, V. Gotte, C. Malgrange, C. Brouder, and C.R. Natoli, *X-ray natural circular dichroism in a uniaxial gyrotropic single crystal of LiIO₃*: J. Chem. Phys., **108**, 6394 (1998)
43. J. Goulon, A. Rogalev, C. Goulon-Ginet, G. Benayoun, L. Paolasini, C. Brouder, C. Malgrange, and P.A. Metcalf, *First Observation of Nonreciprocal X-ray Gyrotropy*: Phys. Rev. Lett. **85**, 4385 (2000)
44. J. Goulon, A. Rogalev, F. Wilhelm, C. Goulon-Ginet, P. Carra, D. Cabaret, and C. Brouder, *X-Ray Magnetochiral Dichroism: A New Spectroscopic Probe of Parity Nonconserving Magnetic Solids*: Phys. Rev. Lett. **88**, 237401 (2002)
45. P. Carra, A. Jerez, and I. Marri, *X-ray dichroism in noncentrosymmetric crystals*: Phys. Rev. B **67**, 045111 (2003)
46. J. Goulon, A. Rogalev, F. Wilhelm, C. Goulon-Ginet, P. Carra, I. Marri, and C. Brouder, *X-ray Optical Activity: Application of Sum Rules*: J. Exp. Theor. Phys., **97**, 402 (2003)
47. J. Stöhr and R. Nakajima, *Magnetic properties of transition-metal multilayers studied with X-ray magnetic circular dichroism spectroscopy*: IBM J. Res. Develop. **42**, 73 (1998)
48. U. Fano, *Spin Orientation of Photoelectrons Ejected by Circularly Polarized Light*: Phys. Rev. **178**, 131 (1969)
49. M. Altarelli, *Orbital-magnetization sum rule for X-ray circular dichroism - a simple proof*: Phys. Rev. B **47**, 597 (1993)
50. J. Igarashi and K. Hirai, *Magnetic circular dichroism at the K-edge of nickel and iron*: Phys. Rev. B **50**, 17820 (1994)
51. A. Ankudinov A. and J.J. Rehr, *Sum rules for polarization-dependent X-ray absorption*: Phys. Rev. B **51**, 1282 (1995)
52. A.F. Starace, *Potential barrier effects in photoabsorption. 1. General theory*: Phys. Rev. **B5**, 1773 (1972)
53. H. Ebert, *Magneto-optical effects in transition metal systems*: Rep. Prog. Phys. **59**, 1665 (1996)
54. P. Carra, H. König, B.T. Thole, and M. Altarelli, *Magnetic X-ray dichroism - general features of dipolar and quadrupolar spectra*: Physica B **192**, 182 (1993)
55. R. Wu, D. Wang and A.J. Freeman, *First principles investigation of the validity and range of applicability of the X-ray magnetic circular dichroism sum rule*: Phys. Rev. Lett. **71**, 3581 (1993)
56. C.T. Chen, Y.U. Idzerda, H.-J. Lin, N.V. Smith, G. Meigs, E. Chaban, G.H. Ho, E. Pellegrin, and F. Sette, *Experimental confirmation of the X-ray magnetic circular dichroism sum rules for iron and cobalt*: Phys. Rev. Lett. **75**, 152 (1995)
57. J. Vogel, A. Fontaine, V. Cros, F. Petroff, J.P. Kappler, G. Krill, A. Rogalev, and J. Goulon, *Structure and magnetism of Pd in Pd/Fe multilayers studied by x-ray magnetic circular dichroism at the Pd L₂,₃ edges*: Phys. Rev. B **55**, 3663 (1997)
58. W. Grange, M. Maret, J.P. Kappler, J. Vogel, A. Fontaine, F. Petroff, G. Krill, A. Rogalev, J. Goulon, M. Finazzi, and N. Brookes, *Magnetocrystalline*

anisotropy in (111) CoPt₃ thin films probed by x-ray magnetic circular dichroism: Phys. Rev. B **58**, 6298 (1998)

59. M. Finazzi, P. Sainctavit, A.M. Dias, J.P. Kappler, G. Krill, J.P. Sanchez, P. Dalmas de Reotier, A. Yaouanc, A. Rogalev, and J. Goulon, X-ray magnetic circular dichroism at the U $M_{4,5}$ absorption edges of UFe_2: Phys. Rev. B **55**, 3010 (1997)

60. A. Rogalev, J. Goulon, C. Goulon-Ginet, and C. Malgrange: Instrumentation Developments for Polarization Dependent X-ray Spectroscopies at the ESRF Beamline ID12A. **Magnetism and Synchrotron Radiation** ed. by E. Beaurepaire, F. Scheurer, G. Krill and J.P. Kappler - Springer-Verlag, 60 (2001)

61. C. Malgrange, L. Varga, C. Giles, A. Rogalev, J. Goulon, Phase plates for X-ray optics: Proc. SPIE **3773**, 326 (1999)

62. A.J.P. Meyer, Propriétés magnétiques de MnAu: C.R. Acad. Sci. Paris **242**, 2315 (1956)

63. J. Goulon, C. Goulon-Ginet, R. Cortés, and J.M. Dubois, On experimental attenuation factors of the amplitude of the EXAFS oscillations in absorption, reflectivity and luminescence measurements: J. Phys. (Paris) **43**, 539 (1982)

64. M. Brown, R.E. Peierls, and E.A. Stern, White lines in X-ray absorption: Phys. Rev. B **15**, 738 (1977)

65. L.F. Mattheiss and R.E. Dietz, Relativistic tight-binding calculation of core-valence transitions in Pt and Au: Phys. Rev. B **22**, 1663 (1980)

66. J. Stöhr and R. Wu: Proc. NATO Advanced Study Institute "New Directions in Research with 3rd Generation Soft X-ray Synchrotron Radiation Sources", ed. by A.S. Slachter and F.J. Wuilleumier (Kluver Academic Publishers, Dordrecht) (1994)

67. C. Brouder and J.P. Kappler, in Magnetism and Synchrotron Radiation, ed. by E. Beaurepaire, B. Carrière and J.P. Kappler, p. 19–32 (Les editions de physique, Les Ulis) (1997)

68. Y. Wu, J. Stöhr, B.D. Hermsmeier, M.G. Samant, and D. Weller, Enhanced orbital magnetic moment on Co atoms in Co/Pd multilayers - a magnetic circular X-ray dichroism study: Phys. Rev. Lett. **69**, 2307 (1992)

69. M. Kuhn and T.K. Sham, Charge redistribution and electronic behavior in a series of Au-Cu alloys: Phys. Rev. B **49**, 1647 (1994)

70. V. Antonov, B. Harmon, and A. Yaresko: **Electronic structure and magneto-optical properties of solids** (Kluwer Academic Publishers, Dordrecht, Boston, London) (2004)

71. R.M. White: Quantum Theory of Magnetism (McGraw-Hill Advanced Physics Monograph Series, New York, 1970)

72. R. Kubo and Y. Obata, Note on the paramagnetic susceptibility and the gyromagnetic ratio in metals: J. Phys. Soc. Japan **11**, 547 (1956)

73. J.A. Seitchik, A.C. Gossard, and V. Jaccarino, Knight shifts and susceptibilities of transition metals: palladium: Phys. Rev. **136**, A1119 (1964)

74. H. Ebert et al. in Magnetism and Synchrotron Radiation ed. by E. Beaurepaire, F.Scheurer, G. Krill and J.P. Kappler Springer-Verlag, 343 (2001)

75. J. Goulon, A. Rogalev, G. Goujon, C. Gauthier, E. Moguiline, A. Sole, S. Feite, F. Wilhelm, N. Jaouen, C. Goulon-Ginet, P. Dressler, P. Rohr, M.-O Lampert, and R. Henck, Advanced detection systems for X-ray fluorescence excitation spectroscopy: J. Synch. Rad. **12**, 57 (2005)

76. P. Poulopoulos, M. Angelakeris, E.T. Papaioannou, N.K. Flevaris, D. Niarchos, M. Nyvlt V. Prosser, S. Visnovsky, C. Mueller, P. Fumagalli, F. Wilhelm, and A. Rogalev, *Structural, magnetic, and spectroscopic magneto-optical properties aspects of Pt-Co multilayers with intentionally alloyed layers*: J. Appl. Phys. **94**, 7662 (2003)

77. M. Angelakeris, P. Poulopoulos, N. Vouroutzis, M. Nyvlt, V. Prosser, S. Visnovsky, R. Krishnan, and N.K. Flevaris, *Structural and spectroscopic magneto-optic studies of Pt-Ni multilayers*: J. Appl. Phys. **82**, 5640 (1997)

78. F. Wilhelm, P. Poulopoulos, G. Ceballos, H. Wende, K. Baberschke, P. Srivastava, D. Benea, H. Ebert, M. Angelakeris, N.K. Flevaris, D. Niarchos, A. Rogalev, and N.B. Brookes, *Layer-resolved Magnetic Moments in Ni/Pt Multilayers*: Phys. Rev. Lett. **85**, 413 (2000)

79. J. Goulon, A. Rogalev, C. Gauthier, C. Goulon-Ginet, S. Paste, R. Signorato, C. Neumann, L. Varga, and C. Malgrange, *Instrumentation developments for x-ray linear and circular dichroism at the ESRF beamline ID12A*: J. Synch. Rad., **5**, 232 (1998)

80. J. Goulon, N.B. Brookes, C. Gauthier, J.B. Goedkoop, C. Goulon-Ginet, M. Hagelstein, and A. Rogalev, *Instrumentation development for ESRF beamlines*: Physica B, **208&209**, 199 (1995)

81. F. Wilhelm, P. Poulopoulos, H. Wende, A. Scherz, K. Baberschke, M. Angelakeris, N.K. Flevaris, and A. Rogalev, *Systematics of the induced magnetic moments in 5d layers and the violation of the third Hund's rule*: Phys. Rev. Lett. **87**, 207202 (2002)

82. G. Schütz, M. Knulle, and H. Ebert, *Magnetic circular X-ray dichroism and its relation to local moments*: Phys. Scr. **T49**, 302 (1993)

83. F. Wilhelm, P. Poulopoulos, H. Wende, A. Scherz, K. Baberschke, M. Angelakeris, N.K. Flevaris, and A. Rogalev, Reply on comment by R. Tyer, G. van der Laan, W.M. Temmerman, and Z. Szotek, Phys. Rev. Lett. **90**, 129702 (2003)

84. N. Jaouen, F. Wilhelm, A. Rogalev, J. Goulon, J.-M. Tonnerre, and S. Andrieu, *Influence of the structure of Fe on the interfacial spin and orbital magnetic moments of Ir in Fe/Ir multilayers*: J. Magn. Magn. Mat. **272–276**, e1615 (2004)

85. A.K. Rajagopal and M. Mochena, *Spin-orbit interactions in the many-body theory of magnetic electron systems*: Phys. Rev. B **57**, 11582 (1998)

Spin-Polarized Photoemission

Jürg Osterwalder

Physik-Institut, Universität Zürich, Winterthurerstr. 190, CH-8057 Zürich,
Switzerland
osterwal@physik.unizh.ch

Abstract. The principles of angle-resolved photoemission spectroscopy (ARPES)
with spin resolution are outlined, with the emphasis on conceptual clarity and
not on completeness. In Sect. 2, the theoretical ingredients of the photoemission
process are discussed, including the single-particle matrix element as well as the
many-body reaction of the solid as reflected in the spectral function. Experimental
parameters for probing defined electron states within a three-dimensional Brillouin
zone are defined. Sections 3 and 4 list several situations and mechanisms where spin
polarization is produced or reduced in photoemission signals, and describe how they
can be measured. Spin-polarized ARPES data from a ferromagnetic Ni(111) surface
and from the spin-orbit split surface state on Au(111) follow as case studies. Finally,
the future prospects of the technique are assessed.

1 Introduction

The spin of the electron is at the heart of magnetic and other electron correlation phenomena in condensed matter physics. There is a multitude of
experimental techniques that probe the collective behaviour of spins in solids
and the interaction of spins with other degrees of freedom. They all project
out some observable of electronic states but measure an ensemble average
for others. Magnetic resonance techniques operate directly in spin space and
provide detailed information about magnetic moments and spin dynamics.
Magnetic x-ray or neutron scattering are highly sensitive to magnetic ordering phenomena and spin structures in real space (see chapters by C. Dufour
and M. Altarelli in this book). Spin-polarized angle-resolved photoemission
(ARPES) is complementary to these techniques and offers unique information especially for itinerant magnetic systems. It measures not only the spin
state but also the energy and momentum of such states and is therefore sensitive to spin structures in reciprocal space, in cases where the direction and
magnitude of the electron spin depends on the wave vector. In this sense, it
is a complete experiment that can provide very detailed information e.g. on
the exchange-split bands of itinerant ferromagnets [1].

Ordered spin structures both in real space and in reciprocal space are becoming increasingly more important in the context of *spintronics* [2], where
functionalities are introduced in electronic devices that are based on the electron spin. Vigorous research is now going on worldwide for designing ways

J. Osterwalder: *Spin-Polarized Photoemission*, Lect. Notes Phys. **697**, 95–120 (2006)
www.springerlink.com   © Springer-Verlag Berlin Heidelberg 2006

and materials to inject spin-polarized electron (or hole) currents into semi-conductors and for finding ways to manipulate them. A device that illustrates the combination of these concepts is the spin field-effect transistor proposed by Datta and Das [3], where ferromagnetic source and drain electrodes represent simple real space spin structures whereas the momentum dependence of the spin in the active channel is necessary for the control of the spin precession by the electric field of the gate electrode. ARPES with spin resolution is likely to play an important role in the materials-related investigation of these spin-dependent phenomena.

Unfortunately, in order to characterize the spin-polarized band structure of a solid, we cannot simply use the spin of the electrons as a tag and measure independently the photoelectron spectra for spin-up and spin-down electrons, e.g. in the equivalent of a Stern-Gerlach experiment. Unlike neutral atoms, electrons are strongly deflected in magnetic fields. By means of sophisticated arguments it can be shown that these Lorentz forces and the uncertainty principle conspire to render impossible the complete spin separation by magnetic fields [4, 5]. In practice, one therefore exploits the spin-dependence in a scattering experiment in order to characterize the *spin polarization* in a beam of electrons that has been preselected according to energy E and wave vector k by an angle-resolving spectrometer. It is defined as

$$P_I(E, k) = \frac{I_\uparrow(E, k) - I_\downarrow(E, k)}{I_\uparrow(E, k) + I_\downarrow(E, k)} , \tag{1}$$

where $I_{\uparrow,\downarrow}(E, k)$ represent the cleanly spin-resolved ARPES spectra for a spin quantization axis defined by the scattering geometry [5]. We refer to Section 4 for more details on how the spin polarization is extracted from left-right scattering asymmetries. At this point we emphasize that the measured quantities are the spin-integrated spectrum $I_M(E, k) = I_\uparrow(E, k) + I_\downarrow(E, k)$ and the spin-polarization $P_I(E, k)$, from which the spin-dependent spectra can be recovered as

$$I_{\uparrow,\downarrow}(E, k) = I_M(E, k)(1 \pm P_I(E, k))/2 . \tag{2}$$

Does the spin-polarized spectrum reflect the true spin polarization of the electronic states in the system under study? Unfortunately, this is not necessarily true. In Sect. 3, several different mechanisms inherent in the photoemission process are discussed that affect the spin polarization that is experimentally observed. One is due to strong correlation effects that often occur in magnetic materials when the photoeffect produces a hole in a localized level. Likewise, spin-dependent electron-electron scattering effects of the photoelectron when leaving the sample surface can reduce one spin channel $I_{\uparrow,\downarrow}(E, k)$ with respect to the other. Moreover, the interaction of the photoelectrons with other elementary excitations in the solid (phonons, magnons) can be different in both spin channels. Finally, chiral measurement geometries can introduce dichroic photoemission matrix elements that affect the two spin channels differently. All these effects can hamper the characterization of spin-resolved bands in the ground state. They are in most cases not well known. On the other hand, they provide interesting opportunities for studying spin-dependent many-body interactions in solids.

2 Angle-Resolved Photoemission Spectroscopy (ARPES) and Fermi Surface Mapping

In order to understand how photoemission spectra relate to the electronic band structure and elementary excitation spectra, and how spin polarization gets transferred from the initial state into the photoelectron final state, we need to establish the theoretical foundations of the method. In a general sense the photoemission process promotes the solid, containing N electrons within the interaction region, from the quantum mechanical N-electron state $\Psi_i(N)$ to the photoexcited $(N-1)$-electron state $\Psi_f(N-1, \boldsymbol{k_i})$ with a hole (or missing electron) in quantum state $\boldsymbol{k_i}$. At the same time, a photoelectron is produced in a free-electron state $\phi_f(\epsilon_f, \boldsymbol{k})$ propagating in vacuum (see below for the relation between the two wave vectors $\boldsymbol{k}$ and $\boldsymbol{k_i}$). From Fermi's Golden Rule the process occurs with a probability

$$ w \sim |\langle \Psi_f(N-1, \boldsymbol{k_i})\phi_f(E, \boldsymbol{k})|\boldsymbol{A} \cdot \boldsymbol{p}|\Psi_i(N)\rangle|^2 \delta(E_f - E_i - h\nu) \,, \qquad (3) $$

where the operator $\boldsymbol{A} \cdot \boldsymbol{p}$ describes the interaction between the vector potential $\boldsymbol{A}$ of the absorbed photon and the electron momentum $\boldsymbol{p}$, and E_i and E_f are the total energies of the system before and after the emission process (including the photoelectron energy ϵ_f) [1]. The matrix element can be factorized into a one-electron matrix element $M_{i,f} = \langle \phi_f(\epsilon_f, \boldsymbol{k_f})|r|\phi_i(\epsilon_i, \boldsymbol{k_i})\rangle$,[1] connecting the photoelectron with the initial one-electron state $\phi_i(\epsilon_i, \boldsymbol{k_i})$, times an overlap integral $\langle \Psi_f(N-1, \boldsymbol{k_i})|\Psi_i(N-1, \boldsymbol{k_i})\rangle$ of $(N-1)$-electron states involved before and after the removal of an electron in quantum state $\boldsymbol{k_i}$. In the process, the $(N-1)$-electron state finds itself suddenly in a changed potential with one electron missing, and in the so-called *sudden approximation* $\Psi_f(N-1, \boldsymbol{k_i})$ is expanded in Eigenstates $\tilde{\Psi}_f(N-1, \boldsymbol{k_i})$ of the perturbed Hamiltonian: $\Psi_f(N-1, \boldsymbol{k_i}) = \sum_s c_s \tilde{\Psi}_{f,s}(N-1, \boldsymbol{k_i})$. Here, the values $|c_s|^2$ describe the probability for finding the $(N-1)$-electron system in the excited state $\tilde{\Psi}_{f,s}(N-1, \boldsymbol{k_i})$ with label s after the removal of electron $\boldsymbol{k_i}$. In a solid, there is a continuum of excited states, including electronic, phononic and also magnetic excitations, and these probabilities are thus described by a continuous spectral function $A(\boldsymbol{k_i}, E)$, where E is the excitation energy of he system (i.e. $E = 0$ corresponds to the ground state of the $(N-1)$-electron system

---

[1] Due to commutation relations, and if the vector potential $\boldsymbol{A}$ varies slowly over atomic dimensions, the operator $\boldsymbol{A} \cdot \boldsymbol{p}$ in these matrix elements can be replaced by the operator r [1].

with total energy E_f^0). For each wave vector $\mathbf{k_i}$ the spectral function describes the excitation spectrum produced by the sudden presence of a hole in state $\mathbf{k_i}$. There are powerful theoretical concepts for the calculation of $A(\mathbf{k_i}, E)$ including electron-electron, electron-phonon, and electron-magnon interaction, although the actual computations are not a trivial task. The photoemission intensity measured along a direction (θ, ϕ) can be written as

$$I(\epsilon_f, \mathbf{k}) \sim |M_{i,f}|^2 \times A(\mathbf{k_i}, E) \times \delta(\mathbf{k_i} - \mathbf{k_f} + \mathbf{G}) \times \delta(\epsilon_f + E_f^0 + E - E_i - h\nu) . \quad (4)$$

The two δ-functions represent energy and momentum conservation. In a one-electron picture where excitations of the many-body system are neglected $(E = 0)$, the total-energy difference $E_f^0 - E_i$ equals the binding energy ϵ_B of the initial-state one-electron wave function and the δ-function simply ensures that $\epsilon_f = h\nu - \epsilon_B$ which is often referred to as the Einstein relation.

The second δ-function connects the wave vectors $\mathbf{k_i}$ and $\mathbf{k_f}$: In the case of itinerant states in crystalline solids, crystal momentum $\hbar \mathbf{k_i}$ is a good quantum number. It is conserved in the photoemission process up to a reciprocal lattice vector $\mathbf{G}$, because the photon momentum can usually be neglected in a typical ARPES experiment.[2] A complication arises due to the fact that the photoelectron is measured in vacuum, i.e. after it has left the potential range of the sample. The detection angles θ_m and ϕ_m, together with a kinetic energy value ϵ_f define a wave vector $\mathbf{k}$ of a free electron in vacuum:

$$\mathbf{k} = \frac{1}{\hbar} \sqrt{2m\epsilon_f} \times (\sin\theta_m \cos\phi_m, \sin\theta_m \sin\phi_m, \cos\theta) , \quad (7)$$

where m is the electron mass. The question of how $\mathbf{k}$ relates to the wave vector $\mathbf{k_f}$ of the photoelectron inside the crystal represents a fundamental difficulty in ARPES, where the photoelectron has to penetrate the surface barrier of the solid. The crystal periodicity parallel to the surface ensures that the wave vector components parallel to the surface plane are conserved: $\mathbf{k_\parallel} = \mathbf{k_{f,\parallel}} + \mathbf{g}$.[3] In the direction perpendicular to the surface, the periodicity of the crystal lattice is truncated abruptly by the surface potential step, where the average potential rises from the inner potential $-V_0$ in the solid to zero in vacuum. In the so-called *free-electron final-state approximation*, where the

---

[2] Convenient conversion formulas from energies E_{kin} or $h\nu$ (in eV) to wave numbers $|\mathbf{k}|$ in Å^{-1} for free electrons and photons, respectively, are:

$$|\mathbf{k}| = 0.5123 \times \sqrt{E_{kin}} \approx 0.51 \times \sqrt{E_{kin}} \quad (5)$$

and

$$|\mathbf{k}| = 0.5068 \times h\nu \times 10^{-3} \approx 0.51 \times h\nu \times 10^{-3} . \quad (6)$$

[3] A surface reciprocal lattice vector $\mathbf{g}$ can appear in cases where the surface is reconstructed, showing a periodicity that is different from the truncated bulk crystal, and leading to so-called surface Umklapp scattering.

photoelectron state in the solid is treated like a single plane wave with wave vector

$$\boldsymbol{k_f} = \frac{1}{\hbar}\sqrt{2m(\epsilon_f + V_0)} \times (\sin\theta\cos\phi_m, \sin\theta\sin\phi_m, \cos\theta) \,, \qquad (8)$$

the consequence is a reduction of the component $\boldsymbol{k_\perp}$ perpendicular to the surface. This leads to a refraction of the photoelectron wave that can be described by Snell's law:

$$\sin\theta = \sin\theta_m \sqrt{\frac{\epsilon_f}{\epsilon_f + V_0}} \,. \qquad (9)$$

Here, the electron energy is referenced to the vacuum level. Typical values for V_0 are of the order of $10\,\mathrm{eV}$, and for low energies refraction angles can be substantial.

This may seem like a crude approximation for a photoelectron propagating through a periodic potential, but it has nevertheless proven to be quite successful.[4]

A fundamental difficulty arises because photoelectrons have a relatively short inelastic mean free path. This means that only those electrons originating from a thin selvage region typically a few atomic layers thick contribute to the true spectrum, while electrons originating from deeper within the crystal constitute a smoothly rising inelastic background at lower kinetic energies. But it also means that $\boldsymbol{k_\perp}$ is no longer well defined but smeared out. This severely limits the precision at which $\boldsymbol{k_f}$ can be measured and has to be carefully considered when interpreting ARPES data. The problem does not occur in two-dimensional electronic states where $\boldsymbol{k_\parallel}$ is the only relevant quantum number.

The two δ-functions in (4), providing a kinematical description of photoemission processes, represent a very stringent condition under which photoemission intensity can be observed, especially when excitations in the many-body system are neglected and $\epsilon_B = E_f^0 - E_i$, where $\epsilon_B(\boldsymbol{k_i})$ represents the single-particle band structure of the solid[5]. The conventional picture of direct

---

[4] In the traditional *three-step model* of photoemission, the process is described as a sequence of three steps: (1) photoexcitation of an electron into an unoccupied state within the sample band structure, (2) propagation of this band state to the surface, undergoing also inelastic losses, and (3) penetration of the surface potential step and coupling to a free-electron state in vacuum. In the more accurate *one-step model* the entire process is coherently described in one single step, with a final state wave function that considers all scattering events of the photoelectron within the surface region, including inelastic processes, and the coupling to the vacuum state in a so-called *time-reversed low-energy electron diffraction (LEED) state*. It appears that the use of a free-electron final state in the three-step model corrects for some inadequacies of the former approach [6].

[5] In principle, many-body effects are included implicitly because the energy E_f^0 is the lowest energy of the $(N-1)$-electron system in the excited state $\tilde{\Psi}_{f,s}(N-$

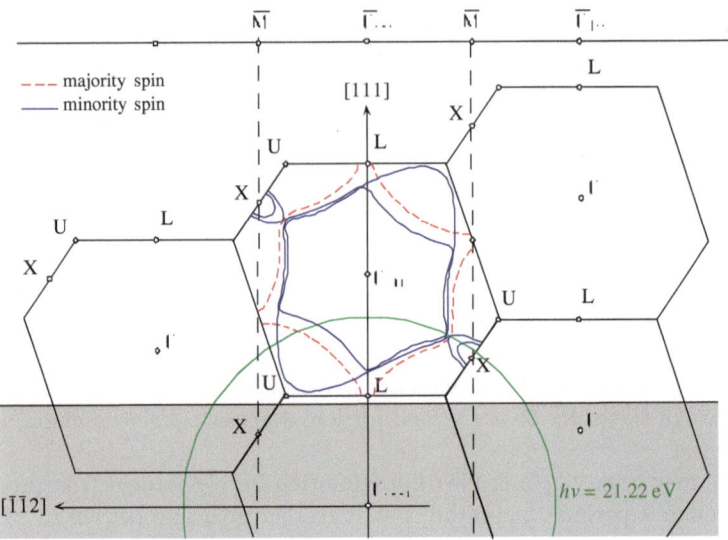

Fig. 1. Schematic diagram for k-space mapping experiments.In the free-electron final-state approximation direct transitions are expected where the final state sphere (large circle centered at Γ_{000}) intersects with an initial state band in the extended zone scheme. $k_\parallel$ of these intersection points is conserved throughout the photoemission process. The grey area in k-space is not accessible due to total internal reflection according to (9) (From [7])

transitions results [1]: in a band structure plotted in the *reduced zone scheme* the conservation of wave vectors up to a reciprocal lattice vector means that an electron is transferred vertically, i.e. at constant crystal momentum, from an initial state band to a final state band $\epsilon_B^i(\mathbf{k_i}) \rightarrow \epsilon_B^f(\mathbf{k_i})$ with energy levels separated by

$$\epsilon_B^f(\mathbf{k_i}) - \epsilon_B^i(\mathbf{k_i}) = h\nu \ , \tag{10}$$

i.e. by the photon energy. In the free-electron final-state approximation, the upper level is connected to the value of $\mathbf{k_i}$ by the simple relation

$$\epsilon_B^f(\mathbf{k_i}) = \epsilon_B^f(\mathbf{k_f} - \mathbf{G}) = \hbar^2 |\mathbf{k_f}|^2 / 2m \ . \tag{11}$$

The geometrical implications of these equations is illustrated in Fig. 1, where a planar section through the reciprocal space of a Ni crystal is plotted. The geometry and the periodicity are defined by the Brillouin zones given in the *repeated zone scheme*. In one of the zones, centered at the reciprocal lattice point $\mathbf{G} = \Gamma_{111}$, the sections through the Fermi surfaces defined by the Ni

---

$1, \mathbf{k_i}$) with electronic charges rearranged in order to screen the photohole $\mathbf{k_i}$. In highly itinerant cases, this effect may be small [1], and $\epsilon_B(\mathbf{k_i})$ is quite close to a single-particle band structure. For more localized hole states the deviations may become substantial (see Sect. 3.2).

band structure are given. These curves are defined by the condition $\epsilon_B^i(\mathbf{k}_i) = \epsilon_F$ where ϵ_F is the Fermi energy. The large circle centered at $\mathbf{G} = \Gamma_{000}$, which we may term the *measurement sphere*, represents the free-electron final states for the photon energy $h\nu = 21.22\,\text{eV}$ with a radius calculated according to Equations (10) and (11). Measurement sphere and Fermi surface sections thus represent states in $\mathbf{k}$-space with the correct energy separation, and whereever they intersect, momentum is conserved. As is obvious from this figure, there are only few points in this plane that qualify for the kinematics according to (4). If other vertical planes, obtained by azimuthal rotations about the surface normal (here the [111] direction), are considered in an equivalent way, it becomes clear that these few points will form continuous lines on the measurement sphere.

This simple kinematical concept is borne out nicely by the data shown in Fig. 2a, which shows a complete *Fermi surface map* measured on a Ni(111) surface at a photon energy of $h\nu = 21.22\,\text{eV}$. In this figure, photoemission intensities at the Fermi energy are plotted as gray scale values for a complete hemispherical set of emission angles ($\theta_m = [0..88°]$, $\phi_m = [0..360°]$), projected parallel onto the $\mathbf{k}_\parallel$-plane. It exhibits a number of well-defined contours which are well reproduced in a theoretical plot (Fig. 2b) that was obtained by plotting all intersections of measurement sphere and Fermi surface in the same projection. The latter was obtained within the standard band structure package Wien2k [8]. The excellent agreement in most details demonstrates

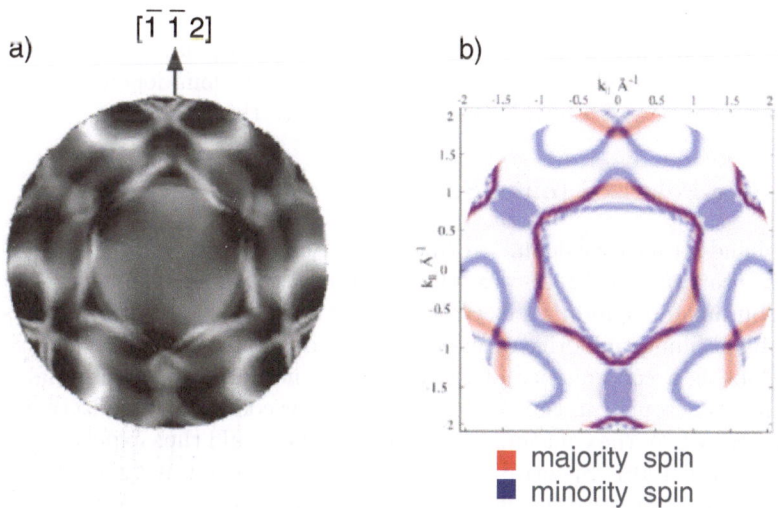

Fig. 2. (a) He Iα excited Fermi surface map ($h\nu = 21.21\,\text{eV}$) from Ni(111). A $\mathbf{k}_\parallel$ projection of the raw data is presented in a linear grey scale, with highest intensities in white, lowest in black. In (**b**) the corresponding spin-polarized band structure calculation is displayed, showing Fermi level crossings at the same $\mathbf{k}_\parallel$ locations as in the measurement (a). Majority spin bands are shown in red, minority spin bands in grey (From [10], with improved experimental data by W. Auwärter)

that this simple kinematical interpretation of ARPES data, including the use of the free-electron final-state model, is a viable concept in order to measure and identify band structures. Refined models using more realistic final states are under investigation, but they have yet to prove their general applicability [9].

The experimental Fermi surface contours of Fig. 2a appear with different brightness values (photoemission intensities) and with different widths. These properties are not explained within the kinematical picture. They have to do with state- and k-dependent matrix elements $M_{i,f}$ (intensities) and with broadening of $k_\perp$ as well as issues related to the spectral function $A(k_i, E)$ (widths) [6].

The basic ingredients in the formation of valence photoemission signals have now been considered. In the next section it will be discussed how a measured spin polarization in such signals relates to that in the initial state, and which processes produce or reduce spin polarization.

3 Sources of Spin Polarization in ARPES Data

The photon of the exciting light does not directly couple to the electron spin, and one might expect that the spin polarization measured in a spin-resolved photoemission experiment should directly reflect the spin polarization of the initial states under study. However, indirect coupling occurs in the process due to several mechanisms involving a spin-dependent reaction of the remaining $(N-1)$-electron system (if it is strongly spin polarized), spin-orbit effects and spin-dependent elastic exchange scattering (spin-polarized photoelectron diffraction) in the photoelectron final state, or finally spin-dependent inelastic scattering processes during electron transport to the surface.

3.1 Spin-Polarized Initial States

The band-structure calculation that formed the basis for plotting the Fermi surface contours in Fig. 2b was fully spin-polarized. In a ferromagnetic material, spin-up and spin-down electrons receive their label ($\uparrow, \downarrow$) from the respective band filling (minority or majority electrons), and the exchange splitting develops as a consequence of the exchange and correlation potentials. The resulting contours can thus be labeled in a color code according to their spin. At places where the contours do not overlap one would thus expect a spin polarization of 100%. Spectral overlap and poorly defined magnetization state of a sample can reduce this value considerably (see Sect. 5.1). A demagnetized sample produces zero spin polarization, because spin-polarization values from

---

[6] In the theoretical contours of Fig. 2b, the width is a consequence of the finite momentum shell around the ideal measuring sphere that had to be defined in order to produce continuous contours at the given sampling density.

different magnetic domains average out in the fixed spin detection geometry. Conversely, such experiments can be used for magnetometry [11].

Spin-splittings of bands can not only arise due to electron exchange interactions, but also from the spin-orbit interaction (see Sect. 5.2), which couples the spin to the symmetry of the lattice. This forms the basis for the magnetic anisotropy in exchange-coupled systems, and it can lead to non-trivial spin structures in spin-orbit coupled systems, where the *direction* and the *magnitude* of the spin-polarization vector is k_i-dependent. In this case, the observation of spin polarization from spin-split bands is not dependent on a net magnetization of the sample. On the other hand, it is necessary to introduce spin polarization as a vectorial quantity, defined as the vector of expectation values of the spin operators for a given state or ensemble of states

$$\boldsymbol{P} = \frac{2}{\hbar}(\langle S_x \rangle, \langle S_x \rangle, \langle S_x \rangle) , \tag{12}$$

where the normalization constant $2/\hbar$ makes sure that the absolute value is $P \leq 1$ [5].

3.2 Effects of Electron Correlation on Spin Polarization

In materials with electronic states that are rather localized and where the Coulomb interactions between the electrons are thus strong, the extraction of one electron can lead to severe many-body excitations of the system. This means that the spectral function appearing in (4) leads to a strong redistribution of photoelectron intensities over the energy axis, and the photoemission spectrum measured at a particular wave vector k_f no longer reflects a single-particle state with sharp energy $\epsilon_B^i(k_i)$. The typical effects are threefold [1,12]: the single-particle peak shifts to lower binding energy and receives a finite width, and at higher binding energy a much broader distribution of spectral weight occurs. The first two effects suggest that the missing electron (hole) can be described as a coherent *quasiparticle* (with defined energy and with momentum k_i) forming the lowest-energy state of the $(N-1)$-electron system (hence the *relaxation shift*), which has a short life time due to the strong interactions (hence the finite width). The broader distribution reflects the occurrence of a continuum of excited many-body states lacking the defined energy-momentum relation of a particle. It is thus termed the *incoherent* part of the spectral function.

In a ferromagnetic material these effects can be strongly spin dependent. This is illustrated in Fig. 3, taken from a recent comparative investigation of photoemission spectra and spectral functions in cobalt metal [13]. Panels (a) and (c) show the results of a spin-resolved band structure, calculated along a line in k_i-space that corresponds to a photoemission data set presented in the same study (not shown). The typical picture of an itinerant ferromagnet is displayed, with two rather similar band structures for the two spin channels

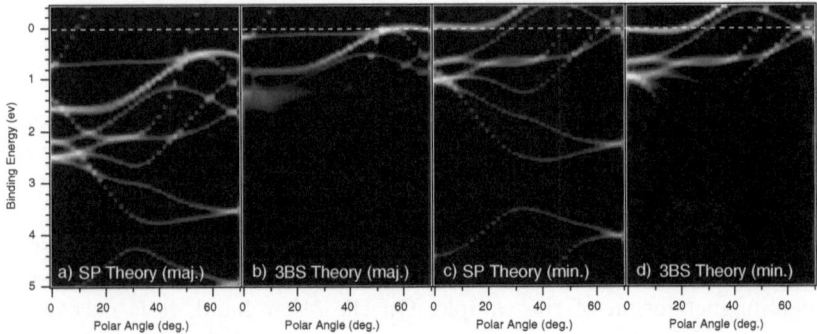

Fig. 3. Comparison of spin-resolved single-particle band structure and many-body quasiparticle spectral function for the same wave vector region in hexagonal close-packed Co. Binding energies are referenced to the Fermi level, and wave vectors are described in terms of polar angles θ_m in a photoemission experiment on a (0001) oriented thick Co film. Panels (**a**) and (**b**) compare single- and quasiparticle spectra for majority ($\downarrow$) electrons, panels (**c**) and (**d**) those for minority ($\uparrow$) electrons (from [13])

and with a large exchange splitting of the order of 1.5 eV between the two. Clearly, one expects large spin polarization values in photoemission from such spin-split bands where they do not overlap. However, the many-body calculation (panels (b) and (d)) conveys a completely different picture. It uses a simple model on top of the detailed band structure calculation where the missing electron interacts strongly with one additional electron-hole pair in the so-called *three-body scattering approximation*. The result shows that there is a severe renormalization of band positions and line widths in the majority channel, while the effects in the minority channel are only minor. In a *strong ferromagnet* like Co, where the majority d band is completely filled, this asymmetry arises for the following reason: electron-hole pairs can only be created in the minority channel (see Fig. 4). A hole produced due to photoemission in the majority channel is accompanied by the full phase space of electron-hole pair excitations in the minority d band, thus rendering the quasiparticle renormalization very strong. For a photohole in the minority d band, the configurations for electron-hole pairs are restricted, because there are now two holes that interact strongly and repulsively. The overall result of this asymmetry is that both spin channels produce spectral functions that are rather similar, which leads to the expectation of strongly reduced spin polarization in photoemission data from cobalt. On the experimental side, spin-integrated spectra measured along the same line in k_i-space [13] are in good agreement with these quasiparticle spectra, and so are spin-resolved photoemission spectra for thick Co(0001) films measured at normal emission ($\theta_m = 0°$) with the same photon energy [14,15].

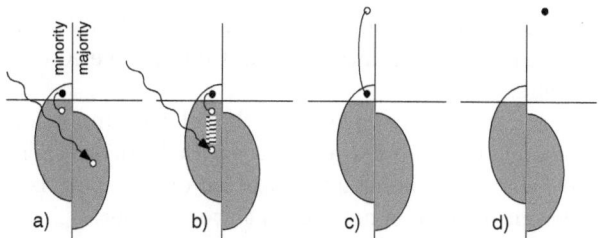

Fig. 4. Schematic representation of the interactions involved in several spin-dependent processes within the d band of a strong ferromagnet. (**a**) photoemission in the majority spin band, (**b**) photoemission in the minority spin band, (**c**) inelastic scattering into the minority channel, and (**d**) inelastic scattering into the majority channel. In (a) and (b), electron-hole pair excitation is possible only in the minority channel. In (d) there is no spin-conserving inelastic scattering into empty d states. Situations (c) and (d) refer to Sect. 3.4

3.3 Spin-Polarization Introduced by the Photoexcitation Process

Several distinct mechanisms have been identified that can introduce or change spin polarization in photoemission intensities through spin-selectivity in the matrix element $M_{i,f}$ of (4). They are most conveniently studied in photoemission from non-magnetic samples where no spin-polarization is present in the initial states. The most robust effects are produced when *circularly polarized light* is used, where spin-polarized emission is quite commonly observed. Optical pumping of spin-orbit split initial state bands leads to asymmetric occupation of final-state spin states [16]. A spin-polarized electron source based on a GaAs photocathode and illuminated by circularly polarized laser light produces electron beams with 43% spin polarization [17]. Also at higher photon energies substantial spin polarization can be observed, especially in the heavier elements [18].

Also excitation by *linearly polarized light* can introduce spin-dependent matrix-element effects and thus produce spin polarization in photoemission from non-magnetic samples. The mechanisms here include spin-dependent transmission through the surface of solids with non-vanishing spin-orbit coupling [19] and more subtle effects that showed up first in photoemission calculations within the *one-step model* [20–22] and were later observed experimentally [23–25]. The mechanisms behind the latter effects are related to spin-orbit induced hybridization of initial-state wave functions, and their occurrence depends strongly on the symmetry of the solid and of the particular surface, as well as the absolute directions of photon incidence and polarization and electron emission. In materials containing ions with finite magnetic moments, intra-atomic multiplet splittings due to final-state $L - S$ term splittings are an important internal source of spin polarization [26], mostly discussed in the context of core levels. Such moments can also give rise to spin-polarized photoelectron diffraction effects [27], both in core and valence photoemission: the scattering potential of the photoelectron within the ionic environment depends on the relative orientation of the electron spin and the magnetic moments.

It is hard to get an intuitive picture of the relative importance of all these mechanisms, but one has to be aware that such effects are there in principle. How much they may blur the characterization of spin polarization and spin structure in the initial state is not clear at the moment.

3.4 Spin-Polarization Induced or Reduced During Electron Transport

The theoretical expression for photoemission intensities given in (4) reflects the so-called *intrinsic* contribution to the measured intensity, including the photoemission process and, in principle, the full multiple-elastic-scattering photoelectron final state. It considers the *extrinsic* inelastic scattering processes that remove photoelectrons from the elastic channel only implicitly by assuming final state wave functions that decay exponentially away from the surface into the crystal. Extrinsic inelastic processes occur along the way as photoelectrons propagate towards the surface, and collected energy losses move them from the *elastic peak position* to a smoothly rising inelastic background on the higher binding energy side, underneath the intrinsic spectrum. The photoelectron inelastic mean free path $\lambda_i(\epsilon_f)$ defines the surface sensitivity of photoemission as a probe for the electronic structure of solids, because the electrons appearing in the inelastic background no longer reveal the spectral information contained in the intrinsic spectrum due to the random nature of the inelastic processes. Inelastic scattering cross sections are very high, and values of $\lambda_i(\epsilon_f)$ are thus of the order of a few Å. As it turns out, inelastic processes can be an important source for photoelectron spin polarization or depolarization as well.

In a ferromagnet, this latter effect can be described simply by considering two different values $\lambda_i^\uparrow(\epsilon_f)$ and $\lambda_i^\downarrow(\epsilon_f)$ for photoelectron spins antiparallel or parallel to the sample magnetization, respectively. Evidence for the inequivalence of $\lambda_i^\uparrow(\epsilon_f)$ and $\lambda_i^\downarrow(\epsilon_f)$ has come from experiments involving thin ferromagnetic overlayers on top of non-magnetic substrates [15, 28]. In spin-resolved detection, a distinctive valence photoemission signal is attenuated more rapidly in the minority spin channel. In the case of Co(0001) films grown on W(110), values of $\lambda_i^\downarrow = 6.6$ Å and $\lambda_i^\uparrow = 9.0$ Å were extracted from the thickness dependent signal attenuation for $\epsilon_f = 14$ eV (referenced to the Fermi energy). This asymmetry is brought about by the different density of empty d-states that are available for scattering in the two spin channels (see Fig. 4c and d). An empirical law by Siegmann relates the spin-asymmetry in the total inelastic scattering cross section to the paramagnetic occupation number n of the d shell and its change Δn in each spin channel in forming the ferromagnetic state [29]. This simple model works quite successfully for the entire series of $3d$ transition metals.

This spin-dependent transport effect was also demonstrated in transmission experiments where electron beams of controlled spin polarization were passed through self-supporting and magnetized Au/Co/Au tri-layers, and has also been termed the *spin- filter effect* [30]. If the spin polarization vector P of the incident electron beam is perpendicular to the magnetization M of the sample, the vector P of the elastically transmitted electron beam points into a different direction when it emerges from the sample. This is due to the elastic and inelastic interactions of the electrons inside the ferromagnet. Due to the elastic interaction mediated by the exchange field, P precesses around the axis of M while it rotates into the direction of M due to inelastic yet spin conserving scattering. The total motion of P depends on the electron energy and is characteristically different with Fe, Co, and Ni [31].

The first experiments with spin-polarized electrons were much less sophisticated and designed to examine whether spin polarization appears at all with photoemitted electrons. Quantum theory had made it clear that the magnetization in Fe, Co, and Ni must be predominantly generated by the spin polarization of the metallic electrons, and that the spin polarization should be preserved in the process of emission. While the first attempts to detect the spin polarization of photoelectrons failed [32,33], it became obvious soon thereafter that with atomically clean surfaces, photoelectrons from all kinds of ferromagnetic materials exhibit sizeable spin polarization [34,35]. However, almost all theories on magnetism predict that the spin polarization of threshold photoelectrons should be negative (i.e. minority in character) in Co and Ni, because the majority spin states must be located well below the Fermi-level, separated by an energy gap called the Stoner gap. The magnitude of the magnetic moment and the number of available d-electrons dictate that the majority states are completely occupied in both cases. Yet threshold photoelectrons show negative spin polarization only with Ni [36] but not with Co [37,38]. Today, this is understood by the spin-filter effect active in transport of the photoelectrons to the surface, removing the minority spins by scattering on the minority spin holes in the 3d-states, and enhancing the positive polarization of the photoemitted electrons. The spin-filter effect is much stronger in Co due to the larger density of minority spin holes [29].

A seminal experiment showing how overwhelming this spin-filter effect can be with low-energy threshold electrons uses the spin polarization of Auger electrons as well as the polarization of low energy cascade electrons from a Fe(100) surface covered with one monolayer of Gd [39]. The $N_{45}N_{67}N_{67}$ Auger electrons excited from the $4f^7$-states of Gd are spin polarized antiparallel to the electrons emitted from the $M_{23}M_{45}M_{45}$ Auger electrons of the Fe-substrate, as expected due to the well known antiparallel coupling of the Gd- and Fe-magnetic moments. In contrast, the low-energy cascade electrons emerging at threshold already show the reversed polarization of the Gd overlayer, proving that most of their formation process occurs in one single Gd layer. This much shorter escape depth of low-energy electrons [40] excludes earlier interpretation of the photoelectron spin polarization in terms of quasi elastic spin flip scattering [41,42]. Rather, the polarization of threshold photoelectrons must be attributed to the spin conserving scattering on the large and spin-polarized density of $5d$ holes of the rare earth metals.

One has thus to be aware that the spin polarization of photoelectrons may be considerably altered in transport to the surface if unoccupied d-states are present. With monolayer or sub-monolayer coatings of, e.g., the alkali metals, the spin polarization is not affected [41], which has been used in a number of interesting photoemission experiments [43,44] to reduce the photoelectric work function.

4 Measurement of Spin Polarization in ARPES

In a spin-resolved ARPES experiment, an electrostatic energy analyzer provides an energy and momentum selected photoelectron beam at the exit aperture while preserving the spin polarization and spin orientation. As pointed out earlier, electrons within this prepared beam cannot be separated according to their spin along a predefined axis like neutral atoms in a Stern-Gerlach experiment. Therefore, rather inefficient spin detection schemes that are based on spin-dependent scattering processes need to be applied in order to determine the spin polarization. Among these, Mott scattering and polarized low-energy electron diffraction (PLEED) are the most frequently used. Two detectors placed symmetrically with respect to the beam axis measure the left-right asymmetry of intensities backscattered from a suitable target:

$$A_x = \frac{(I_L - I_R)}{(I_L + I_R)} . \tag{13}$$

The scattering asymmetry results from the spin-orbit interaction in the target region and depends on the polarization component perpendicular to the scattering plane (here x). The asymmetry measured for a 100% polarized electron beam depends on the electron energy, the target material and thickness, and the scattering angle; it is called the Sherman function S [5]. Usually it is calibrated experimentally, and it can then be used to determine the spin polarization P_x of an electron beam along a defined axis from the measured scattering asymmetry A_x in the related scattering plane (here the yz plane):

$$P_x = A_x/S . \tag{14}$$

Sherman functions can take values of almost up to 0.5 theoretically, and for ultimately thin target foils that avoid multiple elastic or inelastic scattering events. On the other hand, only a small fraction I/I_0 of the electrons in the beam are backscattered into the detectors. For judging the efficiency of a spin detector, the figure of merit $\epsilon = (I/I_0)S^2$ reflects both of these aspects. Figures of merit for typical spin polarimeters are in the range of 10^{-4} to

10^{-3}. It is the relevant quantity for establishing the statistical error in a polarization measurement:

$$\Delta P_x = 1/\sqrt{\epsilon I_0} \ . \tag{15}$$

In deciding upon the particular choice of spin detector, the figure of merit ϵ is not the only all-important factor. Another criterion is temporal stability. The level of complexity of a typical photoemission experiment, including sample preparation, is such that one favours a polarimater that produces reproducible absolute spin polarization values without frequent recalibration or repeated target preparation. Although the PLEED detectors can have high figures of merit, the low electron energies of typically 100 eV make the Sherman function dependent on the condition of the target surface that can deteriorate over the period of one or several photoemission measurements. The target surface needs to be periodically reestablished. This is the main reason why the rather inefficient high-energy Mott detectors, where the scattering occurs at 50–120 keV, are used in most cases. Here, the high-energy electrons penetrate the target foil of several hundred nanometers thickness, which usually consists of Au or some other heavy element providing a strong spin-orbit interaction. These devices can thus be operated over weeks and months under stable conditions.

In ARPES, and specifically in Fermi surface mapping experiments, the need arises for full three-dimensional spin polarimetry. In order to access arbitrary locations in momentum space, the sample needs to be brought into the corresponding orientations relative to the spectrometer. For a magnetized sample, the magnetization vector, and thus the spin quantization axis, is rotated relative to the spin detector. Likewise, for spin-orbit induced spin structures, where the spin vector depends on the electron momentum, the spin polarization vector can have components in all three spatial directions. Moreover, since the total spin polarization vector need not have unity length, one cannot deduce a third component from measuring the other two.

In a single Mott polarimeter, two pairs of detectors can be placed in two orthogonal scattering planes, thus measuring the two spin components transverse to the beam direction. By combining two of these devices in an othogonal geometry, with an electrostatic beam switcher in front of them (see Fig. 5), a fully three-dimensional spin polarimeter has been built [7]. The scattering planes are placed such that the two Mott polarimeters share a common spin quantization axis (z axis). The redundancy along this axis provides useful consistency checks. The beam is switched at typically 1 Hz between the two devices. The three components of the spin polarization vector thus measured in the coordinate frame of the polarimeter can be transformed into a spin polarization vector in the sample coordinate frame by means of a rotation matrix containing the emission angles θ_m and ϕ_m selected in order to reach the specific point $\boldsymbol{k_i}$ in momentum space [7].

For spin-polarized ARPES experiments on exchange-split bands from magnetized samples, the proper control of the magnetization state represents

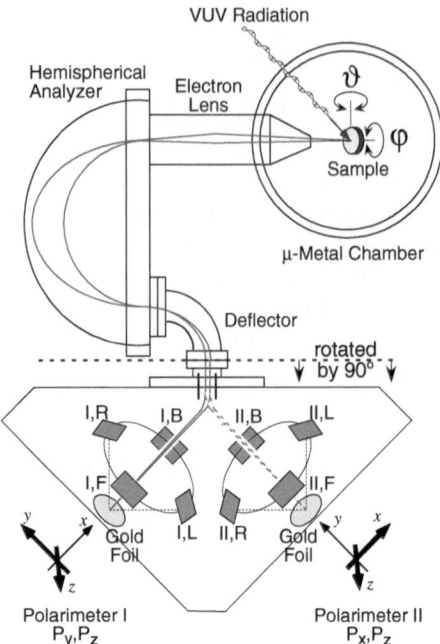

Fig. 5. Schematic view of the three-dimensional polarimeter [7]. Electrons that are photoemitted from a sample by ultraviolet radiation are energy and angle selected by an electrostatic analyzer and detected in two orthogonal Mott polarimeters (named polarimeter I and II). In an electrostatic beam deflection system the spin direction is conserved and polarimeter I measures the polarization components P_y and P_z, while polarimeter II measures P_x and P_z. The beam is switched between the two in order to allow quasi-simultaneous data collection. In the figure, the polarimeter system is shown rotated by $90°$ for graphical clarity, i.e. in reality the z axis is directed straight to the left and parallel to the electron lens of the spectrometer

a further experimental difficulty to overcome. In order to measure the spin-polarization from such bands quantitatively, it has to be established that the probing spot is entirely within a single magnetic domain of known magnetization direction. In order to preserve the angular distribution of the relatively slow photoelectrons, such measurements are usually done with the sample in remanence, and preferably with in-plane magnetization if the magnetic surface so permits, in order to avoid stray fields. Best results are obtained with single crystal samples cut out in the so-called picture frame geometry [45]. A coil wound around the bottom of the frame is used to change the direction of the sample magnetization at the top, i.e. at the surface under study. Upon passing a current through the coil, essentially all magnetic flux is contained within the crystal, and the magnetization can be easily switched back and forth in the measurement position. Alternatively, the sample can be brought into an external field coil that is preferably removed from the photoemission stage.

Control of the sample magnetization brings the further advantage that purely instrumental asymmetries in the Mott scattering, caused by slightly different detector geometries or sensitivities, or by target non-uniformities, can be cancelled by forming cross asymmetries with reversed sample magnetizations [5] (see following section).

5 Case Studies

5.1 Spin Polarization of Nickel Bulk Bands Measured on Clean Ni(111)

The Fermi surface scan in Fig. 2 showed an excellent agreement between measured contours and those obtained from a spin-polarized band structure calculation. In the latter, the spin character of the various bands is known in terms of majority and minority spin, irrespective of the actual magnetization direction. From the comparison it is straightforward to identify the respective spin states of the measured contours. Nevertheless, it is good to verify the assignment by direct measurement of the spin polarization. This is demonstrated in Fig. 6 showing spin-polarized ARPES data measured along a circular arc centered at the $[\overline{1}\overline{1}2]$ direction and following the upper rim of the plot in Fig. 2a. The top curve gives the measured intensity, showing four rather narrowly spaced peaks near the center and two larger peaks at azimuthal angles of $\sim \pm 30°$, corresponding to the "goggle"-like pattern in Fig. 2a. In the bottom curve, the experimental asymmetries are given for the same scan, measured with the two detectors of one Mott polarimeter that probe the P_z component. For this nearly grazing emission geometry ($\theta_m = 78°$), the largely in-plane magnetization vector has a strong component along the z-axis of the Mott polarimeter system when the sample azimuth is chosen appropriately [7]. The curve represents, in fact, the cross asymmetry measured along the z-axis, evaluated with two consecutive measurements with switched magnetization direction (indicated by $\oplus$ and $\ominus$):

$$A^{\otimes} = \frac{(I_L^{\oplus} + I_R^{\ominus}) - (I_R^{\oplus} + I_L^{\ominus})}{(I_L^{\oplus} + I_R^{\ominus}) + (I_R^{\oplus} + I_L^{\ominus})} . \tag{16}$$

This procedure removes any purely instrumental asymmetry [5] and thus establishes a dependable zero line in the asymmetry measurement[7]. One notes

---

[7] As a matter of fact, the cross asymmetry with arithmetical mean values as described in (16) is used in situations where one of the channels I_L or I_R is suspected to have a constant offset in count rate. More often the two detector channels have slightly different detection sensitivities, in which case the cross asymmetry based

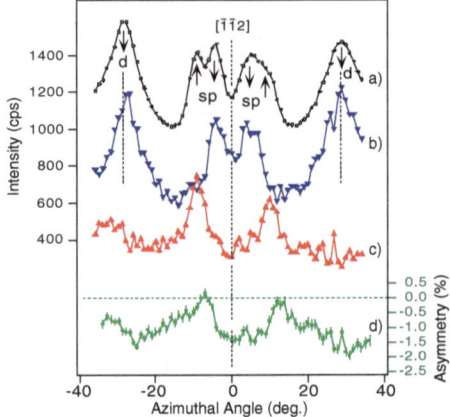

Fig. 6. (a) He Iα excited and spin-integrated intensity scan measured at the Fermi energy on Ni(111) along a path near the upper edge of Fig. 2a, producing an azimuthal momentum distribution curve. The peaks are annotated by their known orbital and spin assignment. Curves **(b)** and **(c)** present the spin-resolved minority and majority spin spectra, respectively, as obtained from curves (a) and **(d)** by applying (2). Curve (d) is the spin asymmetry for the polarization component measured along the sample magnetization direction (from [7]). See text for a discussion of quantitative aspects

immediately that the asymmetry values are negative (corresponding to minority spin character) throughout most of the scan, with the exception of two narrow regions where the majority *sp* band peaks are seen in the intensity curve. In between each of the two *sp* spin pairs the rapid sign change of the asymmetry manifests the different spin character of the two peaks. The assignment based on the band structure calculation (Fig. 2b) is thus correct.

From the curves a) and d) in Fig. 6 one can, in principle, calculate the spin-resolved momentum distribution curves for majority and minority spins. Before Equation 2 can be applied, one needs to translate asymmetry values into spin-polarization values. The Sherman function of the device has not been precisely measured for lack of an electron beam of known polarization. From comparison to Mott polarimeters of the same design it is known to be in the range of $S = 15$–25%. With these numbers the asymmetry contrast of $\Delta A_z \approx 1.2\%$ seen in Fig. 6 translates into polarization values of the order of 6% which is very low, much lower than expected from these clean exchange-split bands. This strong reduction in measured spin asymmetries

---

on geometrical mean values leads to cancellation of the unknown sensitivity factor:

$$\tilde{A}^{\otimes} = \frac{\sqrt{I_L^{\oplus} I_R^{\ominus}} - \sqrt{I_R^{\oplus} I_L^{\ominus}}}{\sqrt{I_L^{\oplus} I_R^{\ominus}} + \sqrt{I_R^{\oplus} I_L^{\ominus}}} . \tag{17}$$

comes from two effects: (i) The sample is only poorly magnetized; in fact the following analysis suggests a degree of magnetization of only about 12%. (ii) The direct transitions from the exchange split bands ride on an unpolarized background (signal-to-background ratio ~0.6) that arises due to quasielastic scattering processes involving phonons, magnons and other electrons. In this particular experiment, which was carried out with a non-monochromatized He discharge lamp, a part of the background results also from photoexcitation with different photon energies $h\nu$ [7].

The second effect can be further assessed by a simple spectral synthesis model as is illustrated in Fig. 7. Fully spin-polarized Lorentzian lines are superposed on a uniform unpolarized background in order to model the intensity scan of Fig. 6a. In the model, the spin-resolved curves can be readily generated. The resulting polarization values show an amplitude of almost 50%, while we measure a much lower value of 6% indicative of a poor magnetization state of the sample. It is interesting to note that this analysis provides a nice confirmation that the uniform background underneath the direct transitions is indeed unpolarized. Any significant background polarization makes the asymmetry curve move with respect to the zero line and deform significantly from the plotted curve that agrees well with the experimental data except for the absolute scale.

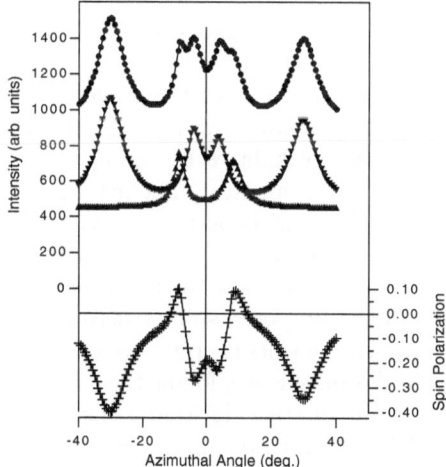

Fig. 7. (a) Synthesis of azimuthal intensity scan, modelling the data of Fig. 6a by using 6 Lorentzian lines of suitable positions, widths and intensities plus a constant background. The spin polarization of the individual Lorentzian lines is assumed to be +100% or −100%, respectively, depending on the majority or minority spin assignment as given in Fig. 6. The background intensity is assumed to be unpolarized. With these assumptions, the curves (**b**) and (**c**) are generated, reproducing the intensity curves for minority and majority spin electrons, respectively. The resulting spin polarization curve as calculated by (1) is shown as curve (**d**)

After the experimental asymmetry curve has been scaled up by roughly a factor of 8 to represent a fully magnetized sample, (2) can be applied to the experimental data in order to generate the spin-resolved spectra (curves b) and c) in Fig. 6). The result is quite gratifying in that it cleanly separates majority and minority bands. Note, however, that this procedure is unsuitable to study temperature-dependent spin-state mixing, which would reflect itself in reduced asymmetries for higher temperatures as the sample magnetization drops.

5.2 Spin-Polarized Surface States on Au(111)

Electronic band states can be spin-split by the spin-orbit interaction if the Kramers degeneracy, expressed as $E^\uparrow(\boldsymbol{k}) = E^\downarrow(\boldsymbol{k})$, is lifted by removing either time-reversal symmetry or inversion symmetry. Time-reversal symmetry alone demands only that $E^\uparrow(\boldsymbol{k}) = E^\downarrow(-\boldsymbol{k})$. For any crystal, inversion symmetry is naturally broken at the surface, and splittings can be expected. A remarkable example for this effect is the Shockley surface state on the Au(111) surface, representing a quasi-two-dimensional nearly-free electron gas. LaShell et al. [46] observed for this state an energy splitting that is proportional to the wave vector. Based on the Rashba term in the Hamiltonian, describing the spin-orbit interaction in propagating electronic states,

$$H_{S.O.} = \frac{\mu_B}{2c^2}(\boldsymbol{v} \times \boldsymbol{E}) \cdot \boldsymbol{\sigma} , \qquad (18)$$

they postulated a characteristic spin structure along the two concentric circular Fermi surfaces of the split states (see Fig. 8). In this equation, μ_B is the Bohr magneton, c the speed of light, $\boldsymbol{v}$ the group velocity of the particular state, and $\boldsymbol{\sigma}$ the vectorial spin operator. The average electric field vector $\boldsymbol{E}$ seen by the surface state electrons is perpendicular to the surface and pointing outside. Due to the vector product, the spin vector is thus expected to be entirely in-plane and perpendicular to the velocity vector, i.e. tangential to the Fermi surface. In the proposed spin structure, the spin vectors of the outer state, i.e. the state that has its energy lowered by the spin-orbit interaction to produce a larger Fermi surface, follow the contour in an anti-clockwise sense, while those of the inner state do so in a clockwise sense.

Spin-polarized ARPES with three-dimensional spin polarimetry has been applied to verify the spin polarization [47] and the spin structure [48] of the spin-orbit split Au(111) surface state. Figure 9a shows angle-resolved energy distribution curves for five different angles probing the dispersion of the state. The energy resolution has been much relaxed with respect to the data shown in Fig. 8d because of the low efficiency of the spin detectors and the very high number of electron counts needed for reliable spin analysis (typically 10^5 electrons). The energy splitting in the spectra is no longer observed in the raw intensity spectra, it is just reflected in an increased line width of the peak at polar angles of $\pm 3.5°$. At $\pm 4.5°$ the inner state

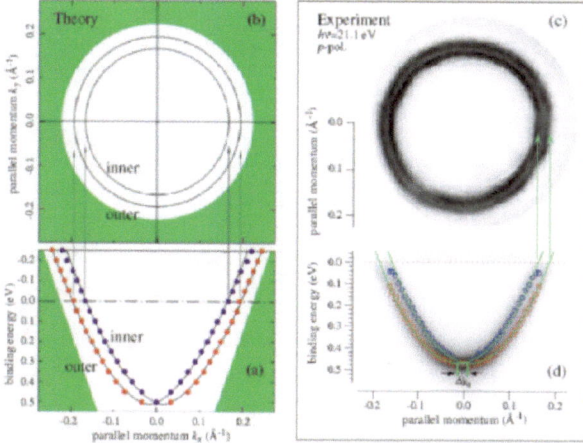

Fig. 8. Spin-orbit splitting of the Au(111) Shockley surface state found in the energy gap near the L-point of the bulk Brillouin zone. (**a**) The dispersion along k_x of the inner and outer surface state, as obtained from a first-principles calculation. The region of bulk bands is shaded in gray. (**b**) The theoretical momentum distribution of the surface states at the Fermi energy. (**c**) The experimental Fermi surface map, as obtained from a spin-integrated high-resolution photoemission experiment. (**d**) The experimental dispersion map in an arbitrary azimuth. The peak positions as determined from peak fitting to the energy distribution curves are indicated by circles, and parabolic fits to these positions are shown by solid lines. In both experimental panels, the measured intensities are shown in a linear gray scale, with black corresponding to the highest intensity (from [49])

has already crossed the Fermi level and the peak is again narrower. The polarization spectra reveal the spin polarization of the split states. The in-plane polarimeter component tangential to the Fermi surface is shown, as well as the out-of-plane component. Whereas there is no significant in-plane polarization at normal emission, where the state is not split, a characteristic polarization pattern develops in both directions with increasing polar angle. Depending on the sign of the polar angle θ_m, the polarization values are first negative or positive and switch sign roughly at the peak position. Maximum observed in-plane polarization values are of the order of 50%. On the other hand, there is no out-of-plane polarization visible in this data.

Unlike in the case of magnetized Ni(111) in the previous section, the polarization values do not depend on the sample magnetization. Except for a contribution from some unpolarized background, one should expect each state to be nearly 100% polarized. Reduced values arise from the spectral overlap of the two split components. Polarization modelling has been done along the lines discussed in the context of Fig. 7. For the observed line width, resulting from both the energy resolution of 120 eV and from the angular resolution of 1.2° that samples more states as the dispersion gets steeper, a

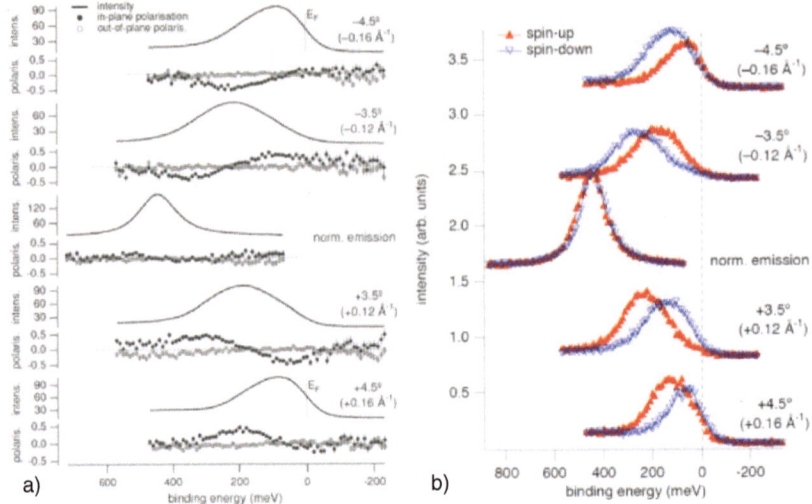

Fig. 9. Spin-polarized photoemission measurements of the Shockley surface state on Au(111) at several emission angles. (**a**) The top curves of each data set show the intensity measured by four channels of one of the Mott polarimeters. The lower curves give the measured polarization values determined from the in-plane and out-of-plane components of the polarization vector, using the relation $P = (1/S_{eff})(I_L - \eta I_R)/(I_L + \eta I_R)$ for the appropriate detector channels, with effective Sherman functions S_{eff} of 0.1–0.15 and relative gain factors $\eta \approx 0.97 - 1.03$ for the two scattering planes. The latter are used in order to compensate for the instrumental asymmetry. The spectrometer resolution was set to 120 meV and 1.2° full width at half maximum, respectively. (**b**) Spin-resolved spectra as derived from the data in a) by using (2) and the in-plane polarization values (from [47])

measured polarization value of 50% is consistent with spin-split states that are essentially ±100% polarized without any significant unpolarized background.

The data of Fig. 9 are also consistent with the spin structure proposed by LaShell et al.: The measurement of the polarimeter channels for P_x and P_y show that the strong in-plane polarization signal is in the direction tangential to the Fermi surface, defined by $P_{tan} = (P_x + P_y)/\sqrt{2}$ [47]. Moreover, when probing the spin polarization of, e.g. the outer state on opposite sides of the Fermi surface, i.e. at $\pm\theta_m$, the in-plane polarization switches sign, so that we have indeed $E^\uparrow(\boldsymbol{k}) = E^\downarrow(-\boldsymbol{k})$. In a similar fashion, the spin structure on a hydrogen-saturated W(110) surface has been shown to circulate in a counterclockwise sense around a hole pocket in the Fermi surface, corresponding here to the "inner state" [50]. The opposite sense of rotation of the inner state is related to the hole-like character of these states.

For further confirmation of this spin structure, the apparatus described in Sect. 4 can produce a complete spin-resolved momentum distribution map of the surface state [48]. This is demonstrated in Fig. 10 which is the first data

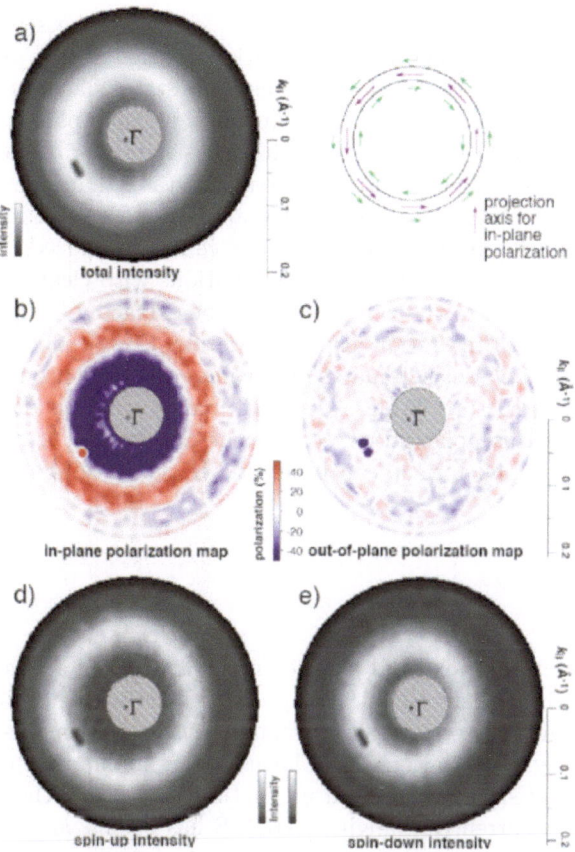

Fig. 10. Measured spin-resolved momentum distribution maps for a binding energy of 170 meV using ultraviolet light of 21.1 eV photon energy. The top panel (**a**) shows the total intensity map and a sketch of the predicted spin structure. Purple arrows indicate the rotating projection axis for in-plane polarization. The center panels (**b**) and (**c**) show the polarization maps for the in-plane and out-of-plane component of the polarization vector. The in-plane polarization (b) is a projection on the tangents to the circular Fermi surface. Red (blue) indicates a counterclockwise (clockwise) spin orientation. The bottom panels show the individual spin-up (**d**) and spin-down (**e**) intensity maps derived from a) and b) by applying (2). Note the different radii of the two Fermi surfaces of the individual bands. These data consisting of 1080 angular settings were measured in 6 h. The shift of the common center of the circles with respect to the center of the graphs is due to a 0.5° misalignment of the sample rotation axis (from [48])

set of its kind. In order to increase the count rate by a factor of two, these data have been measured at a binding energy of 170 meV, where the Fermi-Dirac distribution function is already saturated. Due to the relaxed energy and angular resolution, the spin-split states are not resolved in the intensity panel a), like in Fig. 9a. Nevertheless, the states can be separated by the measurement of the strong in-plane polarization signal (Fig. 10b). It should be noted that, since motion along the circular state is achieved by rotating the sample about the surface normal (angle ϕ_m), the in-plane polarization is always measured directly as P_{tan}, which is thus the natural coordinate frame for this measurement. This is no longer true for states that are non-symmetric with respect to rotations about ϕ_m. By applying (2) to all individual data points, the spin-resolved momentum maps can be produced (panels d) and e)). They show circular momentum distributions with clearly different radii that correspond well to the values that can be extracted for inner and outer states from the high-resolution dispersion data of Fig. 8d. On the other hand, the out-of-plane polarization map shows values scattered around 0% without any significant structure. From a detailed analysis of these data, systematic modulations of this polarization component that would be consistent with the symmetry of the surface can be excluded at a level of 5% polarization amplitude [49].

6 Outlook

Up to this date, spin-polarized photoemission experiments have been carried out by a small number of dedicated groups. This is mainly due to the low efficiency of current spin detectors, making such experiments difficult and time consuming. Applications have been mainly in the field of surface and interface magnetism and ultrathin magnetic films. The author believes that the situation is currently changing. ARPES has been a major player in the study of strongly correlated electron systems and has brought significant advances e.g. in the understanding of high-temperature superconductivity or colossal magnetoresistance. With very few exceptions [51, 52], none of the hundreds of experimental studies have directly exploited the electron spin, which is a key feature in electron correlation physics. As soon as spin-resolved photoemission spectrometers become more readily available, there will be a strong scientific case to use them. Indeed, manufacturers of high-resolution spectrometers now offer complete solutions for adding spin detection to their systems. If, one day in the future, the spin-detectors based on electron scattering asymmetries could be replaced by much more efficient devices based e.g. on spin-dependent electron absorption [53], the technique would experience a tremendous boost.

Acknowledgement

The author has benefitted greatly from the work of and discussions with Moritz Hoesch, Thomas Greber, Matthias Hengsberger, Jorge Lobo-Checa and Jürgen Henk. He is grateful to Hans-Christoph Siegmann for critically reading the manuscript, and for contributing a large part of Sect. 3.4.

References

1. S. Hüfner: *Photoelectron Spectroscopy: Principles and Applications*, 3rd edn. (Springer, Berlin Heidelberg New York 2003)
2. S. A. Wolf, D. D. Awschalom, R. A. Buhrman, J. M. Daughton, S. von Molnar, M. L. Roukes, A. Y. Chtchelkanova, D. M. Treger, Science **294**, 1488 (2001)
3. S. Datta, B. Das: Appl. Phys. Lett. **56**, 665 (1990)
4. N. F. Mott, H. S. W. Massey: *The Theory of Atomic Collisions* (Clarendon, Oxford 1965), Chap. IX
5. J. Kessler: *Polarized Electrons*, 2nd edn. (Springer, Berlin Heidelberg 1985)
6. R. Matzdorf: Surf. Sci. Rep. **30**, 153 (1998)
7. M. Hoesch, T. Greber, V. N. Petrov, M. Muntwiler, M. Hengsberger, W. Auwärter, J. Osterwalder: J. Electron Spectrosc. Relat. Phenom. **124**, 263 (2002)
8. K. Schwarz, P. Blaha, G. K. H. Madsen: Comput. Phys. Commun. **147**, 71 (2002)
9. V. N. Strocov, R. Claessen, G. Nicolay, S. Hüfner, A. Kimura, A. Harasawa, S. Shin, A. Kakizaki, P. O. Nilsson, H. I. Starnberg, P. Blaha: Phys. Rev. Lett. **81**, 4943 (1998)
10. T. J. Kreutz, T. Greber, P. Aebi, J. Osterwalder: Phys. Rev. B **58**, 1300 (1998)
11. H. C. Siegmann: Surf. Sci. **307**, 1076 (1994)
12. M. Grioni: A Photoemission Primer. In: *Magnetism and Synchrotron Radiation*, Lecture Notes in Physics, vol 565, ed by E. Beaurepaire, F. Scheurer, G. Krill, J.-P. Kappler (Springer, Berlin Heidelberg New York 2001) pp 109–129
13. S. Monastra, F. Manghi, C. A. Rozzi, C. Arcangeli, E. Wetli, H.-J. Neff, T. Greber, J. Osterwalder: Phys. Rev. Lett. **88**, 236402 (2002)
14. U. Alkemper, C. Carbone, E. Vescovo, W. Eberhardt, O. Rader, W. Gudat: Phys. Rev. B **50**, 17496 (1994)
15. M. Getzlaff, J. Bansmann, J. Braun, G. Schönhense: J. Magn. Magn. Mat. **161**, 70 (1996)
16. D. T. Pierce, F. Meier: Phys. Rev. B **13**, 5484 (1976)
17. D. T. Pierce, R. J. Celotta, G.-C. Wang, W. N. Unertl, A. Galejs, C. E. Kuyatt, S. R. Mielczarek: Rev. Sci. Instrum. **51**, 478 (1980)
18. U. Heinzmann, G. Schönhense: Spin-Resolved Photoemission from Nonmagnetic Metals and Adsorbates. In: *Polarized Electrons in Surface Physics*, ed by R. Feder (World Scientific, Singapore 1985) pp 467–512
19. H. P. Oepen, K. Hünlich: J. Kirschner, Phys. Rev. Lett. **56**, 496 (1986)
20. E. Tamura, W. Piepke, R. Feder: Phys. Rev. Lett. **59**, 934 (1987)
21. E. Tamura, R. Feder: Europhys. Lett. **16**, 695 (1991)
22. J. Henk, R. Feder: Europhys. Lett. **28**, 609 (1994)
23. B. Schmiedeskamp, B. Vogt, U. Heinzmann: Phys. Rev. Lett. **60**, 651 (1988)
24. N. Irmer, R. David, B. Schmiedeskamp, U. Heinzmann: Phys. Rev. B **45**, 3849 (1992)
25. N. Irmer, R. Frentzen, S.-W. Yu, B. Schmiedeskamp, U. Heinzmann: J. Electron Spectrosc. Relat. Phenom. **78**, 321 (1996)

26. C. S. Fadley, D. A. Shirley, A. J. Freeman, P. S. Bagus, J. V. Mallow: Phys. Rev. Lett. **23**, 1397 (1969); C. S. Fadley, D. A. Sirley, Phys. Rev. A **2**, 1109 (1970)

27. B. Sinkovic, B. Hermsmeier: C. S. Fadley, Phys. Rev. Lett. **55**, 1227 (1985)

28. D. P. Pappas, K.-P. Kämpfer, B. P. Miller, H. Hopster, D. E. Fowler, C. R. Brundle, A. C. Luntz, Z.-X. Shen: Phys. Rev. Lett. **66**, 504 (1991)

29. H. C. Siegmann: J. Electron Spectrosc. Relat. Phenom. **68**, 505 (1994)

30. D. Oberli, R. Burgermeister, S. Riesen, W. Weber, H. C. Siegmann: Phys. Rev. Lett. **81**, 4228 (1998)

31. W. Weber, S. Riesen and H. C. Siegmann: Science **291**, 1015 (2001)

32. R. L. Long, Jr., V. W. Hughes, J. S. Greenberg, I Ames, and R. L. Christensen: Phys. Rev. A **138**, 1630 (1965)

33. A. B. Baganov, and D. B. Diatroptov: Zh Eksp. Teor. Fiz. **54**, 1333 (1968) (Sov. Phys.-JETP **27**, 1733 (1968))

34. G. Busch, M. Campagna, P. Cotti, H. C. Siegmann: Phys. Rev. Lett. **22**, 597 (1969)

35. U. Baenninger, G. Busch, M. Campagna, H. C. Siegmann: Phys. Rev. Lett. **25**, 585 (1970)

36. W. Eib and S. F. Alvarado: Phys. Rev. Lett. **37**, 444 (1976)

37. G. Busch, M. Campagna, D. T. Pierce, H. C. Siegmann: Phys. Rev. Lett **28**, 611 (1972)

38. J. C. Gröbli, A. Kündig, F. Meier, and H. C. Siegmann: Physica B **204**, 359 362 (1995)

39. M. Taborelli, R. Allenspach, G. Boffa, and M. Landolt: Phys. Rev. Lett **56**, 2869 (1986)

40. O. M. Paul: Ph. D Thesis. ETH Zürich Nr. **9210** (1990)

41. F. Meier, D. Pescia, M. Baumberger: Phys. Rev. Lett. **49**, 747 (1982)

42. F. Meier, G. L. Bona, S. Hüfner: Phys. Rev. Lett. **52**, 1152 (1984)

43. M. Aeschlimann, M. Bauer, and S. Pawlik: Phys. Rev. Lett. **79**, 5158 (1997)

44. A. Vaterlaus, F. Milani, and F. Meier: Phys. Rev. Lett. **65**, 3041 (1990)

45. M. Donath: Surf. Sci. Rep. **20**, 251 (1994)

46. S. LaShell, B. A. McDougall, E. Jensen: Phys. Rev. Lett. **77**, 3419 (1996)

47. M. Muntwiler, M. Hoesch, V. N. Petrov, M. Hengsberger, L. Patthey, M. Shi, M. Falub, T. Greber, J. Osterwalder: J. Electron Spectrosc. Relat. Phenom. **137–140**, 119 (2004)

48. M. Hoesch, M. Muntwiler, V. N. Petrov, M. Hengsberger, L. Patthey, M. Shi, M. Falub, T. Greber, J. Osterwalder: Phys. Rev. B **69**, 241401(R) (2004)

49. J. Henk, M. Hoesch, J. Osterwalder, A. Ernst, P. Bruno: J. Phys. Condens. Matter **16**, 7581 (2004)

50. M. Hochstrasser, J. G. Tobin, E. Rotenberg, S. D. Kevan: Phys. Rev. Lett. **89**, 216802 (2002)

51. N. B. Brookes, G. Ghiringhelli, O. Tjernberg, L. H. Tjeng, T. Mizokawa, T. W. Li, A. A. Menovsky: Phys. Rev. Lett. **87**, 237003 (2001)

52. J. H. Park, E. Vescovo, H. J. Kim, C. Kwon, R. Ramesh, T. Venkatesan: Phys. Rev. Lett. **81**, 1953 (1998)

53. M. Erbudak, G. Ravano: J. Appl. Phys. **52**, 5032 (1981)

Band-Structure Theory of Dichroism

Alexander Yaresko[1], Alexander Perlov[2], Victor Antonov[3], and Bruce Harmon[4]

[1] Max Planck Institute for the Physics of Complex Systems, 01187 Dresden, Germany
yaresko@mpipks-dresden.mpg.de
[2] University of Munich, 81377, Munich, Germany
[3] Institute of Metal Physics, National Academy of Sciences of Ukraine, 03142 Kiev, Ukraine
[4] Ames Laboratory, Iowa State University, Iowa, 50011 USA

Abstract. A theoretical approach to magneto-optical effects in visible and x-ray regions on the basis of an itinerant description for the underlying electronic structure is presented. The macroscopic description of the magneto-optical effects in terms of the dielectric constant of the media is outlined. Then, it is sketched how to calculate the dielectric constant and magneto-optical spectra on the basis of the underlying band structure. Simple illustrations are given to show that the magneto-optical effects appear as a result of symmetry lowering caused by the interplay of magnetic ordering and spin-orbit coupling.

1 Introduction

In 1845 Faraday discovered [1] that the polarization vector of linearly polarized light is rotated upon transmission through a sample that is exposed to a magnetic field parallel to the propagation direction of the light. About 30 years later, Kerr [2] observed that when linearly polarized light is reflected from a magnetic solid, its polarization plane also rotates over a small angle with respect to that of the incident light. This discovery has become known as the magneto-optical (MO) Kerr effect. Since then, many other magneto-optical effects, as for example the Zeeman, Voigt and Cotton-Mouton effects [3], have been discovered. These effects have in common that the interaction of polarized light with a magnetic solid depends on its polarization.

The quantum mechanical understanding of the Kerr effect began as early as 1932 when Hulme [4] proposed that the Kerr effect could be attributed to spin-orbit (SO) coupling (see, also Kittel [5]). The symmetry between left- and right-hand circularly polarized light is broken due to the SO coupling in a magnetic solid. This leads to different refractive indices for the two kinds of circularly polarized light, so that incident linearly polarized light is reflected with elliptical polarization, and the major elliptical axis is rotated by the so called Kerr angle from the original axis of linear polarization. The first systematic study of the frequency dependent Kerr and Faraday effects was developed by Argyres [6] and later Cooper presented a more general theory

A. Yaresko et al.: *Band-Structure Theory of Dichroism*, Lect. Notes Phys. **697**, 121–141 (2006)
www.springerlink.com   © Springer-Verlag Berlin Heidelberg 2006

using some simplifying assumptions [7]. The very powerful linear response techniques of Kubo [8] gave general formulas for the conductivity tensor which are now widely used. A general theory for the frequency dependent conductivity of ferromagnetic (FM) metals over a wide range of frequencies and temperatures was developed in 1968 by Kondorsky and Vediaev [9]. The first ab initio calculation of MO properties was made by Callaway and co-workers in the middle of the 1970s [10, 11]. They calculated the absorption parts of the conductivity tensor elements σ_{xx} and σ_{xy} for bulk Fe and Ni and obtained rather good agreement with experiment.

In 1975 the theoretical work of Erskine and Stern showed that the x-ray absorption could be used to determine the x-ray magnetic circular dichroism (XMCD) in transition metals when left- and right-circularly polarized x-ray beams are used [12]. In 1985 Thole et al. [13] predicted a strong magnetic dichroism in the $M_{4,5}$ x-ray absorption spectra of magnetic rare-earth materials, for which they calculated the temperature and polarization dependence. A year later this XMCD effect was confirmed experimentally by van der Laan et al. [14] at the Tb $M_{4,5}$-absorption edge of terbium iron garnet. The next year Schütz et al. [15] measured the K-edge absorption of iron with circularly polarized x-rays, and found an asymmetry in absorption of about 10^{-4}. This was shortly followed by the observation of magnetic EXAFS [16]. A theoretical description for the XMCD at the Fe K-absorption edge was given by Ebert et al. [17] using a spin-polarized version of relativistic multiple scattering theory. In 1990 Chen et al. [18] observed a large magnetic dichroism at the $L_{2,3}$ edge of nickel metal. Full multiplet calculations for $3d$ transition metal $L_{2,3}$ edges by Thole and van der Laan [19] were confirmed by several measurements on transition metal oxides. First considered as a rather exotic technique, XMCD has now developed as an important measurement technique for local magnetic moments. Whereas optical and MO spectra are often swamped by too many transitions between occupied and empty valence states, x-ray excitations have the advantage that the core state has a localized wave function, which offers the specificity of the element, site and the symmetry. Recent progress in devices for circularly polarized synchrotron radiation have now made it possible to explore routinely the polarization dependence of magnetic materials. The results of the corresponding theoretical investigations can be found, e.g., in a review paper by Ebert [20] and a book by Antonov et al. [21].

The aim of this contribution is to give an overview of the basic ideas behind the band-structure theory of the magneto-optical effects.

2 Classical Description of Magneto-Optical Effects

Magneto-optical effects refer to various changes in the polarization state of light upon interaction with materials possessing a net magnetic moment, including rotation of the plane of linearly polarized light (Faraday, Kerr rotation), and the complementary differential absorption of left and right circularly polarized light (circular dichroism). In the visible spectral range these effects result from electron excitations in the conduction bands. Near x-ray absorption edges, or resonances, magneto-optical effects can be enhanced by transitions from well-defined atomic core levels to conduction states selected by transition symmetry. Using straightforward symmetry considerations it can be shown that all MO phenomena are caused by the symmetry reduction caused by magnetic ordering [22], in comparison to the paramagnetic state. For the optical properties this symmetry reduction has only consequences when the SO coupling is included. To calculate MO properties one has therefore to account for magnetism and SO coupling simultaneously in the electronic structure of materials.

The interaction of the electromagnetic radiation with a magnetic medium is described classically by Maxwell's equations (in GSM or Gaussian units) [23]:

$$\nabla \times \boldsymbol{E} + \frac{1}{c}\frac{\partial \boldsymbol{B}}{\partial t} = 0\,, \tag{1}$$

$$\nabla \cdot \boldsymbol{B} = 0\,, \tag{2}$$

$$\nabla \times \boldsymbol{H} - \frac{1}{c}\frac{\partial \boldsymbol{D}}{\partial t} = \frac{4\pi}{c}\boldsymbol{J}\,, \tag{3}$$

$$\nabla \cdot \boldsymbol{D} = 4\pi\rho\,, \tag{4}$$

where $\boldsymbol{D}$ is the electric displacement, a quantity that is related to the electric field $\boldsymbol{E}$ caused by polarization $\boldsymbol{P}$ of the medium, $\boldsymbol{B}$ is the magnetic induction, $\boldsymbol{H}$ the macroscopic magnetic field and ρ and $\boldsymbol{J}$ the macroscopic charge and current densities.

The Maxwell equations are supplemented by so-called material equations which introduce specific properties of the material such as the electrical conductivity σ, the dielectric permittivity or dielectric constant ε, and magnetic permeability μ

$$\boldsymbol{D} = \varepsilon_0 \varepsilon \cdot \boldsymbol{E} = \boldsymbol{E} + 4\pi\boldsymbol{P}\,, \tag{5}$$

$$\boldsymbol{B} = \mu_0 \mu \cdot \boldsymbol{H} = \boldsymbol{H} + 4\pi\boldsymbol{M}\,, \tag{6}$$

$$\boldsymbol{J} = \sigma \cdot \boldsymbol{E}\,, \tag{7}$$

where $\boldsymbol{M}$ is the magnetization, ε_0 is the vacuum permittivity and μ_0 is the vacuum permeability. In general the dielectric constant is a function of both spatial and time variables that relates the displacement field $\boldsymbol{D}(\boldsymbol{r}, t)$ to the electric field $\boldsymbol{E}(\boldsymbol{r}', t')$

$$\boldsymbol{D}(\boldsymbol{r}, t) = \int \int_{-\infty}^{t} \varepsilon(\boldsymbol{r}, \boldsymbol{r}', t')\boldsymbol{E}(\boldsymbol{r}', t')\mathrm{d}t'\mathrm{d}\boldsymbol{r}'\,. \tag{8}$$

In the following we neglect the spatial dependence of the dielectric constant and consider only its frequency dependence $\varepsilon(\omega)$. Usually, the effect of the magnetic permeability tensor $\boldsymbol{\mu}(\omega)$ on optical phenomena is small and we assume that $\boldsymbol{\mu}(\omega) = \mu_0 \boldsymbol{I}$ where $\boldsymbol{I}$ is a unit tensor. It should be stressed that for high electric and magnetic fields ε and μ may depend on the field strength. In such cases higher order terms in a Taylor expansion of the material parameters lead to appearance of non-linear effects [24–26].

The dielectric constant $\varepsilon(\omega)$ and optical conductivity $\boldsymbol{\sigma}(\omega)$ are complex quantities

$$\varepsilon = \varepsilon_1 + \mathrm{i}\varepsilon_2 , \tag{9}$$

$$\sigma = \sigma_1 + \mathrm{i}\sigma_2 , \tag{10}$$

and are related to each other through [27]

$$\varepsilon = 1 + \frac{4\pi\mathrm{i}}{\omega}\sigma . \tag{11}$$

In the absence of a spontaneous magnetisation or external magnetic field and in a general case of an anisotropic material ε is a symmetric tensor, which, after transformation to its principal axes, has three components. For simplicity, let us consider a material of cubic structure with a magnetization $\boldsymbol{M}$ directed along its z axis. Above the Curie temperature T_C the crystal has a cubic symmetry and all three components of the dielectric tensor are equal so that

$$\varepsilon(\omega) = \varepsilon \boldsymbol{I} . \tag{12}$$

When the magnetization $\boldsymbol{M}\|[001]$ appears below T_C the symmetry is lower and $\varepsilon(\omega)$ becomes [22]

$$\varepsilon(\boldsymbol{M},\omega) = \begin{pmatrix} \varepsilon_{xx} & \varepsilon_{xy} & 0 \\ -\varepsilon_{xy} & \varepsilon_{xx} & 0 \\ 0 & 0 & \varepsilon_{zz} \end{pmatrix} . \tag{13}$$

The remaining symmetry of the system depends on the orientation of the magnetization. For cubic systems, for example, the effective symmetry is tetragonal, trigonal or orthorhombic depending on whether the magnetization is aligned along the [001]-, the [111]- or the [110]-axis, respectively [28].

The components of the dielectric tensor depend on the magnetization and satisfy the following Onsager relations

$$\varepsilon_{\alpha\beta}(-\boldsymbol{M},\omega) = \varepsilon_{\beta\alpha}(\boldsymbol{M},\omega) , \tag{14}$$

where $\alpha, \beta = x, y$ or z. The Onsager relations mean that the diagonal components of the dielectric tensor are even functions of $\boldsymbol{M}$, whereas the nondiagonal ones are odd functions of $\boldsymbol{M}$. In the lowest order in $\boldsymbol{M}$

$$\varepsilon_{xy} \sim M , \quad \varepsilon_{zz} - \varepsilon_{xx} \sim M^2 . \tag{15}$$

In the absence of an external current ($\boldsymbol{J} = 0$) and free charges ($\rho = 0$) (1)–(7) reduce to

$$\nabla \times \boldsymbol{E} = -\mu_0 \frac{1}{c} \frac{\partial \boldsymbol{H}}{\partial t}, \tag{16}$$

$$\nabla \times \boldsymbol{H} = \varepsilon_0 \varepsilon \frac{1}{c} \frac{\partial \boldsymbol{E}}{\partial t}. \tag{17}$$

After substitution of $\boldsymbol{E}$ and $\boldsymbol{H}$ in a form of plane waves

$$\boldsymbol{E} = \boldsymbol{E}_0 \exp[-\mathrm{i}(\omega t - \boldsymbol{q} \cdot \boldsymbol{r})], \tag{18}$$

$$\boldsymbol{H} = \boldsymbol{H}_0 \exp[-\mathrm{i}(\omega t - \boldsymbol{q} \cdot \boldsymbol{r})], \tag{19}$$

where ω is a frequency and $\boldsymbol{q}$ is a wave vector of light, one arrives to a secular equation

$$\begin{pmatrix} N^2 - \varepsilon_{xx} & -\varepsilon_{xy} & 0 \\ \varepsilon_{xy} & N^2 - \varepsilon_{xx} & 0 \\ 0 & 0 & \varepsilon_{zz} \end{pmatrix} \begin{pmatrix} E_x \\ E_y \\ E_z \end{pmatrix} = 0, \tag{20}$$

where $\boldsymbol{N}$ is a unit vector directed along $\boldsymbol{q}$. When the light propagates along z direction, i.e., along $\boldsymbol{M}$, $E_z = 0$, while from the condition for nonzero solutions of E_x and E_y, one finds the eigenvalues

$$n_{\pm}^2 = \varepsilon_{xx} \pm \varepsilon_{xy}. \tag{21}$$

This means that the normal modes of the light are

$$D_+ = \varepsilon_0 n_+^2 (E_x + \mathrm{i}E_y), \tag{22}$$

$$D_- = \varepsilon_0 n_-^2 (E_x - \mathrm{i}E_y), \tag{23}$$

i.e. a right and a left circularly polarized light wave with complex refractive indices of n_+ and n_-, respectively.

2.1 Faraday Effect

Linearly polarized light, which propagates along z direction, can be resolved into two circularly polarized light waves with an equal amplitude. In a ferromagnet, the left-hand and right-hand circularly polarized lights propagate generally with different refractive indices or different velocities c/n_+ and c/n_-. After travelling through a sample of thickness l, there appears a phase difference between two circularly polarized lights. When the two transmitted light waves are combined at the exit surface of the sample, they yield again a linearly polarized light, but its plane of polarization is rotated by the so-called Faraday angle θ_F given by [29] (see Fig. 1)

$$\theta_F = \frac{\omega l}{2c} \mathrm{Re}(n_+ - n_-). \tag{24}$$

The direction of the rotation depends on the relative orientation of the magnetization and the light propagation. When the transmitted light is reflected

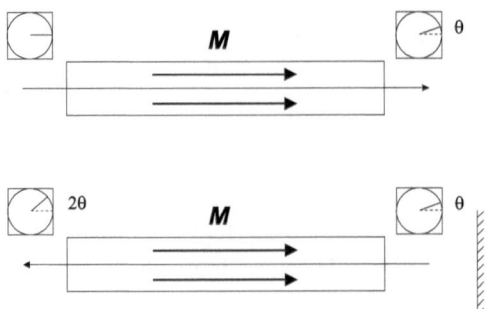

Fig. 1. Faraday rotation

by a mirror and propagates through the ferromagnet to the starting point as in the lower panel of Fig. 1, the total rotation angle is doubled to $2\theta_F$. This is known as the nonreciprocal property of light propagation in ferromagnets.

If two circularly polarized waves attenuate at different rates, then after travelling through the sample, their relative amplitude change. Therefore the transmitted light becomes elliptically polarized, with the ellipticity defined as

$$\eta_F = -\frac{\omega l}{2c}\text{Im}(n_+ - n_-).\tag{25}$$

The ellipticity η_F corresponds to the ratio of the minor to the major axes of the polarization ellipsoid, and is related to the magnetic circular dichroism, which is defined by the difference of the absorption coefficient μ between the right and left circularly polarized light

$$\Delta\mu(\omega) = \mu_+(\omega) - \mu_-(\omega) = -\frac{4\eta_F(\omega)}{l}.\tag{26}$$

In the general case the Faraday rotation and ellipticity per unit length are related to the components of the dielectric tensor by the following expressions

$$\theta_F(\omega) = \frac{\omega}{2c}\frac{k\varepsilon_{xy}^{(1)} - n\varepsilon_{xy}^{(2)}}{n^2 + k^2},\tag{27}$$

$$\eta_F(\omega) = -\frac{\omega}{2c}\frac{n\varepsilon_{xy}^{(1)} + k\varepsilon_{xy}^{(2)}}{n^2 + k^2},\tag{28}$$

were the refractive index n and extinction coefficient k are defined by

$$\varepsilon_{xx}^{(1)} = n^2 - k^2, \qquad \varepsilon_{xx}^{(2)} = 2nk.\tag{29}$$

For a transparent ferromagnet $k = 0$ and (27) and (28) simplify to

$$\theta_F(\omega) = -\frac{\omega}{2c}\frac{\varepsilon_{xy}^{(2)}}{\left(\varepsilon_{xx}^{(1)}\right)^{\frac{1}{2}}}, \qquad \eta_F(\omega) = -\frac{\omega}{2c}\frac{\varepsilon_{xy}^{(1)}}{\left(\varepsilon_{xx}^{(1)}\right)^{\frac{1}{2}}}.\tag{30}$$

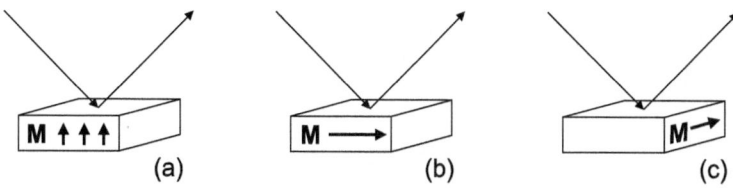

Fig. 2. Three different geometries for the MO Kerr effect: (**a**) the polar Kerr effect, (**b**) the longitudinal Kerr effect, (**c**) the transversal, or equatorial Kerr effect

This means that θ_F and η_F are proportional to the imaginary part and the real part of the nondiagonal component of the dielectric tensor ε, respectively.

2.2 MO Kerr Effect

The magneto-optical Kerr effect was discovered by the Scottish physicist John Kerr in 1888. He observed that when linearly polarized light is reflected at normal incidence from the polished pole of an electromagnet, it becomes elliptically polarized with the major axis of the ellipse rotated with respect to the plane of polarization of the incident beam [2].

Depending on the orientation of the magnetization vector relative to the reflective surface and the plane of incidence of the light beam, three types of the magneto-optical effects in reflection are distinguished: polar, longitudinal (meridional), and transverse (equatorial) effects (Fig. 2). In the polar geometry (Fig. 2(a)) the magnetization vector M is oriented perpendicularly to the reflective surface and parallel to the plane of incidence. In the longitudinal Kerr effect the magnetization M is parallel both to the sample surface and to the plane of incidence (Fig. 2(b)). In both these effects the reflection from the magnetic material causes a rotation of the plane of polarization and an appearance of ellipticity of the reflected linearly polarized light. The transverse (equatorial) Kerr effect (Fig. 2(c)) is observed when the magnetization vector is oriented perpendicularly to the plane of incidence of the light. In this case the change of intensity and phase of the linearly polarized light reflected by the magnetized medium are detected [30].

In the polar Kerr geometry (Fig. 2(a)) the normal modes in the sample are two circularly polarized light. Outside the sample these are also possible modes and the reflection coefficients can be easily calculated using the Fresnel relations. For linearly polarized incident light the reflected light will in general be elliptically polarized. The relation between the complex polar Kerr angle and and the complex refraction indices is given by [31]

$$\frac{1 + \tan \eta_\mathrm{K}}{1 - \tan \eta_\mathrm{K}} e^{2i\theta_\mathrm{K}} = \frac{1 + n_+}{1 - n_+} \frac{1 - n_-}{1 + n_-} . \tag{31}$$

From (31) it can be seen that the maximal observable rotation θ_K is $\pm 90°$ and the maximal achievable Kerr ellipticity η_K is $\pm 45°$.

For most materials the Kerr rotation and ellipticity are less than 1°. When $|\varepsilon_{xy}| \ll |\varepsilon_{xx}|$ and for small θ_K and η_K the above exact expression can be approximated by [32]

$$\theta_K + i\eta_K \approx \frac{-\varepsilon_{xy}}{(\varepsilon_{xx} - 1)\sqrt{\varepsilon_{xx}}} \, . \tag{32}$$

From (32) it follows that the Kerr rotation and ellipticity are proportional to the nondiagonal component of the dielectric tensor. However, the denominator of (32) is a complex function of ε_{xx}. When it becomes small a resonant amplification of the Kerr rotation can be observed as, for example, in CeSb for which the record value of $\theta_K \approx 17°$ was reported [33–35].

3 Linear Response Theory

In the previous section the response of the medium to electromagnetic waves was described in a phenomenological manner in terms of the frequency dependent complex dielectric constant and conductivity. Microscopic calculations of the optical conductivity tensor are based on the Kubo linear response formalism in which the conductivity is given in terms of current-current correlation functions [8, 27]

$$\sigma_{\alpha\beta}(\omega) = \frac{1}{\hbar\omega V} \int_{-\infty}^{0} d\tau e^{-i\omega\tau} \langle [J_\beta(\tau), J_\alpha(0)] \rangle \,, \tag{33}$$

where $J(\tau)$ is the electric current operator. Unfortunately this formula is unmanageable for practical implementation as it involves the many-body ground state wave function. An expression for $\boldsymbol{\sigma}(\omega)$ that is feasible for calculations using band structure methods was derived by Callaway [10, 36]

$$\sigma_{\alpha\beta}(\omega) = \frac{ie^2}{m^2 \hbar V} \sum_{\boldsymbol{k}} \sum_{nn'} \frac{f(\epsilon_{n\boldsymbol{k}}) - f(\epsilon_{n'\boldsymbol{k}})}{\omega_{nn'}(\boldsymbol{k})} \frac{M^\alpha_{n'n}(\boldsymbol{k}) M^\beta_{nn'}(\boldsymbol{k})}{\omega - \omega_{nn'}(\boldsymbol{k}) + i\gamma} \,. \tag{34}$$

It relates the macroscopic optical conductivity to the sum of interband transitions between Bloch states $\Psi^{n\boldsymbol{k}}$ and $\Psi^{n'\boldsymbol{k}}$ with energies $\epsilon_{n\boldsymbol{k}}$ and $\epsilon_{n'\boldsymbol{k}}$ with n and n' being the band indices. In (34), V is the unit cell volume, $f(\epsilon_{n\boldsymbol{k}})$ is the Fermi function, $\hbar\omega_{nn'}(\boldsymbol{k}) = \epsilon_{n\boldsymbol{k}} - \epsilon_{n'\boldsymbol{k}}$, and $\gamma = 1/\tau$ is a phenomenological relaxation time parameter that takes into account the finite lifetime of the excited electronic states. Finally, $M^\alpha_{nn'}(\boldsymbol{k})$ are matrix elements which will be considered in detail in Sect. 6.

Because of causality $\boldsymbol{\sigma}(\omega)$ is analytical in the upper half of the complex ω-plane. As a consequence its real and imaginary parts are connected by the following Kramers-Kronig relations [37]

$$\sigma^{(2)}(\omega) = -\frac{2\omega}{\pi}\mathcal{P}\int_o^\infty \frac{\sigma^{(1)}(\omega')}{\omega'^2 - \omega^2}\mathrm{d}\omega',\tag{35}$$

$$\sigma^{(1)}(\omega) = \frac{2}{\pi}\mathcal{P}\int_o^\infty \frac{\omega'\sigma^{(2)}(\omega')}{\omega'^2 - \omega^2}\mathrm{d}\omega',\tag{36}$$

where $\mathcal{P}$ is the principal part of the integral. In practice it is often more convenient to calculate the absorptive parts of the conductivity tensor components using expressions

$$\sigma_{\alpha\alpha}^{(1)}(\omega) = \frac{\pi e^2}{m^2\hbar\omega V}\sum_{\boldsymbol{k}}\sum_{n'}^{\mathrm{occ}}\sum_{n}^{\mathrm{unocc}}|M_{n'n}^\alpha(\boldsymbol{k})|^2\delta(\omega - \omega_{nn'}(\boldsymbol{k})),\tag{37}$$

$$\sigma_{\alpha\beta}^{(2)}(\omega) = \frac{\pi e^2}{m^2\hbar\omega V}\sum_{\boldsymbol{k}}\sum_{n'}^{\mathrm{occ}}\sum_{n}^{\mathrm{unocc}}\mathrm{Im}[M_{n'n}^\alpha(\boldsymbol{k})M_{nn'}^\beta(\boldsymbol{k})]\delta(\omega - \omega_{nn'}(\boldsymbol{k})),\tag{38}$$

obtained from (34) by taking $\gamma \to 0$ limit. Then, the corresponding dispersive parts can be calculated using the relations (35) and (36).

Alternatively, the absorption part of the optical conductivity can be expressed in terms of the one-particle Green function $G(E)$ [38]

$$\sigma_{\alpha\beta}^{(1)}(\omega) = \frac{1}{\pi\omega}\int_{E_F-\hbar\omega}^{E_F}\mathrm{d}E\,\mathrm{tr}[\hat{j}_\alpha\mathrm{Im}G(E)\hat{j}_\beta\mathrm{Im}G(E + \hbar\omega)],\tag{39}$$

where $\mathrm{Im}G(E)$ stands for the anti-Hermitian part of the Green function, $\hat{\boldsymbol{j}}$ is the electron current operator, and E_F is the Fermi level. The one-particle Green function in (39) is defined as a solution of the equation

$$[\hat{H}_0 + \Sigma(E) - E]\hat{G}(E) = \hat{I},\tag{40}$$

where $\hat{H}_0$ is the one-particle Hamiltonian including the kinetic energy, the electron-ion Coulomb interaction and the Hartree potential, while the self-energy $\Sigma(E)$ describes all static and dynamic effects of electron-electron exchange and correlations. This approach allows to incorporate the self-energy obtained by many-body technics into band-structure calculations of the optical properties.

Equations (37)–(39) are derived in the $q = k' - k = 0$ limit, i.e., they include only the contribution from direct interband transitions to the optical conductivity. The missing intraband contributions are quite difficult to account for because in principle they depend on lattice imperfections of the system as well as on the temperature. For that reason they are, in general, represented by the phenomenological Drude contribution to the diagonal components of the optical conductivity [27]

$$\sigma^{(1)}(\omega) = \frac{\sigma_0}{1 + \omega^2\tau_D^2},\tag{41}$$

where σ_0 is a *dc* conductivity and τ_D is a relaxation time parameter. The intraband contribution to the off-diagonal optical conductivity is very small and usually neglected in the case of magnetically ordered materials [31, 32].

It is important to note that the symmetry reduction due to the presence of spontaneous magnetization, that leads to the appearance of nonzero off-diagonal components of the optical conductivity tensor, occurs only if both the spin-polarization and the spin-orbit coupling are simultaneously taken into account in the calculations. If one switches off the spin-orbit coupling for a spin-polarized system its properties no longer depend on the orientation of the magnetization. For this reason its symmetry remains the same as the symmetry of the corresponding paramagnetic state and the off-diagonal components of $\boldsymbol{\sigma}(\omega)$, e.g. $\varepsilon_{xy}(\omega)$ in (13), vanish. Thus, although the symmetry of the magnetic system is entirely determined by the orientation of the magnetization, the magnitude of the off-diagonal components of the conductivity tensor depends on the magnitude of the magnetization and the strength of the spin-orbit coupling.

4 MO Effects in the X-ray Region

In recent years the study of magneto-optical effects in the soft x-ray range has gained great importance as a tool for the investigation of magnetic materials [20, 21].

The attenuation of the x-ray intensity when passing through a sample of thickness d

$$I(d) = I_0 e^{-\mu_{q\lambda}(\omega)d} \tag{42}$$

is determined by the absorption coefficient $\mu_{q\lambda}(\omega)$ which in general depends of the wavevector $\boldsymbol{q}$, energy $\hbar\omega$, and polarization λ of the radiation. Because $\mu_{q\lambda}(\omega)$ is related to the absorptive part of the dielectric function $\varepsilon_{q\lambda}$ or the optical conductivity $\sigma_{q\lambda}$ via [20]

$$\mu_{q\lambda}(\omega) = \frac{\omega}{c}\varepsilon_{q\lambda}^{(2)}(\omega) = \frac{4\pi}{c}\sigma_{q\lambda}^{(1)}(\omega) \tag{43}$$

it can be evaluated using the expression similar to (37)

$$\mu_{q\lambda}(\omega) = \frac{\pi c^2}{\hbar\omega m V} \sum_i^{\text{occ}} \sum_f^{\text{unocc}} |M_{if}^{q\lambda}|^2 \delta(\hbar\omega - E_f + E_i). \tag{44}$$

In contrast to (37) in which the matrix elements of the electron-photon interaction are evaluated between two Bloch states, the matrix elements $M_{if}^{q\lambda}$ are calculated between an extended final state f and a well localized initial core state i. The sum over the initial states i is usually restricted to one core shell which corresponds to a fine-tuning to a particular absorption edge. This important property makes x-ray absorption an element specific probe.

If x-rays are absorbed by a magnetic solid the absorption coefficients for left and right circularly polarized photons are in general different so that

$$\Delta\mu(\omega) = \mu^{q+}(\omega) - \mu^{q-}(\omega) \neq 0\,. \qquad (45)$$

This effect can be experimentally detected [15] and is called the X-ray Magnetic Circular Dichroism (XMCD). Similar to the other MO effects considered in Sect. 2 XMCD is the consequence of the symmetry reduction caused by magnetic ordering, which can be accounted for only if the spin-polarization and the spin-orbit coupling are included into the calculation. Comparing the magneto-optical effects in visible and x-ray spectral ranges one can note an important difference. For the former the spin-polarization and the spin-orbit coupling strength of the initial and final states are equally important. For the latter, however, because of the large spin-orbit splitting of the initial states, e.g. $p_{1/2,3/2}$ core states, an intense XMCD signal at $L_{2,3}$- or $M_{2,3}$-edges can be observed even for light magnetic atoms. The spin splitting of the initial core states is usually small and can be neglected. Based on these considerations, Erskine and Stern [12] were the first to predict a very large XMCD for the $M_{2,3}$-edge of Ni.

During recent years important magneto-optical sum rules were derived [39–41] that relate the integrated intensity of absorption XMCD spectra to the expectation values of the operators σ_z and l_z, and to the spin and orbital magnetic moments μ_s and μ_l, respectively. This makes XMCD a powerful tool for studying magnetic properties of solids. The XMCD sum rules and their applications are discussed in details in other contributions to this collection.

5 Spin Polarized Relativistic Calculations

As was mentioned above the magneto-optical effects can be calculated using band structure methods only if the spin-polarization and spin-orbit coupling are simultaneously taken into account.

Rajagopal and Callaway [42] pointed out that treating magnetism in a proper relativistic way, by extending the original Hohenberg-Kohn-Sham density functional formalism [43, 44], leads to a current density formalism with the expectation value of the four-current density operator as the central quantity. However, as this approach is too difficult to implement an extension of the non-relativistic spin-density-functional formalism suggested by MacDonald and Vosko [45] and Ramana and Rajagopal [46] is used in order to describe relativistic effects and spontaneous magnetism on an equal footing. The corresponding Dirac Hamiltonian has the form

$$\hat{H} = \hat{H}_0 + \hat{H}_m = \frac{c}{i}\boldsymbol{\alpha}\nabla + \frac{c^2}{2}(\beta - I) + V(\boldsymbol{r}) + \mu_{\mathrm{B}}\beta\boldsymbol{\sigma}\boldsymbol{B}(\boldsymbol{r})\,, \qquad (46)$$

where $\boldsymbol{\alpha}$ and β are Dirac matrices

$$\alpha = \begin{pmatrix} 0 & \boldsymbol{\sigma} \\ \boldsymbol{\sigma} & 0 \end{pmatrix}, \quad \beta = \begin{pmatrix} I & 0 \\ 0 & -I \end{pmatrix}, \tag{47}$$

$\boldsymbol{\sigma}$ is the vector of Pauli matrices

$$\sigma_x = \begin{pmatrix} 0 & 1 \\ 1 & 0 \end{pmatrix}, \quad \sigma_y = \begin{pmatrix} 0 & -i \\ i & 0 \end{pmatrix}, \quad \sigma_z = \begin{pmatrix} 1 & 0 \\ 0 & -1 \end{pmatrix}, \tag{48}$$

and I is a 2×2 unitary matrix. The spin-independent potential $V(\boldsymbol{r}) = V_{\mathrm{H}}(\boldsymbol{r}) + V_{\mathrm{xc}}(\boldsymbol{r})$ in (46) includes the Hartree potential $V_{\mathrm{H}}(\boldsymbol{r})$ and the spin-averaged part of the exchange-correlation potential

$$V_{\mathrm{xc}}(\boldsymbol{r}) = \frac{\partial E_{\mathrm{xc}}}{\partial \rho}(\boldsymbol{r}). \tag{49}$$

The effective magnetic field $\boldsymbol{B}(\boldsymbol{r})$ in $\hat{H}_m = \mu_{\mathrm{B}} \beta \boldsymbol{\sigma} \boldsymbol{B}(\boldsymbol{r})$ is defined as

$$\boldsymbol{B}(\boldsymbol{r}) = \frac{\partial E_{\mathrm{xc}}}{\partial \boldsymbol{m}}(\boldsymbol{r}), \tag{50}$$

with $\rho(\boldsymbol{r})$ and $\boldsymbol{m}(\boldsymbol{r})$ being the electron density and spin-magnetization density, respectively. The magnetic term $\hat{H}_m$ does not contain the explicit coupling of the magnetic field $\boldsymbol{B}(\boldsymbol{r})$ to the orbital degree of freedom of the electron, which would be present in a general current density functional formalism. However, this coupling *is provided* indirectly via the spin-orbit interaction which is contained in $\hat{H}_0$, and the Hamiltonian (46) accounts properly for the fact that time-reversal symmetry is broken by the magnetic ordering.

In the spherically symmetric potential a solution of the spin-independent part of the Dirac Hamiltonian $\hat{H}_0$ can be written as

$$\Phi_{\kappa\mu}(\varepsilon, \boldsymbol{r}) = \begin{pmatrix} g_\kappa(\varepsilon, r) \chi_{\kappa\mu}(\hat{r}) \\ i f_\kappa(\varepsilon, r) \chi_{-\kappa\mu}(\hat{r}) \end{pmatrix}, \tag{51}$$

where the radial dependence is provided by the so-called large $g_\kappa(r)$ and small $f_\kappa(r)$ components of the radial solution and the angular dependence is determined by bi-spinors $\chi_{\kappa\mu}$ which can be expressed as a sum of products of spherical harmonics $Y_{lm}(\hat{r})$ and spin functions χ_s weighted by Clebsch-Gordan coefficients $C^{j\mu}_{lm,1/2s}$ [47]

$$\chi_{\kappa\mu}(\hat{r}) = \sum_s C^{j\mu}_{l\mu-s,\frac{1}{2}s} Y_{l\mu-s}(\hat{r}) \chi_s. \tag{52}$$

κ and μ (or m_j) in (51) are relativistic quantum numbers that are defined by

$$-(1 + \boldsymbol{\sigma} \cdot \hat{\boldsymbol{l}}) \chi_{\kappa\mu} = \kappa \chi_{\kappa\mu}, \tag{53}$$

$$\hat{j}_z \chi_{\kappa\mu} = \mu \chi_{\kappa\mu}, \tag{54}$$

and, depending on the value of the total momentum j of the electron, can have the following values

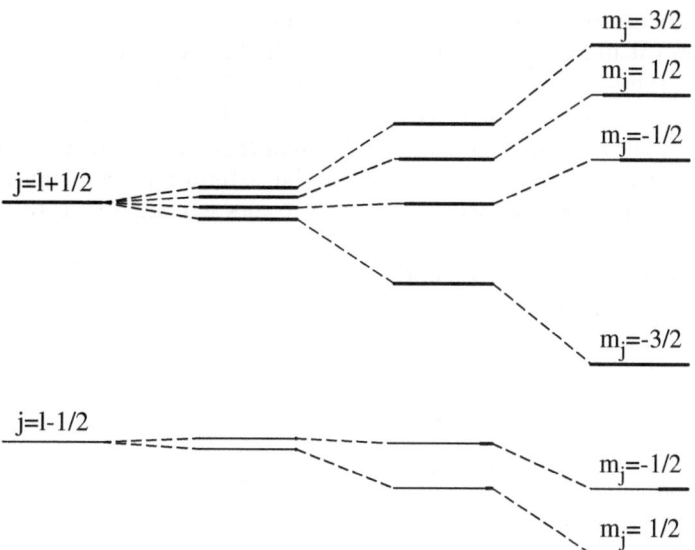

Fig. 3. Lifting of the degeneracy of $l - 1/2$ and $l + 1/2$ levels with $l = 1$ by the effective magnetic field. The relative strength of the magnetic field increases from left to right. The contribution of $l + 1/2$ states to the wave function is denoted by the length of thick horizontal lines

$$\kappa = \begin{cases} -l - 1, & \text{for } j = l + 1/2 \\ l, & \text{for } j = l - 1/2 \end{cases}, \quad \text{and} \quad -j \le \mu \le j . \tag{55}$$

If the magnetization is directed along the z axis the spin-dependent part of (46) becomes $\hat{H}_m = \mu_B \beta \sigma_z B(r)$. Then, using (51)–(52) and taking into account that

$$\langle \chi_{\kappa\mu} | \sigma_z | \chi_{\kappa'\mu} \rangle = \begin{cases} -2\mu/(2\kappa + 1), & \kappa = \kappa' \\ -\sqrt{1 - 4\mu^2/(\kappa - \kappa')^2}, & \kappa = -\kappa' - 1 \end{cases} \tag{56}$$

one can easily verify that the matrix elements $\langle \Phi_{\kappa\mu} | \hat{H}_m | \Phi_{\kappa'\mu} \rangle$ between the states with the same μ and different κ, κ' do not vanish if $l - l' = 0, \pm 2$, that leads to an infinite set of coupled equations [48–50]. However, it is commonly accepted [20,51] that the coupling of the states with $\Delta l = \pm 2$ is less important and can be neglected. This means that the only good quantum number is μ whereas the states with $j = l - 1/2$ and $j = l - 1/2$ are mixed by $\hat{H}_m$.

Figure 3 illustrates the lifting of the degeneracy of two- and four-fold degenerate $p_{1/2}$ and $p_{3/2}$ states by the magnetic field of increased strength. It should be noted that the energy splitting produced by the effective spin-only coupling magnetic field is not proportional to μ. The relation $\Delta E = g M_j$, well known from the theory of the Zeeman effect [52] is however recovered if the term proportional to $\hat{l} \cdot \boldsymbol{B}$ is added to the Hamiltonian.

In principle one can set up a formalism for spin-polarized relativistic band structure calculations, in particular using the LMTO method [53, 54], starting from the solution of the full Hamiltonian (46) [55, 56]. In practice it is, however, more convenient to construct the LMTO basis functions from the solution of the spin-independent Hamiltonian $\hat{H}_0$ and then include the matrix elements of $\hat{H}_m$ at the variational step. The latter approach is much simpler to implement while the accuracy of the results is comparable to the more rigorous one [55, 57].

Without going into detail of the spin-polarized relativistic LMTO method we write down only the final expression for the Bloch wave function

$$\Psi^{nk}(r) = \sum_{t\kappa\mu} \left[A_{t\kappa\mu}^{nk} \Phi_{\nu\kappa\mu}(r_t) + B_{t\kappa\mu}^{nk} \dot{\Phi}_{\nu\kappa\mu}(r_t) \right] , \tag{57}$$

where A^{nk} and B^{nk} are matrices that depend on the eigenvectors calculated by solving the LMTO eigenvalue problem, index t numbers atomic spheres centred at R_t in the unit cell, and $r_t = r - R_t$. $\Phi_{\nu\kappa\mu}(r_t)$ is the solution (51) of the Dirac equation in the atomic sphere surrounding atom t taken at a fixed energy ε_ν, and $\dot{\Phi}_{\nu\kappa\mu}(r_t)$ is its energy derivative.

6 Matrix Elements of the Electron–Photon Interaction

The interaction of an electromagnetic wave represented by a vector potential $A_{q\lambda}(r)$ with the wave vector q and polarization λ with an electron is described by the electron-photon interaction operator

$$\hat{H}_{\text{el-ph}} = -\frac{1}{c} J A_{q\lambda}(r) = -\frac{1}{c} J a_\lambda A e^{iqr} , \tag{58}$$

where J is the electronic current density operator

$$J = -ec\alpha . \tag{59}$$

The components of the polarization vector for linearly polarized light are given by

$$a_x = \begin{pmatrix} 1 \\ 0 \\ 0 \end{pmatrix} , \quad a_y = \begin{pmatrix} 0 \\ 1 \\ 0 \end{pmatrix} , \quad a_z = \begin{pmatrix} 0 \\ 0 \\ 1 \end{pmatrix} . \tag{60}$$

For q pointing along the z axis, left $(+)$ and right $(-)$ circularly polarized lights are represented by the polarization vector

$$a_\pm = \frac{1}{\sqrt{2}} \begin{pmatrix} 1 \\ \pm i \\ 0 \end{pmatrix} . \tag{61}$$

In order to study optical or magneto-optical phenomena one needs to calculate matrix elements of the form

$$M_{q\lambda}^{\text{fi}} = \langle \Psi_{\text{f}} | \hat{H}_{\text{el}-\text{ph}} | \Psi_{\text{i}} \rangle . \tag{62}$$

Depending on the kind of spectroscopy and photon energy the initial and final states represented by their wave functions Ψ_{i} and Ψ_{f} can be either Bloch states or tightly bound core states. In the LMTO and many other band structure methods $\Psi_{\text{i,f}}$ are calculated in the form of the expansion in terms of the functions $\Phi_{\kappa\mu}(E, r)$ defined by (51) and it is sufficient to consider the matrix elements (62) between these functions having a unique spin-angular character $\Lambda = (\kappa, \mu)$.

It is generally argued that in the frequency range of conventional optics the amplitude of the vector potential varies only slowly on a microscopic scale. This implies that it is sufficient to expand the exponential factor in (58) [58]

$$e^{i\boldsymbol{q}\boldsymbol{r}} = 1 + i\boldsymbol{q}\boldsymbol{r} - \frac{1}{2}(\boldsymbol{q}\boldsymbol{r})^2 + \dots \tag{63}$$

and retain just the first constant term, in which case only the electric dipole interaction is accounted for. For x-ray regime the next term in the expansion that represents the magnetic dipole and electric quadrupole interactions may also be important. However, Arola et al. [59] showed that the contribution from the magnetic dipole and electric quadrupole interactions to the anomalous scattering cross section at the iron K edge is two orders of magnitude smaller that the electric dipole contribution.

Dropping the constant factor ec the matrix elements of $\hat{H}_{\text{el}-\text{ph}}$ in the dipole approximation are given by

$$
\begin{aligned}
M_{\kappa\mu,\kappa'\mu'}^{\lambda} &= \langle \Phi_{\kappa\mu} | \boldsymbol{\alpha} \boldsymbol{a}_{\lambda} | \Phi_{\kappa'\mu'} \rangle \\
&= \mathrm{i}(R_{\kappa\kappa'}^{(1)} A_{\kappa\mu,-\kappa'\mu'}^{\lambda} - R_{\kappa\kappa'}^{(2)} A_{-\kappa\mu,\kappa'\mu'}^{\lambda}) ,
\end{aligned}
\tag{64}
$$

with the radial matrix elements

$$R_{\kappa\kappa'}^{(1)} = \int g_{\kappa}(r, \varepsilon_{\kappa}) f_{\kappa'}(r, \varepsilon_{\kappa'}) r^2 \mathrm{d}r , \tag{65}$$

$$R_{\kappa\kappa'}^{(2)} = \int f_{\kappa}(r, \varepsilon_{\kappa}) g_{\kappa'}(r, \varepsilon_{\kappa'}) r^2 \mathrm{d}r , \tag{66}$$

and the angular matrix elements

$$A_{\kappa\mu,\kappa'\mu'}^{\lambda} = \langle \chi_{\kappa\mu} | \boldsymbol{\sigma} \boldsymbol{a}_{\lambda} | \chi_{\kappa'\mu'} \rangle . \tag{67}$$

For various polarization states λ the latter are found to be

$$A_{\kappa\mu,\kappa'\mu'}^{\pm} = 2C_{l\mu\mp\frac{1}{2},\frac{1}{2}\pm\frac{1}{2}}^{j\mu} C_{l\mu\pm\frac{1}{2},\frac{1}{2}\mp\frac{1}{2}}^{j'\mu} \delta_{ll'} \delta_{\mu,\mu'\pm1} \tag{68}$$

$$A_{\kappa\mu,\kappa'\mu'}^{x} = \frac{1}{\sqrt{2}}(A_{\kappa\mu,\kappa'\mu'}^{-} + A_{\kappa\mu,\kappa'\mu'}^{+}) \tag{69}$$

$$A^y_{\kappa\mu,\kappa'\mu'} = \frac{i}{\sqrt{2}}\left(A^-_{\kappa\mu,\kappa'\mu'} - A^+_{\kappa\mu,\kappa'\mu'}\right) \tag{70}$$

$$A^z_{\kappa\mu,\kappa'\mu'} = \left(C^{j\mu}_{l\mu-\frac{1}{2},\frac{1}{2}\frac{1}{2}}C^{j'\mu}_{l\mu-\frac{1}{2},\frac{1}{2}\frac{1}{2}} - C^{j\mu}_{l\mu+\frac{1}{2},\frac{1}{2}-\frac{1}{2}}C^{j'\mu}_{l\mu+\frac{1}{2},\frac{1}{2}-\frac{1}{2}}\right)\delta_{ll'}\delta_{\mu\mu'}. \tag{71}$$

Taking the coupling of major and minor components in (64) into account, this gives rise to the well known dipole selection rules

$$l - l' = \pm1$$
$$j - j' = 0, \pm1$$
$$\mu - \mu' = \begin{cases} \pm1 & \text{for} \quad \lambda = \pm1 \\ 0 & \text{for} \quad \lambda = z \\ \pm1 & \text{for} \quad \lambda = x, y \end{cases}$$

for the matrix elements $M^\lambda_{\kappa\mu,\kappa'\mu'}$. If the magnetic dipole and electric quadrupole interactions are taken into account transitions between the states with $l - l' = 0$ or ±2 and $j - j' = \pm2$ become also allowed.

Finally, after substitution of the one-centre expansion (57) for the LMTO wave function into the (62) for $M^\lambda_{\kappa\mu,\kappa'\mu'}$ one arrives at the expression for the matrix elements of the electron-photon interaction between two Bloch states with band indexes n and n' that can be used to calculate $\boldsymbol{\sigma}(\omega)$ according to (37), (38) or $\mu_{q\lambda}(\omega)$ according to (44)

$$\begin{aligned} M^\lambda_{nn'}(\boldsymbol{k}) &= \langle \Psi^{n\boldsymbol{k}}(\boldsymbol{r})|\boldsymbol{\alpha}a_\lambda|\Psi^{n'\boldsymbol{k}}(\boldsymbol{r})\rangle \\ &= \sum_t \sum_{\kappa\mu} \sum_{\kappa'\mu'} \Big[A^{n\boldsymbol{k}}_{t\kappa\mu}\langle\Phi_{\nu\kappa\mu}|\boldsymbol{\alpha}a_\lambda|\Phi_{\nu\kappa'\mu'}\rangle A^{n'\boldsymbol{k}}_{t\kappa'\mu'} + \ldots \\ &\quad + B^{n\boldsymbol{k}}_{t\kappa\mu}\langle\dot{\Phi}_{\nu\kappa\mu}|\boldsymbol{\alpha}a_\lambda|\dot{\Phi}_{\nu\kappa'\mu'}\rangle B^{n'\boldsymbol{k}}_{t\kappa'\mu'} \Big], \end{aligned} \tag{72}$$

where we omitted the mixed terms containing Φ_ν and $\dot{\Phi}_\nu$. The integrals $\langle\ldots\rangle$ in (72) are evaluated using the expression (64) with the radial integration being restricted to the atomic sphere t. In the case of optical spectra the sum over κ' runs over the values from $-l^t_{\max} - 1$ to $l^t_{\max}$ where $l^t_{\max}$ is the highest l included into the decomposition of the LMTO wave function inside the atomic sphere t. For x-ray spectra this sum is usually restricted to two values of κ' corresponding to the core states with $j = l - 1/2$ and $j = l + 1/2$, e.g., $p_{1/2}$ and $p_{3/2}$ in the case of $L_{2,3}$ absorption spectra, whereas the sum over t includes only the sites occupied by atoms of the same element. Because of the dipole selection rules the sums over κ and μ contain only the terms with $\kappa = \kappa' \pm 1$ and $\mu = \mu' \pm 1$.

It should be pointed out that the expressions for $\boldsymbol{\sigma}(\omega)$ and $\mu_{q\lambda}(\omega)$ contain products of the matrix elements $M^\lambda_{nn'}(\boldsymbol{k})$. As the wave function of a core level disappears outside the corresponding atomic sphere, the x-ray absorption coefficient is given by the sum of the contributions from different atoms. On the contrary, $\boldsymbol{\sigma}(\omega)$ cannot be easily decomposed into the sum of atomic contributions due to the appearance of interference terms caused by the delocalized character of the wave function of initial states.

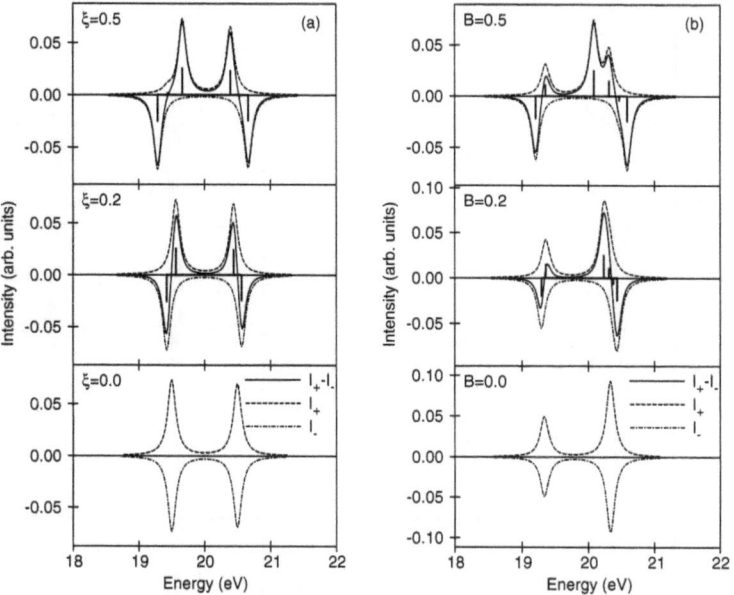

Fig. 4. Model K-edge XMCD spectra calculated for the cases of strong spin-polarization ($B = 1$) and increasing SO coupling (**a**) and strong SO coupling ($\xi = 1$) and increasing spin polarization (**b**) of the "valence" p electrons

7 "Toy" XMCD Spectra

To illustrate the effect of the symmetry breaking, caused by the simultaneous presence of the-spin orbit coupling and spin polarization and the matrix elements described in the previous section, we present in Fig. 4 model K-edge XMCD spectra. These calculations are performed assuming that the absorption is determined by transitions between occupied "core" s states and completely empty "valence" p states. The spin splitting of the s states is neglected. All radial matrix elements are assumed to be equal. The spectra are obtained by Gaussian broadening of individual contributions from the transitions between the discrete atomic-like states. The length of vertical bars in Fig. 4 is proportional to the difference of the squared angular matrix elements for left and right circularly polarized photons.

When either the spin-orbit coupling strength ξ or the effective magnetic field B is equal to zero, as in the lower panels of Fig. 4, p states with the quantum numbers $|\mu|$ and $-|\mu|$ are degenerate. The transitions with $\Delta\mu = \pm 1$ occur at the same energy and the absorption spectra for left and right circularly polarized x-rays are the same that leads to the absence of the dichroic signal. If both the SO coupling and magnetic field are switched on the degeneracy of the $\pm|\mu|$ states is lifted. Then, the transitions with $\Delta\mu = 1$ and $\Delta\mu = -1$ have different energies and no longer compensate each other

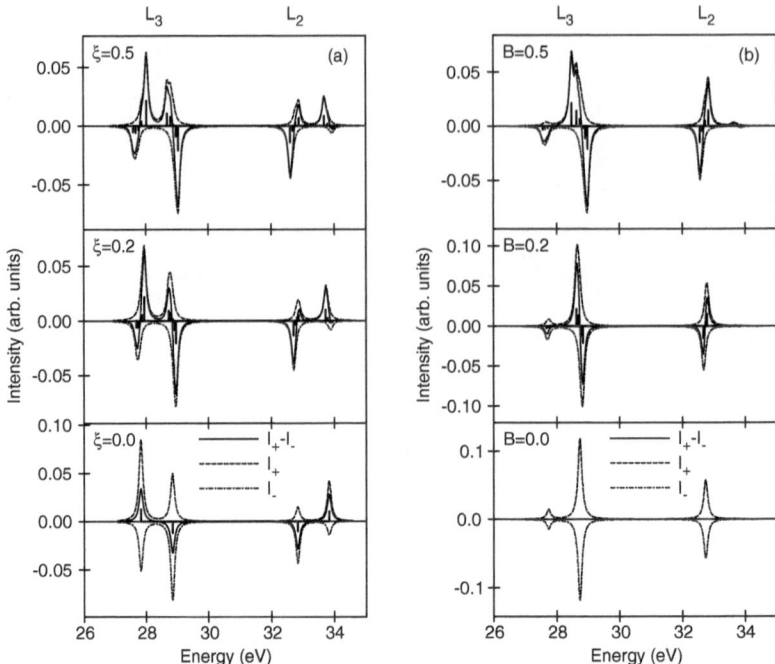

Fig. 5. Model $L_{2,3}$ XMCD spectra calculated for the cases of strong spin-polarization ($B = 1$) and increasing SO coupling (**a**) and strong SO coupling ($\xi = 1$) and increasing spin polarization (**b**) of the d electrons

which is reflected in the increasing magnitude of the XMCD spectra. If the p states were partially occupied the transition to the occupied states would be prohibited and the low frequency part of the spectra up to the energy corresponding to the transition to the first unoccupied p state would become zero.

The dependence of the model $L_{2,3}$ XMCD spectra on ξ and B is illustrated by Fig. 5. In this example "core" p states are split by strong SO interaction into $p_{1/2}$ and $p_{3/2}$ sub-shells whereas their exchange splitting is set to zero. As a result a non-zero XMCD signal appears even if the spin-orbit coupling of "valence" d states is neglected and only their spin-polarization is taken into account (lower panel in Fig. 5(a)). This situation is typical for magnetic $3d$ metal compounds.

When the SO coupling of the d states is strong they are also split into well separated $d_{3/2}$ and $d_{5/2}$ sub-shells (Fig. 5(b)). One can note that because of the difference of the corresponding angular matrix elements the L_3 spectrum is mainly formed by $p_{3/2} \rightarrow d_{5/2}$ transitions; the $p_{3/2} \rightarrow d_{3/2}$ contribution being much weaker. Similar to the case of the model K spectra switching on the magnetic field B lifts the degeneracy of $d_{\kappa,\mu}$ and $d_{\kappa,-\mu}$ states which, in turn, leads to the appearance of non-zero XMCD.

Numerous results of realistic band structure calculations of magneto-optical (see, e.g., [20, 21] and references therein) and XMCD spectra can be found in literature. For example, XMCD spectra were calculated for $3d$ transition metals in [60–63], their alloys in [64–76], for multilayers in [77–79], for hcp Gd in [80], for Uranium compounds in [81–87].

References

1. M. Faraday: Phil. Trans. R. Soc. **136**, 1 (1846)
2. J. Kerr: Philos. Mag. **3**, 321 (1877)
3. M. J. Freiser: IEEE Trans. Magn. **4**, 1 (1968)
4. H. R. Hulme: Proc. R. Soc. (London) Ser. A **135**, 237 (1932)
5. C. Kittel: Phys. Rev. **83**, A208 (1951)
6. P. N. Argyres: Phys. Rev. **97**, 334 (1955)
7. B. R. Cooper: Phys. Rev. **139**, A1504 (1965)
8. R. Kubo: J. Phys. Soc. Japan **12**, 570 (1957)
9. A. E. Kondorsky and A. V. Vediaev: J. Appl. Phys. **39**, 559 (1968)
10. C. S. Wang and J. Callaway: Phys. Rev. B **9**, 4897 (1974)
11. M. Singh, C. S. Wang, and J. Callaway: Phys. Rev. B **11**, 287 (1975)
12. J. L. Erskine and E. A. Stern: Phys. Rev. B **12**, 5016 (1975)
13. B. T. Thole, G. van der Laan, and G. A. Sawatzky: Phys. Rev. Lett. **55**, 2086 (1985)
14. G. van der Laan, B. T. Thole, G. A. Sawatzky, et al.: Phys. Rev. B **34**, 6529 (1986)
15. G. Schütz, W. Wagner, W. Wilhelm, et al.: Phys. Rev. Lett. **58**, 737 (1987)
16. G. Schütz, R. Frahm, P. Mautner, et al.: Phys. Rev. Lett. **62**, 2620 (1989)
17. H. Ebert, P. Strange, and B. L. Gyorffy: J. Appl. Phys. **63**, 3055 (1988)
18. C. T. Chen, F. Sette, Y. Ma, et al.: Phys. Rev. B **42**, 7262 (1990)
19. G. van der Laan and B. T. Thole: Phys. Rev. B **43**, 13401 (1991)
20. H. Ebert: Rep. Prog. Phys. **59**, 1665 (1996)
21. V. Antonov, B. Harmon, and A. Yaresko: *Electronic structure and magneto-optical properties of solids* (Kluwer Academic Publishers, Dordrecht, Boston, London, 2004)
22. W. H. Kleiner: Phys. Rev. **142**, 318 (1966)
23. J. D. Jackson: *Classical Electrodynamics* (J. Willey, New York, 1975)
24. N. Bloembergen: *Nonlinear Optics* (Benjamin, New York, 1965)
25. P. N. Butcher and D. Cotter: *The Elements of Nonlinear Optics* (Cambridge University Press, Cambridge, 1991)
26. P. Mills: *Nonlinear Optics* (Springer-Verlag, Berlin, 1991)
27. M. Dressel and G. Griner: *Electrodinamics of Solids* (Cambridge University Press, Cambridge, 2001)
28. A. Cracknell: J. Phys. C **2**, 1425 (1969)
29. A. K. Zvezdin and V. A. Kotov: *Modern Magnetooptics and Magnetooptical Materials* (Institute of Physics Publishing, Bristol and Philadelphia, 1997)
30. K. H. J. Buschow: In: *Ferromagnetic Materials*, vol 4, ed by E. P. Wohlfarth and K. H. J. Buschow (North-Holland, Amsterdam, 1988), p 588
31. W. Reim and J. Schoenes: In: *Ferromagnetic Materials*, vol 5, ed by E. P. Wohlfarth and K. H. J. Buschow (North-Holland, Amsterdam, 1990), p 133

140   A. Yaresko et al.

32. J. Schoenes: In: *Electronic and Magnetic Properties of Metals and Ceramics*, vol 3A of *Materials Science and Technology*, ed by R. W. Cahn, P. Haasen, and E. J. Kramer (Verlag Chemie, Weinheim, 1992), p 147, volume editor: K. H. J. Buschow
33. R. Pittini, J. Schoenes, O. Vogt, et al.: Phys. Rev. Lett. **77**, 944 (1996)
34. F. Salghetti-Drioli, K. Mattenberger, P. Wachter, et al.: Solid State Commun. **109**, 687 (1999)
35. A. N. Yaresko, P. M. Oppeneer, A. Y. Perlov, et al.: Europhys. Lett. **36**, 551 (1996)
36. J. Callaway: *Quantum Theory of the Solid State* (Academic, New York, 1974)
37. H. S. Bennet and E. A. Stern: Phys. Rev. **137**, A448 (1965)
38. A. Perlov, S. Chadov, and H. Ebert: Phys. Rev. B **68**, 245112 (2003)
39. B. T. Thole, P. Carra, F. Sette, et al.: Phys. Rev. Lett. **68**, 1943 (1992)
40. P. Carra, B. T. Thole, M. Altarelli, et al.: Phys. Rev. Lett. **70**, 694 (1993)
41. A. Ankudinov and J. J. Rehr: Phys. Rev. B **51**, 1282 (1995)
42. A. K. Rajagopal and J. Callaway: Phys. Rev. B **7**, 1912 (1973)
43. P. Hohenberg and W. Kohn: Phys. Rev. B **136**, 864 (1964)
44. W. Kohn and L. J. Sham: Phys. Rev. A **140**, 1133 (1965)
45. A. H. MacDonald and S. H. Vosko: J. Phys. C **12**, 2977 (1979)
46. M. Ramana and A. Rajagopal: Adv. Chem. Phys. **54**, 231 (1983)
47. D. A. Varshalovich, A. N. Moskalev, and V. K. Khersonskii: *Quantum Theory of Angular Momentum* (World Scientific, Singapure, 1989)
48. R. Feder, F. Rosicky, and B. Ackermann: Z. Phys. B **52**, 31 (1983)
49. P. Strange, J. Staunton, and B. L. Gyorffy: J. Phys. C **17**, 3355 (1984)
50. P. Cortona, S. Doniach, and C. Sommers: Phys. Rev. A **31**, 2842 (1985)
51. A. Jenkins and P. Strange: J. Phys.: Condens. Matter **6**, 3499 (1994)
52. L. D. Landau and E. M. Lifshitz: *Quantum Mechanics*, vol 3 of *Course of Theoretical Physics* (Pergamon, Oxford, 1977)
53. O. K. Andersen: Phys. Rev. B **12**, 3060 (1975)
54. V. V. Nemoshkalenko, A. E. Krasovskii, V. N. Antonov, et al.: Phys. status solidi (b) **120**, 283 (1983)
55. H. Ebert: Phys. Rev. B **38**, 9390 (1988)
56. A. E. Krasovskii: Phys. Rev. B **60**, 12788 (1999)
57. V. N. Antonov, A. Y. Perlov, A. P. Shpak, et al.: J. Magn. Magn. Materials **146**, 205 (1995)
58. M. E. Rose: *Relativistic Electron Theory* (Willey, Amsterdam, 1961)
59. E. Arola, P. Strange, and B. Gyorffy: Phys. Rev. B **55**, 472 (1997)
60. J. Igarashi and K. Hirai: Phys. Rev. B **50**, 17820 (1994)
61. R. Wu, D. Wang, and A. Freeman: J. Appl. Phys. **75**, 5802 (1994)
62. J. Igarashi and K. Hirai: Phys. Rev. B **53**, 6442 (1996)
63. H. Ebert: Solid State Commun. **100**, 677 (1996)
64. S. Ostanin, V. Popescu, H. Ebert, et al.: J. Phys.: Condens. Matter **11**, 9095 (1999)
65. I. Galanakis, S. Ostanin, M. Alouani, et al.: Phys. Rev. B **61**, 4093 (2000)
66. I. Galanakis, S. Ostanin, M. Alouani, et al.: Phys. Rev. B **61**, 599 (2000)
67. W. Grange, I. Galanakis, M. Alouani, et al.: Phys. Rev. B **62**, 1157 (2000)
68. S. Ostanin, V. Popescu, and H. Ebert: J. Phys.: Condens. Matter **13**, 3895 (2001)
69. V. N. Antonov, B. N. Harmon, and A. N. Yaresko: Phys. Rev. B **64**, 024402 (2001)

70. J. Kunes, P. Oppeneer, H. Mertins, et al.: Phys. Rev. B **64**, 174417 (2001)
71. Y. Zhao, W. Geng, K. Park, et al.: Phys. Rev. B **64**, 035207 (2001)
72. I. Galanakis, M. Alouani, and H. Dreysse: J. Magn. Magn. Materials **242**, 27 (2002)
73. V. N. Antonov, B. N. Harmon, and A. N. Yaresko: Phys. Rev. B **67**, 024417 (2003)
74. M. Takahashi and J. Igarashi: Phys. Rev. B **67**, 245104 (2003)
75. V. Antonov, B. Harmon, L. Bekenov, et al.: Phys. Rev. B **71**, 174428 (2005)
76. V. Antonov, H. Dürr, Y. Kucherenko, et al.: Phys. Rev. B **72**, 054441 (2005)
77. G. Guo, H. Ebert, W. Temmerman, et al.: Phys. Rev. B **50**, 3861 (1994)
78. C. Ederer, M. Komelj, M. Fahnle, et al.: Phys. Rev. B **66**, 094413 (2002)
79. J. Minar, A. Perlov, H. Ebert, et al.: J. Phys.: Condens. Matter **17**, 5785 (2005)
80. H. Ebert, G. Schutz, and W. Temmerman: Solid State Commun. **76**, 475 (1990)
81. J. Kunes, P. Novak, M. Divis, et al.: Phys. Rev. B **63**, 205111 (2001)
82. M. Kucera, J. Kunes, A. Kolomiets, et al.: Phys. Rev. B **66**, 144405 (2002)
83. V. N. Antonov, B. N. Harmon, and A. N. Yaresko: Phys. Rev. B **68**, 214424 (2003)
84. V. N. Antonov, B. N. Harmon, O. V. Andryushchenko, et al.: Phys. Rev. B **68**, 214425 (2003)
85. A. N. Yaresko, V. N. Antonov, and B. N. Harmon: Phys. Rev. B **68**, 214426 (2003)
86. V. Antonov, B. Harmon, O. Andryushchenko, et al.: Low Temp. Phys. **30**, 305 (2004)
87. A. Yaresko, P. de Reotier, A. Yaouanc, et al.: J. Phys.: Condens. Matter **17**, 2443 (2005)

Hitchhiker's Guide to Multiplet Calculations

Gerrit van der Laan

Daresbury Laboratory, Warrington WA4 4AD, UK
g.vanderlaan@dl.ac.uk

Abstract. Core level spectra from localized magnetic systems, such as $3d$ transition metal compounds or rare earths, obtained with high resolution synchrotron radiation can be simulated theoretically by calculating the transitions from the ground state to the allowed final states, e.g. $3d^n \rightarrow 2p^5 3d^{n+1}$. Presented here is the angular momentum coupling theory, Racah-Wigner algebra, point group theory and multipole moment expansion behind such calculations for the dichroic spectra of x-ray absorption, photoemission and x-ray resonant magnetic scattering. The first part of this chapter reviews the standard results of multiplet theory and surveys the Cowan-Butler-Thole approach. The last part presents some more advanced topics, such as the x-ray optical activity, which is induced by interference between different electric and magnetic multipole fields, and the method of statistical moment analysis, which relates the moment distribution of the dichroic spectra to effective operators.

1 Introduction

The recent developments in high resolution monochromatization of synchrotron radiation have provided the opportunity to obtain detailed core level spectra using a variety of techniques such as x-ray absorption, photoemission, Auger spectroscopy, x-ray resonant magnetic scattering, etc. In localized materials these spectra can be simulated theoretically by calculating all possible transitions from a given ground state to the allowed final states, e.g. the near edge structure of the $2p$ absorption spectrum of a $3d$ transition metal compound can be obtained by calculating the transition probabilities $3d^n \rightarrow 2p^5 3d^{n+1}$ [1, 2]. This chapter presents the angular momentum coupling theory, Racah-Wigner algebra, point group theory and multipole moment expansion that is behind such calculations. The approach has several advantages over the usual infinitesimal operator approach, the most important one is that the results are valid even if the JM basis is not used. Some relevant classic texts on angular momentum coupling are [3–34].

The first part (Sects. 2–17) of this chapter gives a review of "standard work" in the field of multiplet calculations including a detailed treatment of crystal field interaction. The second part presents the more advanced subjects of electric and magnetic multipole fields and x-ray optical activity (Sects. 18–20) and statistical moment analysis (Sects. 21–22).

G. van der Laan: *Hitchhiker's Guide to Multiplet Calculations*, Lect. Notes Phys. **697**, 143–199 (2006)
www.springerlink.com

The detailed outline is as follows: Section 2 introduces the interplay between the two main interactions. *Spin-orbit and electrostatic interactions* result in jj and LS coupling, respectively, and the combined presence leads to intermediate coupling. Section 3 shows that the *multiplet structure* depends on only a few parameters; those of the Coulomb and exchange interactions and the spin-orbit interaction. We discuss the structure of *Thole's multiplet program* for atomic multiplet calculations in the presence of crystal field, magnetic field and hybridization. References are given to systematic studies of x-ray absorption, magnetic x-ray dichroism and photoemission from rare earths, transition metals and actinides. The main observations in the case of the rare earths $M_{4,5}$ absorption edges are summarized. Section 4 presents the *Wigner-Eckart theorem*, which states that matrix elements of operators with the same rank differ only in the *reduced matrix elements*. Some expressions for useful *operators* are given. Section 5 presents the *selection rules for the electric-dipole transitions* in LS coupling from which the transition probabilities are obtained. Section 6 treats the *angular momentum coupling* using *Clebsch-Gordan coefficients* and $3j$ *symbols*. Momentum coupling is an essential ingredient to calculate many-electron systems. Expressions are given for the reduced matrix element in the case of *coupled and uncoupled basis functions*. Section 7 discusses the *coefficients of fractional parentage*, required to make the wave function antisymmetric in the case of equivalent electrons. This allows us to define the *many-electron unit tensor operators* $U^{(k)}$ and $V^{(k1)}$. Section 8 discusses the *Coulomb and exchange interactions*, where the Slater integrals F^k and G^k give a splitting in states of different energy. The *Pauli principle*, which requires that the total wavefunction must be antisymmetric, is responsible for lowering the energy of the high-spin state. As an example we show the atomic spectra for the Ni $d^8 \rightarrow p^5 d^9$ transition under continuous change from LS to jj coupling. Section 9 emphasizes the difference between the non-interacting particle approach and the *Hartree-Fock approach*. Some limitations of the Hartree-Fock approach are mentioned. Section 10 shows that itinerant metals, such as Fe, Co and Ni, should be calculated using an *Anderson impurity approach*, by taking both multiplet structure and *configuration interaction* into account. Section 11 introduces *second quantization* as an alternative to include the Pauli principle, where the use of creation and annihilation operators removes the need for coefficients of fractional parentage. Section 12 presents the *spin-orbit interaction* in LS coupling, the Landé interval rule and the transformation matrix between the LS and jj coupling scheme. Section 13 discusses the influence of an external magnetic field, or *Zeeman interaction* on the matrix elements. Section 14 treats the important subject of the implementing the *crystal field interaction* in the multiplet theory. Starting from the quantities calculated in spherical symmetry one can branch to subsequent lower *point group symmetries*. The *branching tables* for SO_3–D_2, SO_3–C_4 and SO_3–SO_2 are compared. The concepts of the $3jm$ symbol, $3j$ phase and $2jm$ phase are explained. It is shown how the Wigner-Eckart theorem and matrix elements are evaluated using crystal field functions and how to compare the coefficients with the conventional parameters $10Dq$, $10Ds$ and $10Dt$. Branchings for the electric-dipole transition operator and the magnetic interaction are given. Section 15 shows that for each type of spectroscopy the point group symmetry determines the *number of fundamental spectra*. Section 16 treats *second-order processes*, such as x-ray resonant magnetic scattering. Expressions for multipole transitions are derived using *vector spherical harmonics*. The *diffraction conditions* are discussed, where screw axes and glide planes can give constructive interference. Section 17 gives the *symmetry properties* of the spherical harmonics and *spherical tensors*. Continuous symmetries arising from homogeneity of space lead to conservation laws, e.g. angular momentum conservation is a consequence of rotational invariance. Like discrete symmetries associated with inversion of space and time they provide selection rules. *Rotation* symmetry, *parity*, *hermiticity* and *time reversal* symmetry are extensively discussed in order to prepare the reader for the next sections where these concepts are explored to examine the physical properties. The second, more advanced, part starts with Sect. 18 where the *electromagnetic interaction* is expressed in terms of *electric and magnetic multipole fields*. The parity and time reversal symmetry discussed in Sect. 19 are of particular interest for the study of materials using x-ray absorption and scattering. We discuss the electric-dipole ($E1$), magnetic-dipole ($M1$) and electric-quadrupole ($E2$) terms. By recoupling the expression for the $E1E2$ cross-section the different terms can be related to physical properties, such as the orbital anapole moment, which is time-odd and parity-odd. This is applied in Sect. 20 to *x-ray optical activity*, which is induced by *interference between multipoles* of different orders in the expansion of the photon wave vector.

Section 21 introduces the *statistical moment analysis*, which relates the moment distribution of the spectrum to *effective operators*, e.g. the zeroth-moment, which is the integrated intensity, corresponds to the sum rule result. Section 22 gives an application of statistical moment analysis for the $2p$ absorption spectra in $3d$ metals using the *independent particle model*. The line shape of the x-ray magnetic dichroism (XMCD) spectrum is primarily determined by the spin polarization in the valence band, with other ground state moments playing a less prominent role. The described model provides insight into the influence of the *ground state properties* on the spectrum. This greatly simplifies any calculational fit of the spectrum since it removes the need for a large number of calculations trying different parameter sets. Finally, Sect. 23 contains *conclusions and outlook*.

2 Spin-Orbit Interaction Versus Coulomb Interaction

In the Hamiltonian used for electronic structure calculations, there are two standard ways to couple the angular momenta of multi-electronic systems:

Russell-Saunders (LS) and jj coupling. In atoms where the spin-orbit coupling is weak compared to the Coulomb and exchange interaction, the orbital angular momenta l of individual electrons are coupled to a total orbital angular momentum L and likewise the spin angular momenta s are coupled to a total S. Then L and S are coupled to form the total angular momentum J. This approach simplifies the calculation of the Coulomb and exchange interactions, which commutate with L, S and J, and hence are diagonal in these quantum numbers. For heavier elements with larger nuclear charge, there is also an important contribution from the spin-orbit interaction. This interaction is diagonal in j and J, but not in L and S. Therefore, in the jj coupling scheme, the spin and orbital angular momenta, s and l, of each electron are coupled to form individual electron angular momenta j and then the different j are coupled to give the total angular momentum J. Both schemes are equally viable. In a many-electron system, the basis states obtained in LS and jj coupling can be changed into each other by using a recoupling transformation matrix (see 39 below). For purposes of numerical calculation of the energy levels and the spectral intensities by diagonalization of the Hamiltonian matrix it is immaterial what pure-coupling representation is used. The solution obtained by exact matrix diagonalization is known as intermediate coupling. Note that for a one-electron system the LS and jj coupling scheme give the same result. However, the one-electron approach commonly used to calculate the electronic structure of materials does not take full advantage of the recoupling techniques that are available for many-electron systems.

For convenience, the atomic ground state is often expressed in a pure-coupling representation. It is known that LS coupling holds quite well for transition metals (in the absence of crystal field) and for rare earth metals. These systems exhibit a Hund's rule ground state with maximum S and L, which are coupled by spin-orbit interaction antiparallel or parallel to each other for less or more than half filled shell, respectively, resulting in $J = |L - S|$ or $J = L + S$, respectively. However, for the $5f$ states of the actinides the spin-orbit interaction is much stronger, giving a significant mixing of the Hund's rule ground state by other LS states with the same J value. Hence, the LS states are less pure and there is a tendency towards the jj coupling limit [35–38]. The choice of the coupling limit has profound implications for the expectation value of the spin-orbit interaction, as well as for any other orbital-related interactions, such as the orbital magnetic moment.

3 Multiplet Structure

The multiplet theory is used to calculate core-level spectra of (more or less) localized materials, e.g. for x-ray absorption the transitions $4f^n \to 3d^9 4f^{n+1}$ [2] and $3d^n \to 2p^5 3d^{n+1}$ [1]. The number of levels for each configuration l^n is equal to the binomial $\binom{4l+2}{n}$, which can become quite large, e.g. for

$4f^6 \rightarrow 3d^9 4f^7$ there are 3,003 initial state and 48,048 final state levels, resulting in matrices of these dimensions that have to be diagonalized. Hence group theory, especially in the form of the powerful Wigner-Racah algebra is a prerequisite for such calculations. The basis states are antisymmetric states. Using Wigner-Racah algebra the matrix elements are calculated from reduced matrix elements (RME) in SO_3 symmetry. In this way only a few parameters are required, i.e. the Slater parameters and the spin-orbit coupling. The book of Cowan [22] provides a thorough and useful guide to this process.

As an example, the ab initio Hartree-Fock values of the parameters in the initial and final state configurations of Co^{2+} and Gd^{3+} are given Tables 1 and 2, respectively. The actual values for the Coulomb and exchange interaction used in the calculation are normally scaled to 80% of these values.

The following general remarks are of relevance.

(1) Only a few parameters ("generators") fix the many multiplet lines in spherical symmetry. The number of parameters increases when going to lower symmetry (e.g. crystal field or point group symmetry).

(2) The Coulomb and exchange interactions are usually much larger than the valence spin-orbit interaction. However, this does not mean that the spin-orbit interaction can be neglected in the ground state. It often splits the lower level, hence changing the symmetry of the ground state. Since at room temperature $kT = 0.026\,\text{eV}$, there might be some population of the higher levels, which we can take into account by a Boltzmann distribution.

(3) When the core spin-orbit splitting is much larger than the Coulomb and exchange interactions the spectrum splits nicely into to two separate manifolds, e.g. the $2p_{1/2}$ (L_2) and $2p_{3/2}$ (L_3).

(4) Hund's rule: empirical rule for LS coupling. Lowest energy term of a configuration with equivalent electrons is that term of maximum S which has the largest value of L. This rule does generally not apply for non-equivalent electrons (such as the excited state). The relative energies of the various LS terms can be qualitatively predicted in simple cases by the following semiclassical argument based on the vector model [22]. For each electron with orbital angular momentum l_1, the electronic charge cloud is concentrated near an orbital plane perpendicular to l_1. For either maximum (l_1 parallel to l_2) or minimum L (l_1 antiparallel to l_2), the two

Table 1. Hartree-Fock parameters (in eV) for Co $3d^7 \rightarrow 2p^5 3d^8$

	ζ_{3d}	(3d, 3d) F^2	F^4	(2p, 3d) ζ_{2p}	F^2	G^1	G^3
Co $3d^7$	0.06	11.6	7.2	-	-	-	-
Co $2p^5 3d^8$	0.08	12.4	7.7	9.7	7.3	5.4	3.1

Table 2. Hartree-Fock parameters (in eV) for Gd $4f^7 \to 3d^9 4f^8$

		$(4f, 4f)$					$(3d, 4f)$			
	ζ_{4f}	F^2	F^4	F^6	ζ_{3d}	F^2	F^4	G^1	G^3	G^5
Gd $4f^7$	0.20	11.6	7.3	5.2	–	–	–	–	–	–
Gd $3d^9 4f^8$	0.23	12.2	7.6	5.5	12.4	7.8	3.6	5.6	3.3	2.3

orbital planes more or less coincide; the two charge clouds thus overlap strongly and there is high repulsion energy. For intermediate L value, on the other hand, the orbital planes lie at an angle to each other, the charge clouds are more separated, and the energy is lower. Thus e.g. in a pd configuration we would expect the D levels to lie lower than the P and F levels.

3.1 Computer Code by Theo Thole

The code developed by Theo Thole is tailored for multiplet calculations, and serves here as an example [39]. The total suite of programs is comprised of several self-contained codes intended for consecutive use. First of all, Cowan's programs RCN31 and RCN2 calculate the initial and final-state wave functions in intermediate coupling using the atomic Hartree-Fock (HF) method with relativistic corrections [40]. After empirical scaling of the output parameters, such as the Slater integrals and spin-orbit parameters, Cowan's program RCN9 calculates the electric dipole and quadrupole transition matrix elements from the initial state to the final-state levels of the specified configurations [22].

Using the group-theoretical program RACER (or its predecessor RACAH) these calculations can be extended to point group symmetry. The chain-of-groups approach starts with the calculation of the reduced matrix elements of all necessary operators in the spherical group, as can be done using Cowan's atomic multiplet program. The Wigner-Eckart theorem is then applied to obtain the reduced matrix elements in any desired point group. Butler's point group program [41] calculates the necessary isoscalar factors, using modern group-theoretical results [42,43] to obtain a consistent set of coefficients. The advantage over standard methods [44] is that all point groups are treated in a uniform way. The program can calculate the transition probabilities between any two configurations in x-ray absorption, photoemission, inverse photoemission, etc.

The program BAND, developed by Thole in collaboration with Kotani and coworkers, extends the atomic calculations with hybridization. This enables the study of the interplay between atomic multiplet structure and solid-state hybridization using a cluster model, Anderson impurity model or Fermi liquid approach. Transitions are calculated between a mixed ground state d^n, $d^{n-1}k$, ..., where k denotes a combination of appropriate symmetry of orbitals on the adjacent sites, and final states $\underline{c}d^n$, $\underline{c}d^{n-1}k$, ... as a function

Table 3. Multiplet calculations of x-ray absorption, magnetic x-ray dichroism (MXD) and photoemission spectra for transition metal, rare earths and actinides over a full row of elements

	Rare earths	
$3d$ x-ray absorption	Thole et al. (1985)	Ref. [2]
$3d$ XMCD	Goedkoop et al. (1988)	Ref. [46]
$3d$, $4d$ XMCD	Imada, Jo (1990)	Ref. [47]
$4f$ polarized photoemission	van der Laan, Thole (1993)	Ref. [48]
$4d$ photoemission	Ogasawara, Kotani, Thole (1994)	Ref. [49]
	Transition metal compounds with crystal field	
$2p$ x-ray absorption	Yamaguchi et al. (1982)	Ref. [44]
$2p$ x-ray absorption	de Groot et al. (1990)	Ref. [50]
$3p$ x-ray absorption	van der Laan (1991)	Ref. [51]
$2p$ XMCD	van der Laan, Thole (1991)	Ref. [1]
$2p$ x-ray absorption	van der Laan, Kirkman (1992)	Ref. [52]
$3d$, $3p$, $3s$ resonant photoemission	Tanaka, Jo (1994)	Ref. [53]
	Actinides	
$4d$, $5d$ XMCD	Ogasawara, Kotani, Thole (1991)	Ref. [54]
$4d$ XMCD	van der Laan, Thole (1996)	Ref. [37]

of the charge-transfer energy, on-site Coulomb interaction, transfer integral and bandwidth. The routine AUGER can be used to calculate Auger and resonant photoemission decay by taking into account radiative transitions to first order and Auger transitions to infinite order [45].

For large matrices a considerable acceleration of several orders of magnitude can be achieved with the program AGSPARSE which is using the Lanczos method, originally introduced to detect the latent roots and the principal axes of a matrix by a process of minimized iterations with least squares. The algorithm is particularly convenient for matrices of very large dimensions, which are in sparse form. Instead of a spectrum containing thousands of lines, the program gives the best fit to this spectrum with a specified number of lines.

Calculational results of multiplet spectra over a full series of transition metal compounds, rare earths and actinides can be found in the reference material that is listed in Table 3.

3.2 Rare Earth X-Ray Absorption Spectra

A systematic study of the $M_{4,5}$ ($3d$) x-ray absorption spectra (XAS) in rare earth metals was presented in [2]. These spectra display strong peak structures due to dipole transitions $3d^{10}4f^n \rightarrow 3d^9 4f^{n+1}$. The large $3d$ spin-orbit interaction splits the final states into a $j = 5/2$ and $3/2$ manifold which have a large energy separation. Comparison of calculated results with high-resolution total-electron-yield measurements revealed the following [2]:

(1) The electrostatic and exchange parameters need to be scaled down to about 80% of their HF value to account for the Coulomb interaction with configurations omitted in the calculation. The $3d$ spin-orbit parameter requires only a scaling between 95 to 100%. This is in agreement with other results. [55]

(2) The intrinsic line width of the core state is smaller than was expected from [56] The Γ values of the M_5 width range from 0.2 eV for La to 0.3 eV for Yb and of the M_4 width range from 0.3 eV for La to 0.6 for Tm.

(3) The absorption length is so short that very strong saturation effects are occurring. The theoretical line shape matches the experimental result when the saturation in the total-electron-yield signal is included using the calculated value of the absorption length. The short length scale, which ranges from 29 Å in La to 126 Å in Tm, has been confirmed by angular dependent photoelectric yield measurements [57].

(4) There are systematic trends in the spectral shape over the lanthanide series. The $3d_{5/2}$ to $3d_{3/2}$ peak intensity ratio is non-statistical. The high-energy peak decreases as the f shell fills up with electrons. This was already observed by Demekhin [58] and the origin is explained in [36].

(5) The electric-dipole selection rules from the Hund's rule ground state strongly limit the number of accessible final states. Compared to the total manifold of final states, the XAS lines are located in a narrow energy region. In light elements, such La, Ce and Pr, the XAS transitions are located at the high-energy side, and in heavy elements, such as Er and Tm, the XAS transitions are located at the low-energy side. This is due to the spin selection rule, as is explained in [36].

(6) Whereas the XAS is composed of hundreds of lines, e.g. the Gd $3d$ contains 1077 lines, there is a bunching of the spectral intensity, often resulting in only a few peaks. These are scars of the selection rules.

(7) Different valencies of a particular rare earth element have usually completely different peak profiles and show a displacement in energy (chemical shift). This allows to separate different valence state as has been applied e.g. to determine the relative concentrations of rare earth ions in terbium mixed oxides [59].

4 Reduced Matrix Elements

An irreducible spherical tensor operator $T_q^{(k)}$ transforms under rotation as

$$\mathcal{D}(R)T_q^{(k)}\mathcal{D}^\dagger(R) = \sum_{q'=-k}^{k} T_{q'}^{(k)} D_{q'q}^{(k)}(R) , \tag{1}$$

where $\mathcal{D}(R)$ is the rotation operator and $D^{(j)}_{mm'}(R) = \langle jm'|D(R)|jm\rangle$. This means that under the coordinate rotation the tensor $T^{(k)}_q$ transforms into a linear combination of the $(2k+1)$-fold set $\{T^{(k)}_{q'}\}$ with expansion coefficients $D^{(k)}_{qq'}(R)$. The Wigner-Eckart theorem gives

$$\langle \alpha JM|T^{(k)}_q|\alpha' J'M'\rangle = (-)^{J-M} \begin{pmatrix} J & k & J' \\ -M & q & M' \end{pmatrix} \langle \alpha J\|T^{(k)}\|\alpha' J'\rangle \,, \qquad (2)$$

so that a matrix element can be written as a product of a $3j$ symbol (or, alternatively, a Clebsch-Gordan coefficient) and a reduced matrix element (RME) $\langle \alpha J\|T^{(k)}\|\alpha' J'\rangle$. The first factor, that does not depend on the nature of the tensor operator, depends only on the geometry, i.e. the polarization and orientation of the system. The properties of the $3j$ symbol give that $(J\,k\,J')$ must satisfy the triangle condition $|J'-k| \le J \le J'+k$ and that $M = M'+q$. Thus matrix elements of tensors with the same rank (k) differ only by their RME, which is a scalar.

The RME can be obtained from traditional methods. Since e.g. it is known that the one-electron matrix element $\langle ll|l_z|ll\rangle \equiv \langle ll|l^{(1)}_0|ll\rangle = l$, we can obtain the RME by substitution in (2), so that

$$\langle l\|l^{(1)}\|l'\rangle = \delta_{ll'}[l(l+1)(2l+1)]^{1/2} \,, \qquad (3)$$
$$\langle s\|s^{(1)}\|s'\rangle = \delta_{ss'}[s(s+1)(2s+1)]^{1/2} = (3/2)^{(1/2)} \,, \qquad (4)$$
$$\langle l\|C^{(k)}\|l'\rangle = (\ \ )^l \, [l,l']^{1/2} \begin{pmatrix} l & k & l' \\ 0 & 0 & 0 \end{pmatrix} - (-)^k \, \langle l'\|C^{(k)}\|l\rangle \,, \qquad (5)$$

where $[l,l']$ is short for $(2l+1)(2l'+1)$. The $3j$ symbol is zero unless $l' + k + l$ is even. However, note that Brink and Satchler [12] use $\langle l\|l^{(1)}\|l'\rangle = \delta_{ll'}[l(l+1)]^{1/2}$.

We also present the example of the matrix element of the orbital moment operator [40]. This is a single-variable operator acting on a coupled state.

$$\langle l^n LSJM|L^{(1)}_q|L'S'J'M'\rangle = (-)^{J-M} \begin{pmatrix} J & 1 & J' \\ -M & q & M' \end{pmatrix} \langle l^n LSJ\|L^{(1)}\|l^n L'S'J'\rangle \,,$$
$$\langle l^n LSJ\|L^{(1)}\|l^n L'S'J'\rangle = (-)^{L+S+J'}[J,J']^{1/2}\langle l^n LS\|L^{(1)}\|l^n L'S'\rangle \,,$$
$$\langle l^n LS\|L^{(1)}\|l^n L'S'\rangle = \delta_{LL'}[L(L+1)(2L+1)]^{1/2} \,. \qquad (6)$$

5 Selection Rules for Electric-Dipole Transitions

The restriction on the magnetic quantum numbers imposes certain conditions (selection rules) on the matrix elements of operators. This is of great utility in discussing the probability of transitions between states. The condition for a non-vanishing angular part is imposed by the $3j$ symbol in the Wigner-Eckart theorem. For electric-dipole transitions, the vector operator $P_q^{(1)}$ gives the selection rules $J - J' = 0, \pm 1$ (with the restriction that $J' = J = 0$ is not allowed) and $q = M - M' = 0, \pm 1$.

Because the electric-dipole operator $P^{(1)}$ does not involve spin coordinates, the matrix element calculation is simplest in LS-coupled basis. From the equations for tensor products it follows that

$$D_{LS} = \langle LSJ \| P_q^{(1)} \| L'S'J' \rangle = \delta_{SS'} D_{\text{line}} \langle LS \| P^{(1)} \| L'S' \rangle , \qquad (7)$$

where the line factor

$$D_{\text{line}} = (-)^{L+S+J'+1} [J, J']^{1/2} \left\{ \begin{matrix} L & S & J \\ J' & 1 & L' \end{matrix} \right\} , \qquad (8)$$

contains the entire dependence of the matrix element on J and J'. Taking note of the delta factor for S and of the triangle conditions that must be satisfied for the $6j$ symbol to be non-zero, we find, in addition to the selection rule on J, the following selection rules: $\Delta S = S - S' = 0$, $\Delta L = L - L' = 0, \pm 1$, with $L \neq 0$ if $L = L'$. Therefore, the relative strengths of the various lines of a multiplet are

$$S_{LS} = (D_{\text{line}})^2 [J, J'] \left\{ \begin{matrix} L & S & J \\ J' & 1 & L' \end{matrix} \right\}^2 . \qquad (9)$$

The above equations are straightforward to apply for one-electron configurations.

6 Angular Momentum Coupling

For a many-electron system the situation becomes more complicated. The electric-dipole operator acts on a single electron so we have to remove one electron from the many-electron system, while we will have to know the quantum numbers of both the extracted electron and the remaining system. The coupling coefficients give the construction of many-particle states of given transformation properties from products of fewer states.

Two angular momenta j_1 and j_2 may be coupled together to give eigenfunctions of the angular momentum $\mathbf{j} = \mathbf{j}_1 + \mathbf{j}_2$ by using Clebsch-Gordan (CG) coefficients

$$|j_1 j_2 j m\rangle = \sum_{m_1 m_2} C_{j_1 m_1 j_2 m_2}^{jm} |j_1 m_1\rangle |j_2 m_2\rangle$$

$$= (-)^{j_1 - j_2 + m} [j]^{1/2} \sum_{m_1 m_2} \begin{pmatrix} j_1 & j_2 & j \\ m_1 & m_2 & -m \end{pmatrix} |j_1 m_1\rangle |j_2 m_2\rangle , \qquad (10)$$

where $m_1 + m_2 + m = 0$ and the triangular condition $\Delta(j_1 j_2 j)$ applies. The CG coefficients are the vector-coupling coefficients. They can also be written in $3j$ symbols, having the benefit of elegant symmetry properties.

From the symmetry properties of the Clebsch-Gordan coefficients it follows that $|j_1 j_2 j m\rangle = (-)^{j_1 + j_2 - j}|j_2 j_1 j m\rangle$, hence the wave function changes sign for $j_1 + j_2 - j$ is odd. For two equivalent electrons ($j_1 = j_2$) the requirement is that j is even (odd) for integer (half-integer) j_1.

The Condon-Shortley phase convention, where $J_{\pm}|jm\rangle = \sqrt{(j \mp m)(j \pm m + 1)}|j\,m\pm 1\rangle$, gives that the C-coefficients are real and fulfil the unitary properties

$$\sum_{m_1} C^{jm}_{j_1 m_1 j_2 m_2} C^{j'm}_{j_1 m_1 j_2 m_2} = \delta_{jj'} , \tag{11}$$

$$\sum_{j} C^{jm}_{j_1 m_1 j_2 m_2} C^{jm'}_{j_1 m'_1 j_2 m'_2} = \delta_{m_1 m'_1} \delta_{mm'} , \tag{12}$$

from which it follows that

$$|j_1 m_1\rangle|j_2 m_2\rangle = \sum_{j} C^{jm}_{j_1 m_1 j_2 m_2}|jm\rangle . \tag{13}$$

Note that the indices 1 and 2 refer to different spaces. In case of the same space we have the coupling rule

$$Y_{l_1 m_1} Y_{l_2 m_2} = \sum_{l} \sqrt{\frac{[l_1, l_2]}{4\pi[l]}} C^{l0}_{l_1 m_1 l_2 m_2} C^{lm}_{l_1 m_1 l_2 m_2} Y_{lm} , \tag{14}$$

for spherical harmonics, where the symmetry properties of $C^{l0}_{l_1 m_1 l_2 m_2}$ lead to the parity rule $l_1 + l_2 + l$ is even.

6.1 Coupled Tensors

The RME of a tensor product of two tensor operators can be evaluated for coupled basis functions as

$$\langle \alpha j \| [T^{(k_1)} \times W^{(k_2)}]^{(K)} \| \alpha' j' \rangle = (-)^{j+K+j'} [K]^{1/2} \sum_{\alpha'' j''} \begin{Bmatrix} k_1 & k_2 & K \\ j' & j & j'' \end{Bmatrix}$$

$$\times \langle \alpha j \| T^{(k_1)} \| \alpha'' j'' \rangle \langle \alpha'' j'' \| W^{(k_2)} \| \alpha' j' \rangle , \tag{15}$$

and for uncoupled basis functions as

$$\langle \alpha_1 j_1 \alpha_2 j_2 j \| [T^{(k_1)} \times W^{(k_2)}]^{(K)} \| \alpha'_1 j'_1 \alpha'_2 j'_2 j' \rangle$$

$$= [j, j', K]^{1/2} \langle \alpha_1 j_1 \| T^{(k_1)} \| \alpha'_1 j'_1 \rangle \langle \alpha_2 j_2 \| T^{(k_2)} \| \alpha'_2 j'_2 \rangle \begin{Bmatrix} j_1 & j_2 & j \\ j'_1 & j'_2 & j' \\ k_1 & k_2 & K \end{Bmatrix} . \tag{16}$$

7 Coefficients of Fractional Parentage

When two or more electrons are equivalent, normalization difficulties are encountered if Clebsch-Gordan expansions are applied to antisymmetrized product functions. These difficulties are avoided by using appropriate linear combinations of coupled simple-product (non-antisymmetrized) functions; the linear expansion is independent of J, M, M_L and M_S. If $|l^{n-1}\overline{\alpha LS}\rangle$ is an antisymmetric function for $n-1$ equivalent electrons in a subshell l^{n-1}, and if $|(l^{n-1}\overline{\alpha LS}, l)LS\rangle$ is a coupled but not antisymmetrized function, then a completely antisymmetric function for l^n can be written as [4, 19, 22, 24]

$$|l^n \alpha LS\rangle = \sum_{\overline{\alpha LS}} |(l^{n-1}\overline{\alpha LS}, l)LS\rangle \, (l^{n-1}\overline{\alpha LS}\} l^n \alpha LS) \,. \qquad (17)$$

Therefore, the wave function describing the state αLS of a group l^n of equivalent electrons is the linear combination of the functions $\psi_{\alpha SL}(l^{n-1}\overline{\alpha LS}, l)$ corresponding to the different initial terms $\overline{\alpha LS}$ of configuration l^{n-1}. Only the well-defined linear combinations in (17) will comply with the Pauli principle. The $(l^{n-1}\overline{\alpha LS}\} l^n \alpha LS)$ are called the coefficients of fractional parentage (cfp). They describe how the state ψ is built up from its possible parents. The cfp's are real and chosen such that the Pauli principle is fulfilled and the wave functions are normalised

$$\sum_{\overline{\alpha LS}} (l^n \alpha LS\{l^{n-1}\overline{\alpha LS})(l^{n-1}\overline{\alpha LS}\} l^n \alpha' LS) = \delta_{\alpha \alpha'} \,. \qquad (18)$$

Even if one-electron RMEs were readily to hand, the process of calculating corresponding many-electron RMEs involves the summation over products of cfp. Therefore, it is convenient to have available tables of the many-electron RMEs. This task is significantly simplified by the fact that the RMEs are just scaling factors in the Wigner-Eckart theorem. We can always choose to factorize the RME scale factors into two parts. We define a unit orbital angular momentum operator $u^{(k)}$ such that $\langle l \| u^{(k)} \| l' \rangle = \delta(lkl')$, i.e. the triangle relation factor, thus $\langle l \| u^{(k)} \| l' \rangle = 1$ for $|l - l'| \le k \le l + l'$, which gives

$$\langle lm | f_q^{(k)} | l'm' \rangle = (-)^{l-m} \begin{pmatrix} l & k & l' \\ -m & q & m' \end{pmatrix} \langle l \| u^{(k)} \| l' \rangle \langle l \| f^{(k)} \| l' \rangle \,. \qquad (19)$$

The corresponding many-electron unit tensor operator, $U^{(k)} = \sum_n u^{(k)}(n)$, for a subshell l^n has reduced matrix element

$$\langle l^n \alpha LS \| U^{(k)} \| l^n \alpha' L'S' \rangle = \delta_{SS'} n \, (-)^{l+L+k} [L, L']^{1/2}$$

$$\times \sum_{\overline{\alpha LS}} (-)^{\overline{L}} (l^n \alpha LS\{l^{n-1}\overline{\alpha LS}) \begin{Bmatrix} l & k & l \\ L & \overline{L} & L' \end{Bmatrix} (l^{n-1}\overline{\alpha LS}\} l^n \alpha' L'S') \,. \qquad (20)$$

The utility of the unit tensor derives from the relationship,

$$\langle\psi\|F^{(k)}\|\psi'\rangle = \langle\alpha j\|f^{(k)}\|\alpha' j'\rangle\langle\psi\|U^{(k)}\|\psi'\rangle \ . \tag{21}$$

Similarly, we have a double (orbital + spin) unit operator $v^{(k1)}$ such that $\langle ls\|v^{(k1)}\|l's'\rangle \equiv \langle l\|u^{(k)}\|l'\rangle\,\langle s\|s^{(1)}\|s'\rangle = (3/2)^{1/2}$.

$$\langle l^n\alpha LS\|V^{(k1)}\|l^n\alpha'L'S'\rangle = (3/2)^{1/2}n\,(-)^{l+L+k}[L,L',S,S']^{1/2}$$

$$\times \sum_{\overline{\alpha LS}}(-)^{\overline{L}+\overline{S}+S+3/2}(l^n\alpha LS\{|l^{n-1}\overline{\alpha LS}) \left\{ \begin{matrix} l & k & l \\ L & \overline{L} & L' \end{matrix} \right\} \left\{ \begin{matrix} s & 1 & s \\ S & \overline{S} & S' \end{matrix} \right\}$$

$$\times (l^{n-1}\overline{\alpha LS}|\}l^n\alpha'L'S') \ . \tag{22}$$

Methods to calculate the cfp were first given by Racah [4](d), and are discussed in the book of Judd [14]. Tables for $U^{(k)}$ and $V^{(k1)}$ for all states of the p^n, d^n and f^n configurations can be found in the book by Nielson and Koster [60].

8 Coulomb and Exchange Interaction

For N electrons in an atom with nuclear charge Ze the non-relativistic Hamiltonian is

$$\mathcal{H} = -\frac{\hbar^2}{2m}\sum_{i-1}^{N}\nabla_i^2 - \sum_{i-1}^{N}\frac{Ze^2}{r_i} + \sum_{i<j}^{N}\frac{e^2}{r_{ij}} \ . \tag{23}$$

The first term describes the kinetic energy of all electrons, the second one gives the potential energy of all electrons in the potential of the nucleus and the third term describes the repulsive Coulomb potential of the electron-electron interaction.

Since the Schrödinger equation for the Hamiltonian with $N > 1$ is not exactly solvable, one makes the approximation that each electron moves independently in a central field build up from the nuclear potential and the average potential of all the other electrons. The electron-electron interaction is taken as a perturbation potential. The matrix elements of this potential

$$\langle\alpha SLJM_J|\sum_{i<j}^{N}\frac{e^2}{r_{ij}}|\alpha'S'L'J'M'_J\rangle = E_{\rm C} + E_{\rm X} \ , \tag{24}$$

are independent of the quantum numbers J and M_J and diagonal in L and S, but not diagonal in α. Here, $E_{\rm C}$ and $E_{\rm X}$ are the Coulomb and exchange energy, respectively. The Coulomb potential can be expanded in Legendre polynomials P_k which can be written using spherical harmonics Y_{kq} as

$$\frac{e^2}{r_{ij}} = e^2 \sum_{k=0}^{\infty} \frac{r_<^k}{r_>^{k+1}} P_k \left(\cos \omega_{ij} \right) = e^2 \frac{r_<^k}{r_>^{k+1}} \left(C_i^{(k)*}(\theta_1, \phi_1) \cdot C_j^{(k)}(\theta_2, \phi_2) \right) ,$$

$$(25)$$

where

$$C_q^{(k)}(\theta, \phi) \equiv \sqrt{\frac{4\pi}{2k+1}} \, Y_{kq}(\theta, \phi) , \qquad (26)$$

$r_<$ and $r_>$ are the lesser and greater of the distance of the electrons i and j to the nucleus and ω gives the angle between these vectors.

8.1 Slater Integrals

For a system with equivalent electrons (those in the same shell, i.e. with same n and l) one obtains as result for the Coulomb energy

$$E_C = \sum_k f_k F^k . \qquad (27)$$

The F^k are known as the Coulomb integrals or Slater radial integrals. The $f_k^{(ij)}$ is the matrix element of

$$\langle l_i \| C^{(k)} \| l_i \rangle \langle l_j \| C^{(k)} \| l_j \rangle \, U_{(i)}^{(k)} \cdot U_{(j)}^{(k)} , \qquad (28)$$

which builds the angular dependent part of the matrix elements. The sum in (28) is constrained by the 3j symbol in $\langle l \| C^{(k)} \| l \rangle$ giving $0 \le k \le 2l$ and k even. Thus the Coulomb part contains only a few k values, e.g. $k = 0, 2, 4$ for d electrons and $k = 0, 2, 4, 6$ for f electrons.

For non-equivalent electrons we have, besides the Coulomb interaction, also an exchange interaction, where two electrons are interchanged in the Slater determinant. The energy is

$$E_X = \sum_k g_k G^k , \qquad (29)$$

with angular part g_k and radial part G^k and $k = |l - l'|, |l - l'| + 2, \dots, l + l'$.

As an example, the pd configuration gives $L = |l_d - l_p|, \dots, l_d + l_p = 1, 2, 3$, indicated as P, D and F states, respectively, and $2S + 1 = 1, 3$ for singlet and triplet. The energies for the terms ^{2S+1}L are expressed in Slater integrals as

$$E(^1D) = F^0 - \tfrac{7}{35}F^2 - \tfrac{3}{15}G^1 + \tfrac{21}{245}G^3 = -1.94 \text{ eV},$$

$$E(^3F) = F^0 + \tfrac{2}{35}F^2 - \tfrac{6}{15}G^1 - \tfrac{3}{245}G^3 = -1.53 \text{ eV},$$

$$E(^3D) = F^0 - \tfrac{7}{35}F^2 + \tfrac{3}{15}G^1 - \tfrac{21}{245}G^3 = -0.54 \text{ eV},$$

$$E(^3P) = F^0 + \tfrac{7}{35}F^2 - \tfrac{1}{15}G^1 - \tfrac{63}{245}G^3 = 0.25 \text{ eV},$$

$$E(^1P) = F^0 + \tfrac{7}{35}F^2 + \tfrac{1}{15}G^1 + \tfrac{63}{245}G^3 = 2.22 \text{ eV},$$

$$E(^1F) = F^0 + \tfrac{2}{35}F^2 + \tfrac{6}{15}G^1 + \tfrac{3}{245}G^3 = 2.24 \text{ eV}.$$

The values on the right-hand side are obtained using the Hartree-Fock values, $F^0 = 0$, $F^2 = 7.72$, $G^1 = 5.79$ and $G^3 = 3.29$ eV, scaled to 80% for Ni $2p^5 3d^9$.

The total wavefunction should be antisymmetric with respect to an interchange of all the coordinates (space and spin) of the two electrons. This means that for two equivalent electrons in the same shell only terms with even $L + S$ are allowed, e.g. for the $3d^2$, we obtain

$$E(^3F) = F^0 - \tfrac{8}{49}F^2 - \tfrac{1}{49}F^4 \; = -1.47 \text{ eV},$$

$$E(^1D) = F^0 - \tfrac{3}{49}F^2 + \tfrac{4}{49}F^4 \; = -0.10 \text{ eV},$$

$$E(^3P) = F^0 + \tfrac{3}{21}F^2 - \tfrac{4}{21}F^4 \; = 0.24 \text{ eV},$$

$$E(^1G) = F^0 + \tfrac{4}{49}F^2 + \tfrac{1}{441}F^4 = 0.81 \text{ eV},$$

$$E(^1S) = F^0 + \tfrac{2}{7}F^2 + \tfrac{2}{7}F^4 \quad = 4.53 \text{ eV}.$$

The values on the right-hand side are obtained using the Hartree-Fock values, $F^0 = 0$, $F^2 = 12.23$ and $F^4 = 7.60$ eV, scaled to 80% for Ni $3d^8$.

One might well wonder why singlet and triplet states of two coupled electrons have different energies, while the operator e^2/r_{12} is spin independent. The answer comes from the spatial distributions of electrons in the singlet and triplet states. For the singlet, the spacial part is a symmetric function, while for the triplet state it is antisymmetric function. Hence if $r_1 \rightarrow r_2$ the probability for parallel spin is zero, which is a manifestation of the Pauli principle or the antisymmetric requirement. Since electrons with parallel spin avoid each other while electrons with opposite spin do not, the Coulomb repulsion will be different in both cases: higher for the singlet state and lower for the triplet state. Thus the difference in energy is electrostatic in origin and not magnetic.

8.2 Atomic Ni $d^8 \rightarrow p^5 d^9$ Absorption Spectra

In Fig. 1 we illustrate the continuous change from LS to jj coupling, where the Ni $d^8 \rightarrow p^5 d^9$ transition probability from the initial states ^3F, ^1D and ^1G is given as function of core-valence electrostatic interaction over core spin-orbit interaction, i.e. $U(p,d)/\zeta_p$. [36] The spectra at the top have $U(p,d)/\zeta_p = 0$ (no electrostatic interactions). This corresponds to jj coupling and without spin-orbit in the ground state the transitions to the $2p_{3/2}$ and $2p_{1/2}$ give an intensity ratio 2:1. The bottom spectra obey LS coupling ($\zeta_p = 0$) and the final states are, in order of increasing energy, ^1D, ^3F, ^3D, ^3P, ^1F and ^1P (c.f. Sect. 8). Although the term ^1D in the $p^5 d^9$ configuration violates Hund's rule, the average energy of the triplets is still lower than that of the singlets. In LS coupling (selection rules: $\Delta S = 0$, and $\Delta L = 0, \pm 1$) we have $^3F \rightarrow \{^3F, ^3D\}$ and $^1D \rightarrow \{^1P, ^1D, ^1F\}$ and $^1G \rightarrow ^1F$. When ζ_p is increased levels might split and forbidden levels might mix in. The continuous change is shown in Fig. 1,

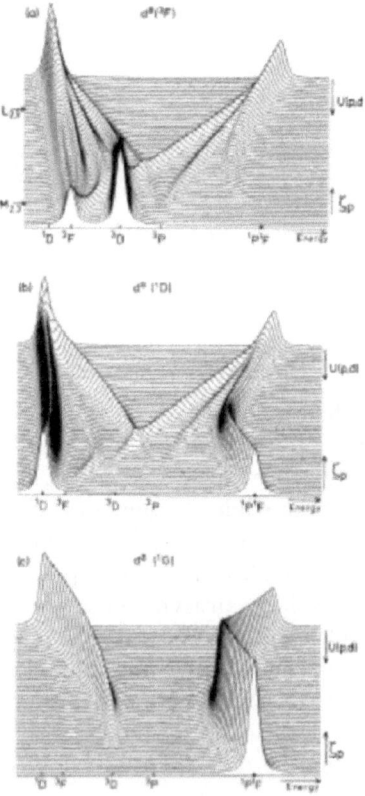

Fig. 1. Calculated transition probability of the transition $d^8 \rightarrow p^5 d^9$ for the initial states ^3F, ^1D and ^1G as a function of $U(p,d)/\zeta_p$ [36]

where in all cases we observe strong changes going from LS to jj coupling. The approximate $L_{2,3}$ and $M_{2,3}$ spectra (neglecting Fano broadening) are indicated in Fig. 1(a). It is interesting to observe the effect on the branching ratio $B = L_3/(L_2 + L_3)$. Compared to the statistical ratio of $2/3$, the value of B is increased for high-spin (triplet) and reduced for low-spin (singlet) ground states. This is because high-spin and low-spin states are at low and high energy, respectively. The relation of B with the spin state is given by a general rule, but it is noted that there is also a further rule stating that B is proportional to the ground state spin-orbit interaction. For simplicity of this particular example, the spin-orbit interaction was set to zero but it presence normally gives a pronounced effect on B [36]. Another interesting feature from Fig. 1 is that it offers an interpretation of the fine structure in the core-hole spectra. It is clear that the shoulders in the spectra near the jj limit are remnants of the peaks in the LS limit. The character of e.g. high-spin LS terms is not only distributed preferably to the $p_{3/2}$ manifold, but within each manifold it goes preferably to the low-energy side.

9 Hartree-Fock Calculations

There are two basic independent-particle approaches that may be classified as "non-interacting" and "Hartree-Fock". We follow the discussion in Weiss-bluth [18]. The two approaches are similar in that each assumes the electrons are uncorrelated except that they must obey the exclusion principle. However, they are different in that Hartree-Fock includes electron-electron Coulomb interaction in the energy, while neglecting the correlation that is introduced in the true wavefunction due to those interactions. In general, "non-interacting" theories have some effective potential that incorporates some effect of the real interaction, but there is no interaction term explicitly included in the effective Hamiltonian. This approach is often referred to as "Hartree" or "Hartree-like" after D.R. Hartree who included an average Coulomb interaction in a rather heuristic way. More to the point of modern calculations, all calculations following the Kohn-Sham method involved a non-interacting Hamiltonian with an effective potential chosen to incorporate exchange and correlation effects approximately.

It is observed that the electrostatic potentials $U(r_i, r_j)$ depend on the magnitudes of r_i and r_j and the angle θ between them, but not on the separate orientations of r_i and r_j, hence the Coulomb and exchange are spherical symmetric. This requires that all magnetic sublevels are equally populated.

The central-field problem is solved using the Hartree-Fock Self-Consistent Field method, which involves iterative procedures on which certain conditions of self-consistency have been imposed. If the spin-orbital are of the central field variety, the Hartree-Fock equations can be expressed in radial functions only, thereby simplifying the problem to one-dimension. This procedure implies that all electrons in the same shell (same nl values) have the same radial function. This is known as the "restricted Hartree-Fock" method. In the un-restricted version, orbitals with different spins are permitted to have different radial functions.

The Coulomb repulsion between two electrons prevents them from ap-proaching too close. This means that the positions of all electrons are corre-lated, *i.e.* all electrons are surrounded by a Coulomb hole. The Hartree-Fock wave function is not exact and the correlation not properly accounted for. However, the error for electrons with parallel spin is much smaller as for op-posite spin. This difference is associated with the antisymmetric properties for the wave function, which correlates the positions of electrons with paral-lel spin. This feature due to the exclusion principle gives raise to the Fermi hole. The difference between the Hartree-Fock energy and the exact energy is known as the correlation energy. The HF energy is about 1% too high.

The ordinary HF approximation uses a single configuration, given by a sin-gle Slater determinant. However to expand an n-electron antisymmetric wave function one needs a complete set of Slater determinants. This method for obtaining improved eigenfunctions and eigenvalues for the n-electron prob-lem is know as "configuration interaction". To reduce the numerical work one often includes only configurations whose energy are close to one another and therefore interact more strongly than the configurations with widely sepa-rated energies.

10 Configuration Interaction

Atomic multiplet calculations give good results for localized systems, such as rare earths metals, even in the case of $4f$ photoemission. Figure 2 shows the calculated Tb $4f^8(^7F_6) \rightarrow 4f^8\varepsilon l$ photoemission spectrum compared with the experimental result [61]. We just mention here the relation between the peak intensity in the MCD and the orbital quantum number in the final state. [62] High L' peaks show positive and low L' peaks show negative dichroism. Furthermore, the intensity of the integrated XMCD is proportional to $\langle L_z \rangle$ as given by a sum rule [62].

While for localized materials, such as rare earths, there is a good agree-ment between many-particle model and experiment (cf. Fig. 2), one would expect itinerant metals, like nickel and iron, to agree with a one-particle model. Perhaps surprizingly, this is not the case for the $2p$ photoemission of $3d$ transition metals. The experimental and calculated MCD in photoemis-sion spectra for Ni metal are compared in Fig. 3. The measured $2p$ photoemis-sion of Ni metal displays an MCD with a complex structure over an energy range of more than $35\,\mathrm{eV}$ [63]. The one-electron spectrum calculated with the spin polarized relativistic Korringa-Kohn-Rostoker (SPR-KKR) Green's function method [64] confirms the shape predicted by the angular momentum coupling. The effective exchange field of $0.3\,\mathrm{eV}$ is much smaller than the sep-aration of $17.5\,\mathrm{eV}$ caused by the $2p$ spin-orbit interaction, so that the signal

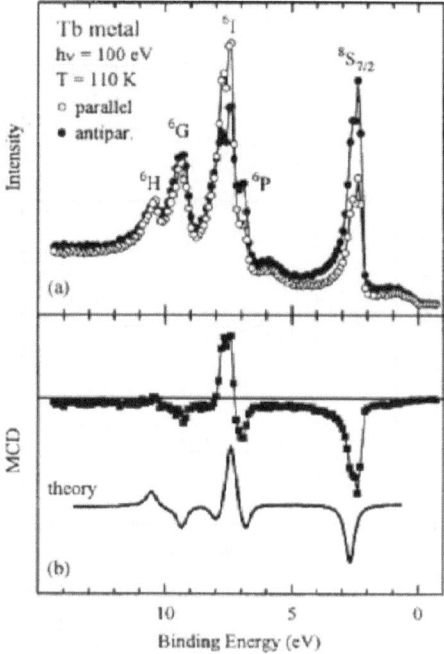

Fig. 2. Tb 4 f photoemission spectrum (**a**) measured with parallel and antiparallel alignment of magnetization and light helicity. (**b**) Resulting difference spectrum, i.e. the MCD, compared with the theoretical prediction, calculated in intermediate coupling. Note that high L' peaks show positive and low L' peaks show negative dichroism. See [61] for experiment and [48] for theory

is centred near the $2p_{3/2}$ and $2p_{1/2}$ main lines. However, the experimental spectrum shows intense structures at binding energies much higher above the main lines. Multi-electron excitations do not show up in the one-particle calculation, although most other effects, such as multiple scattering are included [64]. Figure 3 also displays the spectra calculated using the final-state impurity model, which takes into account configuration interaction (CI). In this case the spectral structure agrees very well with the experimental results. The main line is due to a well screened state, while the satellites are due to poorly screened and states. Details of this model can be found elsewhere [63, 65].

Also for iron metal with its more itinerant character there is a remarkable difference between the experiment and the one-particle calculation. As shown in [66] the leading $2p_{3/2}$ structure consists of a sharp negative peak similar to the one-electron calculation, however, the positive peak at the high BE side is smeared out over a broad energy range. In fact, the measured MCD signal remains positive over the entire energy region between the $2p_{3/2}$ and $2p_{1/2}$ main lines. Since angular momentum algebra shows that the

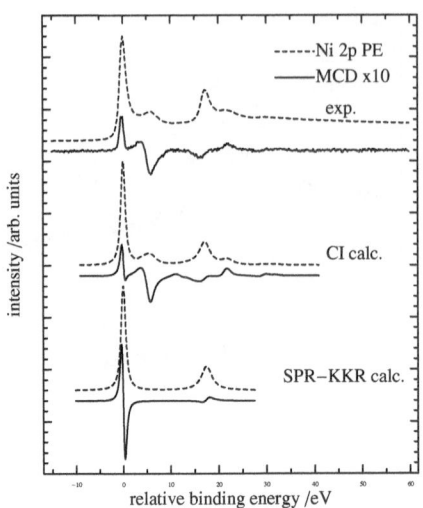

Fig. 3. Experimental and calculated Ni $2p$ photoemission of nickel metal at a photon energy of $1100\,\mathrm{eV}$ using wide angle detection. The MCD spectra (*dashed lines*) are given on a $10 \times$ enhanced intensity scale with respect to the helicity averaged spectra (*drawn lines*). The calculations, performed using the spin-polarized relativistic Korringa-Kohn-Rostoker (SPR-KKR) Green's function method, include multiple scattering but exclude electron correlation effects [64]. The impurity model calculation, indicated by CI, takes into account the configuration interaction between the states with d^8, d^9 and d^{10} Reference [63]

one-particle model for a deep core hole can only lead to the centred line shape as displayed, the strongly different spectrum that is observed must have its origin in the presence of correlation effects. However, an impurity calculation is more complicated to perform for iron than for nickel metal, where the number of number of final-state configurations is limited. Also the situation for nickel is relatively clear cut because the various final-state configurations are well separated in energy, whereas in iron metal there is a strong mixing between the different configurations. Similar results as for iron have also been reported for the Co $2p$ photoemission of cobalt metal [67, 68].

11 Second Quantization

The Pauli principle can alternatively be taken into account using creation and annihilation operators ($a^\dagger$ and a, respectively) using the method of second quantization. We first look how this is done in first quantization, in which case Slater determinants are used.

Basis wavefunctions are constructed for an n-electron atom from linear combinations of n one-electron spin-orbitals. If angular momenta are not coupled together, antisymmetrization can be accomplished by forming determinantal functions, e.g. for a two-particle state

$$\psi_{jk}(q_1, q_2) = \frac{1}{\sqrt{2}} \begin{vmatrix} \psi_j(q_1) & \psi_k(q_1) \\ \psi_j(q_2) & \psi_k(q_2) \end{vmatrix} , \tag{30}$$

$$\langle \psi_{ij}(q_1, q_2) | \psi_{kl}(q_1, q_2) \rangle = \delta_{ik}\delta_{jl} - \delta_{il}\delta_{jk} . \tag{31}$$

In second quantization for electrons (fermions) the wave function is $|jk\rangle = a_j^\dagger a_k^\dagger |0\rangle$, which leads to the anticommutation relations

$$\{a_i^\dagger a_j^\dagger\} \equiv a_i^\dagger a_j^\dagger + a_j^\dagger a_i^\dagger = 0 ,$$
$$\{a_i a_j\} \equiv a_i a_j + a_j a_i = 0 ,$$
$$\{a_i a_j^\dagger\} \equiv a_i a_j^\dagger + a_j^\dagger a_i = \delta_{ij} , \tag{32}$$

and $a_k|0\rangle = \langle 0|a_k^\dagger = 0$, since there are no electrons in the vacuum state $|0\rangle$.

$$\langle ij|kl \rangle = \langle 0|a_j a_i a_k^\dagger a_l^\dagger \rangle = \delta_{ik}\delta_{jl} - \delta_{il}\delta_{jk} . \tag{33}$$

Thus first and second quantization gives equivalent results.

The one-particle operator is

$$F = \sum_{i=1}^{N} f(\mathbf{r}_i) \Longrightarrow F = \sum_{ij} f_{ij} a_i^\dagger a_j , \tag{34}$$

with $f_{ij} = \langle i|f|j \rangle$.

The two-particle operator is

$$G = \frac{1}{2} \sum_{i \neq 1}^{N} g(\mathbf{r}_i) \Longrightarrow G = \frac{1}{2} \sum_{ij} g_{ijkl} a_i^\dagger a_j^\dagger a_k a_l , \tag{35}$$

with $g_{ijkl} = \langle ij|g|kl \rangle$.

The use of creation and annihilation operators removes the need for coefficients of fractional parentage [16]. To manipulate the operator products one can use Wick's theorem, which provides a formalism for contraction of arbitrary creation and annihilation operators [16, 24].

12 Spin-Orbit Interaction

The Hamiltonian of the spin-orbit interaction is [40]

$$\mathcal{H}_{SO} = \sum_{i=1}^{N} \zeta(r_i)\, s_i^{(1)} \cdot l_i^{(1)} \,, \tag{36}$$

with radial part $\zeta(r_i)$ and angular part $\mathbf{s}_i \cdot \mathbf{l}_i$. The spin-orbit interaction is diagonal in J and independent of M_J, but is not diagonal in α, S and L, so that states with different L and S are coupled. The matrix elements can be calculated using

$$\langle \alpha L S J M | s_i^{(1)} \cdot l_i^{(1)} | \alpha' L' S' J' M' \rangle = \delta_{JJ'} \delta_{MM'} (-)^{L'+S+J}$$
$$\times \begin{Bmatrix} L & S & 1 \\ S' & L' & J \end{Bmatrix} [(3/2)L(L+1)(2L+1)]^{1/2} \langle \alpha S L \| V^{11} \| \alpha' S' L' \rangle \,. \tag{37}$$

The dependence of the interaction on J is given by the $6j$ symbol, while the dependence on the other quantum numbers is in the reduced matrix element. The $6j$ symbol in (37) gives the selection rule $\Delta S = -\Delta L = 0, \pm 1$, which is also immediately clear by expanding the spin-orbit operator in step operators: $l^{(1)} \cdot s^{(1)} = \sum_q (-)^q l_q s_q = l_0 s_0 - l_{+1} s_{-1} - l_{-1} s_{+1}$.

For diagonal terms ($\Delta S = \Delta L = 0$) the splitting follows the Landé interval rule, since

$$\begin{Bmatrix} L & S & 1 \\ S & L & J \end{Bmatrix} = \frac{J(J+1) - L(L+1) - S(S+1)}{2[L(L+1)(2L+1)S(S+1)(2S+1)]^{1/2}} \,. \tag{38}$$

Transformation matrices from one coupling scheme to another ($LS \leftrightarrow jj$) may be computed with the aid of the recoupling equation

$$\langle [(l_1 l_2)L, (s_1 s_2)S]J \mid [(l_1 s_1)j_1, (l_2 s_2)j_2]J \rangle = [L, S, j_1, j_2]^{1/2} \begin{Bmatrix} l_1 & l_2 & L \\ s_1 & s_2 & S \\ j_1 & j_2 & J \end{Bmatrix} \,. \tag{39}$$

13 Zeeman Interaction

An external field H along the z-axis gives a Zeeman term

$$\mathcal{H}_{\mathrm{mag}} = \mu_B H (L_0^{(1)} + g_s S_0^{(1)}) = \mu_B H [J_0^{(1)} + (g_s - 1) S_0^{(1)}] \,, \tag{40}$$

where $\mu_B = e\hbar/2mc$ is the Bohr magneton and $g_s \approx 2$ is the gyromagnetic ratio. Application of the Wigner-Eckart theorem gives [22]

$$\langle b|\mathcal{H}_{\mathrm{mag}}|b'\rangle = \delta_{bb'}\mu_{\mathrm{B}}HM - \delta_{\alpha LSM,\alpha'L'S'M'}\mu_{\mathrm{B}}H(g_s-1)(-)^{L+S+M}[J,J']$$

$$\times \{S(S+1)(2S+1)\}^{1/2} \begin{pmatrix} J & 1 & J' \\ -M & 0 & M' \end{pmatrix} \begin{Bmatrix} L & S & J \\ 1 & J' & S \end{Bmatrix} . \qquad (41)$$

Thus the magnetic matrix is diagonal in all quantum numbers except J; non-zero matrix elements exist only between levels of the same LS term of the same configuration, and then only for $\Delta J = 0, \pm 1$ ($J = J' = 0$ not allowed).

For diagonal matrix elements, the $3j$ and $6j$ symbols can be evaluated analytically to

$$\langle \alpha LSJM|\mathcal{H}_{\mathrm{mag}}|\alpha'L'S'J'M'\rangle = \mu_{\mathrm{B}}g_{LSJ}HM ,$$

$$g_{LSJ} = 1 + (g_s - 1)\frac{J(J+1) + S(S+1) - L(L+1)}{2J(J+1)} , \qquad (42)$$

where g_{LSJ} is the Landé g factor. For sufficiently small values of H, the off-diagonal matrix elements of $\mathcal{H}_{\mathrm{mag}}$ that connect basis states of different J will be negligible compared with the energy differences provided by the Coulomb and spin-orbit interactions. Mixing of basis states of different J can then be neglected. The energy matrix breaks up into blocks according to J value, just as in the field free case. However, note that a crystal field can mix the different J states.

14 Crystal Field Interaction

It is a commonplace use of symmetry to assert that a matrix element $\langle \psi_i|O_k|\psi_j\rangle$ vanishes unless the triple direct product of the representations, $\Gamma(\psi_i^*) \times \Gamma(O_k) \times \Gamma(\psi_j)$, contains the totally symmetric representation (here $\times$ indicates the Kronecker product; a non-vanishing triple Kronecker product is called a *triad*). However, it is clear that the conclusion is not particularly exhaustive: it does not provide any quantitative answers. The aim of this section is to put this straight by using the method described in the book by Butler [41].

Under the effect of an operator O_R of the group G the functions $|\Gamma\gamma\rangle$ transform according to

$$O_R|\Gamma\gamma\rangle = \sum_{\gamma'} |\Gamma\gamma'\rangle\langle\Gamma\gamma'|O_R|\Gamma\gamma\rangle = \sum_{\gamma'} |\Gamma\gamma'\rangle|\Gamma(R)_{\gamma'\gamma} . \qquad (43)$$

The coefficients, written in two ways here, form the elements of a matrix. One first calculates the quantities in spherical symmetry (O_3 or SO_3) and then branch to subsequent lower symmetries. This needs to be done for the wave functions of the initial and final state as well as for the operators (Hamiltonian $\mathcal{H}$, magnetic field H, electric-multipole operator Q, crystal field operators X, configuration interaction V, etc.).

The partners in SO_3–O–D_4–D_2–C_2 basis and in SO_3–O–D_4–C_4–C_2 basis are given in Tables 4 and 5, respectively, together with their relation with the $|JM\rangle$ partners and with x, iy, z basis functions. The examples illustrate two different kind of choices for the partners fixed by the subgroup scheme. It is clear that the chain determines the specific basis functions.

Table 4. The partners in the SO_3–O–D_4–D_2–C_2 basis in Mulliken and Butler notation and their relation with the $|JM\rangle$ partners, x, iy, z basis functions and spectroscopic shorthand notation

SO_3–O–D_4–D_2–C_2		SO_3–SO_2	x, iy, z basis	Sp.
Mulliken	Butler	$|JM\rangle$		not.
$|0\ A_1\ A_1\ A_1\ \Gamma_1\rangle \equiv |0\ 0\ 0\ 0\ 0\rangle = |0\ 0\rangle$				
$|1\ T_1\ A_2\ B_1\ \Gamma_1\rangle \equiv |1\ 1\ \tilde{0}\ \tilde{0}\ 0\rangle = |1\ 0\rangle$			$= |z\rangle$	$T_1 z$
$|1\ T_1\ E\ B_2\ \Gamma_2\rangle \equiv |1\ 1\ 1\ 1\ 1\rangle = -\frac{1}{\sqrt{2}}[|1\ 1\rangle + |1\ \text{-}1\rangle]$			$= |iy\rangle$	$T_1 y$
$|1\ T_1\ E\ B_3\ \Gamma_2\rangle \equiv |1\ 1\ 1\ \tilde{1}\ 1\rangle = \frac{1}{\sqrt{2}}[|1\ 1\rangle - |1\ \text{-}1\rangle]$			$= |-x\rangle$	$T_1 x$
$|2\ E\ A_1\ A_1\ \Gamma_1\rangle \equiv |2\ 2\ 0\ 0\ 0\rangle = -|2\ 0\rangle$			$= |-\frac{1}{2}(2z^2 - x^2 - y^2)\rangle$	$E\theta$
$|2\ E\ B_1\ A_1\ \Gamma_1\rangle \equiv |2\ 2\ 2\ 0\ 0\rangle = \frac{1}{\sqrt{2}}[|2\ 2\rangle + |2\ \text{-}2\rangle]$			$= |\frac{1}{\sqrt{2}}(x^2 - y^2)\rangle$	$E\epsilon$
$|2\ T_2\ E\ B_2\ \Gamma_2\rangle \equiv |2\ \tilde{1}\ 1\ 1\ 1\rangle = \frac{1}{\sqrt{2}}[-|2\ 1\rangle + |2\ \text{-}1\rangle]$			$= |xz\rangle$	$T_2 \eta$
$|2\ T_2\ E\ B_3\ \Gamma_2\rangle \equiv |2\ \tilde{1}\ 1\ \tilde{1}\ 1\rangle = \frac{1}{\sqrt{2}}[|2\ 1\rangle + |2\ \text{-}1\rangle]$			$= |-iyz\rangle$	$T_2 \xi$
$|2\ T_2\ B_2\ B_1\ \Gamma_1\rangle \equiv |2\ \tilde{1}\ \tilde{2}\ \tilde{0}\ 0\rangle = \frac{1}{\sqrt{2}}[|2\ 2\rangle - |2\ \text{-}2\rangle]$			$= |ixy\rangle$	$T_2 \zeta$

We can transform the SO_3–SO_2 partners, i.e. the spherical harmonics $|JM\rangle$, into the crystal field functions $|J\Gamma\gamma\rangle$ of the point group symmetry branch SO_3–G–g,

$$|J\Gamma\gamma\rangle = \sum_M |JM\rangle\langle JM|J\Gamma\gamma\rangle , \qquad (44)$$

where Γ is an irrep of G, and γ is a subspecies label needed when Γ is degenerate, i.e. γ is an irrep of the subgroup g. For clarity one can choose to write the corresponding symmetry group in parentheses behind the irrep: $|J(SO_3)M(SO_2)\rangle \equiv |JM\rangle$ and $|J(SO_3)\Gamma(G)\gamma(g)\rangle \equiv |J\Gamma\gamma\rangle$. The basis kets of an irrep space are called the *partners* of the irrep. The functions which are partners simultaneously of the groups in the chain SO_3–O–D_4–D_2 are different from those for SO_3–O–D_4–C_4.

In D_4 the basis functions are not uniquely fixed, because for the two-dimensional irrep $E(D_4)$ we could choose different linear combinations of x and y as bases. However, branching further to either D_2 or C_4 fixes the basis to the partners (omitting any constants) $\{x, iy\}$ or $\{x + iy, x - iy\}$, respectively (see Tables 4 and 5). These two sets are related by a unitary transformation. Note that under C_2 (twofold rotation about z axis) which is a subgroup of

Table 5. The partners in the SO_3–O–D_4–C_4–C_2 basis in Mulliken and Butler notation and their relation with the $|JM\rangle$ partners and x, iy, z basis functions

SO₃–O–D₄–C₄–C₂		SO_3–SO_2	x, iy, z basis
Mulliken	Butler	$|JM\rangle$	
$\|0\ A_1\ A_1\ \Gamma_1\ \Gamma_1\rangle = \|0\,0\,0\,0\,0\rangle$		$= \|0\,0\rangle$	
$\|\frac{1}{2}\ E'\ E'\ \Gamma_5\ \Gamma_3\rangle = \|\frac{1}{2}\,\frac{1}{2}\,\frac{1}{2}\,\frac{1}{2}\,\frac{1}{2}\rangle$		$= \|\frac{1}{2}\,\frac{1}{2}\rangle$	$= \|\alpha\rangle$
$\|\frac{1}{2}\ E'\ E'\ \Gamma_6\ \Gamma_4\rangle = \|\frac{1}{2}\,\frac{1}{2}\,\frac{1}{2}\,\text{-}\frac{1}{2}\,\frac{1}{2}\rangle$		$= \|\frac{1}{2}\,\text{-}\frac{1}{2}\rangle$	$= \|\beta\rangle$
$\|1\ T_1\ A_2\ \Gamma_1\ \Gamma_1\rangle = \|1\,1\,\tilde{0}\,0\,0\rangle$		$= \|1\,0\rangle$	$= \|z\rangle$
$\|1\ T_1\ E\ \Gamma_3\ \Gamma_2\rangle = \|1\,1\,1\,1\,1\rangle$		$= -\,\|1\,1\rangle$	$= \|\frac{1}{\sqrt{2}}(x+iy)\rangle$
$\|1\ T_1\ E\ \Gamma_4\ \Gamma_2\rangle = \|1\,1\,1\,\text{-}1\,1\rangle$		$= -\,\|1\,\text{-}1\rangle$	$= \|-\frac{1}{\sqrt{2}}(x-iy)\rangle$
$\|\frac{3}{2}\ U'\ E'\ \Gamma_5\ \Gamma_3\rangle = \|\frac{3}{2}\,\frac{3}{2}\,\frac{1}{2}\,\frac{1}{2}\,\frac{1}{2}\rangle$		$= \|\frac{3}{2}\,\frac{1}{2}\rangle$	
$\|\frac{3}{2}\ U'\ E'\ \Gamma_6\ \Gamma_4\rangle = \|\frac{3}{2}\,\frac{3}{2}\,\frac{1}{2}\,\text{-}\frac{1}{2}\,\text{-}\frac{1}{2}\rangle$		$= -\,\|\frac{3}{2}\,\text{-}\frac{1}{2}\rangle$	
$\|\frac{3}{2}\ U'\ E''\ \Gamma_8\ \Gamma_4\rangle = \|\frac{3}{2}\,\frac{3}{2}\,\frac{3}{2}\,\frac{3}{2}\,\text{-}\frac{1}{2}\rangle$		$= \|\frac{3}{2}\,\frac{3}{2}\rangle$	
$\|\frac{3}{2}\ U'\ E''\ \Gamma_7\ \Gamma_3\rangle = \|\frac{3}{2}\,\frac{3}{2}\,\frac{3}{2}\,\text{-}\frac{3}{2}\,\frac{1}{2}\rangle$		$= \|\frac{3}{2}\,\text{-}\frac{3}{2}\rangle$	
$\|2\ E\ A_1\ \Gamma_1\ \Gamma_1\rangle = \|2\,2\,0\,0\,0\rangle$		$= -\,\|2\,0\rangle$	$= \|-\frac{1}{2}(2z^2 - x^2 - y^2)\rangle$
$\|2\ E\ B_1\ \Gamma_2\ \Gamma_1\rangle = \|2\,2\,2\,2\,0\rangle$		$= -\frac{1}{\sqrt{2}}\,[\|2\,2\rangle + \|2\,\text{-}2\rangle]$	$= \|-\frac{1}{\sqrt{2}}(x^2 - y^2)\rangle$
$\|2\ T_2\ E\ \Gamma_3\ \Gamma_2\rangle = \|2\,\tilde{1}\,1\,1\,1\rangle$		$= -\,\|2\,1\rangle$	$= \|\frac{1}{\sqrt{2}}(ixz - yz)\rangle$
$\|2\ T_2\ E\ \Gamma_4\ \Gamma_2\rangle = \|2\,\tilde{1}\,1\,\text{-}1\,1\rangle$		$= \|2\,\text{-}1\rangle$	$= \|\frac{1}{\sqrt{2}}(xz - iyz)\rangle$
$\|2\ T_2\ B_2\ \Gamma_2\ \Gamma_1\rangle = \|2\,\tilde{1}\,\tilde{2}\,2\,0\rangle$		$= \frac{1}{\sqrt{2}}\,[\|2\,2\rangle - \|2\,\text{-}2\rangle]$	$-\,\|ixy\rangle$

both D_2 and C_4, the x and y axes transform identically, i.e. have equivalent irreducible representations (irreps).

The irreps -1, 0, 1 in the cyclic group C_4 correspond to the magnetic components $M = -1, 0, 1$ of $L = 1$ in SO_2, where the 1 and -1 are conjugate irreps ($M^* = -M$).

14.1 3jm Symbol

In point group symmetry the $3j$ symbol is refereed to as the $3jm$ symbol, which is defined using the vector coupling coefficient by [69]

$$(a\alpha, b\beta|rc\gamma) \equiv |c|^{1/2} \begin{pmatrix} c \\ \gamma \end{pmatrix} \begin{pmatrix} a & b & c^* \\ \alpha & \beta & \gamma^* \end{pmatrix}^r , \qquad (45)$$

where $|(ab)rc\gamma) = \sum_{\alpha\beta} |a\alpha\rangle|b\beta\rangle(a\alpha, b\beta|rc\gamma)$ and r is the multiplicity factor. More generally, but equivalently, the $3jm$ can be defined in terms of the expansion coefficients for the invariant part of the triple direct product of three irreps [10, 69]

$$|((ab)c)r00\rangle \equiv \sum_{\alpha\beta\gamma} \begin{pmatrix} a & b & c \\ \alpha & \beta & \gamma \end{pmatrix}^{*r} |a\alpha\rangle |b\beta\rangle |c\gamma\rangle . \tag{46}$$

It follows that the Wigner-Eckart theorem is written as

$$\langle a\alpha|T_q^{(k)}|b\beta\rangle = \begin{pmatrix} a \\ \alpha \end{pmatrix} \sum_r \begin{pmatrix} a^* & k & b \\ \alpha^* & q & \beta \end{pmatrix}^r \langle a\|T^{(k)}\|b\rangle_r . \tag{47}$$

14.2 $3j$ Phase

Cyclic (even) interchanges of columns leave the value of the $3jm$ coefficient unchanged and noncyclic (odd) interchanges multiply the coefficient by the $3j$ phase $\{abcr\}$, which is defined by

$$\begin{pmatrix} a & b & c \\ \alpha & \beta & \gamma \end{pmatrix}^r = \{abcr\} \begin{pmatrix} b & a & c \\ \beta & \alpha & \gamma \end{pmatrix}^r = \{abcr\} \begin{pmatrix} c & b & a \\ \gamma & \beta & \alpha \end{pmatrix}^r = \{abcr\} \begin{pmatrix} a & c & b \\ \alpha & \gamma & \beta \end{pmatrix}^r . \tag{48}$$

$\{abcr\}$ can have values ± 1.

Closely related is the $2j$ phase, which is defined as $\{a\} \equiv \{aa^*0\}$.

14.3 $2jm$ Phase

The $2jm$ phase $\begin{pmatrix} a \\ \alpha \end{pmatrix}$ is used to relate the transformation properties of $|a\alpha\rangle^\dagger \equiv \langle a\alpha|$ to those of $|a\alpha\rangle$. The $2jm$ phase, which is ± 1, is given by

$$\begin{pmatrix} a \\ \alpha \end{pmatrix} = |a|^{1/2} \begin{pmatrix} a & a^* & 0 \\ \alpha & \alpha^* & 0 \end{pmatrix} = \{a\} \begin{pmatrix} a^* \\ \alpha^* \end{pmatrix} , \tag{49}$$

where the *invariant product*, $|(ab)00\rangle = \sum_{\alpha\beta} |a\alpha\rangle |b\beta\rangle (a\alpha, b\beta|00)$, defines the $3jm$ symbol.

Equation (45) can be regarded as a special case of the Wigner-Eckart theorem in the branching SO_3-SO_2 with $j^* = j$ and $m^* = -m$, $|j| = 2m+1$, $|m| = 1$, the $2jm$ phase $\begin{pmatrix} j \\ m \end{pmatrix} = (-)^{j-m}$, and the total symmetric representations are $j = 0$ and $m = 0$. For SO_3 the $3j$ phase is $\{j_1 j_2 j_3\} = (-)^{j_1+j_2+j_3}$ and the $2j$ phase is $\{j\} = (-)^{2j}$.

14.4 Evaluation of the Matrix Elements

It suffice to have the matrix elements of unit tensors $U^{k\Gamma\gamma}$, since all tensors of the same rank only differ by a RME. For the matrix element of the unit tensor in SO_3–G, we can write

$$\langle Ja\Gamma|U^{ka''\Gamma''}|J'a'\Gamma'\rangle_{\mathrm{G}} = \sum_r \frac{1}{|\Gamma|^{1/2}} \langle J\|U^{(k)}\|J'\rangle_{\mathrm{SO}_3} \begin{pmatrix} J \\ a \\ \Gamma \end{pmatrix} \begin{pmatrix} J & k & J' \\ a^* & a'' & a' \\ \Gamma^* & \Gamma'' & \Gamma' \end{pmatrix}_{\mathrm{rG}}^{\mathrm{SO}_3} ,$$

(50)

where the $\langle J\|U^{(k)}\|J'\rangle$ is the RME in SO$_3$ symmetry, that can be obtained from (Cowan's) Hartree-Fock calculations. The $2jm$ and $3jm$ coefficients in (50) are tabulated by Butler [43]. The $2jm$ factor is a phase factor of ± 1, which relates the transformation properties of the spherical harmonics to their complex conjugates, or the transformation properties of kets to bras. The bras Γ^* and γ^* are the conjugate irreps of Γ and γ. The a gives the branching multiplicity when the irrep Γ occurs r times in J. The $|\Gamma|$ denotes the dimension of irrep Γ and appears in the normalization in the above equation due to the definition of the unit tensor. Racah's factorization lemma gives the important result that a property $A(\mathrm{SO}_3\text{–G–g})$ can be factorized as $A(\mathrm{SO}_3\text{–G}) * A(\mathrm{G–g})$. This provides a simple way to continue the factorization in the chain, which is normally ended upon reaching a totally symmetric representation for $ka\Gamma$.

14.5 Crystal Field Parameters

The crystal field can be written as a multipole expansion over reduced spherical harmonics $C_q^{(k)}$ with conventional crystal field parameters B_q^k,

$$\mathcal{H}_{\mathrm{cf}} = \sum_{k=0}^{2l} \sum_{q=-k}^{k} B_q^k C_q^k(\theta,\phi) = \sum_{ka} |k|^{1/2} X^{ka0} U^{ka0} ,$$

(51)

where at the right hand side, $U^{k\Gamma\gamma}$ are the operators with coefficients $X^{k\Gamma\gamma}$ as parameters for the branch SO$_3$–G–g. The dimension factor in (51) is Butler's convention [41], which gives that

$$X^{ka0} = |l||k|^{1/2} \begin{pmatrix} l & k & l \\ 0 & 0 & 0 \end{pmatrix} \sum_q B_q^k \langle ka0(\mathrm{G})|kq(\mathrm{SO}_2)\rangle .$$

(52)

Since $0 \leq k \leq 2l$ and k is even, (51) reduces for $3d$ electrons to

$$\mathcal{H}_{\mathrm{cf}} = X^{0000} U^{0000} + \sum_{\Gamma\gamma} X^{2\Gamma\gamma 0} U^{2\Gamma\gamma 0} + \sum_{\Gamma\gamma} X^{4\Gamma\gamma 0} U^{4\Gamma\gamma 0} ,$$

(53)

where the sum runs over all irreps in the group chain that are invariant (total symmetric) in the lowest group.

A Hartree-Fock program, such as Cowan's code, calculates multipole operator reduced matrix electrons between multielectron states expressed into one-electron operators acting within one shell (for static crystal field) or between two shells (for hybridization). The one-electron operators are normalized such that their SO$_3$ RMEs are 1, thus are unit tensor operators

$\langle l\|u^k\|l'\rangle = 1$. This holds for static crystal field as well as for hybridization. In practise $l = l'$. We will describe how to obtain matrix elements in point-group symmetry, given the coefficients X^k of the unit operators U^k.

In order to compare the crystal field coefficients X^k to other parameters, such as the familiar Dq, Ds and Dt parameters, we will compare their effects on the d functions. Generally, the best way to specify the strength of the crystal functions is simply to give the energy separations of the d states. The parameter values are often artificial and uninformative. The unit crystal field operators given by RCG9 are scaled such that $\langle d\|U^{(k)}\|d\rangle = 1$ for the allowed values of k. For $k = 2$ and 4, in Butler notation, $\langle 2\|4\|2\rangle_{SO3} = 1$ and $\langle 2\|2\|2\rangle_{SO3} = 1$. In the octahedral group (SO$_3$–O) the only total symmetric operators are U^{00} and U^{40}, so that $\mathcal{H}_{cf} = X^{00}U^{00} + X^{40}U^{40}$. The operator U^{00} only shifts all irreps by an equal amount. For 2(O) (E in Mulliken notation) the energy of the two degenerate components d_{z^2} and $d_{x^2-y^2}$ is

$$\left\langle \begin{matrix} 2 & 4 & 2 \\ 2 & 0 & 2 \end{matrix} \right\rangle_O^{SO3} = |2|_O^{-1/2}\langle 2\|4\|2\rangle_{SO3} \begin{pmatrix} 2 \\ 2 \end{pmatrix} \begin{pmatrix} 2 & 4 & 2 \\ 2 & 0 & 2 \end{pmatrix}_O^{SO3} = \frac{1}{\sqrt{2}}\cdot 1\cdot 1\cdot\frac{1}{\sqrt{15}} = \frac{1}{\sqrt{30}},$$

where the group label is attached as a subscript to avoid confusion. For $\tilde{1}$(O) (T$_2$ in Mulliken notation) the energy of the three degenerate components d_{xy}, d_{yz} and d_{xz} is

$$\left\langle \begin{matrix} 2 & 4 & 2 \\ \tilde{1} & 0 & \tilde{1} \end{matrix} \right\rangle_O^{SO3} = |\tilde{1}|_O^{-1/2}\langle 2\|4\|2\rangle_{SO3} \begin{pmatrix} 2 \\ \tilde{1} \end{pmatrix} \begin{pmatrix} 2 & 4 & 2 \\ \tilde{1} & 0 & \tilde{1} \end{pmatrix}_O^{SO3} = \frac{1}{\sqrt{3}}\cdot 1\cdot 1\cdot\frac{-\sqrt{2}}{3\sqrt{5}} = \frac{-2}{3\sqrt{30}}.$$

Thus a strength of 1 for the operator U^{40} gives an energy separation of $5/(3\sqrt{30}) \approx 0.304$. Therefore, $10Dq \approx 0.304\,X^{40}$ and an energy splitting of $10Dq = 1\,\text{eV}$ is obtained for $X^{40} \approx 3.286$.

Analogously, in tetrahedral symmetry the energies may be expressed in terms of the strength of the unit operators $\{Dq, Ds, Dt\}$ or $\{\Delta, \epsilon, \tau\}$. The SO$_3$–O–D$_4$ crystal field Hamiltonian is $X^{000}U^{000} + X^{400}U^{400} + X^{420}U^{420} + X^{220}U^{220}$. Using the $2jm$ and $3jm$ symbols tabulated by Butler, we obtain from the matrix $\varepsilon(\Gamma) = \langle\Gamma|\sum_k U^k|\Gamma\rangle X^k$ the values for the energies ε of the irreps $\Gamma = a_1$, b_1, e and b_2 as

$$\varepsilon(a_1) = \frac{1}{\sqrt{30}}X^{400} + \frac{1}{\sqrt{42}}X^{420} + \frac{2}{\sqrt{70}}X^{220} \equiv 6Dq - 2Ds - 6Dt\,,$$

$$\varepsilon(b_1) = \frac{1}{\sqrt{30}}X^{400} - \frac{1}{\sqrt{42}}X^{420} - \frac{2}{\sqrt{70}}X^{220} \equiv 6Dq + 2Ds - Dt\,,$$

$$\varepsilon(e) = -\frac{2}{3\sqrt{30}}X^{400} - \frac{2}{3\sqrt{42}}X^{420} + \frac{1}{\sqrt{70}}X^{220} \equiv -4Dq - Ds + 4Dt\,,$$

$$\varepsilon(b_2) = -\frac{2}{3\sqrt{30}}X^{400} + \frac{4}{3\sqrt{42}}X^{420} - \frac{2}{\sqrt{70}}X^{220} \equiv -4Dq + 2Ds - Dt\,,$$

where at the right hand side we give the usual crystal field parameters $Dq \equiv \frac{1}{10}\Delta$, $Ds \equiv \frac{1}{7}(\epsilon+\tau)$ and $Dt \equiv \frac{1}{35}(3\epsilon-4\tau)$, with the energy differences defined as $\Delta \equiv \varepsilon(b_1) - \varepsilon(b_2)$, $\epsilon \equiv \varepsilon(b_1) - \varepsilon(a_1)$ and $\tau \equiv \varepsilon(b_2) - \varepsilon(e)$, which gives the relation between Butler's parameters and the usual parameters as

$$X^{400} = 6\sqrt{30}Dq - \tfrac{7}{2}\sqrt{30}Dt \ ,$$

$$X^{420} = -\tfrac{5}{2}\sqrt{42}Dt \ ,$$

$$X^{220} = -\sqrt{70}Ds \ ,$$

$$Dq = \frac{1}{6\sqrt{30}}X^{400} - \frac{7}{30\sqrt{42}}X^{420},$$

$$Ds = -\frac{1}{\sqrt{70}}X^{220},$$

$$Dt = -\frac{2}{5\sqrt{42}}X^{420}.$$

Thus in general we can write the energies in matrix notation as $\varepsilon = \mathbf{AX}$. By inverting this matrix: $\mathbf{X} = \mathbf{A}^{-1}\varepsilon$, we obtain the values of $\mathbf{X}$, given the values for $\langle \Gamma | U | \Gamma \rangle$. Here, $\varepsilon(\Gamma)$ is the energy shift for orbital Γ, or mixing for symmetry Γ.

It is often convenient to make new operators V_Γ which have only matrix elements between levels of symmetry Γ, the others being zero: $\langle \Gamma | V_\Gamma | \Gamma' \rangle = \delta_{\Gamma\Gamma'}$. We can calculate $V_{\Gamma_1} = \mathbf{X}_{\Gamma_1}\mathbf{U} = \mathbf{X}_{\Gamma_1}\mathbf{A}^{-1}\varepsilon_{\Gamma_1}$. As an example

$$\mathbf{X}_{b_1} = \mathbf{A}^{-1}\begin{pmatrix} 1 \\ 0 \\ 0 \\ 0 \end{pmatrix} = \begin{pmatrix} \frac{1}{\sqrt{5}} \\ \frac{3}{10}\sqrt{30} \\ \frac{3}{14}\sqrt{42} \\ -\frac{1}{7}\sqrt{30} \end{pmatrix},$$

so that

$$V_{b_1} = \frac{1}{\sqrt{5}} U^{000} + \frac{3}{10}\sqrt{30}\, U^{400} + \frac{3}{14}\sqrt{42}\, U^{420} - \frac{1}{7}\sqrt{30}\, U^{220}.$$

To use an operator V with predescribed matrix elements $\langle \Gamma | V | \Gamma \rangle$ we simply take $V = \sum_\Gamma \langle \Gamma | V | \Gamma \rangle V_\Gamma$.

14.6 Electric-Dipole Transition Operator

The electric-dipole transition operator is given by

$$\hat{\epsilon} \cdot \mathbf{r} = e_x x + e_y y + e_z z = \sum_{q=-1,0,1} (-)^q C^{(1)}_{-q}(\hat{\epsilon})\, r C^{(1)}_q(\hat{r}) \ , \qquad (54)$$

where $r C^{(1)}_0(\hat{r}) = z$ and $r C^{(1)}_{\pm 1}(\hat{r}) = \mp \frac{1}{\sqrt{2}}(x \pm iy)$.

In SO_3–O–D_4–C_4 symmetry the electric-dipole operator can be written as (c.f. Table 7)

$$\hat{e} \cdot \mathbf{r} = e_z T^{11\bar{0}0} + \frac{1}{\sqrt{2}}(e_x - ie_y)T^{1111} - \frac{1}{\sqrt{2}}(e_x + ie_y)T^{111-1} . \qquad (55)$$

A vector $\mathbf{r} = \mathbf{e}x$ is invariant with respect to transformations of the coordinate axes, since the unit vectors $\mathbf{e}$ and the coordinates $\mathbf{x}$ are contragredient. Also the set of states $|a\alpha\rangle$ and $\langle a\alpha| \equiv |a\alpha\rangle^{\dagger}$ are contragredient. We can distinguish between ambivalent groups (e.g. SO_3, D_4, O), where $a^* = a$ for all irreps and nonambivalent group (e.g. C_4 and C_3) where $b = a^* \neq a$.

14.7 Magnetic Field

Rotation inversion groups $G_i = G \times C_i$ (e.g. $O_3 = SO_3 \times C_i$ and $O_h = O \times C_i$) distinguish axial vectors (even), J, and polar (odd) vectors, r.

In the presence of a magnetic field, but without crystal field, it is sufficient to branch to SO_2 using the group-subgroup chain in Table 6.

Table 6. The irreps of the Hamiltonian, magnetic field and electric-dipole operator Q in the group-subgroup chain O_3-SO_3-SO_2

Operator	O_3	SO_3	SO_2
$\mathcal{H}_0$	0^+	0	0
H_{mag}	1^+	1	0
$Q_0^{(1)}$	1^-	1	0
$Q_{-1}^{(1)}$	1^-	1	-1
$Q_{+1}^{(1)}$	1^-	1	1

The Hamiltonian $\mathcal{H}_0$ is diagonal in J, therefore it has to be total symmetric with even parity in all groups, i.e. 0^+ in parity conserving groups and 0 otherwise. The magnetic field is given by a vector 1, hence it can mix states with $\Delta J = 0, \pm 1$. The magnetic field is an axial vector, like J, and therefore has even parity. With the field along the z axis, H_{mag} is diagonal in SO_2 symmetry. The electric-dipole operator $Q^{(1)}$ has odd parity, 1^- in O_3. It splits in z-component, left- and right-circular component in SO_2. The electric-quadrupole operator (not listed) has even parity, 2^+ in O_3 and splits in -2, -1, 0, 1, 2 in SO_2. Another example is given for D_4 crystal field in Table 7. In the presence of a magnetic field we have to branch to the cyclic group C_4.

Table 7. The irreps of the Hamiltonian, magnetic field, crystal-field parameters X and electric-dipole operator Q in the group-subgroup chain O_3-O_h-D_{4h}-D_4 -C_4

Operator	O_3	O_h	D_{4h}	D_4	C_4
$\mathcal{H}_0$	0^+	0^+	0^+	0	0
$\mathcal{H}_{\mathrm{mag}}$	1^+	1^+	$\tilde{0}^+$	$\tilde{0}$	0
X^{400}	4^+	0^+	0^+	0	0
X^{420}	4^+	2^+	0^+	0	0
X^{220}	2^+	2^+	0^+	0	0
$Q_0^{(1)}$	1^-	1^-	$\tilde{0}^-$	$\tilde{0}$	0
$Q_{+1}^{(1)}$	1^-	1^-	1^-	1	1
$Q_{-1}^{(1)}$	1^-	1^-	1^-	1	-1

15 Fundamental Spectra

We know that in the cylindrical group, SO_2, where J is split into $2J + 1$ different M levels we can expect besides the isotropic spectrum also a linear dichroism and circular dichroism spectrum. How will this be in an arbitrary point group [70]?

In the presence of a crystal field the different m levels in the ground state are mixed, thus $m \neq m'$. Since the basis functions $l\Gamma\gamma$ can have multiplicity, we have to branch them to low enough group to obtain orthonormal one dimensional functions. The fundamental spectra are given by

$$
I^{x\Gamma\gamma 0} = n_{lx}^{-1} \sum_{f\Gamma_1\gamma_1\Gamma_2\gamma_2\Gamma\gamma\kappa_1} \langle g| l^\dagger_{\Gamma_1\gamma_1\kappa_1} |f\rangle \langle f| l_{\Gamma_2\gamma_2\kappa_1} |g\rangle
\begin{pmatrix} l \\ \Gamma_1 \\ \gamma_1 \\ \kappa_1 \end{pmatrix}
\begin{pmatrix} l & x & l \\ \Gamma_1^* & \Gamma & \Gamma_2 \\ \gamma_1^* & \gamma & \gamma_2 \\ \kappa_1^* & 0 & \kappa_1 \end{pmatrix}
\begin{matrix} SO_3 \\ G \\ G' \\ g \end{matrix} ,
$$

(56)

where for brevity we have suppressed the multiplicity numbers. The fundamental spectra, like all physical properties, must be invariant under all operations of the group g, or else they are zero. Therefore, x must branch to the total symmetric representation $0(g)$ and the number of fundamental spectra is equal to the number of branches of x. For each point group the number of fundamental spectra is given in Table 8. $x = 0$, which gives the isotropic spectrum, branches in every point group to the totally symmetric representation. There is circular dichroism, branching from $x = 1$, when the degeneracy with the complex conjugate states (Kramer's degeneracy) is lifted by breaking of the time reversal symmetry, which can occur in the cyclic groups C_n when a magnetic field and spin-orbit interaction is present. The presence of the magnetic field not only lowers the symmetry $SO_3 \rightarrow SO_2$ or $D_n \rightarrow C_n$ but also breaks the time reversal symmetry. Linear dichroism ($x = 2$) due to crystal field and/or magnetic alignment is present in the cyclic and dihedral groups, but not in the cubic and icosahedral groups. The multiple

Table 8. Number of fundamental spectra in point group g. The $x = 0$, 1, 2 indicate the isotropic spectrum, circular dichroism and linear dichroism, respectively [70]

g	$x = 0$	$x = 1$	$x = 2$
SO_3, O_3, K, K_h, O, O_h T, T_h	1	0	0
D_∞, $D_{\infty h}$, D_6, D_{6h}, D_5, D_{5h}, D_4, D_{4h},D_3, D_{3d}	1	0	1
D_2, D_{2h}	1	0	2
C_∞, SO_2, $C_{\infty h}$, C_6, C_{6h}, C_5, C_{5i}, C_4, C_{4h},C_3, C_{3i}	1	1	1
C_2, C_{2h}	1	1	3
C_1, C_i	1	3	5

fundamental spectra can be labeled using the irreps in the chain of groups. For instance, in D_2 there are two linear dichroism spectra labeled by 2200 (z^2) and 2220 ($x^2 - y^2$), whereas in C_2 there are 22000 (z^2), 22200 ($x^2 - y^2$) and $2\bar{1}\bar{2}00$ (xy). All nine spectra are present in C_1 and C_i and the number of fundamental spectra for each x is $2x + 1$, i.e. one isotropic spectrum, three circular dichroism spectra and five linear dichroism spectra.

16 Second-Order Processes

In the kinetic energy operator $\frac{1}{2m} \left[\mathbf{p} - \frac{e}{c} \mathbf{A}(\mathbf{r}_i) \right]^2$ of the non-relativistic Hamiltonian for the photon-matter interaction, the first-order term of the squared vector potential, $\mathbf{A}^2$, gives rise to the familiar Thomson scattering, while a second-order term in $\mathbf{p} \cdot \mathbf{A}$ describes a resonant process. For excitation of a core electron into an empty valence state, the latter term will usually be dominant. The matrix element for the second-order processes is

$$ f = \frac{e^2}{m^2 c^2} \sum_n \frac{\langle g|\mathbf{A}_{\text{out}} \cdot \mathbf{p}|n\rangle\langle n|\mathbf{p} \cdot \mathbf{A}_{\text{in}}|g\rangle}{E_n - E_g - \hbar\omega - i\Gamma/2} \ . \tag{57} $$

At resonance, the intermediate state $|n\rangle$ makes the process selective to the environment (element, orientation, magnetism).

Following the book of Varshalovich et al. [27] a plane wave can be expanded in a series of vector spherical harmonics [71]

$$ \mathbf{A}(\mathbf{k}, \mathbf{r}) = \varepsilon(\mathbf{k}) \exp(i\mathbf{k} \cdot \mathbf{r}) = 4\pi \sum_{LM} \sum_{l=L-1}^{L+1} i^l j_l(kr) \left\{ \varepsilon(\mathbf{k}) \cdot \mathbf{Y}_{LM}^{l*}(\hat{\mathbf{k}}) \right\} \mathbf{Y}_{LM}^l(\hat{\mathbf{r}}) \ , $$
$$ \tag{58} $$

where $\varepsilon(\mathbf{k})$ is the unit polarization vector, $\hat{\mathbf{k}} = (\vartheta_k, \varphi_k)$ is the unit vector of photon momentum, $j_l(kr)$ is the spherical Bessel function and $\mathbf{Y}_{LM}^l(\vartheta, \varphi) = \sum_L \{Y_m^l, e_\rho^1\}_M^L$ is the vector spherical harmonic, where the coupling is defined using the Clebsch-Gordan coefficient $C_{l'm'l''m''}^{lm}$ as

$$\{A^{l'}_{m'}, B^{l''}_{m''}\}^l_m \equiv \sum_{m'm''} C^{lm}_{l'm'l''m''} A^{l'}_{m'} B^{l''}_{m''} . \tag{59}$$

The vector spherical harmonics $\mathbf{Y}^l_{LM}$ ($l = L - 1, L, L + 1$) can be transformed into the complete set of orthogonal vector spherical harmonics $\mathbf{Y}^{(\lambda)}_{LM}$ ($\lambda = -1, 0, 1$) [72] as

$$\mathbf{Y}^{(1)}_{LM} \equiv \sqrt{\frac{L+1}{2L+1}} \mathbf{Y}^{L-1}_{LM} + \sqrt{\frac{L}{2L+1}} \mathbf{Y}^{L+1}_{LM} = \frac{1}{\sqrt{L(L+1)}} \nabla_\Omega Y_{LM} ,$$

$$\mathbf{Y}^{(0)}_{LM} \equiv \mathbf{Y}^L_{LM} = \frac{\hat{\mathbf{L}}}{\sqrt{L(L+1)}} Y_{LM} ,$$

$$\mathbf{Y}^{(-1)}_{LM} \equiv \sqrt{\frac{L}{2L+1}} \mathbf{Y}^{L-1}_{LM} - \sqrt{\frac{L+1}{2L+1}} \mathbf{Y}^{L+1}_{LM} = \mathbf{n} Y_{LM} , \tag{60}$$

where the orbital momentum operator $\hat{\mathbf{L}} = -\mathrm{i}\mathbf{n} \times \nabla_\Omega$, $\mathbf{n} \equiv \mathbf{r}/r$ and ∇_Ω denotes the angular part of the vector differential operator ∇.

The $\mathbf{Y}^{(1)}_{LM}$ and $\mathbf{Y}^{(0)}_{LM}$ are transverse with respect to $\mathbf{n}(\vartheta, \varphi)$, while $\mathbf{Y}^{(-1)}_{LM}$ is longitudinal, i.e. $\mathbf{n} \cdot \mathbf{Y}^{(1)}_{JM} = \mathbf{n} \cdot \mathbf{Y}^{(0)}_{LM} = 0$ and $\mathbf{n} \times \mathbf{Y}^{(-1)}_{LM} = 0$. The $\mathbf{Y}^{(1)}_{LM}$ and $\mathbf{Y}^{(0)}_{LM}$ are the *electric* and *magnetic multipoles*, respectively. The longitudinal contribution, $\mathbf{Y}^{(-1)}_{LM}$, vanishes due to the transversality of the light [$\mathbf{k} \cdot \varepsilon(\mathbf{k}) = 0$]. Equation (60) shows that the parity is $(-)^L$ and $(-)^{L+1}$ for electric transitions and magnetic transitions, respectively.

Assuming that the interaction radius is small compared to the photon wavelength, $kr \ll 1$ and the Bessel function $j_L(kr) \to (kr)^L/(2L+1)!!$, the major contribution to the sum over L comes from the term with the lowest L value. Here, $(2L+1)!! \equiv (2L+1)(2L-1)\cdots 1$. The matrix elements can be separated in a geometric and physical (dynamic) part, e.g. for the electric multipole transition,

$$\langle n|\mathbf{A}(\mathbf{k}, \mathbf{r}_i) \cdot \mathbf{p}|g\rangle \propto \sum_{LM} \left\{ \varepsilon(\mathbf{k}) \mathbf{Y}^{(1)*}_{LM}(\hat{\mathbf{k}}) \right\} \langle n|J^L_M|g\rangle$$

$$\propto \sum_{LM} \left\{ \varepsilon(\mathbf{k}), Y^{L-1}(\hat{\mathbf{k}}) \right\}^L_M \langle n|J^L_M|g\rangle . \tag{61}$$

where in the physical part the *current operator*

$$J^L_M = -\frac{4\pi \mathrm{i}^L k^L}{(2L+1)!!} \sqrt{\frac{L+1}{L}} \sum_i e r^L_i Y^L_M(\mathbf{r}_i) , \tag{62}$$

excites an electron from an occupied (core) to an empty (valence) state. With the Wigner-Eckart theorem the spherical harmonic Y^L_M gives the selection rules $-L \le \Delta M \le L$.

For the electric 2^L-pole transition in spherical symmetry we obtain the multipole expansion [73, 74]

$$f^{EL}(\omega) = 4\pi\lambda \sum_{z=0}^{2L} \left\{ T^z(\varepsilon_f^*, \hat{\mathbf{k}}_f; \varepsilon_i, \hat{\mathbf{k}}_i)_{EL}, \langle g|F^z(\omega)_{EL}|g\rangle \right\}^0 , \quad (63)$$

with physical part

$$F_\zeta^z(\omega)_{EL} = \sum_n \frac{1}{2\lambda} \left[\frac{(J^L)^\dagger|n\rangle\langle n|J^L}{E_n - E_g - \hbar\omega - i\Gamma/2} \right]_\zeta^z , \quad (64)$$

and geometric part

$$T_\zeta^z(\varepsilon_f^*, \hat{\mathbf{k}}_f; \varepsilon_i, \hat{\mathbf{k}}_i)_{EL} = \frac{2L+1}{L+1} \left[\left\{ \varepsilon_f^*, Y^{L-1}(\hat{\mathbf{k}}_f) \right\}^L, \left\{ \varepsilon_i, Y^{L-1}(\hat{\mathbf{k}}_i) \right\}^L \right]_\zeta^z . \quad (65)$$

For electric-dipole transitions ($L = 1$) the dependence on $\hat{\mathbf{k}}_i$ and $\hat{\mathbf{k}}_f$ disappears and in cylindrical symmetry ($\zeta = 0$) we obtain [75]:

$$\frac{2}{3\lambda} f^{E1}(\omega) = 4\pi \sum_{z=0,1,2} T_0^z(\varepsilon_f^*; \varepsilon_i)_{E1} \langle g|F_0^z(\omega)_{E1}|g\rangle$$

$$= -\frac{1}{\sqrt{3}} \varepsilon_f^* \cdot \varepsilon_i F_0^0(\omega) - i\frac{1}{\sqrt{2}} (\varepsilon_f^* \times \varepsilon_i) \cdot \mathbf{z} F_0^1(\omega)$$

$$+ \frac{1}{\sqrt{30}} [3(\varepsilon_f^* \cdot \mathbf{z})(\varepsilon_i \cdot \mathbf{z}) - \varepsilon_f^* \cdot \varepsilon_i] F_0^2(\omega) . \quad (66)$$

where $\mathbf{z}$ is a direction along the cylinder axis and $F_0^z(\omega)$ is short for $\langle g|F_0^z(\omega)_{E1}|g\rangle$.

16.1 Diffraction Conditions

The diffracted amplitude generated by the whole crystal is $\sum_s f_s^{EL} e^{i\mathbf{q}\cdot\mathbf{R}_s}$ with $\mathbf{q} = \mathbf{k}_i - \mathbf{k}_f$ the scattering wave vector and s running over all sites. Thus a necessary condition for reflection is that the total phase factor in $\sum_s f_s^{EL} e^{i\mathbf{q}\cdot\mathbf{R}_s}$ does not vanish. If all sites have the same f^{EL} then we have $f^{EL} \sum_s e^{i\mathbf{q}\cdot\mathbf{R}_s}$, which is the result for normal Bragg scattering. However, if different sites have different orientations, e.g. an angular rotation by $2\pi/n$ of the spherical harmonics, with φ dependence $e^{i\varphi}$, gives a phase shift $e^{im2\pi/n}$, we obtain $f^{EL} \sum_s e^{im2\pi/n_s + i\mathbf{q}\cdot\mathbf{R}_s}$. Screw axes and glide planes can give constructive interference [74]. For instance, when $\mathbf{q}$ is along an n_j-fold screw axis parallel to $\mathbf{c}$ (i.e. with rotation $2\pi/n$ and translation $j\mathbf{c}/n$, the spherical harmonics change by a factor $e^{im2\pi/n}$ and the Bragg factor $e^{i\mathbf{q}\cdot\mathbf{R}}$ changes by a phase shift $e^{i\mathbf{q}\cdot\mathbf{c}j/n}$. Since this factor must reduce to unity, the reflection is allowed only when $\mathbf{q} \cdot \mathbf{c}/2\pi = (nk - m)/j$, with k an integer. The application is greatly facilitated if one is familiar with the symmetry properties of the spherical harmonics, as described in Sect. 17.

17 Symmetry Properties of Spherical Tensors

Before proceeding to the more advanced part in Sects. 18 and 20, which treats the electric and magnetic multipole fields and the x-ray optical activity, it is useful to recollect several symmetry properties. Continuous symmetries arising from homogeneity of space lead to conservation laws, e.g. the angular momentum conservation is a consequence of the rotational invariance of a system (c.f. Noether's theorem [30]). Like discrete symmetries that are associated with inversion of space and time do not lead to conservation laws and observables they lead to selection rules.

17.1 Rotation

Under an arbitrary rotation $\mathcal{D}$ of the coordinate system described by the Euler angles (α, β, γ) the spherical harmonics transform as

$$\mathcal{D}(\alpha, \beta, \gamma) Y_{lm'}(\vartheta, \varphi) = Y_{lm'}(\vartheta', \varphi') = \sum_m Y_{lm}(\vartheta, \psi) D^l_{mm'}(\alpha, \beta, \gamma)$$

$$= e^{im'\alpha} \sum_m Y_{lm}(\vartheta, \psi) e^{im\gamma} d^l_{mm'}(\beta) . \tag{67}$$

For instance, a rotation φ about the z axis takes Y_{lm} to $e^{im\varphi}Y_{lm}$, a rotation by π about the x axis takes it to $(-)^l Y_{l-m}$ and a rotation about the y axis takes it to $(-)^{l+m}Y_{l-m}$.

Wave functions transform under unitary transformation (defined with $U^\dagger = U^{-1}$) as $\Psi' = U\Psi$, whereas operators transform as $T' = UTU^\dagger$, e.g. for the Hamiltonian $UHU^{-1} = H$ or $[U, H] = 0$.

17.2 Parity

Under inversion (i.e. reflection through the origin) the *parity operator* P changes the *covariant* variables $u = (x, y, z) = (r, \pi, \varphi)$ into the *contravariant* variables $\bar{u} = (-x, -y, -z) = (r, \pi - \vartheta, \varphi + \pi)$. The spherical harmonics transform as

$$PY_{lm}(u) = Y_{lm}(\bar{u}) = (-)^l Y_{lm}(u) . \tag{68}$$

Thus spherical harmonics are functions with parity $(-)^l$. The parity π of an operator A is defined by $PAP = \pi A$. Even parity operator $(\pi = +1)$

include the angular momentum $\mathbf{J} = \mathbf{r} \times \mathbf{p}$ (polar vector) and the magnetic dipole $\mu \cdot \mathbf{B}$. Odd parity operators ($\pi = -1$) include the position vector $\mathbf{r}$, the velocity $\frac{d\mathbf{r}}{dt}$, and hence the momentum $\mathbf{p}$, the electric dipole moment $\mathbf{D} = e\mathbf{r}$, and the helicity vector (chirality).

17.3 Hermiticity

The *complex conjugate* of the spherical harmonics is

$$K_0 Y_{lm}(u) \equiv Y_{lm}^*(u) = (-)^{-m} Y_{l-m}(u) \,, \tag{69}$$

where K_0 is the antiunitary operator for complex conjugation. The complex conjugate of the rotation matrix elements is

$$D_{mn}^{j*}(\alpha, \beta, \gamma) = D_{nm}^j(-\gamma, -\beta, -\alpha) = (-)^{m-n} D_{-m-n}^j(\alpha, \beta, \gamma) \,, \tag{70}$$

The hermitian conjugate of the spherical harmonic, $Y_{lm}(u) = \langle u | lm \rangle$, is [25]

$$Y_{lm}^\dagger(u) \ = Y_{lm}^*(\overline{u}) = (-)^l Y_{lm}^*(u) \ = (-)^{l-m} Y_{l-m}(u) \,,$$
$$\text{or} \quad \langle lm | u \rangle = \langle \overline{u} | lm^* \rangle = (-)^l \langle u | lm^* \rangle = (-)^{l-m} \langle u | l - m \rangle \,, \tag{71}$$

where the last line gives the same expression in the Dirac bra–ket notation. This shows that $Y_{lm}^\dagger$ transforms *contragrediently* to Y_{lm}, i.e.

$$|Y_{lm}^\dagger\rangle = (-)^{l-m} |Y_{l-m}\rangle = \langle Y_{lm}| \,. \tag{72}$$

Therefore, at most only the $m = 0$ components can be hermitian, i.e. $Y_{l0}^\dagger(\theta, \varphi) = Y_{l0}(\theta, \varphi) = \sqrt{\frac{2l+1}{4\pi}} P_l(\cos \theta)$ is real, while the m components have azimuthal dependence $e^{im\varphi}$.

For an operator A we have $K_0 A K_0^{-1} = \pm A$, where the upper (lower) sign is for a real (pure imaginary) operator. The *hermitian conjugate* or *adjoint* of a matrix element for a linear operator is defined as the transposed complex conjugate

$$A_{ij}^\dagger = (A_{ji})^* = A_{ji}^* \,, \quad \text{i.e.} \quad \langle i | A^\dagger | j \rangle = \langle j | A | i \rangle^* \,. \tag{73}$$

$A^\dagger$ transforms contragrediently to A. An operator is called hermitian or self-adjoint if $A^\dagger = A$, in which case $A_{ji}^* = A_{ij}$. A necessary and sufficient condition for this is that all the eigenvalues are real. For an antihermitian operator we have $A^\dagger = -A$, in which case $A_{ji}^* = -A_{ij}$, so that the matrix is imaginary. We can combine both cases by taking $A^\dagger = (-)^\tau A$ and $A_{ji}^* = (-)^\tau A_{ij}$, for an operator which is either hermitian (τ is even) or antihermitian (τ is odd). Alternatively, we can define the operator as $A^\tau = i^\tau A$, where A is hermitian.

To extend the notion of hermitian operators to spherical tensors, $T_q^{(k)}$, one defines

$$T_q^{(k)\dagger} = (-)^p T_q^{(k)*} = (-)^{p-q} T_{-q}^{(k)} \quad \forall \; T = \text{hermitian} \; . \tag{74}$$

The phase $(-)^q$ is essential to preserve the correct rotational properties [c.f. the spherical harmonic property $C_q^{(k)*} = (-)^q C_{-q}^{(k)}$]. The choice of p is arbitrary and varies in the literature. Condon-Shortley and Schwinger use $p = 0$, Edmonds and Fano-Racah use $p = k$ (= the parity), see e.g. Brink and Satchler [12]. Consequently, operators that fulfil $T_q^{(k)\dagger} = (-)^{p+1} T_q^{(k)*}$ are antihermitian. We can combine hermitian tensors (τ is even) and anti-hermitian tensors (τ is odd) by writing

$$T_q^{(k,\tau)\dagger} = (-)^{p-\tau} T_q^{(k,\tau)*} = (-)^{p-\tau-q} T_{-q}^{(k,\tau)} \; . \tag{75}$$

Using the contragredient phase convention for the wave function $|jm^\dagger\rangle = \langle jm| = (-)^{j-m}|j-m\rangle$ the matrix element form of (75) is

$$\langle jm|T_q^{(k,\tau)\dagger}|j'm'\rangle = (-)^{p-\tau}\langle j'm'|T_q^{(k,\tau)}|jm\rangle^* = (-)^{p-\tau-q}\langle jm|T_{-q}^{(k,\tau)}|j'm'\rangle \; . \tag{76}$$

This gives the hermitian and conjugation properties for the reduced matrix elements as [12]

$$\langle j\|T^{(k,\tau)\dagger}\|j'\rangle = (-)^{p-\tau}\langle j'\|T^{(k,\tau)}\|j\rangle^* = (-)^{j-j'-p+\tau}\langle j\|T^{(k,\tau)}\|j'\rangle \; . \tag{77}$$

Thus the reduced matrix element is real (pure imaginary) if $j - j' + \tau$ is even (odd).

The tensor $C_q^{(k)} f_l(r)$ expressed in terms of a modified spherical harmonic $C_q^{(k)}$ and a radial function $f_l(r)$, carries no current ($\tau = 0$) and we can take $p = k$, so that

$$C_q^{(k)\dagger} = (-)^k C_q^{(k)*} = (-)^{k-q} C_{-q}^k \; , \tag{78}$$

with $C_q^{(k)}$ is hermitian and the matrix elements that are real for all k.

Tensors describing physical interactions have integer k, while spinors $\chi_{s\sigma}$ have half-integer s values.

17.4 Time Reversal

Time reversal is described by an antiunitary operator $\Theta = K_0 U = K_0 e^{i\pi S_y}$, where U is a unitary operator and $e^{i\pi S_y} = \sum_m d_{m'm}^j(\pi) = \sum_m (-)^{j-m} \delta_{m',-m}$ for the y component of the spin operator. For the wave function $\Psi_{jm} = |jm\rangle$ the choice

$$\Theta|jm\rangle = (-)^{j-m}|j-m\rangle \; , \tag{79}$$

is invariant with respect to angular momentum coupling. It gives the same phase as a rotation about the y axis, as can be verified with (67). The time reversal Θ has the matrix representation

$$\mathbf{C}^{(j)} = \begin{pmatrix} 0 & \cdots & 0 & 0 & 1 \\ 0 & \cdots & 0 & -1 & 0 \\ 0 & \cdots & 1 & 0 & 0 \\ & \cdots & & & \cdot \end{pmatrix}, \tag{80}$$

which has elements $\mathbf{C}^{(j)}_{m'm} = (-)^{j-m}\delta_{m',-m}$. Hence $\mathbf{C}^{(j)}$ is symmetric for integral j and skew symmetric for half-integral j [11].

The usual spherical harmonics with Y_{l0} is real [so that $\Theta Y_{lm} = Y^*_{lm} = (-)^m Y_{lm}$] will not have the desired time reversal property (79) but the functions $\mathcal{Y}_{lm}(u) \equiv i^l Y_{lm}(u)$ do since

$$\Theta \mathcal{Y}_{lm} \equiv \Theta i^l Y_{lm} = (-)^l i^l Y^*_{lm} = (-)^{l-m} i^l Y_{l-m} = (-)^{l-m} \mathcal{Y}_{l-m}, \tag{81}$$

and the coupled wavefunctions

$$\Psi_{jm} = f_l(r) \sum_m C^{jm}_{lms\sigma} \mathcal{Y}_{lm} \chi_{s\sigma}, \tag{82}$$

with the radial dependence $f_l(r)$, have correct time reversal properties (79).

The time reversal $\Theta|jm\rangle = (-)^{p-m}|j-m\rangle$ with the integer p not specified implies that $\Theta^2|jm\rangle = (-)^{2m}|jm\rangle = (-)^{2j}|jm\rangle = \pm|jm\rangle$, with the upper and lower sign for j is integer and half-integer, respectively. For $\Theta^2|jm\rangle = -|jm\rangle$, the state $|jm\rangle$ is degenerate with its time-reversed state. The two-fold degeneracy in energy of an atomic state with an odd number of electrons (hence half-odd j) is referred to as *Kramer's degeneracy*. The degeneracy is lifted by a velocity dependent field such as the magnetic field $\mathbf{B}$.

Time-even operators ($\tau = 0$) include $\mathbf{r}$, ∇, the Hamiltonian H and the electric field $\mathbf{E}$. Time-odd operators ($\tau = 1$) include $\mathbf{p} = -i\hbar\nabla$, the wave vector $\mathbf{k}$, $\mathbf{L}$, $\mathbf{S}$, $\mathbf{J}$, the magnetic field $\mathbf{B}$, the vector potential $\mathbf{A}$ and the velocity $\mathbf{v}$. This applies to operators in coordinate representation. In momentum representation, $\mathbf{p}$ is real and therefore time-even, while $\mathbf{r}$ is pure imaginary and time-odd. $\mathbf{J}$ is time-odd in either representation.

17.5 Time Reversal Phase

The rule for time reversal transformation of matrix elements is [26]

$$\Theta\left[\langle jm|T|j'm'\rangle\right] = (-)^\tau \langle jm|T|j'm'\rangle^* = (-)^p \langle j'm'|T^\dagger|jm\rangle. \tag{83}$$

Taking the contragredient state as the time reversed state, $\Theta|jm\rangle = \langle jm| = (-)^{l-m}|j-m\rangle$, we can write

$$\theta T^{(k,\tau)}_q \theta^{-1} = (-)^p T^{(k,\tau)\dagger}_q = (-)^\tau T^{(k,\tau)*}_q = (-)^{\tau+q} T^{(k,\tau)}_{-q}, \tag{84}$$

for an operator that is either hermitian ($\tau = 0$) or antihermitian ($\tau = 1$).

It can be easily verified that a spherical tensor $T^{(k)}_q = i^\tau C^{(k)}_q f_l(r)$, with $C^{(k)}_0$ and $f_l(r)$ real, transforms under time reversal as (84). This clarifies the origin of the phase factors: the $(-)^\tau$ arises from the explicit time reversal symmetry of the operator in the specific representation and the $(-)^q$ arises from the rotation symmetry. The operator $T^{(k,\tau)}$ is hermitian (antihermitian) and the matrix elements are real (pure imaginary) for τ is even (odd). Below we give some examples for τ is 0 and 1 in spherical coordinates.

The position vector $\mathbf{r}$ can be described by a spherical tensor $r_q = rC_q^{(1)}$ of rank $k = 1$, which is time even (τ is even). The covariant spherical coordinates r_q are defined as $r_{\pm 1} = \mp r \sin \vartheta \, e^{\pm i\varphi} \sqrt{2} = \mp (x \pm iy)/\sqrt{2}$ and $r_0 = r \cos \vartheta = z$, so that $r_q^* = (-)^q r_{-q}$. The contravariant spherical coordinates are given by $r^q = r_q^*$ [27]. The time reversed operator is

$$\theta r_q \theta^{-1} = (-)^p r_q^\dagger = r_q^* = (-)^q r_{-q} \, . \tag{85}$$

Since $r_q^\dagger = (-)^p r_q^*$, the tensor is hermitian and the matrix elements are real. Thus for the unit vector of the light polarization with $q = \{-1, +1\}$ we have $\hat{\varepsilon}_q = -\hat{\varepsilon}_{-q}^*$ and $\hat{\varepsilon}^q = \hat{\varepsilon}_q^*$ and $\theta \hat{\varepsilon}_q \theta^{-1} = \hat{\varepsilon}_q^*$.

For a current operator $J_q = i^\tau r C_q^{(1)}$ in spherical coordinate representation the matrix elements are $\langle jm'|J_z|jm \rangle = m\delta_{m'm}$ and $\langle jm'|J_\pm|jm \rangle = \sqrt{(j \mp m)(j \pm m + 1)}\delta_{m',m\pm 1}$, where $J_\pm \equiv J_x \pm iJ_y$. With $J_{\pm 1} \equiv \mp J_\pm / \sqrt{2} = \mp (J_x \pm iJ_y)/\sqrt{2}$, we have $J_q^* = (-)^q J_{-q}$. The current operator $\mathbf{J}$ has rank $k = 1$ and is time-odd ($\tau = 1$),

$$\theta J_q \theta^{-1} = (-)^p J_q^\dagger = -J_q^* = -(-)^q J_{-q} \, . \tag{86}$$

Since $J_q^\dagger = (-)^{p+1} J_q^*$, the tensor J_q is antihermitian and the matrix elements are imaginary ($J_0 = -i\frac{\partial}{\partial \varphi}$).

18 Electric and Magnetic Multipole Fields

The electromagnetic interaction is a one-body operator. We follow the derivation of Rose & Brink [5, 6, 12, 15, 17], which circumvents the use of vector spherical harmonics evoked in Sect. 16. The Hamiltonian of a single particle in the presence of a classic electromagnetic field is

$$\mathcal{H} = \frac{1}{2m} \left[\mathbf{p} + \frac{e}{c} \mathbf{A}(\mathbf{r}) \right]^2 + \mu \cdot \mathbf{H}(\mathbf{r}) + e\phi(\mathbf{r}) \, , \tag{87}$$

where e is the charge and $\mu = g\beta \mathbf{S} = \frac{1}{2}g\beta\sigma$ is the spin magnetic moment of the particle and $\beta = e\hbar/2mc$ is the Bohr magneton, $\mathbf{A}$ and ϕ are the vector and scalar electromagnetic potentials, respectively. In the following we will use natural units $\hbar = 1$. The gauge condition is taken such that $\phi = 0$ and we neglect the two-photon process given by $\mathbf{A}^2$ term. This leads to a kinetic energy term $\mathcal{H}_0 = \mathbf{p}^2/(2m)$ and an interaction Hamiltonian, comprising a charge and spin term,

$$\mathcal{H}_{\text{int}} = 2g_l\beta\mathbf{A}\cdot\mathbf{p} + g_s\beta\mathbf{S}\cdot(\nabla\times\mathbf{A}) , \tag{88}$$

where g_l and g_s are the orbital and spin g-factors. The aim is to expand the transverse plane wave $\mathbf{A}$ in a series of spherical vector fields $\mathbf{A}_{LM}^E$ and $\mathbf{A}_{LM}^M$, thereby expressing the interaction Hamiltonian in a series of tensor operators $\mathbf{T}_{LM}^E$ and $\mathbf{T}_{LM}^M$, which are the products of momentum $\mathbf{p}$ and spin $\mathbf{S}$ with the fields $\mathbf{A}_{LM}^E$ and $\mathbf{A}_{LM}^M$.

We define the electric and magnetic multipole fields

$$\mathbf{A}_{LM}^M = \frac{1}{\sqrt{L(L+1)}}\,\mathbf{L}\,\Phi_{LM}, \tag{89}$$

$$\mathbf{A}_{LM}^E = \frac{1}{k}\nabla\times\mathbf{A}_{LM}^M = \frac{1}{k\sqrt{L(L+1)}}\,\nabla\times\mathbf{L}\,\Phi_{LM} , \tag{90}$$

where the scalar functions

$$\Phi_{LM} = \mathrm{i}^L(2L+1)j_L(kr)C_{LM}(\theta,\phi) , \tag{91}$$

satisfy the wave equation $\nabla^2\Phi + k^2\Phi = 0$ and form a complete set. The expansion for the circularly polarized vector plane wave with component $q = \pm 1$ can be written as

$$\mathbf{A} = \mathbf{e}_q\exp(\mathrm{i}\mathbf{k}\cdot\mathbf{r}) = -\frac{1}{\sqrt{2}}\sum_{LM}(q\mathbf{A}_{LM}^M + \mathbf{A}_{LM}^E)\mathcal{D}_{Mq}^L(R)$$

$$= -\frac{1}{\sqrt{2}}\sum_{LM\pi}q^\pi\mathbf{A}_{LM}^\pi\mathcal{D}_{Mq}^L(R) , \tag{92}$$

where R represents the rotation which takes the z axis into the direction of the vector $\mathbf{k}$ and the index π in the compact expression runs over E ($\pi = 0$) and M ($\pi = 1$).

In the long wavelength approximation, $kr \ll 1$, the spherical Bessel functions obtain the approximate value $j_L(kr) \approx (kr)^L/(2L+1)!!$, so that the scalar potential Φ_{LM} takes the simplified form $\Phi_{LM} = r^L C_{LM}(\mathrm{i}k)^L/(2L-1)!!$. This gives for the interaction Hamiltonian

$$\mathcal{H}_{\text{int}} = -\frac{1}{\sqrt{2}}\sum_{LM\pi}\mathcal{D}_{Mq}^L(R)\left[2g_l\beta q^\pi\mathbf{A}_{LM}^\pi\cdot\mathbf{p} + g_s\beta q^\pi\mathbf{S}\cdot(\nabla\times\mathbf{A}_{LM}^\pi)\right] . \tag{93}$$

The charge and spin interaction of the electric and magnetic multipole fields result in a total of four terms in the interaction Hamiltonian,

$$\mathcal{H}_{\text{int}} = -\sum_{LM}\frac{(\mathrm{i}k)^L}{(2L-1)!!}\sqrt{\frac{L+1}{2L}}\mathcal{D}_{Mq}^L(R)\left[Q_{LM} + Q'_{LM} - \mathrm{i}q(M_{LM} + M'_{LM})\right] , \tag{94}$$

with the electric (Q) and magnetic (M) multipole operators

$$Q_{LM} = eg_l(r^L C_{LM}) = -2g_l\beta(\mathrm{i}/k)\nabla(r^L C_{LM})\cdot\mathbf{p}\,,$$
$$M_{LM} = 2\beta g_l\nabla(r^L C_{LM})\cdot\mathbf{L}/(L+1)\,,$$
$$Q'_{LM} = -k\beta g_s\mathbf{L}(r^L C_{LM})\cdot\mathbf{S}/(L+1)\,,$$
$$M'_{LM} = \beta g_s\nabla(r^L C_{LM})\cdot\mathbf{S}\,, \tag{95}$$

where the unprimed (primed) operators result from the charge (spin) interaction. The Hamiltonian can be cast in a simplified form using the *interaction multipole operators* T^π_{LM}, with $\pi = 0$ for E and $\pi = 1$ for M, which are sets of operators all having the same transformation properties,

$$H_{\text{int}} = -\sum_{LM\pi}\mathcal{D}^L_{Mq}(R)q^\pi T^\pi_{LM}\,,$$
$$T^E_{LM} = \alpha^E_L(Q_{LM} + Q'_{LM})\,,$$
$$T^M_{LM} = \alpha^E_M(M_{LM} + M'_{LM})\,,$$
$$\alpha^E_L = \frac{(\mathrm{i}k)^L}{(2L-1)!!}\sqrt{\frac{L+1}{2L}}\,, \text{ and } \alpha^M_L = -\mathrm{i}\alpha^E_L\,, \tag{96}$$

specifically we have $\alpha^E_1 = \mathrm{i}k$, $\alpha^M_1 = k$ and $\alpha^E_2 = -k^2/2\sqrt{3}$.

We have the parity rules $P\Phi_{LM} = (-)^L\Phi_{LM}$, $PA^E_{LM} = (-)^L A^E_{LM}$ and $PA^M_{LM} = (-)^L A^M_{LM}$. The parity of $\mathbf{A}^E\cdot\mathbf{p}$ and $\mathbf{S}\cdot(\nabla\times\mathbf{A}^E)$ is $(-)^L$, whereas the parity of $\mathbf{A}^M\cdot\mathbf{p}$ and $\mathbf{S}\cdot(\nabla\times\mathbf{A}^M)$ is $(-)^{L+1}$. Therefore, the electric and magnetic 2^L-pole interaction operators T^π_{LM} connect states that differ in parity by a factor $(-)^{L+\pi}$. Thus $E1$ has odd parity, while $M1$ and $E2$ have even parity.

Table 9 shows the terms in the expansion of $\exp(\mathrm{i}k\cdot\mathbf{r})$ separated in geometric and dynamic factors. The different orders in the expansion are alternatingly imaginary and real. The parity operator gives $P(\mathrm{i},\hat{\varepsilon},\hat{k},\mathbf{p},\mathbf{r}) = (\mathrm{i},-\hat{\varepsilon},-\hat{k},-\mathbf{p},-\mathbf{r})$.

Table 9. The n-th order geometric and dynamic factors in the $\exp(\mathrm{i}k\cdot\mathbf{r})$ expansion giving the different electric and magnetic multipole terms (n is the number of times that $\hat{k}$ appears). α^π_L is the prefactor. The parity of the geometric and dynamic factors

n	Term	α^π_L	Geometric Factor	Dynamic Factor	Parity
0	$E1$	$\mathrm{i}k$	$\hat{\varepsilon}^{(1)}$	$\mathbf{p}$	-1
$1\{$	$M1$	k	$\{\hat{\varepsilon},\hat{k}\}^{(1)}$	$\{\mathbf{p},\mathbf{r}\}^{(1)}$	$+1$
	$E2$	$-k^2/2\sqrt{3}$	$\{\hat{\varepsilon},\hat{k}\}^{(2)}$	$\{\mathbf{p},\mathbf{r}\}^{(2)}$	$+1$
$2\{$	$M2$	$\mathrm{i}k^2/2\sqrt{3}$	$\{\hat{\varepsilon},\hat{k}^{(2)}\}^{(2)}$	$\{\mathbf{p},\mathbf{r}^{(2)}\}^{(2)}$	-1
	$E3$	$-\mathrm{i}k^3\sqrt{6}/45$	$\{\hat{\varepsilon},\hat{k}^{(2)}\}^{(3)}$	$\{\mathbf{p},\mathbf{r}^{(2)}\}^{(3)}$	-1
$L\{$	ML	$\frac{-\mathrm{i}^{L+1}k^L}{(2L-1)!!}\sqrt{\frac{L+1}{2L}}$	$\{\hat{\varepsilon},\hat{k}^{(L)}\}^{(L)}$	$\{\mathbf{p},\mathbf{r}^{(L)}\}^{(L)}$	$(-)^{L+1}$
	$E(L+1)$	$\frac{\mathrm{i}^{L+1}k^{L+1}}{(2L+1)!!}\sqrt{\frac{L+2}{2L+2}}$	$\{\hat{\varepsilon},\hat{k}^{(L)}\}^{(L+1)}$	$\{\mathbf{p},\mathbf{r}^{(L)}\}^{(L+1)}$	$(-)^{L+1}$

The time-reversal operator gives $\Theta(i, \hat{\varepsilon}, \hat{k}, \mathbf{p}, \mathbf{r}) = (-i, \hat{\varepsilon}^*, -\hat{k}, -\mathbf{p}, \mathbf{r})$, hence the time dependent part $i^{L+1}\hat{k}^L\mathbf{p}$ of the interaction operator T^π_{LM} is always time reversal even.

The zeroth-order contribution $(n = 0)$ gives the $E1$ term $\hat{\varepsilon} \cdot \mathbf{p} \propto \omega\hat{\varepsilon} \cdot \mathbf{r}$. The geometric factor $\hat{\varepsilon}$ is time-even and parity-odd.

The first-order contribution $(n = 1)$ can be factorized in a geometric factor $\{\hat{\varepsilon}, \mathbf{k}\}^{(z)}$ and dynamic factor $\{\mathbf{p}, \mathbf{r}\}^{(z)}$, which are both parity-even. For $z = 0$, the geometric factor vanishes due to transversality, i.e. $\hat{\varepsilon} \cdot \mathbf{k} = 0$. The $z = 1$ term gives the magnetic dipole contribution $(M1)$ with a geometric factor $\{\hat{\varepsilon}, \mathbf{k}\}^{(1)} \propto (\mathbf{k} \times \hat{\varepsilon})$ and dynamic factor $\{\mathbf{p}, \mathbf{r}\}^{(1)} \propto (\mathbf{p} \times \mathbf{r}) \propto \mathbf{L}$. The $z = 2$ term yields the electric quadrupole contribution $(E2)$ with a dynamic factor $\{\mathbf{p}, \mathbf{r}\}^{(2)} \propto \mathbf{L}^2 \propto$ the charge quadrupole.

Thus the leading terms in the transition amplitude can be written as

$$-\frac{m}{\hbar}(E_f - E_i)\left[\langle f|\hat{\varepsilon} \cdot \mathbf{r}|i\rangle + \frac{i}{2}\langle f|(\hat{\varepsilon} \cdot \mathbf{r})(\mathbf{k} \cdot \mathbf{r})|i\rangle\right] - \langle f|(\mathbf{k}\times\hat{\varepsilon})\cdot(\mathbf{L}+g\mathbf{S})|i\rangle , \quad (97)$$

where the first, second and third terms represent the electric dipole $(E1)$, electric quadrupole $(E2)$ and magnetic dipole $(M1)$ transition amplitudes, respectively. The transition probabilities of the $E2$ and $M1$ are $(Z_{\text{eff}}/137)^2$ times smaller than that of the $E1$. The $M1$ operator, with the monopole selection rules $\Delta J \leq 1$, $\Delta L = 0$, $\Delta S = 0$ and $\Delta n = 0$, does not contain a radial variable, therefore, its matrix element vanishes if the radial part of the initial and final states are orthogonal. The $M1$ transition occurs only at low excitation energies and can be neglected in core-level spectroscopy, since it requires configuration interaction. The explicit angular dependence of the $E1$ and $E2$ transitions can be found in [76].

According to the Golden Rule of quantum mechanics the absorption cross section $\sigma(\omega)$ is given by the square modulus of the matrix element of the interaction between initial and final state, summed over the final states with the same energy

$$\sigma(\omega) = \frac{4\pi^2\hbar\alpha}{m^2\omega}\sum_f |\langle f|\mathcal{H}_{\text{int}}|i\rangle|^2 \, \delta(E_f - E_i - \hbar\omega) , \quad (98)$$

which includes the interference terms $\sum_f \langle i|\mathcal{H}^*_{n'}|f\rangle\langle f|\mathcal{H}_n|i\rangle$.

19 Symmetry of Multipole Transitions

The $ELEL'$ absorption cross section can be written as

$$\sigma_{ELEL'} = \sum_{b=|L-L'|}^{L+L'} \sum_{\beta=-b}^{b} (-)^\beta \sum_{\tau=0,1} \mathbf{T}^{(b,\tau)}_\beta \sigma^{(b,\tau)}_{-\beta} , \quad (99)$$

where the tensors $\mathbf{T}_\beta^{(b,\tau)}$ and $\sigma_{-\beta}^{(b,\tau)}$ give the geometric and dynamic factor, respectively, of rank b and time reversal phase factor $(-)^\tau$. The integrated absorption cross sections are related by sum rules to the expectation values of ground state tensors $w_{-\beta}^{(b,\tau)}$. The geometric factor of the squared matrix elements in (98) is (c.f. Table 9)

$$\mathbf{T}_\beta^{(b,\tau)} = \{\{\hat{\varepsilon}^*, \hat{k}^{(L-1)}\}^{(L)}, \{\hat{\varepsilon}, \hat{k}^{(L'-1)}\}^{(L')}\}_\beta^{(b,\tau)} , \qquad (100)$$

with parity condition $(-)^{\Pi} = (-)^{L+L'}$. The time reversal symmetry can be either even ($\tau = 0$) or odd ($\tau = 1$), which cannot be differentiated from the tensor form in (100). To obtain the time reversal symmetry we need to recouple the geometric factor to the form

$$\mathbf{X}^{(b,\tau)} = \{\{\hat{\varepsilon}^*, \hat{\varepsilon}\}^{(a)}, \hat{k}^{(N)}\}^{(b,\tau)} , \qquad (101)$$

where $N = L + L' - 2$ is the number of $\hat{k}$ vectors appearing in the geometric factor. Thus the parity condition is $(-)^{\Pi} = (-)^N$. The tensor $\mathbf{X}$ has the triangle condition, $|N - a| \le b \le N + a$. The time reversal condition is

$$(-)^\tau = (-)^{a+N} = (-)^{a+\Pi} . \qquad (102)$$

For $a = 0$, where $\hat{\varepsilon}^* \cdot \hat{\varepsilon} = 1$, the geometric factor $\hat{k}^{(N)}$ is polarization independent, and we have $(-)^\tau = (-)^b = (-)^N = (-)^{\Pi}$. Thus these indices must be either all even or all odd.

For $a = 1$, we have $(\hat{\varepsilon}^* \times \hat{\varepsilon}) \propto i\hat{k}$. This gives a *chiral effect*, defined as one in which the intensity contribution is reversed by conjugating the polarization vectors [77]. Since $\hat{k} \times \hat{k}^{(N)} = 0$ we have $b = N \pm 1$, and using $(-)^\tau = (-)^{N+1}$ we obtain $b + \tau$ is even. Equation (102) leads to two different possibilities for the chiral effect. $\Pi = 0$ and $\tau = 1$, which gives magnetic circular dichroism (MCD) with b odd and N even. The other possibility is $\Pi = 1$ and $\tau = 0$, which gives the natural circular dichroism (NCD) with b even and N odd (c.f. Table 10).

For $a = 2$ we have $(-)^\tau = (-)^{\Pi} = (-)^N$ while b can have different values.

In the case that $L = L'$, the parity is even so that $(-)^\tau = (-)^a$, hence τ is even (odd) for a is even (odd). In the special case $N = 0$ (i.e. $E1E1$) we have $\mathbf{T}^{(b,\tau)} = \{\hat{\varepsilon}^*, \hat{\varepsilon}\}^{(a)}$, hence $a = b$. For the isotropic spectrum ($a = 0$) the sum rule gives the number of particles n [78]. For the x-ray magnetic circular dichroism (XMCD) ($a = 1$) the sum rule gives the orbital moment $\mathbf{L}$ [79]. For the x-ray magnetic linear dichroism (XMLD) ($a = 2$) the sum rule gives the quadrupole moment $Q \propto \mathbf{L}^2$ [80, 81].

All interference terms will have N odd, so that $(-)^\tau = (-)^{a+1}$, i.e. $\tau + a$ is odd, hence τ is even (odd) for a is odd (even).

The effective operators which form the sum rule results can be built from the triad of mutually orthogonal vector operators [82, 83]: $\mathbf{n} = \mathbf{r}/r$, which is a time-even, polar vector, typically associated with the electric dipole moment. The orbital angular momentum $\mathbf{L}$, which is a time-odd, axial vector.

The toroidal vector $\boldsymbol{\Omega} = \frac{1}{2}[(\mathbf{n} \times \mathbf{L}) - (\mathbf{L} \times \mathbf{n})] = \frac{1}{2}\mathrm{i}[\mathbf{n}, \mathbf{L}^2]$, which is odd with respect to both inversion and time reversal and proportional to the orbital anapole moment.

The $E1E2$ interference has a geometric factor $\mathbf{T}_\beta^{(b,\tau)} = \{\hat{\varepsilon}^*, \{\hat{\varepsilon}, \hat{k}\}^{(2)}\}_\beta^{(b,\tau)}$ and the multipole selection rules impose a mixing (hybridization) between l and $l+1$ valence levels. The following results have been reported [82, 83]:

The x-ray magnetochiral dichroism (XMχD) sum rule involves the ground state expectation value of the toroidal vector $\boldsymbol{\Omega}^{(1,\tau=1)}$ with a geometric factor $\mathbf{T}_0^{(1,\tau=1)} = -\frac{1}{2}\sqrt{\frac{3}{5}}\hat{k}$, associated with the Stokes component S_0.

The XNCD sum rule gives the expectation value of the time-even pseudo-deviator $\mathbf{N}^{(2,\tau=0)} = \{\mathbf{L}, \boldsymbol{\Omega}\}^{(2)}$ and $\mathbf{T}^{(2,\tau=0)} = \frac{1}{2}\sqrt{3}\{\{\hat{\varepsilon}^*, \hat{\varepsilon}\}^{(1)}, \hat{k}\}_0^{(2)}$.

For the non-reciprocal XMLD the effective operator must be a time-odd pseudodeviator $\mathbf{W}_{\pm 2}^{(2,\tau=1)} = \{\mathbf{L}, \mathbf{n}\}^{(2)}$ and $\mathbf{T}_{\pm 2}^{(2,\tau=1)} = \frac{1}{2}\{\{\hat{\varepsilon}^*, \hat{\varepsilon}\}, \hat{k}\}_{\pm 2}^{(3)}$.

An effective operator $\Gamma^{(3,\tau=1)} = \{\{\mathbf{L}, \mathbf{L}\}^{(2)}, \boldsymbol{\Omega}\}^{(3)}$ is involved in the sum rules for both XMχD and nonreciprocal XMLD.

20 X-ray Optical Activity

The results from Sect. 19 allow us to give a simple physical description of direct optical effects. Optical activity induced by pure transitions must have even parity. The only way of inducing natural (field independent) optical activity is by interference between multipoles of different order in the k-expansion of the photon wave vector. Only the interference of a real and an imaginary term T_{LM}^π can lead to odd parity. The $E1M1$ interference in the visible region allows the detection of natural circular dichroism (NCD) and optical rotation (OR), such as in the well-known example of the sugar solution. Although in the visible region the $E2$ transitions are negligibly small, their magnitude increases with photon energy, so that for harder x-rays the $E1E2$ interference can observed. We can distinguish the different cases, given in Table 10, for the properties of the optical activity tensor under space inversion and time reversal. Each one corresponds to a different optical effect and a different way to measure the dichroism.

As a general rule, dichroism can only exist if there is no symmetry that reverses one measurable observable but leaves the rest of the system unchanged. In the Faraday effect the time reversal operator reverses both the magnetization and the direction of the light, thereby leaving the total physical system invariant. In natural circular dichroism, P reverses the rotation angle (screw sense) of both the medium and the light, again leaving the total system invariant. Non-zero effects show up only in substances that have the appropriate symmetry class. Non-reciprocal optical effects invoke magnetic moments, since they are time non-invariant. They are characterized by a reversal of the phase rotation when the light propagates in the opposite direction, such as in the Faraday rotation in transmission or the Kerr

Table 10. The different optical effects. The phase factors for parity (P) and time reversal (T) symmetry are $(-)^{\Pi}$ and $(-)^{\tau}$, respectively. The electric field, **E**, is a polar (i.e. parity-odd) time-even vector and the magnetic field, **M**, is an axial (i.e. parity-even) time-odd vector [84]

Π	τ	Optical Effect	Difference Signal
0	0	Birefringence	Magnetic linear dichroism (MLD)
0	1	Faraday rotation	Magnetic circular dichroism (MCD)
1	0	Optical rotation	Natural circular dichroism (NCD)
1	1	Magneto-electricity	Non-reciprocal linear dichroism

effect in reflection. Also magnetic circular dichroism (MCD), the difference in absorption for left and right circularly polarized light in the presence of an external magnetic field, belongs to this category. Non-reciprocal optical effects are not restricted to media possessing a net magnetic moment, such as ferro- and ferrimagnets. Magnetoelectric antiferromagnets belong to the special class of magnetic ordered materials where there is no net magnetic moment but, in addition to time reversal, also the parity symmetry is broken, while the combined symmetry operation is conserved. Goulon et al. [85] reported a non-reciprocal transverse anisotropy in the low-temperature antiferromagnetic insulating phase of a Cr-doped V_2O_3 crystal in which one single antiferromagnetic domain was grown by magnetoelectric annealing.

For a macroscopic description of the optical activity of molecular systems in terms of spatial dispersion, see [86–88].

21 Statistical Moment Analysis

Moment analysis offers a method which is not based on the calculation of all the individual line positions and intensities in the spectrum but that obtains physical properties from the global intensity distribution [89–92]. The method analyses spectra in terms of the moments of the distribution of their intensity over the energy. The n-th statistical moment of an intensity distribution I is defined as $I^{(n)} = \int I(\omega)\omega^n d\omega$, where the integral is taken over a specified part of the spectrum. The essential idea is that the n-th moment of a spectrum is proportional to the ground state expectation value of an operator $T^{\dagger}\mathcal{H}^n T$, where T is the transition operator and $\mathcal{H}$ is the final state Hamiltonian containing all possible one and two-electron operators of the core and valence shells. Operators acting on a core level can be removed and so we obtain the ground state expectation value of a valence shell operator.

Sum rules have been derived for the circular and linear dichroism signals integrated only over a single core level absorption edge, where the integrals are proportional to the ground state expectation values of the orbital magnetic moment [79] and the quadrupole moment [80,81], respectively. The sum rules use the integral of the spectrum, which is its zeroth statistical moment. Results with the same validity as the sum rules can be obtained for higher statistical moments. In x-ray absorption of deep core levels the first moment of the spectrum is immediately related to the branching ratio of the spin-orbit split edge, where the branching ratio changes with the ground state expectation value of spin-orbit coupling and spin [35, 36].

Moment analysis is especially useful for difference spectra, such as circular or linear dichroism in photoemission which have a zero integral, i.e. they have a vanishing zeroth moment. In this case the first moment is independent of the zero of energy and a quantitative analysis is possible.

Consider a spectrum with intensities given by matrix elements of an operator T between a ground state $|g\rangle$ and a set of final states $|f\rangle$. The intensity is given by

$$I(\omega) = \sum_f \langle g|T^\dagger|f\rangle\langle f|T|g\rangle \delta(E_f - E_g - \omega) . \tag{103}$$

We take the energy of $|g\rangle$ as zero. We consider cases where T is of such a form that it can only reach a limited set of states $|f\rangle$ from $|g\rangle$. Since the states $|f\rangle$ are eigenstates of $\mathcal{H}$ they form a complete set. The moments of the spectrum are defined by

$$I^{(n)} \equiv \int_0^\infty d\omega\, \omega^n I(\omega) = \sum_f \langle g|T^\dagger|f\rangle\langle f|T|g\rangle E_f^n$$
$$= \sum_{ff'} \langle g|T^\dagger|f\rangle\langle f|\mathcal{H}^n|f'\rangle\langle f'|T|g\rangle = \langle g|T^\dagger \mathcal{H}^n T|g\rangle \equiv \langle T^\dagger \mathcal{H}^n T\rangle . \tag{104}$$

Thus we will now be able to find operators that are equivalent to $T^\dagger \mathcal{H}^n T$. For $n = 0$ we measure $\langle T^\dagger T\rangle$, which gives the known "sum rules" for dichroism and spin polarization in x-ray absorption [79, 93, 94] and photoemission [62]. For $n = 1$ we obtain new results measuring $\langle T^\dagger \mathcal{H} T\rangle$, where $\mathcal{H}$ contains a set of terms such as Coulomb, exchange and spin-orbit interactions, crystal fields and magnetic fields.

It is important to note that we do not have to specify what kind of wave functions $|g\rangle$ and $|f\rangle$ refer to. The derivations can be read in more than one way. In a many-body approach the whole system is described by one complicated many-body wave function. Transitions are made to a huge set of many-body final states. In a one-particle model $|g\rangle$ denotes a distribution of the electrons over the valence band and the final states are all possible states with one hole in the core level and one extra electron in the valence band. The ground state expectation value of an operator takes the sum over all occupied or unoccupied levels. Note that $|g\rangle$ does not mean the core level and $|f\rangle$ is not a valence band level. On the contrary, in a one-electron model it is often more appropriate to work with holes and then in XAS $|g\rangle$ denotes a hole in the valence band and $|f\rangle$ a hole in the core level.

22 Spectral Moments for Ground State Tensors

The ground state of a level l can be characterized by coupled tensors moments $\langle \underline{w}^{xyz} \rangle$, where the orbital moment x and the spin moment y are coupled to a total moment z [95]. Moments with even x describe the shape of the charge distribution and moments with odd x describe orbital motion. An underline signifies that the moment is not taken over the electrons but over the *holes*, as required for e.g. x-ray absorption. Thus $\langle \underline{w}^{000} \rangle = n_h$ gives the number of holes. The $3d$ shell contains the spin-orbit coupling $\langle \underline{w}^{110} \rangle = -\sum_i l_i \cdot s_i$, orbital magnetic moment $\langle \underline{w}^{101} \rangle = \frac{1}{2} L_z$, spin magnetic moment $\langle \underline{w}^{011} \rangle = 2S_z$, magnetic dipole term $\langle \underline{w}^{211} \rangle = \frac{7}{2} T_z$, and quadrupole moment $\langle \underline{w}^{202} \rangle = -\frac{1}{2} Q_{zz}$. The w-tensors have the normalization $\langle \underline{w}^{xyz} \rangle = (-)^z$ for a ground state with a single hole, i.e. d^9.

For an electron from a core level $c, \frac{1}{2}, j$ with orbital, spin and angular components γ, σ, m the transition probability into an incompletely filled valence shell l with orbital components λ using q-polarized electric-dipole radiation along the magnetization direction is given by an operator

$$T_q = \sum_{\gamma\sigma} (-)^{l+j-\lambda-m} \begin{pmatrix} l & 1 & c \\ -\lambda & q & \gamma \end{pmatrix} \begin{pmatrix} j & \frac{1}{2} & c \\ -m & \sigma & \gamma \end{pmatrix} l^\dagger_{\lambda\sigma} j_{m\sigma} , \qquad (105)$$

where $l^\dagger_{\lambda\sigma}$ is a creation operator, $j_{m\sigma}$ is an annihilation operator. Reduced matrix elements leading only to an overall scaling have been omitted. This gives the x-ray absorption spectrum with q-polarized light of frequency ω for a ground state $|g\rangle$ and final states $|f\rangle$ as

$$I_q(\omega) = \langle g|T_q^\dagger|f\rangle\langle f|T_q|g\rangle\delta(E_f - E_g - \omega) , \qquad (106)$$

and the XMCD signal is $I = I_1 - I_{-1}$.

It is straightforward to express the dichroic signal of the jm final states as a linear combination of ground state moments [91]

$$I_{jm} = \sum_{xyz} \langle \underline{w}^{xyz} \rangle C_{jm}^{xyz} , \qquad (107)$$

where the coefficients C_{jm}^{xyz} give the probability to create a core hole with quantum numbers jm for a ground state moment $\langle \underline{w}^{xyz} \rangle$ equal to unity.

To expand the signal of a j level in multipole moments r we write

$$C_{jm}^{xyz} = \sum_r C_j^{xyzr} u_{jm}^r , \qquad (108)$$

where the m-dependence is contained in

$$u_{jm}^r \equiv (-)^{j-m} n_{jr}(2r+1) \begin{pmatrix} j & r & j \\ -m & 0 & m \end{pmatrix} , \qquad (109)$$

where $n_{jr} \equiv \begin{pmatrix} j & r & j \\ -j & 0 & j \end{pmatrix}$. This normalization has been chosen such that $\sum_m u^r_{jm} = \delta_{r0}$ and $\sum_{r=0}^{2j} u^r_{jj} = 1$. Substitution of (108) in (107) gives

$$I_{jm} = \sum_{xyz} \langle \underline{w}^{xyz} \rangle \sum_r C_j^{xyzr} u^r_{jm} , \qquad (110)$$

so that the coefficients C_j^{xyzr} give the probability to create a core hole with multipole moment r in the j level for unit ground state moments $\langle \underline{w}^{xyz} \rangle$. Substitution of (109–110) in the definition of the r-th moment of the j level, $I_j^{(r)}$, allows us to write the latter as a function of ground state moments

$$I_j^{(r)} \equiv n_{jr}^{-1} \sum_m I_{jm}(-)^{j-m} \begin{pmatrix} j & r & j \\ -m & 0 & m \end{pmatrix} = \sum_{xyz} \langle \underline{w}^{xyz} \rangle C_j^{xyzr} . \qquad (111)$$

The intensities of the jm sublevels can be calculated from (105) and give the coefficients C_{jm}^{xyz} for each d-shell moment $\langle \underline{w}^{xyz} \rangle$ in the $p \rightarrow d$ XMCD which are shown in Table 11, together with the expressions for $\sum_r C_j^{xyzr} u^r_{jm}$ obtained using (108). From Table 11 it is clear that ground moments with even (odd) z induce only odd (even) moments r. The sum rules are also immediately obvious. Since summation over m gives $u^0_j = 1$ and $u^{r \neq 0}_j = 0$, the signal integrated over both edges is $I_{3/2}^{(0)} + I_{1/2}^{(0)} = 3\langle \underline{w}^{101} \rangle$ and the weighted difference is $I_{3/2}^{(0)} - 2I_{1/2}^{(0)} = \langle \underline{w}^{011} \rangle + 2\langle \underline{w}^{211} \rangle$.

22.1 Independent Particle Model

So far we have not made any assumptions in the spectral moment analysis. However, for the line shape of the spectra we need to know the final state

Table 11. Coefficients C_{jm}^{xyz} of the jm sublevels and resulting expressions $\sum_r C_j^{xyzr} u^r_{jm}$ of the j levels for *unit* ground state moments $\langle \underline{w}^{xyz} \rangle$ in the $p \rightarrow d$ XMCD. The w-tensors are also expressed in traditional notation [92]

(j,m)	$\langle \underline{w}^{000} \rangle$ n_h	$\langle \underline{w}^{110} \rangle$ $-l \cdot s$	$\langle \underline{w}^{101} \rangle$ $\frac{1}{2}L_z$	$\langle \underline{w}^{011} \rangle$ $2S_z$	$\langle \underline{w}^{211} \rangle$ $\frac{7}{2}T_z$	$\langle \underline{w}^{202} \rangle$ $-\frac{1}{2}Q_{zz}$
$(\frac{3}{2},-\frac{3}{2})$	$-\frac{1}{4}$	$-\frac{1}{5}$	$\frac{3}{5}$	$\frac{1}{4}$	$\frac{1}{5}$	$\frac{-1}{2}$
$(\frac{3}{2},-\frac{1}{2})$	$-\frac{1}{12}$	$-\frac{1}{15}$	$\frac{2}{5}$	$-\frac{1}{12}$	$\frac{2}{15}$	$-\frac{1}{6}$
$(\frac{3}{2},+\frac{1}{2})$	$\frac{1}{12}$	$\frac{1}{15}$	$\frac{2}{5}$	$-\frac{1}{12}$	$\frac{2}{15}$	$\frac{1}{6}$
$(\frac{3}{2},+\frac{3}{2})$	$\frac{1}{4}$	$\frac{1}{5}$	$\frac{3}{5}$	$\frac{1}{4}$	$\frac{1}{5}$	$\frac{1}{2}$
$(\frac{1}{2},+\frac{1}{2})$	$\frac{1}{6}$	$-\frac{1}{6}$	$\frac{1}{2}$	$-\frac{1}{6}$	$-\frac{1}{3}$	$\frac{1}{3}$
$(\frac{1}{2},-\frac{1}{2})$	$-\frac{1}{6}$	$\frac{1}{6}$	$\frac{1}{2}$	$-\frac{1}{6}$	$-\frac{1}{3}$	$-\frac{1}{3}$
$j=\frac{3}{2}$	$\frac{5}{9}u^1$	$\frac{4}{9}u^1$	$2u^0+\frac{2}{5}u^2$	$\frac{1}{3}u^0+\frac{2}{5}u^2$	$\frac{2}{3}u^0+\frac{2}{15}u^2$	$\frac{10}{9}u^1$
$j=\frac{1}{2}$	$\frac{1}{3}u^1$	$-\frac{1}{3}u^1$	u^0	$-\frac{1}{3}u^0$	$-\frac{2}{3}u^0$	$\frac{2}{3}u^1$

Hamiltonian $\mathcal{H}$ in order to obtain the energy and intensity distribution over the jm levels. Assuming an independent-particle model we will give an example for the $2p$ XMCD of $3d$ transition metals [96]. Each j level is split by an effective exchange field H_s into $2j + 1$ sublevels with relative energy positions

$$E_{jm} = H_s m \frac{j(j+1) + s(s+1) - l(l+1)}{2j(j+1)} . \tag{112}$$

Since the spectral distribution is $I_{jm}(\omega) = I_{jm}\delta(E_{jm} - E_g - \omega)$ with a constant energy spacing $\frac{1}{3}H_s$ of the sublevels in both j levels, the conversion from multipole moments to spectral moments is straight forward. The main thing to watch is that the energy sequence of the m sublevels in the $j = 1/2$ is reversed compared to that in the $j = 3/2$ level, so that odd moments of the $j = 1/2$ level reverse in sign. The different contributions to the spectra can be obtained directly from Table 11 and are shown in Fig. 4 normalized per unit $(-)^z \langle \underline{w}^{xyz} \rangle$. Operators with odd z give a symmetric signal for each j level, whereas operators with even z give an antisymmetric signal. The moments with $z = 1$ are of special interest because they match the moment transferred by the photon in circular dichroism. As expected from the sum rules the operator $\langle \underline{w}^{101} \rangle$ gives a statistical distribution over the j levels ($\propto 2j+1$). The operators $\langle \underline{w}^{011} \rangle$ and $\langle \underline{w}^{211} \rangle$ will not change the total intensity but transfer intensity between the two j levels. From Fig. 4 it can be checked that the $j = 1/2$ signal disappears when $\langle \underline{w}^{101} \rangle = \langle \underline{w}^{011} \rangle = \langle \underline{w}^{211} \rangle$. This is in agreement with the jj coupled sum rule [97] which states that the $p_{1/2}$ signal is proportional to the component of the ground state $d_{3/2}$ level, which is zero under the given condition. Operators with $z \neq 1$ only shift the intensity *within* a j level. All moments give a negative signal at the low-energy side of the $j = 3/2$ peak.

When the values of the d shell moments are known, the $2p$ spectrum can be obtained from (107) by adding up the different contributions with their relative weights. In Fig. 4 it is seen that, if all moments would have the same value, the shape of the spectrum is strongly dominated by the orbital moment $\langle \underline{w}^{101} \rangle$. However, in a solid the orbital moment is strongly quenched, so that other moments will become more important. In itinerant $3d$ magnets, $\langle \underline{w}^{110} \rangle$ and $\langle \underline{w}^{101} \rangle$ are often an order of magnitude smaller than $\langle \underline{w}^{000} \rangle$ and $\langle \underline{w}^{011} \rangle$. The moments $\langle \underline{w}^{211} \rangle$ and $\langle \underline{w}^{202} \rangle$ vanish in cubic systems, but can become relatively important at the surface where the symmetry is broken [98]. Therefore, the shape of the dichroism spectrum is primarily determined only by the $3d$ spin polarization $P_s \equiv (n_\uparrow - n_\downarrow)/(n_\uparrow + n_\downarrow) = -\langle \underline{w}^{011} \rangle / \langle \underline{w}^{000} \rangle$. For $P_s = 0$ the spectrum is equal to that of $\langle \underline{w}^{000} \rangle$ which is given in Fig. 4 and which exhibits completely antisymmetric peaks. An additional contribution from $\langle \underline{w}^{011} \rangle$ will result in more symmetric peaks.

As an example Fig. 5 shows the separate contributions to the Fe $2p$ XMCD of Fe metal resulting from the different ground state moments. The top spectrum shows the sum of all contributions. In agreement with the sum rules

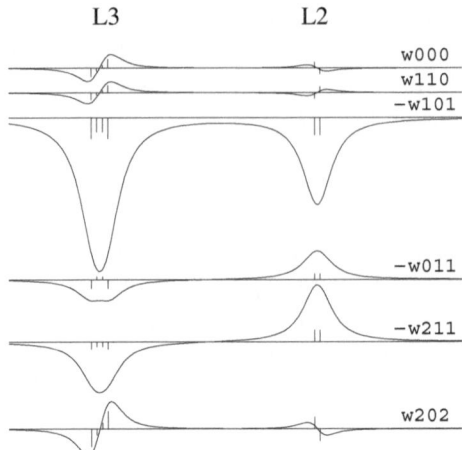

Fig. 4. Separate contributions to the $L_{2,3}$ edges of the $2p$ XMCD for the $3d$ ground-state moments $(-)^z \langle \underline{w}^{xyz} \rangle$. Each signal is given for a moment equal to one. The core spin-orbit parameter is $\zeta(2p) = 8\,\mathrm{eV}$ and the exchange field is $H_s = 0.9\,\mathrm{eV}$. The signals of the jm levels, given by the sticks, have been convoluted with a Lorentzian of $\Gamma = 0.9\,\mathrm{eV}$ [92]

only L_z contributes to the intensity summed over both edges, while L_z, S_z and T_z contribute to the separate L_3 and L_2 intensities. The main contributions to the line shape arise from S_z and n_h. The n_h and to a lesser degree $l \cdot s$ are responsible for the asymmetry in the L_3 XMCD signal. As mentioned the result is for a one-electron model. In the presence of Coulomb interaction and correlation effects, (112) will only hold approximately and the energy levels and intensity will be distributed differently.

23 Conclusions and Outlook

In this chapter, I collected the essential ingredients of what in the synchrotron radiation community is loosely known as multiplet calculations. The basic theoretical principles of atomic spectra were already published in 1935 by Condon and Shortley in their classic book *The Theory of Atomic Spectra* [3] that appeared within a decade of the birth of quantum mechanics. The group-theoretical methods of Giulio Racah in the forties [4] and their reinterpretation by Brian Judd in the sixties [14] brought deeper understanding and enabled the application of atomic theory to complex systems of current interest. These group-theoretical methods facilitated the development of automated calculational methods as by Bob Cowan's code for radiative transition probabilities, which is still today, after more than 40 years, a code of major importance for a wide range of applications [22, 40]. For applications in the area of soft x-ray magnetism Thole and coworkers [39] combined Cowan's code

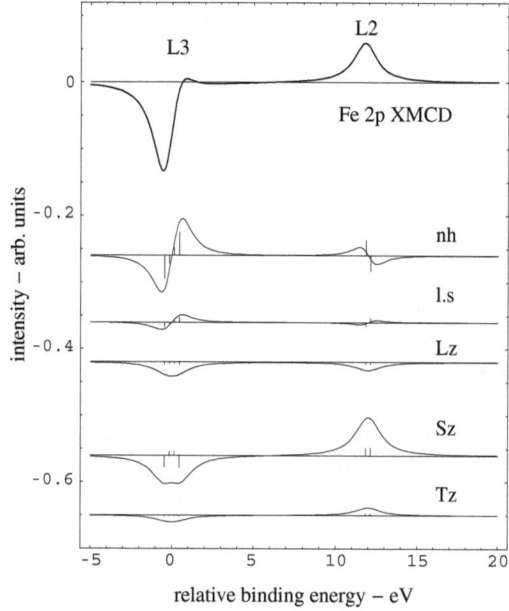

Fig. 5. Relative contributions to the Fe $2p$ XMCD spectrum from the different ground state moments (c.f. Fig. 4) for Fe metal. The top spectrum is the sum of the different contributions and resembles the experimental Fe spectrum [92]

with the point-group symmetry code of Phil Butler [41]. Calculations done with this code give an unprecedented agreement with experimental results for soft x-ray core level spectra of localized materials. In parallel, experimental work in the soft x-ray spectral region strongly benefited from the advent of synchrotron radiation beamlines with an instrumental resolution of similar order of magnitude as the core hole life time (a few hundred meV) allowing to resolve the multiplet structure. Synchrotron radiation further offers a number of unique advantages, such as element-specificity by tuning the x-rays to the resonance energy of the core to valence transition. Due to the selection rules, electric dipole and quadrupole transitions from the ground state reach only a limited subset of final states, thereby providing a fingerprint for the specific ground state. The x-ray transitions involving a deep core state are more straightforward to calculate than transitions in the optical region and the mixing between core and valence states can be neglected. The angular dependent part (containing photon and magnetic polarization) can be separated from the physical part using the Wigner-Eckart theorem.

Recent developments in spectral moment analysis have offered a novel theoretical approach [89, 90]. This method offers an elegant derivation of the sum rules for spin-orbital ground state tensors. When the moments up to the n-th moment are zero, the $(n+1)$-th moment of the experimental spectrum can be analysed. The benefits are large especially in theoretical studies. Spectral moment analysis offers an approach that is model independent, so that it can be used to describe the conversion between localized and itinerant spectra. In the case of the independent-particle model the method provides insight in the origin of the line shape of core-level spectra [91, 92]. It also gives an explanation for the presence of higher-order multipole contributions in resonant enhanced scattering [99, 100].

The effect of magnetic x-ray dichroism was first discovered at LURE in 1986 [101]. Over the timespan of two decades this technique has evolved into a powerful standard technique to separate spin and orbital contributions to the magnetic moments, giving insight into the microscopic origin of anisotropic magnetic properties, such as the magnetocrystalline effect, easy direction of magnetization, magnetostriction and coercivity. Also in photoemission spectroscopy, magnetic dichroism has become a powerful probe to measure different kinds of correlations between the angular moments of core and valence electrons [96, 102]. Chirality is not only introduced through the helicity vector of the light but also by the experimental geometry spanned by the directions of light polarization, sample magnetization and photoemission detection.

Recently, scientific interest has moved towards second-order processes; with the intermediate state created by either XAS or XPS, followed by a decay process. The polarized (or aligned) intermediate state breaks the selection rules for direct transitions and allows "forbidden" transitions, e.g. spin-flip, and gives sensitivity to the local environment. Moreover, the decay process acts as a core-hole clock (stopwatch), allowing a determination of the time scale of the electron screening and spin-dependent screening. Examples of second-order processes are found in "photon in–electron out" techniques, such as resonant photoemission and "photon in–photon out" techniques, such as Raman scattering or resonant inelastic x-ray scattering (RIXS). Sum rules for the magnetic circular dichroism in resonant processes can give information about the higher multipole moments in the ground state [95, 103, 104].

Orbital ordering, which manifests itself in the spatial distribution of the outermost valence electrons, is an important topic in current research of transition-metal oxides, as the magnetic and transport properties are closely related to the orbital and charge degrees of freedom. Transition metal oxides are examples of strongly correlated electron systems, displaying new, unusual, and unexpected behaviour due to nanoscale features in the quantum realm. Advanced properties such as magnetism, colossal magnetoresistance and superconductivity underpin the development of advanced materials. It is now possible to control the electronic and magnetic phases of correlated electron materials in unconventional ways, in some cases with ultra-fast response times. Such control offers the prospect that correlated electron systems may provide a basis for novel future electronics. X-ray diffraction is mainly sensitive to the rather isotropic electron distribution, but tuning the photon energy to an absorption edge gives enhanced sensitivity to valence states. Resonant x-ray diffraction, involving virtual excitations from core to valence states, probes the anisotropic valence charge density allowing forbidden diffraction peaks to appear. Soft x-ray resonant diffraction from single-crystals allows to probe directly the transition metal $3d$ states during orbital ordering using the enhanced sensitivity at the $L_{2,3}$ edges. The resonance process, therefore provides a direct probe of the orbital ordering and gives information on the electronic configuration and the underlying mechanism of the orbital order [105].

The possibilities of the magnetic resonant scattering technique are extended by the use of coherent x-rays. While scattering with incoherent x-ray is only sensitive to statistical averages of the structure, speckle patterns obtained with coherent radiation are sensitive to the particular configuration of the random sample. Magnetic speckle in resonant scattering allows to study static and dynamic magnetic disorder on length scales relevant for nanomagnetism [106]. Time-resolved dichroic studies, performed by subjecting the sample to a magnetic pulse with a constant and adjustable time relationship to the synchrotron pulses, provide an element-specific probe of magnetization dynamics. It allows to study the time dependence of magnetic relaxation effects, such as in read and write times of magnetic media, which are of great technological interest. X-ray optical activity effects, such as x-ray natural circular dichroism (XNCD), observed by Goulon and coworkers [82], offer element-specific access to the absolute configuration of chiral centers in inorganic and organometallic materials.

The near future holds great promises in store with the advent of several new third-generation synchrotron facilities. Currently, XMCD has a sensitivity that allows to perform measurements on very small system, such as magnetically ordered Co chains corresponding to 0.01 monolayer [107]. This paves the way for a wide range of exciting experiments related to exploring magnetism in systems of reduced dimensionality and self-assembled molecular magnets in either 2D (monolayer) or 1D (step edge) arrays. Via novel crystal engineering protocols the dimensionality and ordering of magnetic systems may be systematically tuned. It is worthwhile noting that the understanding of the interplay of structure and magnetism in nanostructured systems is still very much in its infancy. XMCD has already entered into other areas apart from physics such as engineering (magneto-electronics), chemistry (organometallic compounds), earth and environmental sciences (e.g. spinels) and biology (metallo-proteins).

References

1. G. van der Laan, B.T. Thole: Phys. Rev. B **43**, 13401 (1991)
2. B.T. Thole, G. van der Laan, J.C. Fuggle, G.A. Sawatzky, R.C. Karnatak, J.M. Esteva: Phys. Rev. B **32**, 5107 (1985)

3. E.U. Condon, G.H. Shortley: *The Theory of Atomic Spectra* (University Press, Cambridge, 1935)
4. G. Racah: Phys. Rev. **61**, 186 (1942); **62**, 438 (1942); **63**, 367 (1943); Phys. Rev. **76**, 1352 (1949)
5. M.E. Rose: *Multipole fields* (John Wiley, New York, 1955)
6. M.E. Rose: *Elementary Theory of Angular Momentum* (John Wiley, New York 1955)
7. A.R. Edmonds: *Angular Momentum in Quantum Mechanics* (Princeton University Press, Princeton 1957)
8. S. Devons, J.B. Goldfarb: *Angular Correlation*, Handbuch der Physik **62** (Springer Verlag, Berlin 1957) pp. 362–554
9. L.D. Landau, E.M. Lifshitz: *Quantum Mechanics (Non-relativistic Theory)* Course of Theoretical Physics Vol. **3**, (Butterworth, Oxford 1958)
10. U. Fano, G. Racah: *Irreducible Tensorial Sets* (Academic Press, New York 1959)
11. E.P. Wigner: *Group Theory and its Application to the Quantum Mechanics of Atomic Spectra* (Academic Press, New York 1959) p 288
12. D.M. Brink, G.R. Satchler: *Angular Momentum* (Clarendon Press, Oxford, 1962)
13. A. de-Shalit, I. Talmi: *Nuclear Shell Theory* (Academic Press, New York 1963)
14. B.R. Judd: *Operator Techniques in Atomic Spectroscopy* (McGraw-Hill, New York 1963)
15. H.J. Rose, D.M Brink: Rev. Mod. Phys. **39**, 306 (1967)
16. B.R. Judd: *Second Quantization and Atomic Spectroscopy* (The John Hopkins Press, Baltimore, 1967)
17. R.D. Gills: *Gamma Ray Angular Correlations* (Academic Press, London, 1975)
18. M. Weissbluth: *Atoms and Molecules* (Academic Press, New York 1978)
19. I.I. Sobelman: *Atomic Spectra and Radiative Transitions* (Springer Verlag, Berlin 1979)
20. E.U. Condon, H. Odabasi: *Atomic Structure* (Cambridge University Press, Cambridge 1980)
21. J.-M Normand: *A Lie Group: Rotations in Quantum Mechanics* (Elsevier North Holland, New York 1980)
22. R.D. Cowan: *The Theory of Atomic Structure and Spectra* (University of California Press, Berkeley, 1981)
23. L.C. Biedenharn, J.C. Louck: *Angular Momentum in Quantum Physics, Theory and Application*, Encyclopedia of Mathematics (Addison-Wesley, Reading, Mass. 1981)
24. I. Lindgren, J. Morrison: *Atomic Many-Body Theory* (Springer Verlag, Berlin 1982)
25. E. Elbaz: *Algèbre de Racah et Analyse Vectorielle Graphiques* (Ellipses, Paris 1985)
26. V.K. Thankappan: *Quantum Mechanics* (John Wiley, New York 1985)
27. D.A. Varshalovich, A.N. Moskalev, V.K. Khersonskii: *Quantum Theory of Angular Momentum* (World Scientific, Singapore 1988)
28. R.N. Zare: *Angular Momentum* (John Wiley, New York 1988)
29. U. Fano, A.R.P. Rau: *Symmetries in Quantum Physics* (Academic Press, New York 1996)
30. E. Elbaz: *Quantum, The Quantum Theory of Particles, Fields, and Cosmology* (Springer Verlag, Berlin 1998)

31. M. Danos, V. Gillet: *Angular Momentum Calculus in Quantum Physics* (World Scientific, Singapore 1990)
32. I. Talmi: *Simple Models of Complex Nuclei* (Harwood, Chur, Switzerland 1993)
33. V. Devanathan: *Angular Momentum Techniques in Quantum Mechanics* (Kluwer, Dordrecht 1999)
34. K.T. Hecht: *Quantum Mechanics* (Springer Verlag, Berlin 2000)
35. G. van der Laan, B.T. Thole: Phys. Rev. Lett. **60**, 1977 (1988)
36. B.T. Thole, G. van der Laan: Phys. Rev. B **38**, 3158 (1988)
37. G. van der Laan, B.T. Thole: Phys. Rev. B **53**, 14458 (1996)
38. G. van der Laan, K.T. Moore, J.G. Tobin, B.W. Chung, M.A. Wall, A.J. Schwartz: Phys. Rev. Lett. **93**, 097401 (2004)
39. G. van der Laan: J. Electron Spectrosc. Relat. Phenomen. **86**, 41 (1997)
40. R.D. Cowan: J. Opt. Soc. Am. **58**, 808 (1968).
41. P.H. Butler: *Point Group Symmetry, Applications, Methods and Tables* (Plenum, New York, 1981)
42. J.R. Derome, W.T. Sharp: J. Math. Phys. **6**, 1584 (1965)
43. P.H. Butler, B.G. Wybourne: Int. J. Quantum Chem. **10**, 581 (1976)
44. T. Yamaguchi, S. Shibuya, S. Suga, S. Shin, J. Phys. C **15**, 2641 (1982)
45. G. van der Laan, E. Arenholz, Z. Hu, A. Bauer, E. Weschke, Ch. Schüssler-Langeheine, E. Navas, A. Mühlig, G. Kaindl, J.B. Goedkoop, N.B. Brookes: Phys. Rev. B **59**, 8835 (1999)
46. J.B. Goedkoop, B.T. Thole, G. van der Laan, G.A. Sawatzky, F.M.F. de Groot, J.C. Fuggle: Phys. Rev. B **37**, 2086 (1988)
47. S. Imada, T. Jo: J. Phys. Soc. Jpn. **59**, 3358 (1990)
48. G. van der Laan, B.T. Thole: Phys. Rev. B **48**, 210 (1993)
49. H. Ogasawara, A. Kotani, B.T. Thole: Phys. Rev. B **50**, 12332 (1994)
50. F.M.F. de Groot, J.C. Fuggle, B.T. Thole, G.A. Sawatzky: Phys. Rev. B **42**, 5459 (1990). Their spectra for d^1, d^2, d^3, d^5, d^7 in the atomic case with zero crystal field contain errors which make them look different from the spectra with a small crystal field
51. G. van der Laan: J. Phys. Condens. Matter **3**, 7443 (1991)
52. G. van der Laan, I.W. Kirkman: J. Phys.: Condens. Matter **4**, 4189 (1992)
53. A. Tanaka, T. Jo: J. Phys. Soc. Jpn. **63**, 2788 (1994); **64**, 2248 (1995)
54. H. Ogasawara, A. Kotani, B.T. Thole: Phys. Rev. B **44**, 2169 (1991)
55. D.W. Lynch, R.D. Cowan: Phys. Rev. B **36**, 9228 (1987)
56. J.C. Fuggle, S.F. Alvarado: Phys. Rev. A **22**, 1615 (1980)
57. G. van der Laan, B.T. Thole: J. Electron Spectrosc. Relat. Phenomen. **46**, 123 (1988)
58. V.F. Demekhin: Fiz. Tverd. Tela (Leningrad) **16**, 1020 (1974) [Sov. Phys. Solid State **16**, 659 (1974)]
59. G. van der Laan, J.C. Fuggle, M.P. van Dijk, A.J. Burggraaf, J.M. Esteva, R.C. Karnatak: J. Phys. Chem. Solids **47**, 413 (1986)
60. C.W. Nielson, G.F. Koster: *Spectroscopic Coefficients for p^n, d^n and f^n Configurations* (M.I.T. Press, Cambridge, Mass. 1964)
61. E. Arenholz, E. Navas, K. Starke, L. Baumgarten, G. Kaindl: Phys. Rev. B **51**, 8211 (1995)
62. B.T. Thole, G. van der Laan: Phys. Rev. Lett. **70**, 2499 (1993)
63. G. van der Laan, S.S. Dhesi, E. Dudzik: Phys. Rev. B **61** 12277 (2000)
64. G. van der Laan, S.S. Dhesi, E. Dudzik, J. Minar, H. Ebert: J. Phys. Condens. Matter **12**, L275 (2000)

65. C. De Nadaï, G. van der Laan, S.S. Dhesi, N.B. Brookes: Phys. Rev. B **68**, 212401 (2003)
66. H.A. Dürr, G. van der Laan, D. Spanke, F.U. Hillebrecht, N.B. Brookes: Europhys. Lett. **40**, 171 (1997)
67. G. van der Laan: J. Electron Spectrosc. Relat. Phenomen. **117–118**, 89 (2001)
68. G. Panaccione, G. van der Laan, H.A. Dürr: J. Vogel, N.B. Brookes: Eur. Phys. J. B **19**, 281 (2001)
69. S.B. Piepho, P.N. Schatz: *Group Theory in Spectroscopy* (J. Wiley & Sons 1983)
70. G. van der Laan: J. Phys. Soc. Jpn **63**, 2393 (1994)
71. Reference [27], p 228
72. Reference [27], p 210
73. J. Luo, G.T. Trammell, J.P. Hannon: Phys. Rev. Lett. **71**, 287 (1993)
74. P. Carra, B.T. Thole: Rev. Mod. Phys. **66**, 1509 (1994)
75. J.P. Hannon, G.T. Trammell, M. Blume, D. Gibbs: Phys. Rev. Lett. **61**, 1245 (1988)
76. Ch. Brouder: J. Phys.: Condens. Matter **2**, 701 (1990)
77. G.E. Stedman: *Diagram techniques in group theory* (Cambridge University Press, Cambridge, 1990)
78. B.T. Thole, G. van der Laan: Phys. Rev. A **38**, 1943 (1988)
79. B.T. Thole, P. Carra, F. Sette, G. van der Laan: Phys. Rev. Lett. **68**, 1943 (1992)
80. P. Carra, H. König, B.T. Thole, M. Altarelli: Physica (Amsterdam) **192B**, 182 (1993)
81. G. van der Laan: Phys. Rev. Lett. **82**, 640 (1999)
82. J. Goulon, A. Rogalev, F. Wilhelm, C. Goulon-Ginet, P. Carra, I. Marri, Ch. Brouder: J. Exp. Theor. Phys. **97**, 402 (2003)
83. P. Carra, A. Jerez, I. Marri: Phys. Rev. B **67**, 045111 (2003)
84. G. van der Laan: J. Synchr. Rad. **8**, 1059 (2001)
85. J. Goulon, A. Rogalev, C. Goulon-Ginet, G. Benayoun, L. Paolasini, C. Brouder, C. Malgrange, P.A. Metcalf: Phys. Rev. Lett. **85**, 4385 (2000)
86. A.D. Buckingham: Adv. Chem. Phys. **12**, 107 (1967)
87. L.D. Barron: *Molecular Light Scattering and Optical Activity* (Cambridge University Press, Cambridge, 1982)
88. R.E. Raab, O.L. de Lange: *Multipole Theory in Electromagnetism* (Clarendon Press, Oxford 2005)
89. B.T. Thole, G. van der Laan, M. Fabrizio: Phys. Rev. B **50**, 11466 (1994)
90. B.T. Thole, G. van der Laan: Phys. Rev. B **50**, 11474 (1994)
91. G. van der Laan: Phys. Rev. B **55**, 8086 (1997)
92. G. van der Laan: J. Phys.: Condens. Matter **9**, L259 (1997)
93. P. Carra, B.T. Thole, M. Altarelli, X. Wang: Phys. Rev. Lett. **69**, 2307 (1993)
94. G. van der Laan: Phys. Rev. B **57**, 112 (1998)
95. G. van der Laan, B.T. Thole: J. Phys.: Condens. Matter **7**, 9947 (1995)
96. G. van der Laan: Phys. Rev. B **51**, 240 (1995)
97. Reference [37], Equation (38)
98. G. van der Laan: J. Phys.: Condens. Matter **10**, 3239 (1998)
99. S.W. Lovesey: J. Phys.: Condens. Matter **9**, 7501 (1997)
100. S.W. Lovesey, E. Balcar, C. Detlefs, G. van der Laan, D.S. Sivia, U. Staub: J. Phys.: Condens. Matter **15**, 4511 (2003)

101. G. van der Laan, B.T. Thole, G.A. Sawatzky, J.B. Goedkoop, J.C. Fuggle, J.M. Esteva, R.C. Karnatak, J.P. Remeika, H.A. Dabkowska: Phys. Rev. B **34**, 6529 (1986)
102. B.T. Thole, G. van der Laan: Phys. Rev. B **44**, 12424 (1991); **48**, 210 (1993); **49**, 9613 (1994)
103. L. Braicovich, G. van der Laan, G. Ghiringhelli, A. Tagliaferri, M.A. van Veenendaal, N.B. Brookes, M.M. Chervinskii, C. Dallera, B. De Michelis, H.A. Dürr, Phys. Rev. Lett. **82**, 1566 (1999)
104. L. Braicovich, A. Tagliaferri, G. van der Laan, G. Ghiringhelli, N.B. Brookes: Phys. Rev. Lett. **90**, 117401 (2003)
105. S.S. Dhesi, A. Mirone, C. De Nadai, P. Ohresser, P. Bencok, N.B. Brookes, P. Reutler, A. Revcolevschi, A. Tagliaferri, O. Toulemonde, G. van der Laan: Phys. Rev. Lett. **92**, 056403 (2004)
106. G. van der Laan: Physica B **345**, 137 (2004)
107. P. Gambardella, A. Dallmeyer, K. Maiti, M.C. Malagoli, W. Eberhardt, K. Kern, C. Carbone: Nature **416**, 301 (2002)

Resonant X-ray Scattering:
A Theoretical Introduction

Massimo Altarelli

European XFEL Project Team, DESY, Notkestr. 85, 22607 Hamburg, Germany
and Abdus Salam International Centre for Theoretical Physics, Strada Costiera
11, 34100 Trieste, Italy
altarell@ictp.it

Abstract. The theoretical basis of resonant x-ray scattering is reviewed. A detailed
discussion of the physical implications of the various terms in the cross section
for elastic scattering is carried out with special emphasis on applications to the
study of the electronic properties of condensed matter. Resonant Inelastic X-ray
Scattering (RIXS) spectroscopy, and its relationship to techniques such as x-ray
resonant fluorescence spectroscopy and fluorescence detection of x-ray absorption
is also discussed. Relevant examples from the experimental literature are used to
illustrate the general properties derived from the theoretical discussion.

1 Introduction

The purpose of this article is to review the basic aspects of resonant x-ray
scattering (also called anomalous scattering), a method to investigate crystallographic, magnetic, charge and orbital structures and excitations with synchrotron light, which complements more traditional and widespread methods
such as non-resonant x-ray scattering and neutron scattering. Our aim is to
give a pedagogical presentation, by providing a step-by-step guide through
the sometimes elaborate calculations of the relevant scattering amplitudes
and cross sections. The emphasis of our presentation is on the recent applications of resonant scattering to the investigation of the electronic properties
of condensed matter.

Elastic anomalous scattering has long been used by crystallographers as
a help in the solution of the phase problem, a crucial step in the determination of crystal structures from scattering data. A particularly important
development of recent years is the discovery that in the anomalous or resonant region (when the x-ray photon energy is close to an absorption edge of
one of the atomic species of the sample), the scattering amplitude often displays a strong dependence on the polarization of the incoming and scattered
beams. This is formally translated in the description of the atomic scattering amplitude as a *tensor* (rather than as a *scalar*) quantity, with important
consequences for the selection rules for the diffracted beams [1,2].

A further important step forward was the realization that the scattering
of polarized x-rays can deliver information not only on the electron density
distribution, but also on the distribution of magnetic moments. Although

M. Altarelli: *Resonant X-ray Scattering: A Theoretical Introduction*, Lect. Notes Phys. **697**,
201–242 (2006)
www.springerlink.com

the application of magnetic x-ray scattering has only recently become popular, thanks to the development of modern synchrotron light sources, the coupling between photons and magnetic moments is predicted by quantum electrodynamics, and in fact it was described as early as 1954 by Low [3] and Gell-Mann and Goldberger [4] in their derivations of the low-energy limit of the Compton cross section. Later, Platzman and Tzoar [5] pointed out that it would be possible to use this effect to investigate magnetic structures.

Due to the very small cross section, it was not until 1981, however, that the first magnetic scattering experiment was carried out by de Bergevin and Brunel [6] on NiO, demonstrating the basic features of non-resonant scattering. The truly heroic apects of this first experiment performed with an x-ray tube were later alleviated by the advent of synchrotron sources, and experiments were performed to take advantage of the attractive features of x-ray magnetic scattering, as compared to neutron scattering, i.e. the very high momentum resolution and the possibility of a separate determination of the spin and of the orbital contributions to the magnetic moment by the different polarization dependence.

A very important development took place in 1988 with the discovery by Gibbs et al. [7] of resonant magnetic scattering (also called resonant exchange scattering), i.e. of an enhancement of several orders of magnitude of the magnetic scattering intensity when the photon energy is close to an absorption edge of the material. A very large number of studies in rare earth, actinide and transition metal systems followed. Although the price to pay for the resonant enhancement is the loss of a direct interpretation of the scattering intensity in terms of spin and orbital magnetic structure factors, many experiments followed, and contributed to clarify many issues on the electronic structure of magnetic materials: as we shall see, the selection rules for optical transitions make the resonant process sensitive to electronic states with specific orbital character, and enhance their contribution to the magnetic properties.

This last remark leads naturally to another recent development, the exploitation of the sensitivity of resonant scattering not only to charge and magnetic order, but also to orbital order, because the atomic scattering amplitude can vary substantially depending on the occupation of selected spin or orbital states and therefore on their availability to serve as intermediate states in the second order scattering process.

Another weak scattering process that greatly benefits from the resonant enhancement of the scattering amplitude is inelastic scattering, which also has a cross section much smaller than the corresponding elastic process. The study of electronic excitations by resonant inelastic x-ray scattering (RIXS) is becoming an increasingly popular technique, and great progress in energy resolution, on the one hand, and in the interpretation of the resulting spectra, on the other, is taking place.

The structure of the article is the following: in Sect. 2 we recall the formalism necessary to set up the Hamiltonian for the interaction between radiation and matter, and to develop a perturbation description of scattering processes. In Sect. 3 the cross section for the non-resonant case is obtained and discussed, and the resonant elastic case is treated in Sect. 4. Section 5 is devoted to its applications to the study of electronic properties, particularly in magnetic and strongly correlated systems. Finally, Sect. 6 is devoted to a discussion of resonant inelastic scattering and of some recent applications to the study of electronic excitations and properties.

2 Interaction of Radiation with Electronic Matter

A discussion of the microscopic electronic properties of matter must necessarily be formulated in the language of quantum mechanics. We also need to consider relativistic effects, if we want to consider magnetic x-ray scattering, because quantities such as the magnetic moment associated to the electron spin appear only in a relativistic theory, and relativistic effects such as the spin-orbit interaction play an essential role in determining the coupling between radiation and magnetic moments, as we shall see.

We therefore expect the reader to be familiar with basic quantum mechanics and its general formalism, including some aspects of advanced topics such as relativistic quantum mechanics and the second quantization formalism.

In discussing the scattering of electromagnetic waves on a material system composed of electrons and nuclei, we shall follow the usual approach and consider the Hamiltonian for the material system, plus the Hamiltonian for the free electromagnetic field, plus an interaction term between the two systems. As it will be clear soon, the scattering from electrons is much more intense than the scattering from nuclei, and we shall therefore consider matter as a system of electrons, interacting with one another and with a set of nuclei in fixed positions, through a potential energy which can be written

$$V(\mathbf{r}_1, \ldots, \mathbf{r}_N) = \sum_{i=1}^{N} V_{nuc}(\mathbf{r}_i) + \sum_{i>j} V_C\left(|\mathbf{r}_i - \mathbf{r}_j|\right) , \tag{1}$$

where the first term represents the interaction with the nuclei, and in the second the Coulomb interaction is $V_C(r) = e^2/r$. The system of electrons and nuclei is a many-body system and not much progress is possible without suitable approximations. A *self-consistent field* approximation is often introduced in which the dependence of (1) on the positions of all electrons is replaced by a one-electron average,

$$V(\mathbf{r}_1, \ldots, \mathbf{r}_N) \simeq \sum_{i=1}^{N} V(\mathbf{r}_i) . \tag{2}$$

The potential energy is the key ingredient that allows us to write the Hamiltonian for the i-th electron, which, in relativistic quantum mechanics, is the Dirac Hamiltonian [8,9]

$$H_{el} = \sum_{i=1}^{N} (c\boldsymbol{\alpha} \cdot \mathbf{p_i} + \beta mc^2 + V(\mathbf{r}_i)) \,, \tag{3}$$

where $\boldsymbol{\alpha}$ and β are the 4×4 Dirac matrices, and $\mathbf{p}$ is the momentum four-vector (we shall soon revert to the more familiar three-dimensional notation, so the less experienced reader should not be intimidated).

The radiation field, on the other hand, is described by the electric and magnetic fields $\mathbf{E}, \mathbf{B}$, which obey Maxwell's equations [10], and which can also be described by introducing a scalar and a vector potential, $\Phi(\mathbf{r})$ and $\mathbf{A}(\mathbf{r}, t)$

$$\mathbf{B} = \nabla \times \mathbf{A}$$

$$\mathbf{E} = -\nabla\Phi - (1/c)\frac{\partial \mathbf{A}}{\partial t} \,, \tag{4}$$

where the vector $\nabla = (\frac{\partial}{\partial x}, \frac{\partial}{\partial y}, \frac{\partial}{\partial z})$.

For given $\mathbf{E}(\mathbf{r})$ and $\mathbf{B}(\mathbf{r})$, the definition of the vector and scalar potentials is not unique; when describing the fields of electromagnetic waves in vacuum, we can use this freedom to chose the *gauge* in such a way that the scalar potential vanishes, and the vector potential $\mathbf{A}$ is divergence free ($\nabla \cdot \mathbf{A} = 0$) [11]. This will turn out to be a convenient choice later. An arbitrary space- and time-dependent vector potential can be expanded in terms of plane waves, which are characterized by a wavevector $\mathbf{k}$ and by one of the two polarization modes labeled by λ. Let us write this expansion in the following form

$$\mathbf{A}(\mathbf{r}, t) = \sum_{\mathbf{k},\lambda} \left(\frac{hc^2}{\Omega\omega_\mathbf{k}}\right)^{1/2} \left[\mathbf{e}_\lambda(\mathbf{k})a(\mathbf{k}, \lambda)e^{i(\mathbf{k}\cdot\mathbf{r}-\omega_\mathbf{k}t)}\right.$$

$$\left. + \mathbf{e}_\lambda^*(\mathbf{k})a^\dagger(\mathbf{k}, \lambda)e^{-i(\mathbf{k}\cdot\mathbf{r}-\omega_\mathbf{k}t)}\right] \,. \tag{5}$$

In this equation, Ω is the volume of the quantization box, and does not appear in any physically meaningful quantity in the following, $\omega_\mathbf{k}$ is just $c|\mathbf{k}|$, $\mathbf{e}_\lambda$ is the polarization vector associated to the mode λ, i.e. one of two orthogonal vectors in the plane normal to $\mathbf{k}$. Furthermore, in a classical description of the field, $a(\mathbf{k}, \lambda)$ and $a^\dagger(\mathbf{k}, \lambda)$ are the amplitude of the corresponding mode of the field and its complex conjugate. However, in the language of the second quantization formalism a and $a^\dagger$ are *operators*, respectively the annihilation and creation operators of a photon with quantum numbers $(\mathbf{k}, \lambda)$. In this formalism the Hamiltonian of the field takes a very simple and appealing form

$$H_{rad} = \sum_{\mathbf{k},\lambda} \hbar\omega_\mathbf{k} \left(a^\dagger(\mathbf{k}, \lambda)a(\mathbf{k}, \lambda) + 1/2\right) \,. \tag{6}$$

Turning now to the modifications of H_{el} in presence of the electromagnetic field, we follow the usual prescription to insert the $\mathbf{A}(\mathbf{r}_i)$ in the Dirac Hamiltonian [8, 9]

$$H'_{el} = \sum_{i=1}^{N} (c\boldsymbol{\alpha} \cdot [\mathbf{p}_i - (e/c)\mathbf{A}(\mathbf{r}_i)] + \beta mc^2 + V(\mathbf{r}_i)) \,, \tag{7}$$

and consider that the processes we are interested in (scattering of x-ray photons with energies at most in the $\sim$10 keV range), always involve energies much smaller than the electron rest energy, $mc^2 \simeq 511$ keV. This authorizes us to adopt the non-relativistic limit of (7), which is considered and derived in great detail in Sect. 15 of [9]. The resulting Hamiltonian, accurate to order $(1/c)^2$ is

$$H'_{el} = \sum_{i=1}^{N} \left[\frac{(\mathbf{p}_i - (e/c)\mathbf{A}(\mathbf{r}_i))^2}{2m} - \mathbf{p}_i^{\,4}/8m^3c^2 + V(\mathbf{r}_i) - (e\hbar/mc)\mathbf{s}_i \cdot \mathbf{B} \right.$$
$$\left. -(e\hbar/2m^2c^2)\mathbf{s}_i \cdot (\mathbf{E} \times (\mathbf{p}_i - (e/c)\mathbf{A}(\mathbf{r}_i))) + (e\hbar^2/8m^2c^2)\nabla \cdot \mathbf{E} \right]. \tag{8}$$

In this equation, the first term on the r.h.s. is the usual modification of the kinetic energy in presence of a field, the second is the relativistic correction to the kinetic energy, which does not involve the field and is therefore not relevant to our discussion; the fourth term is the interaction of the electron spin $\mathbf{s} = (1/2)\boldsymbol{\sigma}$ with the magnetic field of the radiation, $\mathbf{B} = \nabla \times \mathbf{A}$; the fifth is the spin-orbit interaction term, with the usual modification of the momentum in presence of the field; and the last is the Darwin correction, which is again independent of the radiation field, because the transversality of electromagnetic waves ($\mathbf{k} \cdot \mathbf{E} = 0$) implies $\nabla \cdot \mathbf{E} = 0$, so that the only contribution to this term comes from the electric field of the charges (the gradient of V). After removing all the relativistic corrections to H'_{el} which are not affected by the radiation field, we are left with the following Hamiltonian for the system of electrons and the radiation field

$$H = H'_{el} + H_{rad}$$
$$= \sum_{i=1}^{N} \left[\frac{(\mathbf{p}_i - (e/c)\mathbf{A}(\mathbf{r}_i))^2}{2m} + V(\mathbf{r}_i) \right.$$
$$\left. -(e\hbar/mc)\mathbf{s}_i \cdot \mathbf{B}(e\hbar/2m^2c^2)\mathbf{s}_i \cdot (\mathbf{E} \times (\mathbf{p}_i - (e/c)\mathbf{A}(\mathbf{r}_i))) \right]$$
$$+ \sum_{\mathbf{k},\lambda} \hbar\omega_{\mathbf{k}} \left(a^\dagger(\mathbf{k},\lambda)a(\mathbf{k},\lambda) + 1/2 \right) \,. \tag{9}$$

We are then in a position to separate all the terms mixing electron and photon variables, that constitute the *interaction Hamiltonian*, H_{int}:

$$H = H_{el} + H_{rad} + H_{int} \,, \tag{10}$$

$$H_{el} = \sum_{i=1}^{N} \left[\frac{\mathbf{p}_i^{\,2}}{2m} + V(\mathbf{r}_i) + (e\hbar/2m^2c^2)\mathbf{s}_i \cdot (\nabla V(\mathbf{r}_i) \times \mathbf{p}_i) \right], \tag{11}$$

$$H_{rad} = \sum_{\mathbf{k},\lambda} \hbar\omega_{\mathbf{k}} \left(a^\dagger(\mathbf{k},\lambda)a(\mathbf{k},\lambda) + 1/2 \right) , \qquad (12)$$

$$H_{int} = \sum_{i=1}^{N} [(e^2/2mc^2)\mathbf{A}^2(\mathbf{r}_i) - (e/mc)\mathbf{A}(\mathbf{r}_i)\cdot\mathbf{p}_i$$
$$-(e\hbar/mc)\mathbf{s}_i\cdot(\nabla\times\mathbf{A}(\mathbf{r}_i))$$
$$+(e\hbar/2m^2c^3)\mathbf{s}_i\cdot[(\partial\mathbf{A}(\mathbf{r}_i)/\partial t)\times(\mathbf{p}_i - (e/c)\mathbf{A}(\mathbf{r}_i))]] \quad (13)$$
$$\equiv H_1' + H_2' + H_3' + H_4' .$$

The total Hamiltonian, to the required order of relativistic corrections, is thus split into the Hamiltonian for electronic matter, (11), for the radiation field, (12) and the Hamiltonian describing the interaction between matter and radiation, (13). In the next section, scattering processes will be described as transitions between the eigenstates of H_{el} and H_{rad} induced by the perturbation H_{int}. This can be done by regarding the $\mathbf{A}$ field as a classical quantity, or alternatively and more elegantly, by considering it as an operator, according to the expansion (5) in terms of annihilation and creation operators.

3 Cross Section for Non-Resonant Elastic Scattering

In developing the expressions for the scattering cross section, we closely follow the lucid discussion by Blume [12], warning the reader that this important paper unfortunately contains many misprints.

In this section, our discussion is restricted to *elastic* scattering, i.e. to processes in which the sample (the system of electrons) is in the same state (for simplicity, let us say the ground state) before and after the scattering event. If we consider the scattering of an incoming photon with polarization $\mathbf{e}_\lambda$ and wavevector $\mathbf{k}$ into an outgoing photon with polarization $\mathbf{e}'_{\lambda'}$ and wavevector $\mathbf{k}'$ (conservation of energy implies $|\mathbf{k}| = |\mathbf{k}'|$), we can describe the initial and final state of the system (sample plus radiation field) as:

$$|i\rangle = |0; \ldots, (\mathbf{e}_\lambda, \mathbf{k}), \ldots\rangle ,$$
$$|f\rangle = |0; \ldots, (\mathbf{e}'_{\lambda'}, \mathbf{k}'), \ldots\rangle . \qquad (14)$$

with an obvious notation labelling the ground state of the electronic system with $|0\rangle$, and the radiation field state with the quantum numbers of the photons present in that state.

It is then clear that the transition consists in the *annihilation* of one photon $(\mathbf{e}_\lambda, \mathbf{k})$, and in the *creation* of one photon $(\mathbf{e}'_{\lambda'}, \mathbf{k}')$. This means that the operator $\mathbf{A}$, which is linear in the creation and annihilation operators, must operate twice. Therefore the lowest-order contributing processes will come from applying second-order perturbation theory to H_2' and H_3', which contain one $\mathbf{A}$ operator, and by first-order perturbation theory applied to H_1' and H_4', which contain two $\mathbf{A}$ operators. As a matter of fact, H_4' contains two terms, respectively proportional to $\partial \mathbf{A}/\partial t \cdot \mathbf{p}$ and to $\partial \mathbf{A}/\partial t \cdot (e/c)\mathbf{A}$. However, we shall later see that the first one produces a negligible effect, so we will drop it and retain the second term only.

According to Fermi's Golden Rule of time-dependent perturbation theory, the number of transitions per unit time is proportional to

$$
w = \frac{2\pi}{\hbar} \left| \langle f|H_1' + H_4'|i\rangle + \sum_n \frac{\langle f|H_2' + H_3'|n\rangle \langle n|H_2' + H_3'|i\rangle}{E_0 - E_n + \hbar\omega_\mathbf{k}} \right|^2 \delta(\hbar(\omega_\mathbf{k} - \omega_{\mathbf{k}'})) . \quad (15)
$$

In the second term, the sum over the complete set of eigenstates $|n\rangle$ of the unperturbed Hamiltonian, $H_{el} + H_{rad}$ appears, referred to as the sum over the intermediate states. The calculation of the matrix elements involves both electron and photon operators and is tedious, but straightforward. For example

$$
\begin{aligned}
\langle f|H_1'|i\rangle &= \frac{hc^2}{\Omega\omega_\mathbf{k}} \frac{e^2}{mc^2} \sum_i \langle 0; (\mathbf{e}'_{\lambda'}, \mathbf{k}')|(\mathbf{e}'^*_{\lambda'} \cdot \mathbf{e}_\lambda)a^\dagger(\mathbf{k}', \lambda')a(\mathbf{k}, \lambda)e^{i(\mathbf{k}-\mathbf{k}')\cdot\mathbf{r}_i}|0; (\mathbf{e}_\lambda, \mathbf{k})\rangle \\
&= \frac{hc^2}{\Omega\omega_\mathbf{k}} \frac{e^2}{mc^2} (\mathbf{e}'^*_{\lambda'} \cdot \mathbf{e}_\lambda) \sum_i \langle 0|e^{i(\mathbf{k}-\mathbf{k}')\cdot\mathbf{r}_i}|0\rangle , \quad (16)
\end{aligned}
$$

after taking the photon annihilation and destruction operator matrix elements according to the usual rules. In this matrix element we recognize the *Thomson scattering* amplitude, with the dot product polarization dependence and the structure factor, expressed by the ground state expectation value of $\sum_i e^{i(\mathbf{k}-\mathbf{k}')\cdot\mathbf{r}_i}$. Notice also that the matrix element magnitude per electron is controlled by the quantity $r_0 \equiv e^2/mc^2$ which has the dimensions of length and is the Thomson radius, $r_0 = 2.818 \times 10^{-13}$ cm.

We are now in a position to confirm that the scattering from the nuclei is negligible. In fact, to obtain the corresponding matrix element for nuclear scattering, we should simply replace the electronic positions with the atomic ones, and also replace e^2/mc^2, with Z^2e^2/Mc^2, where Z and M are the nuclear charge and mass. However, M is roughly equal to $2Zm_n$, where m_n is the nucleon mass, and the mass ratio m_n/m is about 1850. Therefore, the scattering matrix element for a nucleus is $\sim Z/(2 \times 1850)$ times smaller than that for an electron, and can be neglected because Z never exceeds 92 (remember that, in addition, there are Z times more electrons than nuclei!).

3.1 Thomson Scattering and Crystallography

Before proceeding to the evaluation of the matrix elements deriving from the other pieces of the interaction Hamiltonian, we briefly consider the implications of the H_1' matrix elements. As a matter of fact, we shall later show that, as long as the photon energy $\hbar\omega_\mathbf{k}$ is well above any of the absorption edges of the atoms in the system, this is the dominant matrix element for the photon scattering process. Consider for example the radiation from a Mo x-ray tube, which allows us to exploit the K_α line, with an energy of 17.4 keV. This is well above all edges of light atoms such as Al, Si, Ca or Ti, which are below 5 keV, and contributions other than H_1' are negligible. In this approximation, (15) simplifies to

$$w = \frac{2\pi}{\hbar} |\langle f|H_1'|i\rangle|^2 \tag{17}$$

We are now ready to replace (16) into (17). However, this gives a number of transitions per unit time which depends on the normalization volume. We would rather have a physically meaningful quantity, i.e. a *cross section*, defined as the number of transitions per unit time, into photon states with energy $\hbar\omega_{\mathbf{k}'} < E < \hbar\omega_{\mathbf{k}'} + dE$, with wavevector $\mathbf{k}'$ in the solid angle dO', divided by the number of incident photons per unit time and area. That is, in differential form

$$\frac{d^2\sigma}{dEdO'} = \frac{w\rho(E)}{c/\Omega} \tag{18}$$

where ρ is the density of photon states (with specified polarization), i.e. the number of wavevectors within dO' satisfying periodic boundary conditions in a box of volume Ω and $\hbar\omega_{\mathbf{k}} \leq \hbar\omega_{\mathbf{k}'} \leq \hbar\omega_{\mathbf{k}} + dE$, i.e.

$$\rho(E)dEdO' = \frac{\Omega}{(2\pi)^3} \frac{E^2}{\hbar^3 c^3} dEdO' \tag{19}$$

Finally, by putting (15), (16), (18) together, and upon multiplying (18) by dE and integrating (remember the Dirac δ in (15)) we obtain the important result

$$\frac{d\sigma}{dO'} = r_0^2 \left| \sum_j \langle 0|e^{i\mathbf{q}\cdot\mathbf{r}_j}|0\rangle \right|^2 (\mathbf{e}_{\lambda'}'^* \cdot \mathbf{e}_\lambda)^2 \tag{20}$$

after defining $(\mathbf{k} - \mathbf{k}') \equiv \mathbf{q}$, the *scattering vector*.

With reference to Fig. 1, define the *scattering plane* as that identified by $\mathbf{k}, \mathbf{k}'$, and introduce a specific basis for the polarization vectors, $\mathbf{e}_\pi$ *parallel* to the scattering plane, and $\mathbf{e}_\sigma$ *perpendicular* to the scattering plane. Define further the *scattering angle* 2θ (the factor 2 is a mere convention!) as the angle between $\mathbf{k}, \mathbf{k}'$. It is easy to see that the polarization factor $(\mathbf{e}_{\lambda'}'^* \cdot \mathbf{e}_\lambda)^2$ forbids σ to π transitions and viceversa, and in other cases is worth

$$\begin{aligned}
(\mathbf{e}_{\lambda'}'^* \cdot \mathbf{e}_\lambda)^2 &= \quad 1, \quad (\sigma \to \sigma) \\
(\mathbf{e}_{\lambda'}'^* \cdot \mathbf{e}_\lambda)^2 &= \cos^2 2\theta, \quad (\pi \to \pi') \ .
\end{aligned} \tag{21}$$

For example, if the photon source is unpolarized, we have to average over the incoming polarizations, and we obtain

$$\frac{d\sigma}{dO'} = (1/2)r_0^2(1 + \cos^2 2\theta) |F(\mathbf{q})|^2 \tag{22}$$

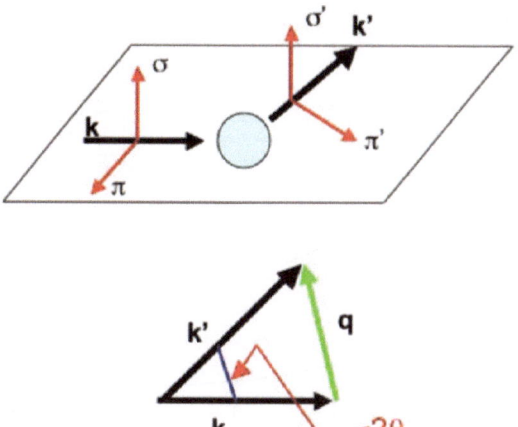

Fig. 1. Scattering and polarization geometry

where we defined

$$F(\mathbf{q}) = \sum_j \langle 0|e^{i\mathbf{q}\cdot\mathbf{r}_j}|0\rangle \ . \tag{23}$$

In full generality, let the scattering object be a system with N electrons, with its ground state $|0\rangle$ described by an antisymmetric wavefunction $\Psi(\mathbf{r}_1, \mathbf{r}_2, \ldots, \mathbf{r}_N)$, from which an electron density is derived as

$$\rho(\mathbf{r}) = N \int d\mathbf{r}_2 d\mathbf{r}_3 \ldots d\mathbf{r}_N \ |\Psi(\mathbf{r}, \mathbf{r}_2, \ldots, \mathbf{r}_N)|^2 \tag{24}$$

It is then easy to see that

$$F(\mathbf{q}) = \int d\mathbf{r} e^{i\mathbf{q}\cdot\mathbf{r}} \rho(\mathbf{r}) \tag{25}$$

so that the scattering cross section with scattering vector $\mathbf{q}$, (20), is proportional to the absolute square of the Fourier transform of the electron density at momentum $\mathbf{q}$. In the particular case in which the N-electron system is an atom, the function $F(\mathbf{q})$ is called the *atomic form factor*. A few examples of form factors are shown in Fig. 2.

Using the convolution properties of Fourier integrals, it is easy to see that

$$|F(\mathbf{q})|^2 = \int d\mathbf{r}' e^{i\mathbf{q}\cdot\mathbf{r}'} \langle \rho(\mathbf{r})\rho(\mathbf{r}+\mathbf{r}')\rangle_{\mathbf{r}} \tag{26}$$

where

$$\langle \rho(\mathbf{r})\rho(\mathbf{r}+\mathbf{r}')\rangle_{\mathbf{r}} = \int d\mathbf{r}\rho(\mathbf{r})\rho(\mathbf{r}+\mathbf{r}') \tag{27}$$

is proportional to the probability of finding two electrons at a distance $\mathbf{r}'$ from each other, a two-particle correlation function, which crystallographers often

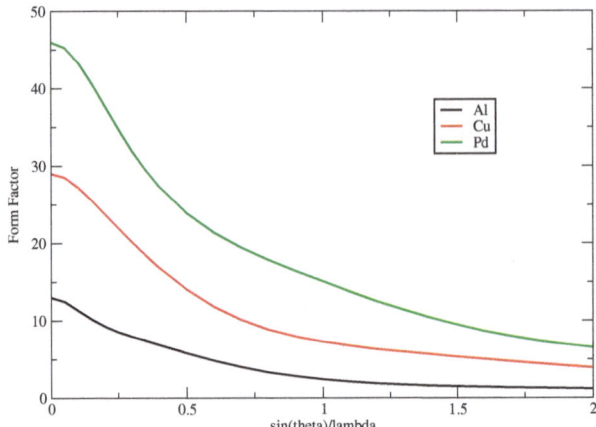

Fig. 2. Atomic form factors of Al, Cu and Pd, calculated from Hartree-Fock electron densities

call the *Patterson* function. This is a special case of the general statement that scattering techniques allow to access correlation functions. Consider now the situation in which the system under investigation is a crystalline solid, for which the electron density is well approximated as

$$\rho(\mathbf{r}) = \sum_{l,m,n} \sum_{i=1}^{I} \rho_i(\mathbf{r} - \mathbf{R}_{l,m,n} - \mathbf{r}_i) \tag{28}$$

which represents a crystal in which I atoms (forming the lattice *basis*) are located at distances $\mathbf{r}_i$, $i = 1, 2, \ldots I$ from every node of the Bravais lattice $\mathbf{R}_{l,m,n} = l\mathbf{a}_1 + m\mathbf{a}_2 + n\mathbf{a}_3$. The Fourier transform of (28) is

$$F(\mathbf{q}) = \sum_{l,m,n} e^{i\mathbf{q}\cdot\mathbf{R}_{l,m,n}} \sum_{i=1}^{I} e^{i\mathbf{q}\cdot\mathbf{r}_i} \int d\mathbf{r} e^{i\mathbf{q}\cdot(\mathbf{r}-\mathbf{R}_{l,m,n}-\mathbf{r}_i)} \rho_i(\mathbf{r} - \mathbf{R}_{l,m,n} - \mathbf{r}_i)$$

$$\equiv \sum_{l,m,n} e^{i\mathbf{q}\cdot\mathbf{R}_{l,m,n}} \sum_{i=1}^{I} e^{i\mathbf{q}\cdot\mathbf{r}_i} f_i(\mathbf{q}) \tag{29}$$

because the integral is independent of l, m, n, and $f_i(\mathbf{q})$ is the atomic form factor of the i-th atom. The sum over l, m, n restricts the permissible $\mathbf{q}$ values to the vectors $\mathbf{G}_{hkl}$ of the *reciprocal lattice* of the Bravais lattice in question. This is the *Laue condition* for crystalline x-ray diffraction and is equivalent to the *Bragg condition*. The intensity of the scattering at each $\mathbf{G}_{hkl}$ (or at each *reflection*), is proportional to the quantities

$$|F(\mathbf{G}_{hkl})|^2 \equiv \left| \sum_{i=1}^{I} e^{i\mathbf{G}_{hkl}\cdot\mathbf{r}_i} f_i(\mathbf{G}_{hkl}) \right|^2 \tag{30}$$

which are called the *structure factors of the unit cell*. A knowledge of the Fourier transform at all wavevectors is of course equivalent to the knowledge of the electron density. However, much to the crystallographers' sorrow, we see that x-ray scattering only delivers the *absolute value* of the Fourier transform. To reconstruct the electron density one should also know the *phase* of each reflection, which, however, is a much more elusive quantity. Notice that the intensity is certainly zero at scattering vectors differing from a reciprocal lattice vector. However, it may also vanish at some $\mathbf{G}_{hkl}$'s if the corresponding structure factor is zero. Consider for instance the case of an elemental lattice, i.e. one in which all atoms are the same. In such case, (30) becomes

$$|F(\mathbf{G}_{hkl})|^2 \equiv |f_a(\mathbf{G}_{hkl})|^2 \left| \sum_{i=1}^{I} e^{i\mathbf{G}_{hkl}\cdot\mathbf{r}_i} \right|^2 . \tag{31}$$

The first factor on the right hand side, the square of the atomic form factor of the element a, contains all the information on the electron density of this atom and is generally nonvanishing (see the examples in Fig. 2). The second factor, however depends only on the *geometry* of the lattice arrangement, and is the square of a complex number which can vanish for some h, k, l combinations. Let us work out the specific example of the diamond lattice, in which C, Si, Ge and gray Sn are found to crystallize. The Bravais lattice is fcc, the basis is given by two atoms at positions $\mathbf{r}_1 = 0$, $\mathbf{r}_2 = (a/4)(\mathbf{x}+\mathbf{y}+\mathbf{z})$, where $\mathbf{x}, \mathbf{y}, \mathbf{z}$ are along the cubic axes, and a is the size of the conventional cubic cell (see Fig. 3). The reciprocal lattice of fcc is the body-centered cubic, bcc, with vectors

$$\mathbf{G}_{hkl} = h\mathbf{b}_1 + k\mathbf{b}_2 + l\mathbf{b}_3 , \tag{32}$$

where

$$\begin{aligned}
\mathbf{b}_1 &= (2\pi/a) \ (-\mathbf{x}+\mathbf{y}+\mathbf{z}) \\
\mathbf{b}_2 &= (2\pi/a) \ (\mathbf{x}-\mathbf{y}+\mathbf{z}) \\
\mathbf{b}_3 &= (2\pi/a) \ (\mathbf{x}+\mathbf{y}-\mathbf{z}) .
\end{aligned} \tag{33}$$

The second factor in (31) is the square of a sum over the two basis atoms and gives

$$\left| \sum_{i=1}^{I} e^{i\mathbf{G}_{hkl}\cdot\mathbf{r}_i} \right|^2 = |1 + e^{i(\pi/2)(h+k+l)}|^2 , \tag{34}$$

which is trivially equal to zero each time $h + k + l = 2\times$ an odd number. Therefore, non-resonant scattering theory predicts a vanishing intensity for reflections such as $(2,2,2)$ or $(6,0,0)$. The two atoms are so located, that, for some reflections, the waves scattered by one interfere destructively with those scattered by the other. This is an example of what crystallographers call *extinction rules* or *systematic absences*, which give important clues as to the position of the atoms in the unit cell. We shall later see how such rules are sometimes violated in the anomalous scattering regime.

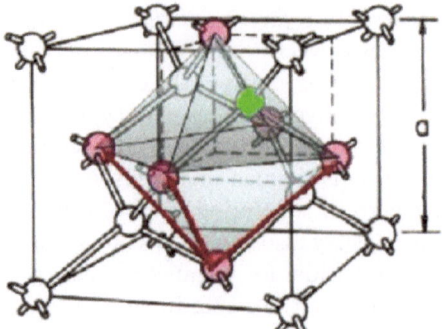

Fig. 3. Conventional cubic cell of the diamond lattice. The shaded area is the primitive unit cell

3.2 Non-resonant Magnetic Scattering

We now resume the systematic exploration of (15), and, after dealing with the matrix elements of H'_1, we consider the remaining terms, which contribute exclusively to *magnetic* scattering.

The next task is the evaluation of $\langle f|H'_4|i\rangle$. Remember that H'_4 contains two terms, respectively proportional to $\partial \mathbf{A}/\partial t \cdot \mathbf{p}$ and to $\partial \mathbf{A}/\partial t \cdot (e/c)\mathbf{A}$. However, we shall soon verify that the second order perturbation on the first term produces a contribution to the cross section which is a factor $(\hbar\omega/mc^2)^2$ smaller than the first order contribution of the second, so we will drop it and retain the second term only. We must first of all determine an expression for the operator $\partial \mathbf{A}/\partial t$ from (5)

$$\partial \mathbf{A}/\partial t = \left(\frac{hc^2}{\Omega\omega_{\mathbf{k}}}\right)^{1/2}[-i\omega_{\mathbf{k}}\mathbf{e}_\lambda(\mathbf{k})a(\mathbf{k},\lambda)e^{i(\mathbf{k}\cdot\mathbf{r}-\omega_{\mathbf{k}}t)}$$
$$+i\omega_{\mathbf{k}}\mathbf{e}^*_\lambda(\mathbf{k})a^\dagger(\mathbf{k},\lambda)e^{-i(\mathbf{k}\cdot\mathbf{r}-\omega_{\mathbf{k}}t)}] . \tag{35}$$

Inserting this expression, the H'_4 matrix element is readily evaluated

$$\langle f|H'_4|i\rangle = -i\left(\frac{e^2}{mc^2}\right)\left(\frac{\hbar\omega_{\mathbf{k}}}{mc^2}\right)\left(\frac{hc^2}{\Omega\omega_{\mathbf{k}}}\right)$$
$$\times \sum_i \langle 0|e^{i(\mathbf{k}-\mathbf{k}')\cdot\mathbf{r}_i}\mathbf{s}_i \cdot (\mathbf{e}^*_{\lambda'}(\mathbf{k}') \times \mathbf{e}_\lambda(\mathbf{k}))|0\rangle . \tag{36}$$

One therefore sees immediately that a term containing the spin operators, i.e. a genuine *magnetic* scattering term appears, and that its magnitude compared to the Thomson term is reduced by the factor $(\hbar\omega_{\mathbf{k}}/mc^2)$. This is a small number, because typically in the x-ray region $\hbar\omega_{\mathbf{k}} \sim 10\,\text{keV}$, while $mc^2 = 511\,\text{keV}$. Additional magnetic information is hidden in the second term in (15), which we now proceed to evaluate. The accessible intermediate states have either no photons, or two photons, and their energy is

$$|n\rangle = |\Psi_n; 0, 0\rangle; E_n = E(\Psi_n) \tag{37}$$

$$|n\rangle = |\Psi_n; (\mathbf{e}_\lambda, \mathbf{k}), (\mathbf{e}'_{\lambda'}, \mathbf{k}')\rangle; E_n = E(\Psi_n) + 2\hbar\omega_\mathbf{k} . \tag{38}$$

The first set of terms (let us call them terms (a)) is reached by the action of the annihilation part of the $\mathbf{A}$ operator on the initial state; the second (terms (b)) by the action of the creation operator part. There is also an additional, important difference between the two kinds of terms: in case (a) the energy denominator can vanish, and give rise to a resonance, when $E_0 - E_n + \hbar\omega_\mathbf{k} = 0$; in case (b) it cannot, because $E_0 - E(\Psi n) - \hbar\omega_\mathbf{k} < 0$ always. To prevent an unphysical divergence of the scattering cross section, we must take into account that the intermediate states $|n\rangle$ are not really stationary, but have a finite lifetime, which is represented by adding a small imaginary part to the eigenvalue, which becomes important only near the resonance condition; i.e. $E(\Psi_n)$ is replaced by $E(\Psi_n) - i\Gamma_n/2$. We want to examine the non-resonant case first, i.e. the case in which $\hbar\omega_\mathbf{k} \gg E(\Psi_n) - E_0$ for all states, or, more precisely, for all states $|n\rangle$ which give an appreciable contribution to the sum in (15). Using the following simple identities for the energy denominators:

$$\frac{1}{E_0 - E(\Psi_n) + \hbar\omega_\mathbf{k} + i\Gamma_n/2}$$
$$= \frac{1}{\hbar\omega_\mathbf{k}} + \frac{E(\Psi_n) - E_0 - i\Gamma_n/2}{\hbar\omega_\mathbf{k}} \frac{1}{E_0 - E(\Psi_n) + \hbar\omega_\mathbf{k} + i\Gamma_n/2}$$
$$\frac{1}{E_0 - E(\Psi_n) - \hbar\omega_\mathbf{k}} = -\frac{1}{\hbar\omega_\mathbf{k}} + \frac{E_0 - E(\Psi_n)}{\hbar\omega_\mathbf{k}} \frac{1}{E_0 - E(\Psi_n) - \hbar\omega_\mathbf{k}} . \tag{39}$$

It is easy to see that in this case the denominators are well approximated by $\pm\hbar\omega_\mathbf{k}$. Substituting (5) into $H'_2 + H'_3$, and paying due attention to the action of photon creation and annihilation operators on the two kinds of intermediate states, we find for type (a) intermediate states

$$\langle f|H'_2 + H'_3|n\rangle\langle n|H'_2 + H'_3|i\rangle$$
$$= \left(\frac{hc^2}{\Omega\omega_\mathbf{k}}\right)\left(\frac{e}{mc}\right)^2 \langle 0| \sum_{j=1}^N \left[\mathbf{e}'^*_{\lambda'} \cdot \mathbf{p}_j - i\hbar(\mathbf{k}' \times \mathbf{e}'^*_{\lambda'}) \cdot \mathbf{s}_j\right] e^{-i\mathbf{k}'\cdot\mathbf{r}_j}|n\rangle$$
$$\langle n| \sum_{j'=1}^N \left[\mathbf{e}_\lambda \cdot \mathbf{p}_{j'} + i\hbar(\mathbf{k} \times \mathbf{e}_\lambda) \cdot \mathbf{s}_{j'}\right] e^{i\mathbf{k}\cdot\mathbf{r}_{j'}}|0\rangle \tag{40}$$

while, for type (b) intermediate states we obtain an expression differing only in that the operators acting between $\langle 0|$ and $|n\rangle$ and between $\langle n|$ and $|0\rangle$ are interchanged. This, together with the fact that the energy denominators, in the non-resonant approximation defined above, are independent of $|n\rangle$ and change sign for the two types of intermediate states, and with the *closure* relationship

$$\sum_n |n\rangle\langle n| = 1 \tag{41}$$

where **1** denotes the unit operator, allows us to write the second term in (15) as the expectation value of a commutator

$$\sum_n \frac{\langle f|H_2' + H_3'|n\rangle\langle n|H_2' + H_3'|i\rangle}{E_0 - E_n + \hbar\omega_\mathbf{k}} \simeq \left(\frac{hc^2}{\Omega\omega_\mathbf{k}}\right)\left(\frac{e}{mc}\right)^2 \langle 0|\,[C', C]\,|0\rangle\,, \quad (42)$$

where

$$C' = \left[\mathbf{e}_{\lambda'}'^* \cdot \mathbf{p}_j - i\hbar(\mathbf{k}' \times \mathbf{e}_{\lambda'}'^*) \cdot \mathbf{s}_j\right] e^{-i\mathbf{k}'\cdot\mathbf{r}_j} \quad (43)$$

$$C = \left[\mathbf{e}_\lambda \cdot \mathbf{p}_j + i\hbar(\mathbf{k} \times \mathbf{e}_\lambda) \cdot \mathbf{s}_j\right] e^{i\mathbf{k}\cdot\mathbf{r}_j}\,. \quad (44)$$

To calculate the commutator is a tedious operation, but is easily performed remembering the basic commutation rules for components of positions, momenta, spin and arbitrary functions of them, referred to the same electron

$$[r_\alpha, p_\beta] = i\hbar\delta_{\alpha\beta}$$
$$[p_\alpha, f(\mathbf{r})] = -i\hbar\partial f/\partial r_\alpha$$
$$[s_\alpha, s_\beta] = i\hbar\epsilon_{\alpha\beta\gamma}s_\gamma\,. \quad (45)$$

Here the antisymmetric tensor $\epsilon_{\alpha\beta\gamma}$ was introduced, and it is worthwhile to remember the expression of the cross product of two vectors in terms of it (summation over repeated indices is implied)

$$(\mathbf{v}_1 \times \mathbf{v}_2)_\alpha = \epsilon_{\alpha\beta\gamma}v_{1\beta}v_{2\gamma}\,. \quad (46)$$

By a careful use of these rules, of the transversality conditions, $\mathbf{e} \cdot \mathbf{k} = 0$ and of a simple vector identity

$$(\mathbf{A} \times \mathbf{B}) \cdot (\mathbf{C} \times \mathbf{D}) \equiv (\mathbf{A} \cdot \mathbf{C})(\mathbf{B} \cdot \mathbf{D}) - (\mathbf{A} \cdot \mathbf{D})(\mathbf{B} \cdot \mathbf{C}) \quad (47)$$

which is applied to the four vectors: $(\mathbf{k} - \mathbf{k}') \equiv \mathbf{q}$, $\mathbf{p}_j$, $\mathbf{e}_{\lambda'}'^*$, $\mathbf{e}_\lambda$, the patient reader should obtain

$$\sum_n \frac{\langle f|H_2' + H_3'|n\rangle\langle n|H_2' + H_3'|i\rangle}{E_0 - E_n + \hbar\omega_\mathbf{k}}$$

$$= -i\left(\frac{hc^2}{\Omega\omega_\mathbf{k}}\right)\left(\frac{e^2}{mc^2}\right)\frac{\hbar\omega_\mathbf{k}}{mc^2}[\langle 0|\sum_j e^{i\mathbf{q}\cdot\mathbf{r}_j}\frac{i\mathbf{q} \times \mathbf{p}_j}{\hbar k^2}|0\rangle(\mathbf{e}_{\lambda'}'^* \times \mathbf{e}_\lambda)$$

$$+\langle 0|\sum_j e^{i\mathbf{q}\cdot\mathbf{r}_j}\mathbf{s}_j|0\rangle[(\mathbf{k}' \times \mathbf{e}_{\lambda'}'^*)(\mathbf{k}' \cdot \mathbf{e}_\lambda) - (\mathbf{k} \times \mathbf{e}_\lambda)(\mathbf{k} \cdot \mathbf{e}_{\lambda'}'^*)$$

$$-(\mathbf{k}' \times \mathbf{e}_{\lambda'}'^*) \times (\mathbf{k} \times \mathbf{e}_\lambda)]]\,. \quad (48)$$

Now that we have the second-order contribution of the $\mathbf{A} \cdot \mathbf{p}$ term in H_2', we can substantiate our claim that the contribution of the $\partial\mathbf{A}/\partial t \times \mathbf{p}$ term of H_4' is negligible. In fact, the magnitude of the latter contribution would be similar to that of the former, which we just evaluated, except for some different prefactors. On the one hand, the time derivative introduces a factor $\omega_{\mathbf{k}}$, on the other, the constant in front of H'_4 introduces, with respect to H'_2, another factor of $\hbar/2mc^2$, so that all in all an extra factor $\hbar\omega_{\mathbf{k}}/2mc^2$ is obtained. This shows that the matrix element of the first part of H'_4 is reduced by $(\hbar\omega_{\mathbf{k}}/2mc^2)^2$ with respect to the Thomson term, and therefore is negligible with respect to the other magnetic scattering terms, which are reduced by $\hbar\omega_{\mathbf{k}}/2mc^2$.

Finally, by putting (15), (16), (36), (48), (18) together, we can complete the cross section expression of (20) to obtain

$$
\frac{d\sigma}{dO'} = r_0^2 \left| \sum_j \langle 0 | e^{i\mathbf{q}\cdot\mathbf{r}_j} | 0 \rangle (\mathbf{e}'^*_{\lambda'} \cdot \mathbf{e}_\lambda) \right.
$$

$$
\left. -i\frac{\hbar\omega_{\mathbf{k}}}{mc^2} \left[\langle 0 | \sum_j e^{i\mathbf{q}\cdot\mathbf{r}_j} \frac{i\mathbf{q} \times \mathbf{p}_j}{\hbar k^2} | 0 \rangle \cdot \mathbf{P}_L + \langle 0 | \sum_j e^{i\mathbf{q}\cdot\mathbf{r}_j} \mathbf{s}_j | 0 \rangle \cdot \mathbf{P}_S \right] \right|^2 (49)
$$

where we introduced the polarization factors

$$
\mathbf{P}_L = (\mathbf{e}'^*_{\lambda'} \times \mathbf{e}_\lambda) \tag{50}
$$
$$
\mathbf{P}_S = [(\mathbf{k}' \times \mathbf{e}'^*_{\lambda'})(\mathbf{k}' \cdot \mathbf{e}_\lambda) - (\mathbf{k} \times \mathbf{e}_\lambda)(\mathbf{k} \cdot \mathbf{e}'^*_{\lambda'}) - (\mathbf{k}' \times \mathbf{e}'^*_{\lambda'}) \times (\mathbf{k} \times \mathbf{e}_\lambda)] . \tag{51}
$$

The indices L and S where adopted for these factors because the second one is attached to the term related to the spin moment, while the first pertains to a term which, as we shall show, is related to the *orbital* moment. In fact, after noting that $|\mathbf{q}| = 2|\mathbf{k}| \sin\theta$, where 2θ is the scattering angle, and defining $\hat{\mathbf{q}} = \mathbf{q}/|\mathbf{q}|$, the relevant quantity can be transformed as follows

$$
\sum_j e^{i\mathbf{q}\cdot\mathbf{r}_j} \frac{i\mathbf{q} \times \mathbf{p}_j}{\hbar k^2} = \frac{i}{\hbar q}(4\sin^2\theta) \sum_j e^{i\mathbf{q}\cdot\mathbf{r}_j} \hat{\mathbf{q}} \times \mathbf{p}_j
$$

$$
= \frac{i}{\hbar q}(4\sin^2\theta)\hat{\mathbf{q}} \times \int d\mathbf{r} e^{i\mathbf{q}\cdot\mathbf{r}} \frac{1}{2} \sum_j [\mathbf{p}_j \delta(\mathbf{r} - \mathbf{r}_j) + \delta(\mathbf{r} - \mathbf{r}_j)\mathbf{p}_j]
$$

$$
= \frac{-im}{e\hbar q}(4\sin^2\theta)\hat{\mathbf{q}} \times \int d\mathbf{r} e^{i\mathbf{q}\cdot\mathbf{r}}\mathbf{j}(\mathbf{r}) = \frac{-im}{e\hbar q}(4\sin^2\theta)\hat{\mathbf{q}} \times \mathbf{j}(\mathbf{q}) \tag{52}
$$

where the electrical current density operator $\mathbf{j}(\mathbf{r}) = (-e/2m)\sum_j [\mathbf{p}_j \delta(\mathbf{r}-\mathbf{r}_j) + \delta(\mathbf{r} - \mathbf{r}_j)\mathbf{p}_j]$ has been expressed in terms of the momentum and the density of electrons multiplied in symmetrized form, because they do not commute.

This current density describes the *microscopic* currents associated to the motion of the electrons, not the *macroscopic* ones, which we can assume to vanish in our system in the absence of external perturbations (remember that all matrix elements in a perturbation calculation refer to the unperturbed system eigenstates). The vanishing of macroscopic currents means that the flux across any surface S internal to the sample vanishes, i.e.

$$\int_S \mathbf{j}(\mathbf{r}) \cdot \mathbf{n}_S dS = 0 \, , \qquad (53)$$

which implies that the microscopic current is divergence-free, i.e. $\nabla \cdot \mathbf{j}(\mathbf{r}) = 0$, and can therefore be expressed as the curl of a vector field. We write this field so that

$$\mathbf{j}(\mathbf{r}) = c[\nabla \times \mathbf{M}_L(\mathbf{r})] \, . \qquad (54)$$

For the purposes of our discussion, we identify $\mathbf{M}_L(\mathbf{r})$ with the density of orbital magnetization. Although a formal identification between operators is analytically involved [13, 14], one can satisfy one's self of the plausibility of (54) by the classical description of magnetic fields in matter; Maxwell's equations for the fields $\mathbf{H}$ and $\mathbf{B} = \mathbf{H} + 4\pi\mathbf{M}$ (no spin magnetization exists in the classical description, so here $\mathbf{M}$ means $\mathbf{M}_L$) prescribe that the microscopic currents are related to the curl of $\mathbf{M}$ by (54) (see for example [15]). Equation (54) implies that $\mathbf{j}(\mathbf{q}) = -ic\mathbf{q} \times \mathbf{M}_L(\mathbf{q})$. Therefore

$$\sum_j e^{i\mathbf{q}\cdot\mathbf{r}_j} \frac{i\mathbf{q} \times \mathbf{p}_j}{\hbar k^2} = \frac{mc}{e\hbar q^2} \mathbf{q} \times [\mathbf{M}_L(\mathbf{q}) \times \mathbf{q}] \qquad (55)$$

We are now ready to collect all the bits and pieces in a formula for the differential cross section

$$\frac{d\sigma}{dO'} = r_0^2 \left| \sum_j \langle 0|e^{i\mathbf{q}\cdot\mathbf{r}_j}|0\rangle (\mathbf{e}'^*_{\lambda'} \cdot \mathbf{e}_\lambda) \right.$$

$$\left. -i\frac{\hbar\omega_\mathbf{k}}{mc^2} \left[\frac{mc}{e\hbar} \langle 0|\hat{\mathbf{q}} \times [\mathbf{M}_L(\mathbf{q}) \times \hat{\mathbf{q}}]|0\rangle \cdot \mathbf{P}_L + \frac{mc}{e\hbar} \langle 0|\mathbf{M}_S(\mathbf{q})|0\rangle \cdot \mathbf{P}_S \right] \right|^2 \quad (56)$$

where the polarization factor $\mathbf{P}_L$ was redefined to include the angular factor

$$\mathbf{P}_L = (\mathbf{e}'^*_{\lambda'} \times \mathbf{e}_\lambda) 4\sin^2\theta \qquad (57)$$

and the Fourier transform of the *spin* magnetization density was introduced

$$\mathbf{M}_S(\mathbf{q}) = \frac{e\hbar}{mc} \sum_j e^{i\mathbf{q}\cdot\mathbf{r}_j} \mathbf{s}_j \, . \qquad (58)$$

We are now ready to obtain from (56) the basic properties of non-resonant magnetic scattering. In a system with an ordered magnetic structure, e.g. an antiferromagnet, the densities of (orbital and spin) magnetization are periodic functions, with Fourier transforms which are non vanishing only for selected $\mathbf{q}$ values corresponding to this periodicity. Some of these vectors may possibly coincide with reciprocal lattice vectors of the crystallographic structure, others will correspond to new reflections (*magnetic* reflections) with nonvanishing intensity below the Néel temperature, where the antiferromagnetic order sets in.

As already noticed, the prefactor $\hbar\omega_{\mathbf{k}}/mc^2$ reduces the intensity of the magnetic terms considerably with respect to the Thomson one. To reinforce this, while all core and valence electrons contribute to Thomson scattering, only electrons in partially filled shells can contribute to magnetic scattering as the orbital and spin moments of filled shells add up to zero. Therefore, apart from the first pioneering experiments [6], the high intensity of sinchrotron light sources is necessary for these experiments.

It is important to notice that $\mathbf{P}_L$ contains the factor $4\sin^2\theta$, and since $|\mathbf{q}| = 2|\mathbf{k}|\sin\theta$, for a given reflection, i.e. for a given $\mathbf{q}$, $\sin\theta$ is proportional to $1/\hbar\omega_{\mathbf{k}}$. Thus, the weight of the orbital part decreases at high photon energies, where spin scattering dominates the magnetic cross section; more generally, the $\omega_{\mathbf{k}}$ dependence of the orbital term is not only in the prefactor.

The different polarization factors $\mathbf{P}_L$, $\mathbf{P}_S$ and the well known polarization properties of synchrotron radiation allow to separate the spin and the orbital contributions to the magnetic moments by changing the experimental geometry. This is a much more direct approach to the separation of the two contributions than it is possible with neutron scattering. This method was applied to rare earth systems such as Ho [7], to actinide systems such as UAs [16, 17] and more recently (see Fig. 4) to 3d antiferromagnets such as NiO [18], and V_2O_3 [19]. Together with the higher momentum resolution allowed by well collimated synchrotron beams, this orbit and spin separation justifies the interest of x-ray scattering for some cases, in spite of the more widespread use of neutron scattering to determine magnetic structures.

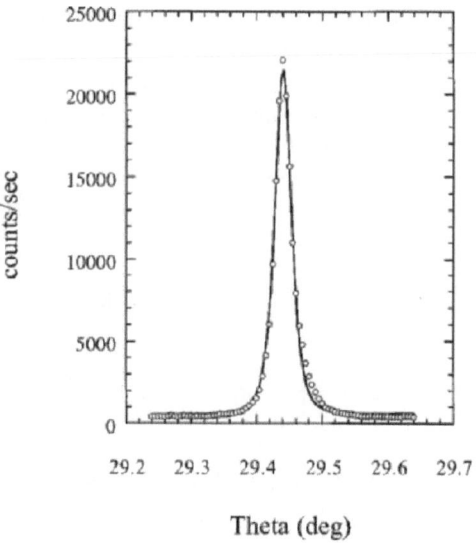

Fig. 4. Rocking curve of the magnetic (3/2,3/2,3/2) reflection in NiO at 300 K (After Fernandez et al. [18])

A further important point to mention about the magnetic terms in (56) is the imaginary prefactor $-i\hbar\omega_{\mathbf{k}}/mc^2$. This means that, upon taking the square modulus, no interference of Thomson and magnetic scattering terms occurs, unless the structure factors:

$$\sum_j \langle 0|e^{i\mathbf{q}\cdot\mathbf{r}_j}|0\rangle \tag{59}$$

are complex (which means that the crystallographic structure is non-centrosymmetric), or that the polarization vectors are complex (corresponding to non-linear, i.e. elliptic or circular polarization). In such cases one has interference terms, and these can be useful in detecting magnetic scattering in ferromagnets [20, 21].

4 Resonant Scattering

We now abandon the assumption of the non-resonant limit and consider the case in which $E(\Psi_n) - E_0 \simeq \hbar\omega_{\mathbf{k}}$, at least for one excited state Ψ_n (normally, in a solid there will be a continuum of states satisfying this condition).

Returning to the expressions of the matrix elements of $H_2' + H_3'$ as written in (40), we want first of all to prove that the contribution of H_2' is always much larger than that of H_3'. To establish this, we begin by remarking that the most important excited states which are resonant with x-ray photons are those in which a core electron in one of the atoms is promoted to an empty one-electron state above the highest occupied orbital. Arguing within an approximate scheme in which the states $|0\rangle$, $|n\rangle$ are reasonably well described by an antisymmetric product of one-electron states, then the matrix elements of the operators H_2' or H_3', which are sums of one-electron operators, can be written [22] in terms of an overlap integral over $N-1$ of the coordinates, multiplied by a one-electron matrix element, i.e.

$$\langle n|H_2' + H_3'|i\rangle$$

$$= \left(\frac{hc^2}{\Omega\omega_{\mathbf{k}}}\right)^{1/2} \left(\frac{e}{mc}\right) \sum_{j=1}^{N} \langle n|\left[\mathbf{e}_\lambda \cdot \mathbf{p}_j + i\hbar(\mathbf{k}\times\mathbf{e}_\lambda)\cdot\mathbf{s}_j\right] e^{i\mathbf{k}\cdot\mathbf{r}_j}|0\rangle$$

$$= \left(\frac{hc^2}{\Omega\omega_{\mathbf{k}}}\right)^{1/2} \left(\frac{e}{mc}\right) \prod_{j=1}^{N-1} \int d\mathbf{r}_j \psi_{vj}^{(n)*}(\mathbf{r}_j)\psi_{vj}^{(0)}(\mathbf{r}_j)$$

$$\times \int d\mathbf{r}_N \psi_{vN}^{(n)*}(\mathbf{r}_N)\left[\mathbf{e}_\lambda \cdot \mathbf{p}_N + i\hbar(\mathbf{k}\times\mathbf{e}_\lambda)\cdot\mathbf{s}_N\right] e^{i\mathbf{k}\cdot\mathbf{r}_N}\psi_c^{(0)}(\mathbf{r}_N) , \tag{60}$$

where ψ_v is a one-electron valence wavefunction, either for the ground or the n-th excited states and ψ_c a core wavefunction which is exponentially decreasing, outside an appropriate core radius r_c. We can then argue that the main contribution to the integral comes from this inner region; and one can see that inside this region $\mathbf{k} \cdot \mathbf{r}_j \ll 1$, for the values of $k = |\mathbf{k}|$ of interest here. This is because at the resonance condition

$$k = \omega/c = E/\hbar c \tag{61}$$

where E is the difference of the core and valence energy, i.e. the core ionization energy. This energy is related to the radius of the core orbital by the approximate hydrogen-like relationship

$$E \simeq \hbar^2/2mr_c^2 \tag{62}$$

whence one finds $r_c \simeq \hbar/\sqrt{2mE}$ and therefore

$$kr_c \simeq \sqrt{E/2mc^2} . \tag{63}$$

The right hand side is always small for all core levels, because $2mc^2$ is about 1 MeV, while the deepest core level (1s in Uranium) has a binding energy of about 116 keV. So, in this most extreme case, $kr_c \simeq 0.34$, and is less for all other core levels. It is therefore legitimate, for $r \leq r_c$, to expand

$$e^{i\mathbf{k}\cdot\mathbf{r}_j} \sim 1 + i\mathbf{k} \cdot \mathbf{r}_j - (\mathbf{k} \cdot \mathbf{r}_j)^2/2 + \dots . \tag{64}$$

and to observe that the terms of the series are rapidly decreasing with increasing order (which is referred to as the multipole order). We can then reach the proof of the statement that H_2' matrix elements dominate over those of H_3', i.e. that the first term in the last integral of (60) dominates over the second. The point is that for given $\psi_{vN}^{(n)}, \psi_c^{(0)}$, the lowest nonvanishing order in the series (64) for the integral of $\mathbf{p}_N$ is lower by one than the lowest nonvanishing order for the second term (which contains the spin, but no $\mathbf{r}_N$ operator). Remember indeed that the selection rules for atomic transitions are the same for $\mathbf{p}$ or for $\mathbf{r}$ matrix elements (a manifestation of the Wigner-Eckart theorem [23]), and since the H_3'-related operator contains the spin but neither $\mathbf{p}$ nor $\mathbf{r}$'s, it is necessary to have one more $\mathbf{r}$ (with respect to the first term) in order to have a nonvanishing integral, i.e. to go to the next order in $\mathbf{k} \cdot \mathbf{r}_N$.

Therefore, near the resonance condition, the resonant terms dominate the cross section, and, among these, only the H_2' matrix elements need to be retained. Equation (15) becomes

$$w = \frac{2\pi}{\hbar} \left| \sum_n \frac{\langle f|H_2'|n\rangle\langle n|H_2'|i\rangle}{E_0 - E_n + \hbar\omega_{\mathbf{k}} + i\Gamma_n/2} \right|^2 \delta(\hbar(\omega_{\mathbf{k}} - \omega_{\mathbf{k}'}))$$

$$= \frac{2\pi}{\hbar} \left| \left(\frac{hc^2e^2}{\Omega\omega_{\mathbf{k}}m^2c^2}\right) \sum_n \frac{\langle 0|\sum_{j=1}^N \mathbf{e}_{\lambda'}'^* \cdot \mathbf{p}_j e^{-i\mathbf{k}'\cdot\mathbf{r}_j}|n\rangle\langle n|\sum_{j'=1}^N \mathbf{e}_\lambda \cdot \mathbf{p}_{j'} e^{i\mathbf{k}\cdot\mathbf{r}_{j'}}|0\rangle}{E_0 - E(\Psi_n) + \hbar\omega_{\mathbf{k}} + i\Gamma_n/2} \right|^2$$

$$\times \delta(\hbar(\omega_{\mathbf{k}} - \omega_{\mathbf{k}'})) \tag{65}$$

As a matter of fact, the above equation contains a contribution that was already taken into account in the non-resonant part; remember (39), where the first piece on the r.h.s. was included in the previous section. Therefore, only the second addendum needs to be considered here and that means that in (65) we must replace

$$\frac{1}{E_0 - E(\Psi_n) + \hbar\omega_{\mathbf{k}} + i\Gamma_n/2} \tag{66}$$

with

$$\frac{E(\Psi_n) - E_0 - i\Gamma_n/2}{\hbar\omega_{\mathbf{k}}} \frac{1}{E_0 - E(\Psi_n) + \hbar\omega_{\mathbf{k}} + i\Gamma_n/2} . \tag{67}$$

4.1 Electric Dipole Approximation

Let us then look in detail into the relevant matrix elements. Consider

$$\langle n|\mathbf{e}_\lambda \cdot \mathbf{p}_j e^{i\mathbf{k}\cdot\mathbf{r}_j}|0\rangle \simeq \langle n|\mathbf{e}_\lambda \cdot \mathbf{p}_j(1 + i\mathbf{k}\cdot\mathbf{r}_j + \ldots)|0\rangle \tag{68}$$

and, for a given $|n\rangle$, consider only the lowest order term for which the matrix element does not vanish. We established already that all higher order terms are negligible in comparison to it. The largest contributions come from those $|n\rangle$'s for which the first term provides a nonvanishing contribution, so that the exponential is simply replaced by 1. These states are said to be accessible by *electric dipole* transitions. In a full quantum electrodynamical formulation, one can see that electric dipole transitions are induced by photons with a total angular momentum of 1. The name electric dipole comes from the fact that in a non-relativistic theory, neglecting the spin-orbit interaction altogether, so that $H_{el} = \sum_j(\mathbf{p}_j^2/2m + V(\mathbf{r}_j))$, one can write

$$\begin{aligned}
\langle n|\mathbf{e}_\lambda \cdot \mathbf{p}_j|0\rangle &= m\langle n|\mathbf{e}_\lambda \cdot \dot{\mathbf{r}}_j|0\rangle \\
&= \frac{-im}{\hbar}\langle n|\mathbf{e}_\lambda \cdot [\mathbf{r}_j, H_{el}]|0\rangle \\
&= \frac{im}{\hbar}[E(\Psi_n) - E(0)]\langle n|\mathbf{e}_\lambda \cdot \mathbf{r}_j|0\rangle .
\end{aligned} \tag{69}$$

In view of this, and neglecting $i\Gamma_n/2$ in the numerator of (67), the sum over intermediate states in (65) becomes

$$\frac{m^2}{\hbar^2}\sum_n \frac{(E(\Psi_n) - E(0))^3}{\hbar\omega_{\mathbf{k}}} \frac{\langle 0|\mathbf{e}'^*_{\lambda'} \cdot \mathbf{R}|n\rangle\langle n|\mathbf{e}_\lambda \cdot \mathbf{R}|0\rangle}{E(\Psi_n) - E(0) + \hbar\omega_{\mathbf{k}} + i\Gamma_n/2} \tag{70}$$

where we defined

$$\mathbf{R} = \sum_j \mathbf{r}_j . \tag{71}$$

In order to make progress and to make contact with the literature [24], we express all vectors in terms of their *spherical* components, i.e. we define

$$R_0 = iR_z, R_{\pm 1} = (\mp i/\sqrt{2})(R_x \pm iR_y) . \tag{72}$$

The definitions of the $0, \pm 1$ components apply to any vector, e.g. to the polarization $\mathbf{e}_\lambda$ as well, and they are clearly inspired from the definition of the spherical harmonics for $l = 1$

$$Y_{1,0} = i\sqrt{\frac{3}{4\pi}}(z/r),\ Y_{1,\pm 1} = \mp i\sqrt{\frac{3}{8\pi}}\left(\frac{x \pm iy}{r}\right),\tag{73}$$

where we adopted the convention for the phases given in [25]. It is easily verified that the scalar product becomes

$$\mathbf{e}_\lambda \cdot \mathbf{R} = \sum_{m=-1}^{1}(-1)^{m-1}e_m R_{-m}.\tag{74}$$

It is then easy to see that

$$\langle 0|\mathbf{e}'^*_{\lambda'}\cdot\mathbf{R}|n\rangle\langle n|\mathbf{e}_\lambda\cdot\mathbf{R}|0\rangle = \sum_{m,m'}(-1)^{m+m'}e'^*_{\lambda'm}e_{\lambda m'}\langle 0|R_{-m}|n\rangle\langle n|R_{-m'}|0\rangle.\tag{75}$$

To simplify this expresssion further, one must take advantage of the symmetry of the physical systemm. The simplest case correspods of course to the highest symmetry, i.e. the spherical symmetry of isolated atoms. Then, the eigenstates $|0\rangle$ and $|n\rangle$ are eigenstates of the angular momentum and of its z-component, and this implies that the sum is restricted to $m = -m'$, because the angular momentum selection rules say that, for the matrix elements

$$\langle 0|R_{-m}|n\rangle \neq 0 \rightarrow -m_0 - m + m_n = 0,$$
$$\langle n|R_{-m'}|0\rangle \neq 0 \Rightarrow -m_n - m' + m_0 = 0,$$
$$\Rightarrow m = -m'\tag{76}$$

The sum in (75) is then simplified and it is worth noticing that

$$\langle 0|R_0|n\rangle\langle n|R_0|0\rangle = -|\langle n|R_0|0\rangle|^2$$
$$\langle 0|R_{-1}|n\rangle\langle n|R_1|0\rangle = |\langle n|R_1|0\rangle|^2$$
$$\langle 0|R_1|n\rangle\langle n|R_{-1}|0\rangle = |\langle n|R_{-1}|0\rangle|^2\tag{77}$$

The first relationship may look surprising, but remember that, because of the factor i in the definition, coming from the chosen convention on the phases of the spherical harmonics, R_0 is an antihermitian operator. Another consequence of that is the fact that if

$$e_{\lambda 0} = ie_{\lambda z}\tag{78}$$

it is also

$$e^*_{\lambda 0} = ie^*_{\lambda z}\tag{79}$$

i.e. the spherical component 0 of the complex conjugate need not be the complex conjugate of the 0 component. With the help of all of the above we can write

$$\langle 0|\mathbf{e}'^{*}_{\lambda'} \cdot \mathbf{R}|n\rangle\langle n|\mathbf{e}_{\lambda} \cdot \mathbf{R}|0\rangle = -e'^{*}_{\lambda'0}e_{\lambda0}|\langle n|R_0|0\rangle|^2 + e'^{*}_{\lambda'1}e_{\lambda-1}|\langle n|R_1|0\rangle|^2$$
$$+e'^{*}_{\lambda'-1}e_{\lambda1}|\langle n|R_{-1}|0\rangle|^2 . \qquad (80)$$

Going back to cartesian coordinates for the polarization vectors, it is possible, with a bit of algebra to recast this expression in the following form

$$e'^{*}_{\lambda'z}e_{\lambda z}|\langle n|R_0|0\rangle|^2 + \frac{1}{2}[e'^{*}_{\lambda'x}e_{\lambda x} + e'^{*}_{\lambda'y}e_{\lambda y}](|\langle n|R_1|0\rangle|^2 + |\langle n|R_{-1}|0\rangle|^2)$$
$$-\frac{i}{2}[e'^{*}_{\lambda'x}e_{\lambda y} - e'^{*}_{\lambda'x}e_{\lambda y}](|\langle n|R_1|0\rangle|^2 - |\langle n|R_{-1}|0\rangle|^2) . \qquad (81)$$

We define

$$F^{e}_{1,m} = m_e \sum_{n} \frac{[E(\Psi_n) - E(0)]^3}{\hbar^3\omega_k} \frac{|\langle n|R_m|0\rangle|^2}{E(0) - E(\Psi_n) + \hbar\omega_{\mathbf{k}} + i\Gamma_n/2} \qquad (82)$$

where the label e and 1 on F remind us that this refers to electric (e) dipole ($l = 1$) contributions, and where we introduced the symbol m_e for the electron mass, to avoid any confusion with the index m, which runs over $0, \pm1$; reinserting the prefactors present in (65) and those allowing to relate w to $d\sigma/dO'$, see (18), we finally obtain

$$\frac{d\sigma}{dO'} = |f_{res}|^2 \qquad (83)$$

where f_{res} is the resonant scattering *amplitude*, given by

$$f_{res} = -r_0 \left[\frac{1}{2}\mathbf{e}'^{*}_{\lambda'} \cdot \mathbf{e}_{\lambda}(F^{e}_{1,1} + F^{e}_{1,-1}) \right.$$
$$-\frac{i}{2}(\mathbf{e}'^{*}_{\lambda'} \times \mathbf{e}_{\lambda}) \cdot \hat{\mathbf{z}}(F^{e}_{1,1} - F^{e}_{1,-1})$$
$$\left. +(\mathbf{e}'^{*}_{\lambda'} \cdot \hat{\mathbf{z}})(\mathbf{e}_{\lambda} \cdot \hat{\mathbf{z}}) \left(F^{e}_{1,0} - \frac{1}{2}F^{e}_{1,1} - \frac{1}{2}F^{e}_{1,-1} \right) \right] \qquad (84)$$

where the unit vector in the z direction (i.e. in the axis of quantization of the angular momenta), $\hat{\mathbf{z}}$, was introduced. Equation (84) was derived in [24] using the relativistic formalism of vector spherical harmonics, soon after the discovery of resonant magnetic scattering by Gibbs et al. [7]. A step-by-step outline of this derivation was reported by Poppe and Fasolino and is available on the website [26].

Before proceeding to discuss the scattering terms beyond the dipole approximation, let us pause briefly to analyze some of the consequences of the results derived so far.

The three terms in (84) describe resonant or anomalous scattering in general, and are rather different in nature. The first is proportional to $(F^e{}_{1,1} + F^e{}_{1,-1})$ and is therefore always present. The second is a genuinely magnetic term, because it originates from the difference between the 1 and the -1 components, which arise only in the presence of a magnetic *preference* for one sense of rotation around the quantization axis. Finally, the last term is nonvanishing for any anisotropic system, a system with a preferential axis, identified either by a crystal anisotropy or by a magnetic moment, which translates into a different occupation for one-electron orbitals with different orientation. The difference in occupation translates into a difference in the value of individual $F^e{}_{1,m}$, which is strongly influenced (see (82)) by the availability of states with the appropriate symmetry, at or near the resonance energy, suitable to play the role of intermediate states.

As an example, consider first an isotropic or cubic environment for an atom. In this case, $F^e{}_{1,m}$ is the same for all m 's and can be called simply $F^e{}_1$. Then

$$f_{res} = -r_0 \frac{1}{2}(\mathbf{e}'^*_{\lambda'} \cdot \mathbf{e}_\lambda)(2F^e{}_1) \tag{85}$$

We can represent the polarization dependence by evaluating the above equation for the various cases arising when $\mathbf{e}'_{\lambda'}, \mathbf{e}_\lambda$ take all possible σ and π orientations and writing the result in terms of a *tensor* scattering amplitude [1,2]

$$f_{res} = \mathbf{e}'^*_{\lambda'} \hat{f}_{res} \mathbf{e}_\lambda \tag{86}$$

The $\hat{f}_{res}$ tensor is represented in matrix form as:

$$\hat{f}_{res} = -r_0 F^e{}_1 \begin{pmatrix} 1 & 0 \\ 0 & \cos 2\theta \end{pmatrix} \tag{87}$$

where the rows correspond to $\mathbf{e}'_{\lambda'} = \sigma$ or π' respectively, and the columns to $\mathbf{e}_\lambda = \sigma$ or π.

Consider next the case in which the atom is in a tetragonally distorted environment, in which z is inequivalent to the x, y directions, so that $2F^e{}_{1,0} - (F^e{}_{1,1} + F^e{}_{1,-1}) \equiv F^e{}_{an} \neq 0$. For a geometry in which the scattering plane is the x, y plane, and the x axis is chosen parallel to the scattering vector $\mathbf{q}$ (see Fig. 5a), corresponding to the experimental situation sketched in Fig. 5b, where the crystal surface and the directions of the incoming and scattered beams are visible, we find that

$$\hat{f}_{res} = -\frac{r_0}{2} \left[(F^e{}_{1,1} + F^e{}_{1,-1}) \begin{pmatrix} 1 & 0 \\ 0 & \cos 2\theta \end{pmatrix} \right.$$

$$\left. + [2F^e{}_{1,0} - (F^e{}_{1,1} + F^e{}_{1,-1})] \begin{pmatrix} 1 & 0 \\ 0 & 0 \end{pmatrix} \right] \tag{88}$$

If the crystal in Fig. 5b is rotated by an angle ϕ about the x axis, that is around the scattering vector $\mathbf{q}$ (*azimuthal scan*), while leaving the scattering

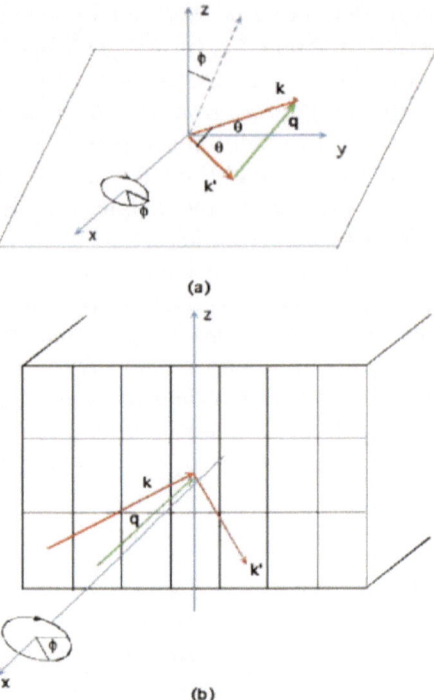

Fig. 5. (a) Sketch of the scattering plane with the scattering angle 2θ, and the azimuthal angle ϕ, describing rotations about the scattering vector **q**; (b) A sketch of the actual experimental geometry corresponding to (a)

geometry, i.e. the directions of **k** and **k'** unchanged (it is not easy to rotate a synchrotron!), the molecular preferred axis is rotated by an angle ϕ as shown by the dashed line in Fig. 5a. The scattering amplitude becomes

$$\hat{f}_{res} = -\frac{r_0}{2}\left[(F^e_{1,1} + F^e_{1,-1})\begin{pmatrix} 1 & 0 \\ 0 & \cos 2\theta \end{pmatrix}\right.$$

$$\left.+[2F^e_{1,0} - (F^e_{1,1} + F^e_{1,-1})]\begin{pmatrix} \cos^2\phi & -\frac{1}{2}\sin\theta\sin 2\phi \\ \frac{1}{2}\sin\theta\sin 2\phi & \sin^2\theta\sin^2\phi \end{pmatrix}\right] \quad (89)$$

This simple example is sufficient to demonstrate how, in the resonant regime, the charge-related scattering may display features such as non-diagonal tensor properties with respect to the polarization of the incoming and scattered beams, and the azimuthal angle dependence, which are absent for the non-resonant Thomson scattering. The tensor nature of the resonant scattering amplitude, which was traditionally written as a single complex number with the notation $f' + if''$ has become increasingly important in recent years.

As we mentioned in passing, (84) also contains magnetic scattering components, as first observed in [7]. The reader may wonder where, in the formulation in terms of electric multipole transitions between the ground and the intermediate states, the sensitivity to magnetic moments may come from. This is a subtle but very important point. In fact, no spin operators appear in the resulting expressions. The sensitivity to magnetic moments comes from the combined action of two ingredients: the Pauli principle and the spin-orbit interaction. The Pauli principle enters because of the already mentioned strong dependence of the scattering amplitude on the availability of states, at or near the resonance energy, suitable to play the role of intermediate states. In a one-electron language, if states with a given spin are predominantly occupied, it is mostly states with the opposite spin which are available to be virtually filled by the promotion of a core electron in the first part of the resonant scattering process. Since the spin is conserved in the optical transition, it is mostly electrons with the same spin as the predominantly available intermediate states which are virtually excited. In the case of all core levels with $l \neq 0$, the spin-orbit interaction is nonvanishing and much larger than in the valence states (for example, the L_2 and L_3 core levels of the rare earths are separated by many hundreds of eV). In a given spin-orbit partner, states with spin up or down have a different orbital character (the spin *polarizes* the orbital state, i.e. it tends to line up the orbital moment along or against the spin direction). Because of the selection rules to the available intermediate states, this orbital polarization translates into a difference between the transition rates for different m's, therefore in an imbalance among the corresponding $F^e_{1,m}$, which is in turn responsible for a nonvanishing magnetic scattering amplitude.

The above qualitative description of resonant magnetic scattering must be modified for s core levels, which have a vanishing spin-orbit interaction. In this case, resonant magnetic scattering is also observable [28], but it must be ascribed to the much weaker spin-orbit interaction of the valence states, which acts to polarize the final states of given spin and to reproduce the same mechanism. We have so far considered the scattering amplitude for a single atom or ion. In order to consider a lattice of atoms, we must perform a coherent superposition of the scattering amplitudes from all atoms, in analogy to (29), in which the amplitude from the atom sitting at $\mathbf{R}_{l,m,n}$ acquires a phase factor $e^{i\mathbf{q} \cdot \mathbf{R}_{l,m,n}}$. If the system displays crystallographic or magnetic order, such that the direction of the preferred axis $\mathbf{z}$ changes from atom to atom, this must also be taken into account, as it will affect the value of the scattering amplitude for given polarization, as visible from (84). In order to illustrate the consequences of these facts, let us consider the case of a basal-plane antiferromagnet, e.g. the rare earths Ho, Tb, and Dy, sketched in Fig. 6. All spins are ferromagnetically aligned in the planes, but their direction rotates by a fixed angle from one plane to the next. It was indeed in Ho, which displays a spiral antiferromagnetic phase in the $20\,\mathrm{K} \leq T \leq 131\,\mathrm{K}$

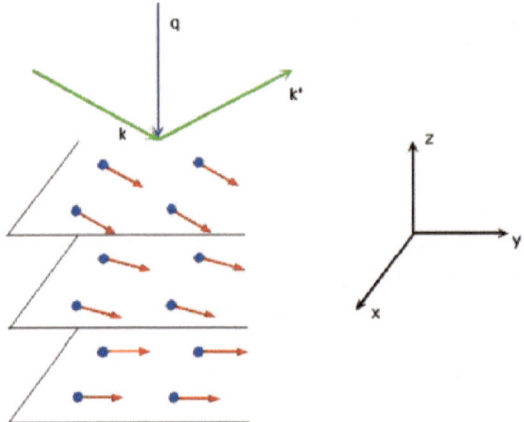

Fig. 6. Schematic view of a basal-plane antiferromagnet and of the scattering geometry

temperature range that the first observations of resonant magnetic scattering took place [7].

With reference to Fig. 6, let us label the atomic positions by a single index $\mathbf{R}_n$, for simplicity, and let us identify the spin direction of the ion n with $\hat{\mathbf{z}}_n$. It is apparent that, with respect to the chosen x, y, z coordinate frame, this vector is given by

$$\hat{\mathbf{z}}_n = (\sin \tau \cdot \mathbf{R}_n, \cos \tau \cdot \mathbf{R}_n, 0) , \qquad (90)$$

where τ is the wavevector associated to the spiral pitch, and if we define for short $(F^e{}_{1,1} + F^e{}_{1,-1}) \equiv F_0$, $(F^e{}_{1,1} - F^e{}_{1,-1}) \equiv F_1$ and $[2F^e{}_{1,0} - (F^e{}_{1,1} + F^e{}_{1,-1})] \equiv F_2$, we are ready to write the scattering amplitude

$$f = \sum_n e^{i\mathbf{q}\cdot\mathbf{R}_n} f_{res}(\mathbf{R}_n, \hat{\mathbf{z}}_n) . \qquad (91)$$

For the case $\sigma \to \sigma'$, the F_1 component does not contribute, and we are left with

$$f = -\frac{r_0}{2} \sum_n e^{i\mathbf{q}\cdot\mathbf{R}_n} [F_0 + F_2 \sin^2 \tau \cdot \mathbf{R}_n] . \qquad (92)$$

Expressing the $\sin^2$ in terms of exponentials, a bit of algebra gives

$$f = -\frac{r_0}{2} \sum_{h,k,l} \left[\left(F_0 + \frac{1}{2}F_2\right) \delta(\mathbf{q} - \mathbf{G}_{hkl}) - \frac{1}{4}F_2\delta(\mathbf{q} - \mathbf{G}_{hkl} \pm 2\tau) \right] . \qquad (93)$$

.

We thus see that the resonant scattering occurs at the crystallographic reciprocal lattice vectors $\mathbf{q} = \mathbf{G}_{hkl}$, with a charge and a magnetic component (F_0 and F_2 respectively) and also at the *second harmonic* magnetic satellite vectors, displaced by $\pm 2\tau$ from each crystallographic reflection. Let us investigate the $\sigma \to \pi'$ scattering channel. In this case, the F_0 component does not contribute, whereas the F_1 does, because it is easy to see that

$$(\mathbf{e}'_{\pi'} \times \mathbf{e}_\sigma) \cdot \hat{\mathbf{z}}_n = \cos\theta \cos\tau \cdot \mathbf{R}_n . \tag{94}$$

Similarly, the F_2 component is readily evaluated, because

$$(\mathbf{e}'_{\pi'} \cdot \hat{\mathbf{z}}_n)(\mathbf{e}_\sigma \cdot \hat{\mathbf{z}}_n) = -\frac{1}{2}\sin\theta \sin 2\tau \cdot \mathbf{R}_n . \tag{95}$$

and finally one gets

$$f = \frac{ir_0}{4}[F_1 \cos\theta \delta(\mathbf{q} - \mathbf{G}_{hkl} \pm \tau)$$
$$+ \frac{F_2}{2}\sin\theta(\delta(\mathbf{q} - \mathbf{G}_{hkl} + 2\tau) - \delta(\mathbf{q} - \mathbf{G}_{hkl} - 2\tau)] . \tag{96}$$

Thus, in the $\sigma \to \pi'$ scattering channel, magnetic reflections are expected again at the second harmonic magnetic satellites, but also at the first harmonic. We are now in a position to consider the actual experimental data by Gibbs et al. [7], some of which are reproduced in Fig. 7.

The most intense resonances occur at an energy of about 8.07 keV, and at wavevectors corresponding to the first and second harmonic magnetic satellites. Most notably, the absence of scattering in the $\sigma \to \sigma'$ channel at the first harmonic satellite is in agreement with the predictions of the analysis reported above. The intermediate states accessible via dipole processes in this energy range must involve promotion of $2p_{3/2}$ electrons to empty $5d$ band states, which are magnetically polarized as a consequence of the exchange interaction with the 4f electrons. In addition, less intense but very visible features resonating at a ~ 6 eV lower energy and occurring not only at the first and second, but also at the third and fourth harmonic satellite positions in $\mathbf{k}$ space are visible, which are not accounted for. In order to explain these features, we have to go beyond the dipole approximation and consider the higher order terms in the expansion of the $e^{i\mathbf{k}\cdot\mathbf{r}}$ exponential. In fact, following [24], we shall see that these features are associated to *quadrupole* transitions.

4.2 Electric Quadrupole Transitions

Our goal is now to derive expressions for electric quadrupole transitions, i.e. for the case in which the resonant transitions are allowed only when the second term in the expansion of $e^{i\mathbf{k}\cdot\mathbf{r}} \simeq 1 + i\mathbf{k} \cdot \mathbf{r} + \cdots$ is retained. The important matrix elements in (65) reduces to

$$\langle n | \mathbf{e}_\lambda \cdot \mathbf{p}_j e^{i\mathbf{k}\cdot\mathbf{r}_j} | 0 \rangle \simeq \langle n | (\mathbf{e}_\lambda \cdot \mathbf{p}_j)(i\mathbf{k} \cdot \mathbf{r}_j) + \cdots | 0 \rangle \tag{97}$$

228   M. Altarelli

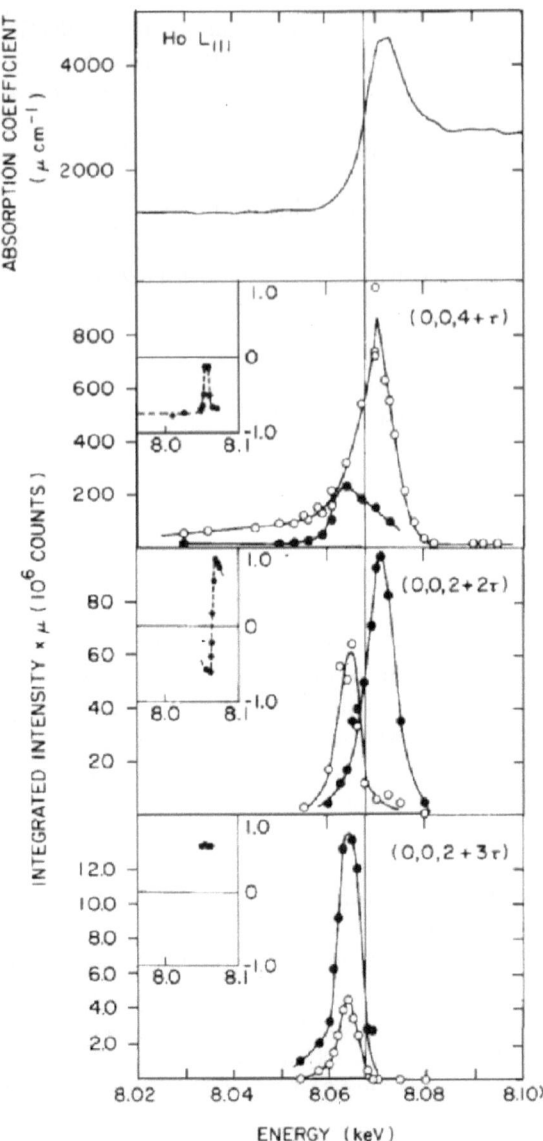

Fig. 7. Dependence on incoming energy of Holmium spectra near the L_3 edge. *Top panel*: absorption spectrum. *Bottom three panels*: intensity scattered at three magnetic satellite reflections; filled circles correspond to σ' and empty circles to π' scattered intensity, for incoming σ polarization. The insets show the degree of outgoing linear polarization defined as $(I_{\sigma'} - I_{\pi'})/(I_{\sigma'} + I_{\pi'})$ (After Gibbs et al. [7])

Using the same approximate trick as in (69), the operator $\mathbf{p}_j$ can be expressed in terms of the commutator of $\mathbf{r}_j$ and H, and allowing H to act on $\langle n|$ and on $\rangle 0|$ by moving it next to them using the commutation rules one finds

$$\langle n|(\mathbf{e}_\lambda \cdot \mathbf{p}_j)(\mathbf{k} \cdot \mathbf{r}_j)|0\rangle = -i\frac{m}{\hbar}(E(0) - E(\Psi_n))\langle n|(\mathbf{e}_\lambda \cdot \mathbf{r}_j)(\mathbf{k} \cdot \mathbf{r}_j)|0\rangle$$
$$-\langle n|(\mathbf{e}_\lambda \cdot \mathbf{r}_j)(\mathbf{k} \cdot \mathbf{p}_j)|0\rangle . \qquad (98)$$

Adding to both sides of this equation the same quantity, namely the left hand side, it becomes

$$2\langle n|(\mathbf{e}_\lambda \cdot \mathbf{p}_j)(\mathbf{k} \cdot \dot{\mathbf{r}}_j)|0\rangle = -i\frac{m}{\hbar}(E(0) - E(\Psi_n))\langle n|(\mathbf{e}_\lambda \cdot \mathbf{r}_j)(\mathbf{k} \cdot \mathbf{r}_j)|0\rangle$$
$$+\langle n|(\mathbf{e}_\lambda \cdot \mathbf{p}_j)(\mathbf{k} \cdot \mathbf{r}_j)|0\rangle - \langle n|(\mathbf{e}_\lambda \cdot \mathbf{r}_j)(\mathbf{k} \cdot \mathbf{p}_j)|0\rangle . \quad (99)$$

Application to the last two terms of this equation of the vector identity (47) finally yields

$$\langle n|(\mathbf{e}_\lambda \cdot \mathbf{p}_j)(\mathbf{k} \cdot \mathbf{r}_j)|0\rangle = -i\frac{m}{2\hbar}(E(0) - E(\Psi_n))\langle n|(\mathbf{e}_\lambda \cdot \mathbf{r}_j)(\mathbf{k} \cdot \mathbf{r}_j)|0\rangle$$
$$+\frac{1}{2}(\mathbf{k} \times \mathbf{e}_\lambda)\langle n|\mathbf{r}_j \times \mathbf{p}_j|0\rangle . \qquad (100)$$

Now it is easy to recognize in the last term the matrix element between the ground and intermediate states of the orbital angular momentum operator (or, in the language of multipole expansions, the matrix element corresponding to *magnetic dipole* transitions). For the transitions resonant with x-ray photons, which involve promotion of a core electron above the Fermi level, the magnetic dipole matrix elements vanish, because of the orthogonality of the *radial* part of core and valence states, as the angular momentum operators only affect the *angular* part of the wavefunctions. Summarizing, one can conclude that the second term in the expansion of the plane-wave exponential produces terms with the matrix elements of products of two components of the position operator $\mathbf{r}_j$ (*electric quadrupole* terms) plus magnetic dipole terms, which are irrelevant in the x-ray range. By analogy with the discussion following (70), define the rank 2 quadrupole moment tensor, with cartesian components $(\alpha, \beta = x, y, z)$

$$Q_{\alpha\beta}^{(2)} = R_\alpha R_\beta - \frac{1}{3}R^2\delta_{\alpha\beta} . \qquad (101)$$

Its spherical components are

$$Q_m^{(2)} = \sqrt{\frac{4\pi}{5}}R^2 Y_m^2(\theta, \phi) \qquad (102)$$

with $m = -2, -1, \ldots, 2$. All matrix elements of importance for quadrupole resonant scattering can be written in terms of the quantities

$$F_{2,m}^e = \frac{m_e}{\hbar^3 c^2} \sum_n [E(\Psi_n) - E(0)]^3 \omega_{\mathbf{k}} \frac{|\langle n|Q_m^{(2)}|0\rangle|^2}{E(0) - E(\Psi_n) + \hbar\omega_{\mathbf{k}} + i\Gamma_n/2} \qquad (103)$$

The scattering amplitude at the quadrupole level comprises 13 different terms (see e.g. [27]) one of which contains no dependence on $\hat{\mathbf{z}}_n$ and is written as

$$-\frac{r_0}{3}(\mathbf{e}'^*_{\lambda'} \cdot \mathbf{e}_\lambda)(\mathbf{k}' \cdot \mathbf{k})[F_{2,2}^e + F_{2,-2}^e] \qquad (104)$$

There are then 2 terms in which there is a linear dependence on $\hat{\mathbf{z}}_n$:

$$-\frac{r_0}{3}[(\mathbf{k}' \cdot \mathbf{k})(\mathbf{e}'^*_{\lambda'} \times \mathbf{e}_\lambda) \cdot \hat{\mathbf{z}}_n + (\mathbf{e}'^*_{\lambda'} \cdot \mathbf{e}_\lambda)(\mathbf{k}' \times \mathbf{k}) \cdot \hat{\mathbf{z}}_n][F_{2,2}^e - F_{2,-2}^e] \quad (105)$$

and this contributes a magnetic scattering reflection at the first harmonic satellite. The other terms have 2, or 3 or 4 factors of $\hat{\mathbf{z}}$ and they give rise to reflections up to the fourth harmonic magnetic satellites, in agreement with the observations of [7] in Ho. The corresponding set of dipole-forbidden, but quadrupole allowed intermediate states is easily identified with the promotiom of $2p_{3/2}$ core electrons to empty $4f$ levels; these resonances probe the huge $4f$ magnetic moment of Ho (about $10\,\mu_B$). This reinforces the strength of these resonances, compensating in part the smaller matrix element for quadrupole transitions in comparison to the dipolar ones. Before concluding this section, it is perhaps worthwhile to emphasize again that the dependence of the polarized radiation scattering amplitude of each atom on the orientation of the local $\hat{\mathbf{z}}_n$ axis can lead to a breakdown of the extinction rules, exemplified by the case of the diamond lattice at the end of subsection III A. In fact, we may no longer be allowed to factor out a constant atomic form factor in (31).

5 Resonant Elastic Scattering: Recent Developments

In the preceding sections, an overview of the theoretical foundations of x-ray magnetic scattering was attempted. The purpose was to a large extent pedagogical, trying to adopt the simplest possible arguments to arrive at an understanding of the information and of the quantities accessible to this experimental technique. In this section, we shall first make a few general comments about the resonant scattering amplitudes derived so far, and then we shall proceed to illustrate some examples from the recent literature.

5.1 Some General Properties of the Scattering Amplitude

We wish to recall some general results of importance for the following. First of all, the connection between the photon scattering amplitude and the optical absorption spectrum of the system is established via the *Optical Theorem* [25] of general scattering theory. This theorem states that:

$$\frac{k}{4\pi}\sigma_t = \mathrm{Im}f(0) \,, \tag{106}$$

where σ_t is the total cross section, comprising elastic and inelastic processes, and f(0) is the *forward* scattering amplitude. In the case of photons in the energy range of interest here, the total cross section σ_t for the interaction with electronic matter is dominated by photoelectric absorption processes (see [29] for typical examples). Therefore, the left-hand side of (106) reproduces, with good approximation, the absorption spectrum. The right-hand side contains the imaginary part of the amplitude of scattering for $\mathbf{k}' = \mathbf{k}$ and $\mathbf{e}'_{\lambda'} = \mathbf{e}_\lambda$, which can be easily obtained as a special case of the general expressions derived above. Application of the theorem to the expression derived at the dipole level, (84), gives the following expression [30] for the absorption cross section

$$\sigma_t = -\lambda r_0 \left[\mathrm{Im}(F^e{}_{1,1} + F^e{}_{1,-1}) - i(\mathbf{e}^*_\lambda \times \mathbf{e}_\lambda) \cdot \hat{\mathbf{z}}\mathrm{Im}(F^e{}_{1,1} - F^e{}_{1,-1}) \right.$$
$$\left. +|\mathbf{e}_\lambda \cdot \hat{\mathbf{z}}|^2 Im(2F^e{}_{1,0} - F^e{}_{1,1} - F^e{}_{1,-1})\right] \,. \tag{107}$$

It is useful to notice that the imaginary part of the expressions (82) is determined by

$$\mathrm{Im}\frac{1}{E(0) - E(\Psi_n) + \hbar\omega_\mathbf{k} + i\Gamma_n/2} = \frac{-\Gamma_n/2}{(E(0) - E(\Psi_n) + \hbar\omega_\mathbf{k})^2 + (\Gamma_n/2)^2} \tag{108}$$

so that the absorption spectrum is described as a sum of Lorentzians, each of which has the broadening parameter corresponding to the inverse lifetime of the corresponding excited state. Notice also that the expression $(\mathbf{e}^*_\lambda \times \mathbf{e}_\lambda)$ vanishes for a real polarization vector (i.e. for linear polarization) and is purely imaginary in other cases. In the traditional crystallographic notation in which the atomic scattering amplitude is a simple scalar quantity, written as the sum of the non-resonant amplitude f_0 and the real and imaginary resonant parts $f'+if''$, the optical theorem identifies f'' as proportional to the optical absorption spectrum, because, in the case of forward scattering, $\mathbf{q} = 0$, f_0 is just $-Zr_0$, where Z is the atomic number. An interesting concept, along the same lines, is that of *interference scattering*, which arises from the fact that the scattering cross section is $|f|^2 = |f_0|^2 + f'^2 + f''^2 + 2Re[f_0(f' + if'')]$. The last term is termed an *interference* between the non-resonant and the resonant scattering, and is important whenever the photon energy is such that resonant and non-resonant components are similar in magnitude. For an application of this idea to magnetic scattering in ferromagnets, see [31]. Another kind, if not of interference, of mixed terms in the cross section arises from the multipolar expansion (64) of the resonant amplitude. In addition to the pure dipole and pure quadrupole terms, considered in Sect. IV, it is clear that mixed dipole-quadrupole terms also arise from the expansion

$$\langle 0| \sum_{j=1}^{N} \mathbf{e}'^{*}_{\lambda'} \cdot \mathbf{p}_j(1 - i\mathbf{k}' \cdot \mathbf{r}_j)|n\rangle\langle n| \sum_{j'=1}^{N} \mathbf{e}_{\lambda} \cdot \mathbf{p}_{j'}(1 + i\mathbf{k} \cdot \mathbf{r}_{j'})|0\rangle$$

$$= \text{dipole-dipole} + \text{quadrupole-quadrupole}$$

$$+ \langle 0| \sum_{j=1}^{N} \mathbf{e}'^{*}_{\lambda'} \cdot \mathbf{p}_j|n\rangle\langle n| \sum_{j'=1}^{N} \mathbf{e}_{\lambda} \cdot \mathbf{p}_{j'} i\mathbf{k} \cdot \mathbf{r}_{j'}|0\rangle$$

$$+ \langle 0| \sum_{j=1}^{N} \mathbf{e}'^{*}_{\lambda'} \cdot \mathbf{p}_j(-i\mathbf{k}' \cdot \mathbf{r}_j)|n\rangle\langle n| \sum_{j'=1}^{N} \mathbf{e}_{\lambda} \cdot \mathbf{p}_{j'}|0\rangle . \tag{109}$$

These terms imply that both dipole and quadrupole matrix elements connect the eigenstates $|0\rangle$ and $|n\rangle$, which implies that inversion symmetry is broken, so that even and odd-parity states are mixed. This can happen for an atom in a non-centrosymmetric site. One can however further doubt the relevance of such terms, given the fact that if dipole-dipole terms are allowed they will dominate over other terms, as previously argued. There are however cases, as we shall see, in which the lattice symmetry is such that even-parity terms, like pure dipole terms, vanish in the summation over atoms, leaving the mixed dipole-quadrupole terms as the largest contribution.

5.2 Some Highlights of the Recent Literature

We return to consider the problem of extinction rules in the diamond lattice, given as an example in the discussion following (31). It has recently been shown [32] that in Ge the forbidden $(6, 0, 0)$ reflection is strongly present when the incoming photon energy is close to the Ge K-edge energy, $11.107 \, \text{keV}$, corresponding to the excitation of $1s$ electrons. In order to understand this observation, which was predicted by Templeton and Templeton [33], we go back to (30), but now consider the resonant scattering amplitude, instead of the atomic form factor. It is important to emphasize that the diamond lattice has inversion symmetry: the center of a bond, connecting two nearest neighbour atoms along the cube diagonal (see Fig. 3), is a center of inversion for the lattice; nonetheless, each atom is not in a centrosymmetric position, and supports therefore mixed terms in the resonant amplitude. If we translate the origin to the midpoint of the two atoms, the coordinates of the basis atoms are $\mathbf{r}_1 = -(a/8)(\mathbf{x} + \mathbf{y} + \mathbf{z})$ and $\mathbf{r}_2 = (a/8)(\mathbf{x} + \mathbf{y} + \mathbf{z})$. With an obvious notation, we can write, starting from (31)

$$|F(\mathbf{G}_{hkl})|^2 \equiv \left| f_{1,res}(\mathbf{G}_{hkl})e^{-i(\frac{\pi}{4})(h+k+l)} + f_{2,res}(\mathbf{G}_{hkl})e^{i(\frac{\pi}{4})(h+k+l)} \right|^2 . \tag{110}$$

If the multipole expansion is now applied to the amplitudes of atom $i, i = 1, 2$

$$f_{i,res} = f_{i,dd} + f_{i,dq} + f_{i,qd} + \cdots . \tag{111}$$

where the dipole-dipole term reduces to the first term of (84), because the T_d symmetry of the diamond lattice does not have a preferential axis and does not allow the other terms. The general form of the mixed terms is

$$
f_{i,dq} + f_{i,qd} \sim i \sum_n \left[\langle 0 | \sum_{j=1}^{N} \mathbf{e}'^*_{\lambda'} \cdot \mathbf{p}_j | n \rangle \langle n | \sum_{j'=1}^{N} \mathbf{e}_\lambda \cdot \mathbf{p}_{j'} \mathbf{k} \cdot \mathbf{r}_{j'} | 0 \rangle \right.
$$

$$
\left. - \langle 0 | \sum_{j=1}^{N} \mathbf{e}'^*_{\lambda'} \cdot \mathbf{p}_j \mathbf{k}' \cdot \mathbf{r}_j | n \rangle \langle n | \sum_{j'=1}^{N} \mathbf{e}_\lambda \cdot \mathbf{p}_{j'} | 0 \rangle \right]
$$

$$
\times \frac{E(\Psi_n) - E_0}{\hbar \omega_\mathbf{k}} \frac{1}{E_0 - E(\Psi_n) + \hbar \omega_\mathbf{k} + i\Gamma_n/2} . \qquad (112)
$$

It is easy to infer, from simple parity arguments, that $f_{1,dd} = f_{2,dd}$, whereas $f_{1,dq} = -f_{2,dq}, f_{1,qd} = -f_{2,qd}$, so that (110) becomes

$$
|F(\mathbf{G}_{hkl})|^2 = \left| 2f_{1,dd}(\mathbf{G}_{hkl}) \cos\left(\frac{\pi}{4}\right) (h + k + l) \right.
$$

$$
\left. - 2i(f_{1,dq} + f_{1,qd})(\mathbf{G}_{hkl}) \sin\left(\frac{\pi}{4}\right) (h + k + l) \right|^2 . \qquad (113)
$$

It is then easy to see that the mixed terms in the second piece of the right-hand side produce a non-vanishing intensity also when $h+k+l = 2\times$ an odd number, thus providing an exception to the extinction rule. The intensity of the scattering is related to the admixture of p-like and d-like wavefunctions in the empty conduction band states, which are accessible from the $1s$ core level via dipole and quadrupole transitions respectively. For a discussion of the information on band structure delivered by this experiment and for a comparison with ab initio calculations, see [34]. This is of course just one of many examples of violation of extinction rules. An example at the level of pure quadrupole terms was discovered in hematite ($\alpha-\mathrm{Fe_2O_3}$) by Finkelstein et al. [35], and discussed theoretically by Carra and Thole [36].

There is at present an ongoing effort to explore new magnetic materials, magnetic microstructures and materials with strong electronic correlations by x-ray scattering, in particular by resonant scattering. This technique has the added value of atomic specificity (an important asset when dealing with systems with more than one magnetic element), and even the specific sensitivity to the electronic shell of the atom, via the selection rules and the choice of the energy. It has originally found applications in the hard x-ray regime, and more recently in the soft x-ray regime, not only in connection to artificial structures (multilayers and superlattices), but also to materials with periods much longer than atomic distances.

Materials with strong electronic correlations, which are very central in contemporary condensed matter research, are characterized by the interplay of structural, electronic and magnetic properties. In some cases, long range ordered arrangements of spins, of charge and of orbital symmetry coexist,

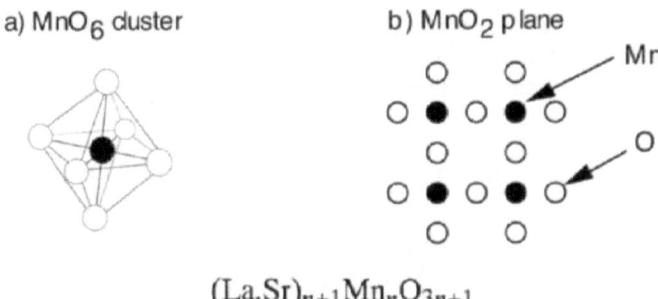

a) MnO$_6$ cluster

b) MnO$_2$ plane

$(La,Sr)_{n+1}Mn_nO_{3n+1}$

Fig. 8. Building blocks of the perovskite structure of the manganites: layers of MnO_6 octahedral separated by La trivalent ions, partly substituted by divalent dopants such as Sr

such as in the case of the manganite family of compounds (see e.g. [37]). Examples of compounds belonging to this family are $La_{1-x}Ca_xMnO_3$ or $(La, Sr/Ca)_{n+1}Mn_nO_{3n+1}$, e.g. for $n = 1$, $La_{0.5}Sr_{1.5}MnO_4$. Resonant scattering from the Mn K-edge can to some extent be sensitive to all these kinds of order parameters and has therefore been used to investigate them [38], although the results on orbital order have generated a considerable controversy [39]. The point is that the degenerate orbitals concerned by the orbital ordering are the Mn d electrons, and dipole resonances at the K-edge involve empty p orbitals in the intermediate states. There are therefore good reasons to believe that the scattering is not especially sensitive to the ordering of the d electrons but rather to the accompanying Jahn-Teller distortions of the MnO_6 octahedra (see Fig. 8). Recently, experimental tests involving the interference scattering of strained manganite films on different substrates [40], confirm that this is indeed the case and ab initio calculations support this conclusion [41].

Another system of interest is V_2O_3, where magnetic and orbital order have been investigated by resonant scattering at the V K-edge [42], and where the interpretation has again been controversial [43–46]. In the case of transition metal oxides presented so far, the K-edges correspond to relatively hard x-rays, with wavelengths in the range of order $\sim$0.1 nm. This wavelength corresponds to the typical order of interatomic distances and is therefore extremely suitable to investigate crystallographic or, possibly, magnetic structures, with periods of the same order. There is however nothing, in our derivations, which would not equally well apply to softer x-rays, and allow the investigation of structures with longer periods. Modern technology for the growth of nanostructures, and in particular of multilayers, has led to the fabrication of structures with characteristic periods in the order of 1 nm or more, which are therefore accessible to investigation by soft x-rays. In multilayers in which 3d transition metals are present as magnetic elements, the $L_{2,3}$ edges, corresponding to dipole allowed transitions from the $2p_{1/2,3/2}$ core levels to the magnetic $3d$ states, have energies ranging from $512\,\mathrm{eV}$ (vanadium) to $932\,\mathrm{eV}$ (copper), corresponding to wavelengths of order $\sim 2\,\mathrm{nm}$ (see for example [47], for an example of the study of Cu/Co superlattices). One major advantage of resonant scattering in the study of multilayer structures is the *element specifity*, which allows us to focus on features relating to one particular element, by selecting the wavelengths appropriate to its edges. In many instances, one can take advantage of such selectivity, as illustrated by examples such as the measurements by Sacchi et al. [48] on V/Fe superlattices, in which it is possible to deduce a ferromagnetic alignment of all the Fe layers, with a small *antiparallel* magnetic moment on the V layers.

But also nature provides for systems with order parameters with periods compatible with soft x-ray wavelengths. As pointed out in [49], orbital order in $La_{0.5}Sr_{1.5}MnO_4$ as well as in other manganites, is within reach of the Mn L-edge resonances, with the advantage that L_2 and L_3 resonances access the $3d$ orbitals in the intermediate states, and should therefore be very sensitive to their orbital and magnetic order. This was confirmed in recent experiments [50–53] exploring the orbital and magnetic order reflections; comparing their dependence on the incoming photon energy with theoretical calculations within an atomic multiplet scheme in a crystal field [54, 55], it turns out that scattering at the L_2 resonance is very sensitive to orbital order, while the intensity at the L_3 resonance is mostly induced by the Jahn-Teller distortion.

6 Resonant Inelastic X-ray Scattering

In this section inelastic processes are considered, in which the state of the material system after the scattering event (the final state) has a different energy than the initial state (for simplicity assumed to be the ground state). In recent years, inelastic scattering with synchrotron radiation has acquired a growing importance. Inelastic processes can take place in the non-resonant as well as in the resonant regime. Here we shall restrict ourselves to the resonant regime, defined as usual by the condition that the incoming photon energy is close to one of the absorption edges of the system. In this case one can consider the dominant resonant terms only, and obtain, in full analogy to (65)

$$w = \frac{2\pi}{\hbar} \sum_f \left| \sum_n \frac{\langle f|H_2'|n\rangle\langle n|H_2'|i\rangle}{E_0 - E_n + \hbar\omega_{\mathbf{k}} + i\Gamma_n/2} \right|^2$$
$$\times \delta(\hbar(\omega_{\mathbf{k}} - \omega_{\mathbf{k}'}) - (E_f - E_i)) \,. \tag{114}$$

Figure 9 shows schematically how an inelastic scattering process can leave the system in a final state with an electron-hole pair in the valence levels or with a pair where the hole is in a core shell. In the latter case the term "resonant x-ray Raman scattering" is used.

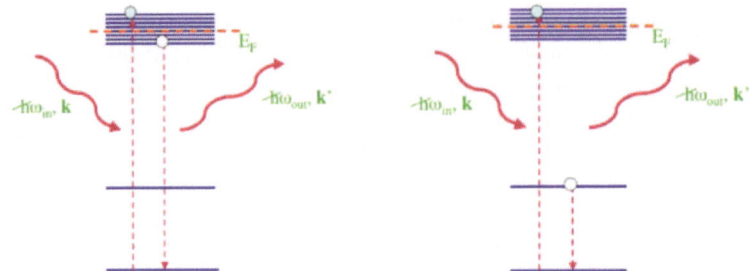

Fig. 9. Schematic description of resonant inelastic inelastic scattering processes

There is a very large intensity difference between elastic and inelastic scattering: this is due to the fact that the intensity of elastic scattering is proportional to N^2, the square of the number of scattering atoms, and the intensity of inelastic scattering is proportional to N. A schematic illustration of the reason [56] is offered in Fig. 10, for a collection of identical non-interacting atoms. The difference is that in elastic scattering virtual excitation of different atoms represent different paths through which the same final state is reached, whereas in the inelastic case the excited electron stays through the final process, together with a hole in a different level: the contributions of processes on different atoms lead therefore to different final states and must be added after squaring.

The argument can be carried over to the more realistic cases in which atoms are interacting, electrons are tunnelling from one to the other and the energy eigenstates are Bloch states, and even electron-hole interactions can be included. Photon transitions, whether virtual or real, conserve momentum, and this restricts the viable intermediate states to a given final state.

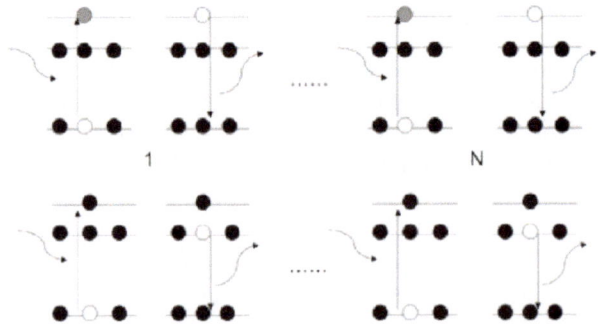

Fig. 10. *Top panel*: elastic scattering, the final state is the same, irrespective of which of the N atoms is virtually excited. The amplitudes for the N different ways to reach the same final state must be first summed and than squared. *Bottom panel*: inelastic scattering: the final state is different, depending on which atom is virtually excited, therefore the amplitudes are first squared and then added

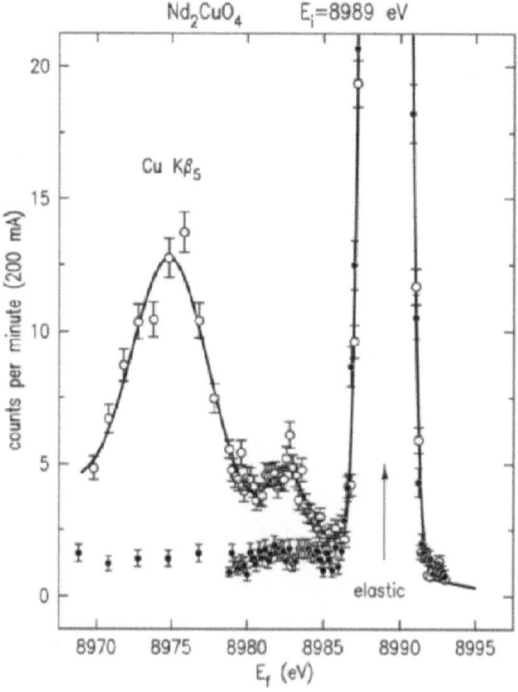

Fig. 11. Scattered intensity as a function of final energy, for a fixed incident energy $E_i = 8989$ eV. The *solid line* is a guide to the eye. *Solid symbols*: Data taken well above resonance ($E_i = 9020$ eV) to characterize inelastic background. After [57]

The large intensity difference between elastic and inelastic scattering is well exemplified by the data [57] reproduced in Fig. 11.

There is an obvious similarity between resonant inelastic x-ray scattering (RIXS) and other techniques such as x-ray fluorescence spectroscopy and absorption spectroscopy in the fluorescence detection mode. The presence of an electron-hole pair in the final state also suggests an analogy with absorption spectroscopy, either in the visible-UV region, for valence holes, or in the soft x-rays for the Raman case. There are however significant differences, because the inelastic scattering process has different selection rules, and the sampling depth of a technique using hard x-rays is always larger than that of soft x-ray spectroscopies. A good illustration of the relationship between inelastic scattering and absorption spectroscopy is provided by the results by Hämäläinen et al. [58] on the Dy L_3 edge in dysprosium nitrate. After quadrupolar excitation from the $2p_{3/2}$ to the $4f$ manifold by the incoming photon, a high resolution analyzer accepts only a narrow band of the outgoing photons corresponding to $3d_{5/2}$ electrons filling the $2p_{3/2}$ holes. The recorded intensity is plotted in Fig. 12, as a function of the incoming photon energy.

238   M. Altarelli

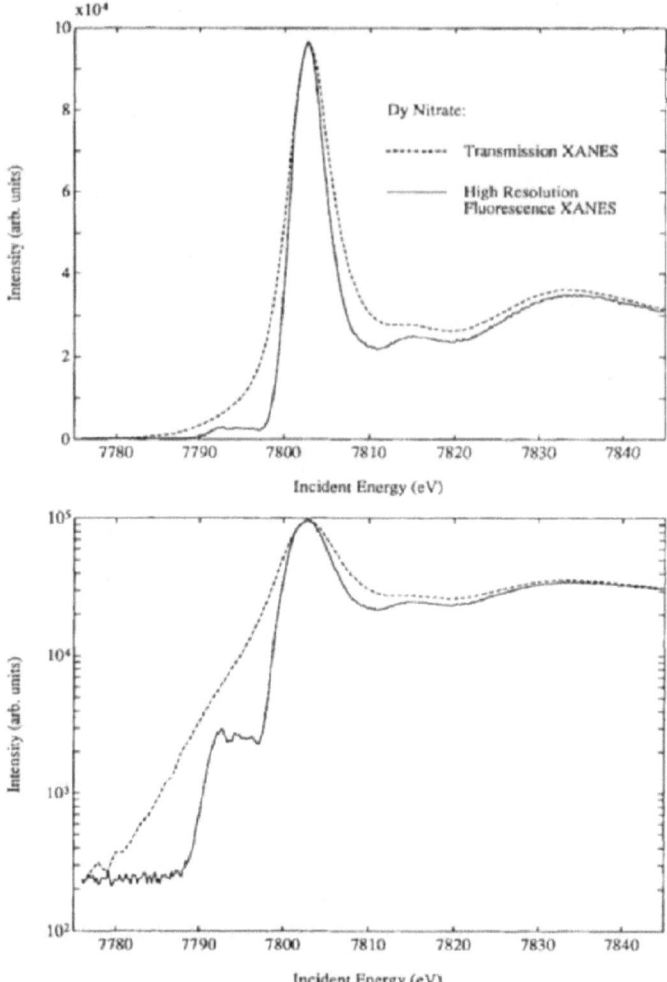

Fig. 12. Inelastic scattering spectrum and absorption (XANES) spectrum of the Dy L_3 edge; both measurements were performed on the same sample with the same energy resolution. After [58]

In general, a wealth of information can be obtained by scanning the incoming photon monochromator and the outgoing photon analyzer in various ways. As pointed out by Carra et al. [59], the possible results of inelastic scattering measurements such as those of [58], can fill a three-dimensional plot, where intensity is recorded as a function of incoming $\omega_{\mathbf{k}}$ and outgoing photon energy $\omega_{\mathbf{k}'}$, or, equivalently, of the incoming and *transferred* energy $\Delta\omega = \omega_{\mathbf{k}} - \omega_{\mathbf{k}'}$ (see Fig. 13).

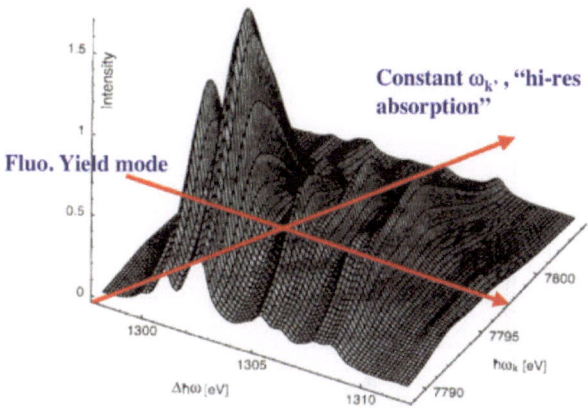

Fig. 13. Calculated inelastic scattering spectrum [59] together with two cuts exemplifying the relationship to other techniques (see text)

From various cuts or slices of this plot we can recover various kind of spectra. If for each value of the incoming energy we integrate over all transferred energies we recover the absorption spectrum in fluorescence yield mode, in which the number of collected outgoing photons, irrespective of their energy, is taken as a measure of the number of absorbed incoming photons of the given energy. The Hämäläinen et al. experiment, on the other hand, in which outgoing photons of the same energy $\omega_{\mathbf{k}'}$ are recorded as a function of $\omega_{\mathbf{k}}$, corresponds to a "diagonal" cut on the plot of Fig. 13, along one of the $\omega_{\mathbf{k}} - \Delta\omega = \text{constant}$ lines. This is related, with the caveats mentioned above, to a soft x-ray absorption spectrum in the region of the M_5 ($3d_{5/2}$) edge. There is, as one can see, much more information in the three-dimensional plot, and therefore in the inelastic scattering experiment, than in just a few cuts and sections, loosely identifiable as corresponding to this or that mode of absorption spectroscopy. The complexity and variety of information contained in RIXS spectra makes theoretical calculations a major challenge. The approaches which have so far been applied most frequently are based either on atomic multiplet models, or on their extension to include the neighbouring ligands via Anderson impurity models or small cluster calculations [60]. On the other hand, from the experimental point of view, this technique, in spite of all the difficulties connected to the low cross-section and the limitations in resolution, can provide rather unique information of importance for many-body physics. A beautiful example is provided by the study [61] of the temperature dependence of Yb valence fluctuations in the Kondo systems YbInCu$_4$ and YbAgCu$_4$. The shift of the resonant energy of the L_3 edge for Yb^{2+} and Yb^{3+} (about 7 eV) is used to distinguish spectral components associated with the two valence states and to assess their relative populations as a function of temperature, in a way that is unambiguous and bulk sensitive (unlike photoemission spectroscopy).

240   M. Altarelli

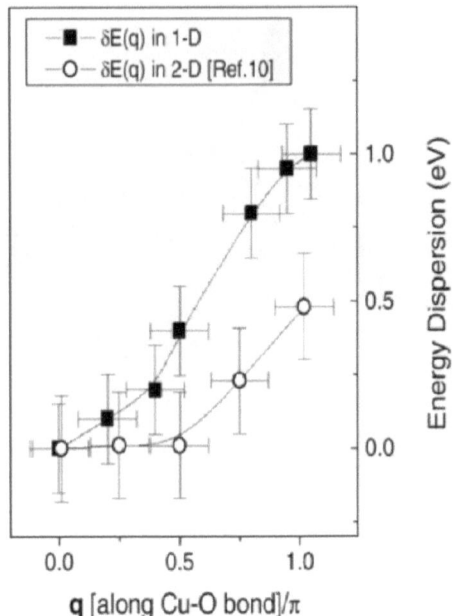

Fig. 14. Momentum dependence of charge excitations in one- and two-dimensional cuprate compounds measured by RIXS (after [63])

A further important piece of information that can be obtained by RIXS is the momentum dependence of electronic excitations, accessible by scanning the experimental geometry (the angle between the incident photon beam and the analyzer direction). This is exemplified by the results [62, 63] on the electronic excitations of low-dimensional cuprates with Mott insulator properties. As shown in Fig. 14, charge excitations display a more marked dispersion in the one-dimensional compounds ($SrCuO_2$, Sr_2CuO_3) than in the two-dimensional compounds ($Ca_2CuO_2Cl_2$). In conclusion, one can say that RIXS is a technique with great promise, as advances in instrumentation should lead to continued improvement in energy resolution.

References

1. D.H. Templeton and L.K. Templeton: Acta Cryst. A**36**, 237 (1980); *ibid.* A**38**, 62 (1982).
2. V.E. Dmitrienko: Acta Cryst. A**39**, 29 (1983).
3. F.E. Low: Phys. Rev. **96**, 1428 (1954).
4. M. Gell-Mann and M.L. Goldberger: Phys. Rev. **96**, 1433 (1954).
5. P. Platzman and N. Tzoar: Phys. Rev. **B2**, 3556 (1970).
6. F. de Bergevin and M. Brunel: Acta Cryst. **A37**, 314 (1981).
7. D. Gibbs, D.R. Harshman, E.D. Isaacs, D.B. McWhan, D. Mills and C. Vettier: Phys. Rev. Lett. **61**, 1241 (1988).

8. V.B. Berestetskii, E.M. Lifshitz and L.P. Pitaevskii: *Quantum Electrodynamics*, 2nd edition (Reed, Oxford, 1982), Sect. 21.
9. A.I. Akhiezer and V.B. Berestetskii: *Quantum Electrodynamics*, (Wiley, New York, 1965), Sects. 12 and 15.
10. J.D. Jackson: *Classical Electrodynamics*, 2nd edition (Wiley, New York, 1975), Sect. 6.3.
11. L.D. Landau and E.M. Lifshitz: *The Classical Theory of Fields*, 4th edition (Reed, Oxford, 1975), Sect. 46.
12. M. Blume: J. Appl. Phys. **57**, 3615 (1985).
13. G. Trammel: Phys. Rev. **92**, 1387 (1953).
14. G. Steinsvoll et al.: Phys. Rev. **161**, 499 (1963).
15. L.D. Landau and E.M. Lifshitz: *Electrodynamics of Continuous Media*, 2nd edition (Reed, Oxford, 1984), Sect. 29, pp. 105–107.
16. D.B. McWhan et al.: Phys. Rev. B**42**, 6007 (1990).
17. S. Langridge et al.: Phys. Rev. B**55**, 6392 (1997).
18. V. Fenandez, C. Vettier, F. de Bergevin, C. Giles and W. Neubeck: Phys. Rev. B**57**, 7870 (1998).
19. L. Paolasini et al.: to be published.
20. M. Brunel, G. Patrat, F. de Bergevin, F. Rousseaux and M. Lemonnier: Acta Cryst. **A39**, 84 (1983).
21. D. Laundy, S.P. Collins and A.J. Rollason: J. Phys.: Condens. Matter **3**, 369 (1991).
22. See e.g. S. Raimes: *Many-Electron Theory*, (North Holland, Amsterdam, 1972).
23. See e.g. A.R. Edmonds: *Angular Momentum in Quantum Mechanics*, (Princeton University Press, Princeton, 1957).
24. J.P. Hannon, G.T. Trammel, M. Blume and D. Gibbs: Phys. Rev. Lett. **61**, 1245 (1988); *ibid.* (E) **62**, 2644 (1989).
25. L.D. Landau and E.M. Lifshitz: *Quantum Mechanics*, 3rd edition (Reed, Oxford, 1977), Appendix c.
26. See the web site http://www-tvs.sci.kun.nl/people/fasolino.html/paper_ps/XRES.ps.
27. J.P. Hill and D.F. McMorrow: Acta Crystallog. **A52**, 10336 (1996).
28. See for example W. Neubeck, C. Vettier, K.B. Lee and F. de Bergevin: Phys. Rev. B**60**, R9912 (1992); A. Stunault, F. de Bergevin, D. Wermeille, C. Vettier, Th. Brckel, N. Bernhoeft, G.J. McIntyre and J.Y. Henry, Phys. Rev. B**60**, 10170 (1999).
29. *X-ray Data Booklet*, edited by the Lawrence Berkeley National Laboratory, Second edition, Jan. 2001 (available on the web at the site http://xdb.lbl.gov/).
30. P. Carra and M. Altarelli: Phys. Rev. Lett. **64**, 1286 (1990).
31. F. de Bergevin et al.: Phys. Rev. B**46**, 10772 (1992).
32. T.L. Lee, R. Felici, K. Hirano, B. Cowie, J. Zegenhagen and R. Colella: Phys. Rev. B**64**, 201316(R) (2001).
33. D.H. Templeton and L.K. Templeton: Phys. Rev. B**49**, 14850 (1994).
34. I.S. Elfimov, N.A. Skorikov, V.I. Anisimov and G.A. Sawatzky: Phys. Rev. Lett. **88**, 015504 (2002) and **88**, 239904 (2002).
35. K.D. Finkelstein, Q. Shen and S. Shastry: Phys. Rev. Lett. **69**, 1612 (1992).
36. P. Carra and B.T. Thole: Rev. Mod. Phys. **66**, 1509 (1994).
37. Y. Tokura, editor, *Colossal Magnetoresistive Oxides*, (Gordon and Breach, New York, 2000).

38. Y. Murakami, H. Kawada, H. Kawata, M. Tanaka, T. Arima, Y. Moritomo, and Y. Tokura: Phys. Rev. Lett. **80**, 1932 (1998).
39. See for example S. Ishihara and S. Maekawa: Phys. Rev. Lett. **80**, 3799 (1998); I.S. Elfimov, V.I. Anisimov and G.A. Sawatzky, Phys. Rev. Lett. **82**, 4264 (1999); M. Benfatto, Y. Joly and C.R. Natoli: Phys. Rev. Lett. **83**, 636 (1999).
40. H. Ohsumi et al.: J. Phys. Soc. Japan **72**, 1006 (2003).
41. N. Binggeli and M. Altarelli: New J. of Physics **6**, 165 (2004).
42. L. Paolasini, C. Vettier, F. de Bergevin, F. Yakhou, D. Mannix, A. Stunault, W. Neubeck, M. Altarelli, M. Fabrizio, P.A. Metcalf, and J.M. Honig: Phys. Rev. Lett. **82**, 4719 (1999).
43. F. Mila, R. Shiina, F.C. Zhang, A. Joshi, M. Ma, V. Anisimov and T.M. Rice: Phys. Rev. Lett. **85**, 1714 (2000).
44. A. Tanaka: J. Phys. Soc. Japan **71**, 1091 (2002).
45. S.W. Lovesey, K.S. Knight and D.S. Sivia: Phys. Rev. B**65**, 224402 (2002).
46. Y. Joly, S. Di Matteo and C.R. Natoli: Phys. Rev. B**69**, 224401 (2004).
47. T.P.A. Hase et al.: Phys. Rev. B**61**, R3792 (2000).
48. M. Sacchi et al.: Phys. Rev. B**60**, R12569 (1999).
49. C.W.M. Castleton and M. Altarelli: Phys. Rev. B**62**, 1033 (2000).
50. S.B. Wilkins, P.D. Spencer, P.D. Hatton, S.P. Collins, M.D. Roper, D. Prabhakaran, and A.T. Boothroyd: Phys. Rev. Lett. **91**, 167205 (2003).
51. S.B. Wilkins, P.D. Hatton, M.D. Roper, D. Prabhakaran, and A.T. Boothroyd: Phys. Rev. Lett. **90**, 187201 (2003).
52. K.J. Thomas et al.: Phys. Rev. Lett. **92**, 237204 (2004).
53. S.S. Dhesi et al.: Phys. Rev. Lett. **92**, 056403 (2004).
54. S.B. Wilkins, N. Stojic, T.W.A. Beale, N. Binggeli, C.W.M. Castleton, P. Bencok, D. Prabhakaran, A.T. Boothroyd, P. Hatton, and M. Altarelli: Phys. Rev. B**71**, 245102 (2005).
55. S.B. Wilkins, N. Stojic, T.A.W. Beale, N. Binggeli, P. Bencok, S. Stanescu, J.F. Mitchell, P. Abbamonte, P.D. Hatton, and M. Altarelli: cond/mat 0412435.
56. V.B. Berestetskii, E.M. Lifshitz and L.P. Pitaevskii: *Quantum Electrodynamics*, 2nd edition (Reed, Oxford, 1982), Sect. 59.
57. J.P. Hill et al.: Phys. Rev. Lett. **80**, 4967 (1998).
58. K. Hämäläinen, D.P. Siddons, J.B. Hastings and L.E. Berman: Phys. Rev. Lett. **67**, 2850 (1991).
59. P. Carra, M. Fabrizio and B.T. Thole: Phys. Rev. Lett. **74**, 3700 (1995).
60. A. Kotani and S. Shin: Rev. Mod. Phys. **73**, 203 (2001).
61. C. Dallera, M. Grioni, A. Shukla, G. Vanko, J.L. Sarrao, J.P. Rueff, and D.L. Cox: Phys. Rev. Lett. **88**, 196403 (2002).
62. M.Z. Hasan et al.: Science **288**, 1811 (2000).
63. M.Z. Hasan et al.: Phys. Rev. Lett. **88**, 177403 (2002).

High Angle Magnetic X-ray Diffraction

Catherine Dufour, Karine Dumesnil, and Thierry Gourieux

Laboratoire Physique des Matériaux, Université de Nancy 1, BP 239, 54506
Vandoeuvre lès Nancy, France
dufour@lpm.u-nancy.fr

Abstract. The magnetic scattering of x-rays was first performed 30 years ago and
was considered as a curiosity. However, due to the advent of high brilliance synchrotron radiation sources, this technique has become a standard microscopic probe
for the investigation of magnetic properties. The magnetic x-ray scattering amplitude contains two terms: the nonresonant term that exists at all photon energies
and the resonant term, induced by multipole electric transitions, that is peaked
near absorption edges. In this paper, the basic concepts of x-ray magnetic scattering are first introduced for both non-resonant and resonant regimes. Then, selected
examples are presented in order to illustrate several specific properties of the technique: i) high resolution in the reciprocal space; ii) ability to separate the orbital
momentum contribution to magnetization; iii) chemical and electronic selectivity;
iv) possibility to investigate weak moment systems and observe resonant signal for
polarized non-magnetic ions; v) multi-**q** structure detection.

1 Introduction

X-ray diffraction has long been interpreted through the Thomson scattering
mechanism, i.e. the interaction between the photons and the charge of the
electrons. However, since x-rays are part of the electromagnetic spectrum,
they are expected to be also sensitive to magnetic distributions.

The sensitivity of x-rays to magnetism has been first considered in analyzing Compton or inelastic scattering. In 1929, Klein and Nishina [1] took the
magnetic scattering into account in their formula for the Compton effect, but
only implicitly since it concerns mean values taken over all spin states. In the
fifties, the cross-section for the scattering of photons by free charges, including magnetic effect was derived by Low [2] and by Gell-Mann and Goldberger
[3] by taking the non-relativistic limit of the Compton cross-section.

For many years, no attempt was made to observe magnetic effects in
diffraction since they are substantively weaker when the scattering is elastic (about a hundredth of the spin dependent inelastic scattering). In 1970,
Platzman and Tzoar [4] carried out the first calculation of the amplitude of
x-rays elastically scattered by a magnetically ordered substance in the relativistic quantum theory. They suggested that this effect could be observed.
De Bergevin and Brunel [5] performed the experimental demonstration two
years later. They used a conventional sealed x-ray tube to measure on a NiO

C. Dufour et al.: *High Angle Magnetic X-ray Diffraction*, Lect. Notes Phys. **697**, 243–273
(2006)
www.springerlink.com

single crystal two superlattice diffraction peaks, which disappeared above the Néel point. The intensities of the magnetic peaks were $4 \cdot 10^{-8}$ smaller than the charge peaks. The counting time for each magnetic peak was 3 days. A few years later, the same authors repeated the calculations of Platzman and Tzoar, made them more explicit and performed magnetic diffraction experiments on ferro- and ferrimagnetic compounds [6].

In 1985, Blume derived the cross-section for x-ray scattering (including magnetic terms) in a way that allows the effect of electron binding to be accounted for [7]. By starting with the non-relativistic Hamiltonian for electrons (to order $(v/c)^2$) and the quantized electromagnetic field, he produced a general formula for the cross-section including virtually all scattering phenomena in appropriate limits (in the "kinematic" or Born approximation), including Thomson, Rayleigh, Bragg, thermal diffuse, Raman and magnetic scattering and anomalous dispersion. In the limit of photon energy large compared to electron binding energy, he recovered the Platzman and Tzoar expression. The results showed resonant magnetic scattering effects when the photon energy is comparable to the inner electron excitation or binding energies. These resonant magnetic scattering effects are related to anomalous dispersion phenomena. However, they exhibit different polarization dependence and disappear in the absence of magnetic order.

The same year, taking advantage of the tunability of the photon energy of the synchrotron x-ray beam, Namikawa et al. [8] reported on the first experiment of Resonant X-ray Magnetic Scattering (RXMS) at the K-edge of nickel.

In 1988, Blume and Gibbs [9] calculated the polarization dependence of magnetic x-ray scattering. They demonstrated that the orbital and spin angular momentum contributions of both ferromagnets and antiferromagnets might be separately measured in a variety of simple geometries. A few months later, the spin and orbital contributions to the x-ray magnetic scattering were experimentally identified in holmium [10]. In the same paper, a fiftyfold resonant enhancement of the magnetic signal and resonant integer harmonics were observed at the L_3 Ho absorption edge.

Simultaneously with these experiments, Hannon et al. [11] suggested an interpretation of RXMS in terms of atomic resonant theory. They predicted the occurrence of large resonant magnetic scattering in the vicinity of the L_2-L_3 and M_4-M_5 absorption edges in the rare earth and actinides and at the K and L edges in the transition metals. These resonances were shown to result from electric multipole transitions, with a sensitivity to magnetization arising from exchange effects. By exchange effects, they meant any effect originating from the anti-symmetry of the wave function under the exchange of electrons (this includes the exclusion principle allowing transitions to unoccupied orbitals as well as the exchange interaction between electrons in different orbitals). For some transitions, the calculated magnetic scattering was shown to be comparable to the charge scattering.

Immediately after this theoretical paper, very large resonance effects were observed in the antiferromagnet UAs [12]: at the U M_4 edge, the intensity of the magnetic peaks was enhanced by 10^7 relative to the non-resonant component far above the edge, and was remarkably 1% of the intensity of the charge peak.

In 1996, Hill and Mc Morrow [13] provided a very useful paper: they reformulated the cross-section for x-ray resonant exchange scattering in terms of linear polarization states perpendicular and parallel to the scattering plane, a basis particularly suitable for synchrotron diffraction experiments.

Over the last 20 years, x-ray magnetic scattering has developed from a scientific curiosity to a routine technique for the investigations of magnetism and magnetic materials. Following the pioneering works described above, a large amount of experiments have been performed, which would have been impossible without the advent of synchrotron sources. X-ray magnetic scattering complements the magnetic neutron scattering technique regarding sample size, Q-resolution, single magnetic domain studies, separation of angular and spin contributions [10, 14–17] to the magnetic moment, and magnetic ordering of highly neutron absorbing compounds [18]. Moreover, the chemical selectivity of x-ray resonant scattering permits to separate the contribution of different elements to the magnetism in compounds, solid solutions [19–22] and superlattices [23, 24]. Finally, the electronic shell selectivity represents a unique method to characterize magnetic interactions from an electronic point of view [25, 26].

We should remark here that dichroism has developed in parallel to scattering. The dichroic effect is proportional to the imaginary part of the scattering factor, evaluated at a zero momentum transfer ($Q = 0$, i.e. absorption only) whereas the scattering contribution is proportional to the sum of the squares of both real and imaginary parts, which can interfere with each other.

This paper is restricted to high angle magnetic x-ray diffraction. In the second paragraph, the basic concept of x-ray magnetic scattering will be introduced. We will start this part with the development of the formalism to evaluate the magnetic scattering cross section. We will calculate separately the non-resonant scattering amplitude (describing magnetic scattering far away from any absorption edge and allowing the spin-orbit separation) and the resonant scattering amplitude (which has to be taken into account when the incident photon energy is tuned to an absorption edge). In the third part, selected examples will be presented, in order to illustrate several specific properties of x-ray magnetic diffraction.

2 Basic Concept of Magnetic X-ray Scattering

2.1 Spin Dependent Cross-Section

We are going to follow the Blume's [7] development of the x-ray scattering cross-section. As mentioned in the introduction, this development contains all terms of magnetic interactions. It described the cross section as a function of the polarization of the incident and diffracted photon beams. The calculations are derived by a perturbative method; they take into account the effects of electron binding and contain the resonant effects.

The non-relativistic Hamiltonian for electrons in a quantized electromagnetic field is:

$$
H = \sum_j \frac{1}{2m} \left(\mathbf{P}_j - \frac{e}{c}\mathbf{A}(\mathbf{r}_j) \right)^2 + \sum_{i,j} V(\mathbf{r}_{ij})
$$

$$
- \frac{e\hbar}{2mc} \sum_j \mathbf{s}_j.\mathbf{B}(\mathbf{r}_j) - \frac{e\hbar}{2(mc)^2} \sum_j \mathbf{s}_j.\mathbf{E}(\mathbf{r}_j) \times \left(\mathbf{P}_j - \frac{e}{c}\mathbf{A}(\mathbf{r}_j) \right)
$$

$$
+ \sum_{q,\lambda} \hbar\omega_q \left(c^+(\mathbf{q},\lambda)c(\mathbf{q},\lambda) + \frac{1}{2} \right)
$$

$$
\tag{1}
$$

$\mathbf{A}(\mathbf{r})$ is the vector potential, $\mathbf{E}(\mathbf{r})$ and $\mathbf{B}(\mathbf{r})$ are respectively the electric and magnetic fields of the radiation. c^+ and c are the operators of photon creation and annihilation. $\mathbf{q}$ is the incident wave vector and the index λ (=1,2) labels the two polarisations of each wave $\mathbf{q} \cdot \mathbf{s}_j$ and $\mathbf{P}_j$ are the spin and momentum operators of the jth electron located at $\mathbf{r}_j$.

The first term in the Hamiltonian (1) describes the kinetic energy of electrons in an electromagnetic field; the second is their Coulomb potential energy. The third and fourth terms are the Zeeman energy and a term including the spin-orbit interaction. The last term gives the energy of the electro-magnetic field.

The vector potential can be written as a linear combination of the photon creation and annihilation operators. Scattering processes conserve the number of photons. Therefore, in a first order perturbation, scattering arises from terms quadratic in $\mathbf{A}$ whereas linear terms contribute to second order perturbation.

Consequently, in the development of the fourth term of (1), which is of order $(v/c)^2$, only the quadratic terms in $\mathbf{A}$ and the independent terms are kept.

Using the Maxwell's equations, the different components of the Hamiltonian can be rewritten $H = H_0 + H_R + H'$ with:

$$
H_0 = \sum_j \frac{1}{2m}\mathbf{P}_j^2 + \sum_{i,j} V(\mathbf{r}_{ij}) + \frac{e\hbar}{2(mc)^2} \sum_j \mathbf{s}_j \cdot (\nabla\phi_j \times \mathbf{P}_j)
$$

$$H_R = \sum_{q,\lambda} \hbar\omega_q \left(c^+(\mathbf{q},\lambda)c(\mathbf{q},\lambda) + \frac{1}{2} \right)$$

$$H' = \frac{e^2}{2mc^2} \sum_j \mathbf{A}^2(\mathbf{r}_j) - \frac{e}{mc} \sum_j \mathbf{A}(\mathbf{r}_j) \cdot \mathbf{P}_j - \frac{e\hbar}{mc} \sum_j \mathbf{s}_j \cdot (\nabla \times \mathbf{A}(\mathbf{r}_j))$$

$$- \frac{e\hbar}{2(mc)^2} \frac{e}{c^2} \sum_j \mathbf{s}_j \cdot \left(\frac{\partial \mathbf{A}(\mathbf{r}_j)}{\partial t} \times \mathbf{A}(\mathbf{r}_j) \right)$$

$$= H'_1 + H'_2 + H'_3 + H'_4$$

$$(2)$$

H_0 is the Hamiltonian of the electrons without radiation, H_R is the Hamiltonian of the radiation and H' the interaction Hamiltonian; ϕ is the Coulomb potential.

Let's assume that, the electrons are initially in a state $|a\rangle$, which is an eigenstate of H_0 with energy E_a, and that there is a single photon in a state $|\mathbf{q}, \lambda\rangle$ of energy $\hbar\omega_q$. The interaction will induce transitions to new eigenstates $|b\rangle$ of energy E_b for the electrons, with a photon in a state $|\mathbf{q}', \lambda'\rangle$ with energy $\hbar\omega'_q$. The probability W of such a transition per time unit is given by the Fermi's golden rule to the second order:

$$W = \frac{2\pi}{\hbar} \left| \langle f|H'_1 + H'_4|i\rangle + \sum_c \frac{\langle f|H'_2 + H'_3|c\rangle \langle c|H'_2 + H'_3|i\rangle}{E_i - E_c} \right|^2 \delta(E_i - E_f)$$

$$= \frac{2\pi}{\hbar} \left| f(\mathbf{q}, \mathbf{q}', \epsilon, \epsilon', \hbar\omega_q, \hbar\omega'_q) \right|^2 \delta(E_i - E_f)$$

$$(3)$$

with $|i\rangle = |a; \mathbf{q}, \lambda\rangle$, $|f\rangle = |b; \mathbf{q}', \lambda'\rangle$, $E_i = E_a + \hbar\omega_q$, $E_f = E_b + \hbar\omega'_q$.

The vectors ϵ and ϵ' are polarization unit vectors of the incident and scattered beams. The states $|c\rangle$ are intermediate states of the system "electrons of the atoms + radiation" and f is the scattering amplitude.

Let's now consider an assembly of N atoms and restrict the problem to the elastic scattering, $\hbar\omega_q = \hbar\omega'_q = \hbar\omega$ and $|a\rangle = |b\rangle$ ($E_a = E_b$). The elastic scattering cross section can be written as:

$$\frac{d\sigma}{d\Omega} = r_0^2 \left| \left\langle a \left| \sum_n e^{i\mathbf{Q}\cdot\mathbf{R}_n} f_n(\mathbf{q}, \mathbf{q}', \epsilon, \epsilon', \hbar\omega) \right| a \right\rangle \right|^2$$

$$(4)$$

$\mathbf{Q} = \mathbf{q} - \mathbf{q}'$. r_0 represents the classical electron radius. The position of an electron j on a site n in a crystal is given by: $\mathbf{R}_j = \mathbf{R}_n + \mathbf{r}_j$ where $\mathbf{R}_n$ is the position of the nth site in the crystal.

f_n is the scattering amplitude of the electrons at site n.

Introducing the Hamiltonian (2) in (3), the scattering amplitude can be expressed:

$$f_n(\mathbf{q}, \mathbf{q}', \boldsymbol{\epsilon}, \boldsymbol{\epsilon}', \hbar\omega) =$$

$$\left\langle a \left| \sum_j e^{i\mathbf{Q} \cdot \mathbf{r}_j} \right| a \right\rangle \boldsymbol{\epsilon}' \cdot \boldsymbol{\epsilon} - i \frac{\hbar\omega}{mc^2} \left\langle a \left| \sum_j e^{i\mathbf{Q} \cdot \mathbf{r}_j} \mathbf{s}_j \right| a \right\rangle \boldsymbol{\epsilon}' \times \boldsymbol{\epsilon} + \frac{1}{m} \sum_c \sum_j$$

$$\left[\frac{\left\langle a \left| [\boldsymbol{\epsilon}' \cdot \mathbf{P}_j - i\hbar(\mathbf{q}' \times \boldsymbol{\epsilon}') \cdot \mathbf{s}_j] e^{-i\mathbf{q}' \cdot \mathbf{r}_j} \right| c \right\rangle \left\langle c \left| [\boldsymbol{\epsilon} \cdot \mathbf{P}_j + i\hbar(\mathbf{q} \times \boldsymbol{\epsilon}) \cdot \mathbf{s}_j] e^{i\mathbf{q} \cdot \mathbf{r}_j} \right| a \right\rangle}{E_a - E_c + \hbar\omega - i\Gamma_c/2} \right.$$

$$\left. + \frac{\left\langle a \left| [\boldsymbol{\epsilon} \cdot \mathbf{P}_j + i\hbar(\mathbf{q} \times \boldsymbol{\epsilon}) \cdot \mathbf{s}_j] e^{i\mathbf{q} \cdot \mathbf{r}_j} \right| c \right\rangle \left\langle c \left| [\boldsymbol{\epsilon}' \cdot \mathbf{P}_j - i\hbar(\mathbf{q}' \times \boldsymbol{\epsilon}') \cdot \mathbf{s}_j] e^{-i\mathbf{q}' \cdot \mathbf{r}_j} \right| a \right\rangle}{E_a - E_c - \hbar\omega} \right]$$

$$(5)$$

In diffraction experiments described in the following sections, the polarizations of incident and scattered beams are linear. Two characteristic directions are used to describe a linear polarization of the radiation: the σ polarization is perpendicular to the scattering plane and the π polarization lies in this plane. During a given experiment, the incoming beam is either σ or π polarized. To perform linear polarization analysis, an analyzer crystal is placed in the diffracted beam. This crystal is chosen because its lattice d-spacing corresponds to $1/2$ of the wavelength of the radiation, so that the charge reflection occurs at $90°$. By rotating the analyzer crystal and the detector by $90°$ around the scattered wave vector, either the σ or the π part of the scattered intensity is measured. σ-σ scattering (respectively σ-π scattering) corresponds to the case of a σ-polarized incident photon for which the polarization is unchanged (respectively rotated).

Using the two unitary vectors parallel to the incident and scattered beams, $\hat{\mathbf{q}} = \mathbf{q}/q$ and $\hat{\mathbf{q}}' = \mathbf{q}'/q$, we can define the coordinate system $(\mathbf{u}_1, \mathbf{u}_2, \mathbf{u}_3)$ as illustrated in Fig. 1:

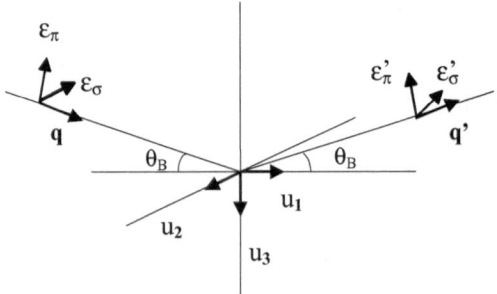

Fig. 1. Coordinate system used. $\mathbf{q}$ and $\mathbf{q}'$ are the incident and scattered wavevectors, θ_B is the Bragg angle, ϵ_σ and ϵ_π are the components of the polarization perpendicular and parallel to the diffraction plane defined by $\mathbf{q}$ and $\mathbf{q}'$. The $\mathbf{u}_i$'s define the basis for the magnetic structure expressed in terms of the incident and scattered wave-vectors

$$\mathbf{u}_1 = (\hat{\mathbf{q}} + \hat{\mathbf{q}}')/2\cos\theta_B, \mathbf{u}_2 = (\hat{\mathbf{q}} \times \hat{\mathbf{q}}')/\sin 2\theta_B, \mathbf{u}_3 = (\hat{\mathbf{q}} - \hat{\mathbf{q}}')/2\sin\theta_B$$

with θ_B the Bragg angle. In this basis, the polarization vectors can be written:

$$\epsilon_\sigma = \epsilon'_\sigma = -\mathbf{u}_2, \epsilon_\pi = -(\cos\theta_B)\mathbf{u}_3 + (\sin\theta_B)\mathbf{u}_1, \epsilon'_\pi = -(\cos\theta_B)\mathbf{u}_3 - (\sin\theta_B)\mathbf{u}_1$$

The dependence of the scattering amplitude on the polarizations of incident and scattered beam can best be expressed using a 2×2 matrix in the base of (σ,π). With this convention, the scalar and vector products of the polarization vectors are:

$$\epsilon' \cdot \epsilon = \begin{pmatrix} 1 & 0 \\ 0 & \cos 2\theta_B \end{pmatrix} \tag{6}$$

$$\epsilon' \times \epsilon = \begin{pmatrix} 0 & \mathbf{u}_1 \cos\theta_B + \mathbf{u}_3 \sin\theta_B \\ -\mathbf{u}_1 \cos\theta_B + \mathbf{u}_3 \sin\theta_B & -\mathbf{u}_2 \sin 2\theta_B \end{pmatrix} \tag{7}$$

The diagonal terms correspond to the non-rotated component of the intensity (σ-σ or π-π), whereas the off-diagonal terms correspond to the rotated component of the intensity (σ-π or π-σ).

Coming back to expression (5) of the scattering amplitude, the first term describes the charge or Thomson scattering. It is directly related to the Fourier transform of the charge density. It is to note that **the charge scattering does not rotate the polarization** (matrix (6) is diagonal). Consequently, during a σ-π or π-σ experiment, no charge contribution should be theoretically measured. The second term of (5) depends on the Fourier transform of the spin density and contributes to the non-resonant magnetic scattering. These first and second terms reflect the scattering process without intermediate state [27].

The third and fourth terms arise from the second order perturbation and contain an energy dependent denominator. The third term corresponds to the case where the initial photon is first absorbed leading to a zero photon intermediate state [27]. It is responsible for resonant anomalous scattering, occurring when the photon energy is tuned to an absorption edge with energy $E_c - E_a$. We take into account the line width of the transition through $i\Gamma_c/2$ where Γ_c is related to the lifetime of the core hole created in the process. The fourth term corresponds to the case where the final photon is first emitted leading to a two photons intermediate state [27]. It is responsible for anomalous scattering far away from resonance. The nominators of the third and fourth terms are Hermitian conjugates. They contain spin dependent contributions arising from $\mathbf{s} \cdot \nabla \times \mathbf{A}$ in H'_3 and interactions between the initial and final states described by $\mathbf{A} \cdot \mathbf{P}$ in H'_2.

Two different cases can be considered depending on the photon energy:

1 if $\hbar\omega$ is far away from any characteristic energy of the system $E_c - E_a$, the denominators scale as $\hbar\omega$ and we obtain the non-resonant magnetic scattering amplitude.

2 if $\hbar\omega$ approaches $E_c - E_a$, the corresponding denominator vanishes giving rise to a resonant contribution.

We are now going to treat successively the resonant and non-resonant regimes.

2.2 Non-resonant X-ray Magnetic Scattering

Far away from any characteristic energy of the system $\hbar\omega \gg E_c - E_a$, the non-resonant magnetic contribution to the scattering factor is reduced to [9]:

$$f_n^{non-res.}(\mathbf{q}, \mathbf{q}', \boldsymbol{\epsilon}, \boldsymbol{\epsilon}', \hbar\omega) = -i\frac{\hbar\omega}{mc^2} \left\langle a \left| \sum_j e^{i\mathbf{Q}\cdot\mathbf{r}_j} \left(i\frac{\mathbf{Q}\times\mathbf{P}_j}{\hbar\mathbf{Q}^2}\cdot\mathbf{A}' + \mathbf{s}_j\cdot\mathbf{B}' \right) \right| a \right\rangle$$

$$= -i\frac{\hbar\omega}{mc^2} \left(\frac{1}{2}\mathbf{L}(\mathbf{Q})\cdot\mathbf{A}'' + \mathbf{S}(\mathbf{Q})\cdot\mathbf{B}' \right) \tag{8}$$

with

$$\mathbf{A}' = \boldsymbol{\epsilon}' \times \boldsymbol{\epsilon}$$
$$\mathbf{A}'' = 2(1 - \hat{\mathbf{q}}\cdot\hat{\mathbf{q}}')(\boldsymbol{\epsilon}' \times \boldsymbol{\epsilon}) + (\hat{\mathbf{q}}' \times \boldsymbol{\epsilon}')(\hat{\mathbf{q}}'\cdot\boldsymbol{\epsilon}) - (\hat{\mathbf{q}} \times \boldsymbol{\epsilon})(\hat{\mathbf{q}}\cdot\boldsymbol{\epsilon}')$$
$$\mathbf{B}' = \boldsymbol{\epsilon}' \times \boldsymbol{\epsilon} + (\hat{\mathbf{q}}' \times \boldsymbol{\epsilon}')(\hat{\mathbf{q}}'\cdot\boldsymbol{\epsilon}) - (\hat{\mathbf{q}} \times \boldsymbol{\epsilon})(\hat{\mathbf{q}}\cdot\boldsymbol{\epsilon}') - (\hat{\mathbf{q}}' \times \boldsymbol{\epsilon}') \times (\hat{\mathbf{q}} \times \boldsymbol{\epsilon})$$

$\mathbf{A}'$, $\mathbf{A}''$ and $\mathbf{B}'$ are geometrical vectors describing the polarization dependence of the scattering. $\mathbf{L}(\mathbf{Q})$ and $\mathbf{S}(\mathbf{Q})$ are the Fourier transforms of the atomic orbital magnetization density and of the spin density.

In the basis $(\mathbf{u}_1, \mathbf{u}_2, \mathbf{u}_3)$ defined above, the scattering amplitude takes the form:

$$f_n^{non-res.} = -i\frac{\hbar\omega}{mc^2}$$
$$\times \begin{pmatrix} S_2 \sin 2\theta_B & -2\sin^2\theta_B\left[\cos\theta_B(L_1 + S_1) \\ & -S_3\sin\theta_B\right] \\ 2\sin^2\theta_B[\cos\theta_B(L_1 + S_1) + S_3\sin\theta_B] & \sin 2\theta_B(2L_2\sin^2\theta_B + S_2) \end{pmatrix} \tag{9}$$

L_i and S_i are the components of $\mathbf{L}(\mathbf{Q})$ and $\mathbf{S}(\mathbf{Q})$ in the basis $(\mathbf{u}_1, \mathbf{u}_2, \mathbf{u}_3)$.

In the case of localized electrons, L_i and S_i are related to the atomic orbital $\boldsymbol{\mu}_L$ and spin $\boldsymbol{\mu}_S$ moments of the atoms through the relations:

$$L_i \equiv f_L(\mathbf{Q})\,\mu_{L_i}/(g_L\mu_B), \ S_i \equiv f_S(\mathbf{Q})\,\mu_{S_i}/(g_S\mu_B)$$

g_L and g_S are the Landé factors, μ_B the Bohr magneton, f_L and f_S the effective orbital and spin form factors and μ_{L_i} and μ_{S_i} the components of $\boldsymbol{\mu}_L$ and $\boldsymbol{\mu}_S$ in the basis $(\mathbf{u}_1, \mathbf{u}_2, \mathbf{u}_3)$.

In the spherical approximation, the spin and orbital form factors are usually expressed as:

$$f_L(\mathbf{Q}) = \langle j_0(\mathbf{Q}) \rangle + \langle j_2(\mathbf{Q}) \rangle \text{ and } f_S(\mathbf{Q}) = \langle j_0(\mathbf{Q}) \rangle$$

where the $\langle j_i(\mathbf{Q}) \rangle$ are appropriate relativistic radial integrals.

The magnetic form factor of an atom decreases more rapidly than the charge form factor because the magnetization density in real space is more diffuse than the charge density. At large Q, this reduces the magnetic contribution even further. The orbital magnetic moment, created by the orbital currents, is more localized than the spin magnetization and therefore its form factor diminishes more rapidly.

From (9), we can note that:

(i)   The non-resonant scattering amplitude is only sensitive to the components of the orbital angular momentum perpendicular to the momentum transfer (L_1 and L_2);

(ii) By measuring the scattered beam intensity dependence on polarization for a given geometry and magnetic structure, the orbital $\mathbf{L}$ and spin $\mathbf{S}$ moments can be estimated separately (the scattering amplitude is only $\mathbf{S}$ dependent in the $\sigma - \sigma$ configuration);

(iii) In contrast to neutron diffraction, the x-ray scattering is sensitive to the magnetization component collinear to the momentum transfer (S_3).

Finally, let us recall that the non-resonant magnetic scattering amplitude is smaller than the charge term by a factor $\hbar\omega/mc^2$ and is further reduced since only unpaired electrons give a contribution to this process whereas the charge scattering amplitude is proportional to the total number of electrons. For 5 keV photons, the ratio is: $\sigma_{mag}/\sigma_{charge} \sim 10^{-6} \langle \mu^2 \rangle$.

2.3 Resonant X-ray Magnetic Scattering (RXMS)

Starting from the third term in Blume's equation (5), containing the resonant contribution, and taking the possible degeneracy of the initial state into account, we can express the resonant scattering amplitude as:

$$f_n^{res.}(\mathbf{q}, \mathbf{q}', \epsilon, \epsilon', \hbar\omega) = -\frac{1}{m} \sum_c \sum_j p_a p_a(c) \frac{E_a - E_c}{\hbar\omega}$$

$$\times \frac{\langle a | [\epsilon' \cdot \mathbf{P}_j - i\hbar(\mathbf{q}' \times \epsilon') \cdot \mathbf{s}_j] e^{-i\mathbf{q}' \cdot \mathbf{r}_j} | c \rangle \langle c | [\epsilon \cdot \mathbf{P}_j + i\hbar(\mathbf{q} \times \epsilon) \cdot \mathbf{s}_j] e^{i\mathbf{q} \cdot \mathbf{r}_j} | a \rangle}{E_a - E_c + \hbar\omega - i\Gamma_c/2}$$

$$(10)$$

p_a gives the statistical probability for the various possible initial states and $p_a(c)$ gives the probability that the intermediate state is empty.

The resonant scattering amplitude (10) can be separated into three distinct terms:

1 $(\epsilon' \cdot \mathbf{P}_j)(\epsilon \cdot \mathbf{P}_j)$ contribution.

The origin of this process arises from the $\mathbf{A} \cdot \mathbf{P}$ contribution of the H_2' term of the interaction Hamiltonian. This term contains classical anomalous scattering contribution and a magnetic term called "exchange resonant scattering" by Hannon [11]. It is usually used to describe the resonant scattering.

2 $(\epsilon' \cdot \mathbf{P}_j)((\mathbf{q} \times \epsilon) \cdot \mathbf{s}_j) - ((\mathbf{q}' \times \epsilon') \cdot \mathbf{s}_j)(\epsilon \cdot \mathbf{P}_j)$ contribution.

This term depends explicitly on the magnetization and is sensitive to its direction. It is smaller by a factor $\hbar\omega/mc^2$ to the Hannon's term and can be neglected. However, it has been used by Namikawa et al. [8] to describe resonant process at the K edge of ferromagnetic Ni because of the absence of spin-orbit splitting.

3 $((\mathbf{q}' \times \epsilon') \cdot \mathbf{s}_j)((\mathbf{q} \times \epsilon) \cdot \mathbf{s}_j)$ contribution.

This term depends only on the spin. It is proportional to S^2 and therefore does not give rise to a pure antiferromagnetic peak. However, it contributes to anomalous scattering.

In the following, we are going to develop in more detail the Hannon's term. An important consequence of this choice is that selection rules for conventional electric multipolar transition can be used. The first contribution to the resonant scattering amplitude is:

$$f_n^{res.}(\mathbf{q}, \mathbf{q}', \epsilon, \epsilon', \hbar\omega) =$$

$$-\frac{1}{m}\sum_c\sum_j p_a p_a(c) \frac{E_a - E_c}{\hbar\omega} \frac{\left\langle a \left| [\epsilon' \cdot \mathbf{P}_j] e^{-i\mathbf{q}'\cdot\mathbf{r}_j} \right| c \right\rangle \left\langle c \left| [\epsilon \cdot \mathbf{P}_j] e^{i\mathbf{q}\cdot\mathbf{r}_j} \right| a \right\rangle}{E_a - E_c + \hbar\omega - i\Gamma_c/2}.$$

$$(11)$$

The $e^{i\mathbf{q}\cdot\mathbf{r}}$ and $\epsilon \cdot \mathbf{P}_i\, e^{i\mathbf{q}\cdot\mathbf{r}}$ terms can be expanded using spherical harmonics [28]. We obtain:

$$e^{i\mathbf{q}\cdot\mathbf{r}_j} \approx 4\pi \sum_{L=0}^{\infty} \sum_{M=-L}^{L} \frac{i^L}{(2L+1)!!}(qr_j)^L Y_{LM}^*(\hat{\mathbf{q}}) Y_{LM}(\hat{\mathbf{r}})$$

$$\epsilon \cdot \mathbf{P}_j e^{i\mathbf{q}\cdot\mathbf{r}_j} \approx mc \sum_{L=0}^{\infty} \sum_{M=-L}^{L} \frac{-4\pi i^L q^L}{(2L+1)!!}\sqrt{\frac{L+1}{L}} r_j^L \epsilon \mathbf{Y}_{LM}^*(\hat{\mathbf{q}}) Y_{LM}(\hat{\mathbf{r}})$$

$$(12)$$

$\mathbf{Y}_{LM}^*(\hat{\mathbf{q}})$ are the electrical vector spherical harmonics.

We neglect here the magnetic contributions of $\epsilon \cdot \mathbf{P}\, e^{i\mathbf{q}\cdot\mathbf{r}}$, which are smaller by a factor $\hbar\omega/mc^2$.

In the case of electric 2^L pole resonance (EL), the resonant scattering amplitude is:

$$f_{EL} = \frac{4\pi}{|\mathbf{q}|} \sum_{M=-L}^{L} F_{LM}(\hbar\omega)\left[\hat{\epsilon}' \cdot \mathbf{Y}_{LM}(\hat{\mathbf{q}}') \mathbf{Y}_{LM}^*(\hat{\mathbf{q}}) \cdot \hat{\epsilon}\right] \qquad (13)$$

with

$$F_{LM}(\hbar\omega_q) = \sum_{a,c} p_a p_a(c) \frac{E_a - E_c}{\hbar\omega} \frac{\Gamma_x(aMc; EL)/\Gamma_c}{x(a,c) - i}$$

$$\Gamma_x(aMc; EL) = 2\frac{4\pi}{((2L+1)!!)^2} \frac{L+1}{L} \left| \langle a | (qr_j)^L Y_{LM}^*(\hat{\mathbf{r}}_j) | c \rangle \right|^2$$

and

$$x(a,c) = \frac{E_a - E_c + \hbar\omega}{\Gamma_c/2} .$$

The $\epsilon' \cdot \mathbf{Y}_{LM}(\hat{\mathbf{q}}')\mathbf{Y}_{LM}^*(\hat{\mathbf{q}}) \cdot \epsilon$ terms determine the geometrical dependence of the scattering amplitude. They represent all scalar invariants that can be formed with the five vectors: ϵ, ϵ', $\mathbf{q}$, $\mathbf{q}'$ and $\mathbf{z}_n$ (direction of magnetic moment $\boldsymbol{\mu}_n$ of atom n). L corresponds to the order of the transition. With the development in electrical spherical harmonics, the dominant processes are the electric dipole ($L = 1$) and electric quadrupole ($L = 2$) transitions.

The factors F_{LM} determine the strength of the resonance and depend on the details of the interaction of the x-ray with the atomic electrons. They can be expressed as a function of the coefficients A_{LM} calculated by Hamrick [29]. He performed the calculations for rare-earth, uranium and transition-metal compounds. The correspondence between F_{LM} and A_{LM} coefficients is given in appendix of [26].

The resonance process corresponds to the virtual emission of one electron from the core level to either the valence band or to an empty state available at the Fermi level, followed by the desexcitation with re-emission of a photon. The sensitivity to the magnetization arises from spin orbit splitting either in the core levels or in the excited states and from the polarization of the excited states. The amplitude of the resonant signal depends on the multipole overlap integrals between the initial and intermediate levels involved in the transition, on the polarization of the x-rays and on the direction of the magnetization. The transition is fast (10^{-16} s) as reflected in the short lifetime of the core hole. This lifetime is correlated to the energy width (1–10 eV) Γ_c including all the desexcitation states of $|c\rangle$. Furthermore, the process is not affected by any magnetic form factor. The scattering occurs if the magnetization density at different sites is correlated. In such a case, the transition contributions add coherently and give rise to Bragg reflections. One has to underline that, in opposition to the non-resonant scattering, which can be related to the scattering of the electromagnetic wave by the orbital and spin moments of the electrons, the resonant magnetic scattering arises from pure electric multipole scattering.

The amplitude ratio between the Thomson scattering and the resonant magnetic scattering is 10^{-2}–10^{-3} at the L_3 edges for rare earth [30].

Dipolar Transitions

For dipolar transitions $E1$, $L = 1$ and $M = 0, \pm 1$, the $\mathbf{q}$ and $\mathbf{q}'$ dependence is negligible. The invariants can be distinguished in terms of their order in

μ_n (order 0, 1 and 2). In the basis $(\mathbf{u}_1, \mathbf{u}_2, \mathbf{u}_3)$, the components of $\mathbf{z}_n$ are z_1, z_2 and z_3. The electric scattering amplitude [13] is:

$$f_{E1} = \boldsymbol{\epsilon}' \cdot \boldsymbol{\epsilon} F_{E1}^{(0)} - i(\boldsymbol{\epsilon}' \times \boldsymbol{\epsilon}) \cdot \mathbf{z}_n F_{E1}^{(1)} + (\boldsymbol{\epsilon}' \cdot \mathbf{z}_n)(\boldsymbol{\epsilon} \cdot \mathbf{z}_n) F_{E1}^{(2)}$$

$$= F_{E1}^{(0)} \begin{pmatrix} 1 & 0 \\ 0 & \cos 2\theta_B \end{pmatrix}$$

$$-iF_{E1}^{(1)} \begin{pmatrix} 0 & z_1 \cos\theta_B + z_3 \sin\theta_B \\ -z_1 \cos\theta_B + z_3 \sin\theta_B & -z_2 \sin 2\theta_B \end{pmatrix}$$

$$+F_{E1}^{(2)} \begin{pmatrix} z_2^2 & -z_2(z_1 \sin\theta_B - z_3 \cos\theta_B) \\ z_2(z_1 \sin\theta_B + z_3 \cos\theta_B) & -\cos^2\theta_B(z_1^2 \tan^2\theta_B + z_3^2) \end{pmatrix}$$

$$(14)$$

$$\text{with: } F_{E1}^{(0)} = \frac{3}{4q}(F_{11} + F_{1-1}) \,,$$

$$F_{E1}^{(1)} = \frac{3}{4q}(F_{11} - F_{1-1})$$

$$\text{and } F_{E1}^{(2)} = \frac{3}{4q}(2F_{10} - F_{11} - F_{1-1}) \,.$$

The first term contributes to the charge peak (it contains no dependence on the magnetic moment direction). The second term is linear in z_i and therefore produces a resonant contribution to pure magnetic Bragg peaks in antiferromagnets. The third term contains z_i to the second power and gives rise to second order diffraction harmonics. Note that there is no $\sigma - \sigma$ scattering for the first order harmonics.

Quadrupolar Transitions

For quadrupolar transitions $E2$, $L = 2$ and $M = 0, \pm 1, \pm 2$, $\mathbf{q}$ and $\mathbf{q}'$ dependences are taken into account. The electric scattering amplitude [13] contains thirteen terms of various orders of the magnetic moments: one zero order term, two first-order terms, five second-order terms, four third-order terms and one fourth-order term. Thus, quadrupolar transitions are characterized by the presence of fourth-order harmonics.

RXMS Amplitudes for Different Absorption Edges

In this paragraph, the amplitudes of the resonant enhancements are qualitatively discussed at the K, L and M edges for some elements relevant for magnetic studies.

K, L₁ or M₁ edges

The magnetic sensitivity at edges involving s states is very weak because of the absence of spin-orbit coupling in the initial states. The observation of magnetic resonance therefore implies the existence of spin-orbit splitting in the intermediate state. In transition metals, the highly polarized $3d$ states are only reached via the $E2$ transition.

$L_{2,3}$ edges

At the $L_{2,3}$ edges of lanthanides and actinides, the initial states are $2p_{1/2}$ and $2p_{3/2}$ orbitals, well separated due to spin-orbit coupling. Dipolar and quadrupolar transitions probe respectively the empty d states and the f states. The polarization of the d band is weak whereas the overlap integrals are large for dipolar transitions. On the contrary, the polarization of the f states is strong and the radial matrix elements are small. Consequently, the resonant magnetic scattering amplitudes are of the same order of magnitude for both dipolar and quadrupolar transitions.

At the $L_{2,3}$ edges, $3d$ metals exhibit a strong resonance. Unfortunately, for a 1.2 nm to 3 nm wavelength, atomic resolution cannot be obtained. However these resonances turn out to be very important for investigations of magnetic films and nanostructures.

$M_{4,5}$ edges

At the $M_{4,5}$ edges of lanthanides and actinides, the initial states are $3d_{3/2}$ and $3d_{5/2}$ orbitals; they are well separated due to spin-orbit splitting. For dipole transitions, the vacant states are $4f$ or $5f$. The spin and orbital moments in these unoccupied levels are correlated due to the spin-orbit coupling in the f orbitals. Moreover, the overlap integral between $3d$ and f states is large.

However in lanthanides, the transition energy is about 1 keV, which corresponds to a 1.2 nm wavelength; so, as for the $L_{2,3}$ edges in $3d$ metals, diffraction experiments on crystal are not possible. For actinides, these edges are around 3.5 keV and several experiments have been performed.

At the $M_{4,5}$ edges, resonant scattering is mainly dipolar because quadrupolar contributions are very weak: they involve transitions to the $6s$, $7s$, $6g$ or $7g$ states and the polarization in each case is weak while the radial integrals are small.

3 Selected Examples

The aim of this section is to illustrate through various examples several specific properties of magnetic x-ray diffraction.

(i)   Magnetic structure determination remains the domain of neutron scat-
      tering. However, because of high resolution in the reciprocal space, mag-
      netic structure refinement is achievable from both non-resonant and res-
      onant magnetic x-ray scattering (3.1).
(ii) Non-resonant x-ray scattering is the only experiment that provides a
      separation of the orbital momentum contribution to magnetization (3.2).
(iii) The richness of the resonant process is very important. By tuning the
      photon energy through absorption edges, it is possible to turn on and
      off the response of given chemical species (3.3). By performing polar-
      ization analysis of the scattered beam, it is further possible to identify
      the nature (d-, f- or even p-electrons) of the magnetic moments (3.4).
      Moreover, intense scattering combined with element selectivity allows
      the investigation of weak magnetic moment systems and the observa-
      tion of a resonant enhancement at non-magnetic (but easily polarizable)
      ions (3.5).

Finally, RXMS can actually detect multi-$\mathbf{q}$ structures, owing to the presence of higher order terms in the development of the scattering amplitude
(3.6).

3.1 Increased Wave-Vector Resolution

The high wave-vector resolution coupled with the ability to separate charge
from magnetic scattering through the polarization dependence has resulted
in a far greater understanding of the magnetic ordering in the rare-earth
series [31]. To illustrate the interest of this increased wave-vector resolution,
we detail, in the following, the pioneering work of Gibbs et al. on Ho [14, 32].

Below its magnetic ordering temperature (138 K), Ho shows a simple c-
axis helical modulation. Neutron scattering measurements suggested that the
magnetic modulation wave vector τ was incommensurate and varied smoothly
with temperature.

Using non-resonant magnetic x-ray scattering measurements, Gibbs et al.
[14] observed that, in certain temperature ranges, the modulation wave-vector
locks to commensurate values. At low temperature, there is a first-order transition between two commensurate periods, namely 2/11 and 1/6, and there is
an indication of a lock-in at 5/27. This lock-in behaviour is explained within
a "spin-slip" model.

In "spin-slip" structures, most of the moments are bunched in pairs about
successive easy directions, but there are, at regular intervals, some single moments aligned along an easy axis (Fig. 2b). These singlets are the "spin-slips"
or spin discommensurations. The formation of these structures is explicable in
terms of the competition between exchange and crystal-field energies. At low
temperature, the hexagonal anisotropy energy is very large and the moments
lie close to the easy axes in the basal plane (Fig. 2a). At high temperature,
the anisotropy is small and the moments are arranged in an undistorted helix

(a)   (b)   (c)

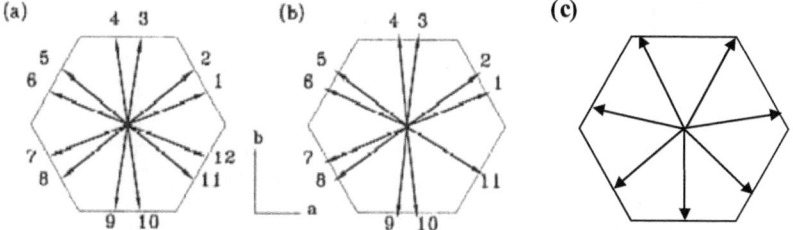

Fig. 2. Basal plane projection of the moments in Ho [33]: (**a**) in the low temperature cone phase; (**b**) in the one-spin-slip phase (wave vector: 2/11 **c***); (**c**) in the high temperature helical phase (wave vector: 0.275 **c***)

that has a wave vector determined by the position of the peak in the exchange function (Fig. 2c). At intermediate temperatures, both energies have similar magnitudes and the compromise arrangements are "spin-slip" structures (Fig. 2b): all the moments are along or close to an easy axis, thereby minimizing the anisotropy energy, and by changing the periodicity of the "spin-slips", an average wave-vector is obtained, that minimizes the exchange energy.

In addition to the magnetic satellites, new reflections were observed, which through their polarization dependence, were found to have a charge character. Such satellites also result from "spin slip" structures. They are due to new crystal modulations arising from the change in magnetoelastic coupling at slip positions.

Subsequently to this pioneering work on Ho, both x-ray and neutron measurements have revealed such "spin-slip" structures to be a common feature of rare-earth systems [31].

3.2 Separation of the Orbital and Spin Moments

As demonstrated in Sect. 2.2, the separation of spin and orbital moments is possible because, in the non-resonant magnetic x-ray scattering cross section, the spin- and the orbital-moment densities have different geometrical prefactors that can be adjusted by changing either the scattering geometry or the x-ray polarization. In the following, the historical example of the spin-orbit separation in Ho will be detailed, then results obtained in an actinide compound (UAs) and in transition metal oxides will be presented.

Rare-Earth: Ho [10, 33]

Ho is a test element for the separation of orbital and spin moments because it exhibits a large magnetic moment and a large orbital one. The first experiment was performed with a poor degree of incident linear polarization (77%). It was however very successful. The Ho magnetic structure has been recalled in Sect. 3.1.

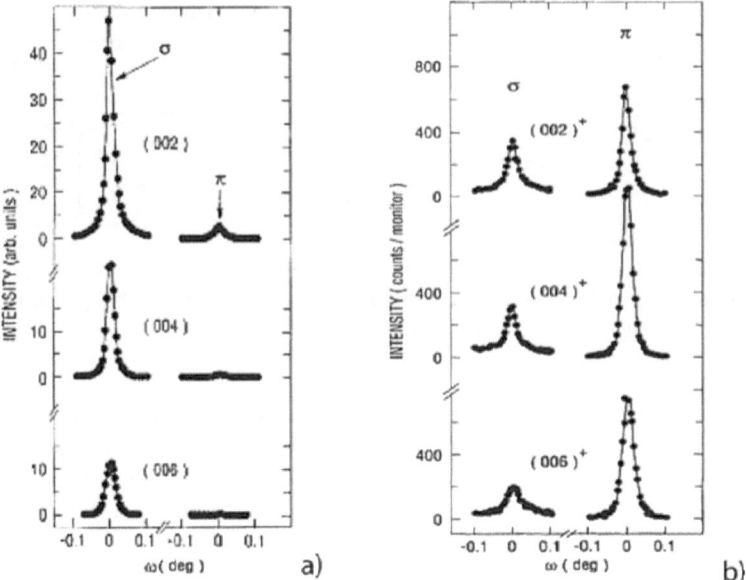

Fig. 3. Rocking curves of the σ- and π-polarized components for Ho [33]: (**a**) charge scattering for the (002), (004) and (006) structural reflections; (**b**) magnetic scattering for the $(0\,0\,2+\tau)$, $(0\,0\,4+\tau)$ and $(0\,0\,6+\tau)$ magnetic reflections, obtained for an incident x-ray energy of 7.847 keV

Figures 3a and 3b show respectively the intensity when rocking the sample at low temperature through each structural peak and each magnetic peak. It is clear that, for each charge reflection, the σ component of the charge scattering is considerably larger than the π component. This behaviour can be understood when considering (9); it reflects the profile of the incident beam, which is predominantly (but not fully) σ polarized. The decrease in intensity observed when increasing the momentum transfer is consistent with the calculation of the structure factor of Ho (usual corrections being included).

In contrast to the result for the charge scattering, the intensity of the π-polarized component of the magnetic scattering dominates the intensity of the σ-polarized component (Fig. 3b): this is an experimental evidence of the rotation of the incident polarization by magnetic scattering.

Starting from (9), this result can be easily interpreted if one considers that the dominance of the π-polarized component reflects the existence of a large orbital moment of the Ho $4f^{10}\,{}^{5}I_8$ ground state.

The results are quantified on Fig. 4 where the degree of linear polarization of the scattered beam $P' = (I'_\sigma - I'_\pi)/(I'_\sigma + I'_\pi)$ is plotted versus the momentum transfer (I' are the scattered intensities). If $P = (I_\sigma - I_\pi)/(I_\sigma + I_\pi)$ is the degree of linear polarization of the incident beam:

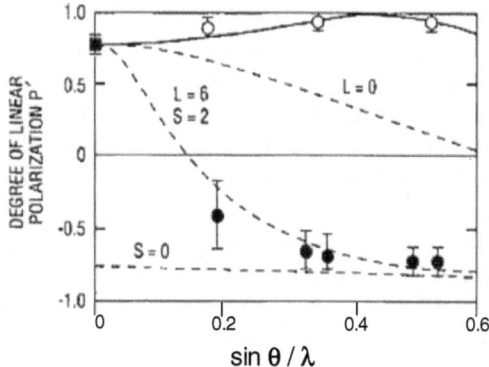

Fig. 4. Degree of linear polarization as a function of $\sin\theta/\lambda$ for holmium [14, 35]: incident beam (*solid square*), charge (*open circles*) and magnetic (*solid circles*) scattering. The *solid line* and the *dashed lines* show the calculated values (from (15) and (16))

– the non resonant degree of charge polarization from the chemical structure is:

$$P'_{charge} = [1 - \cos^2\theta_B + P(1 + \cos^2\theta_B)]/[1 + \cos^2\theta_B + P(1 - \cos^2\theta_B)] \quad (15)$$

– the scattered intensities in the case of a magnetic spiral are [35]:

$$I'_\sigma \propto (1 + P) + (1 - P)(1 + g^2)\sin^2\theta_B$$
$$I'_\pi \propto (1 + P)(1 + g)^2\sin^2\theta_B + (1 - P)[1 + 2g\sin^2\theta_B]^2 \quad (16)$$

$g(\mathbf{Q})$ is the ratio of the orbital and spin magnetization densities, each projected along the total angular momentum; for Ho, $g(\mathbf{Q})$ can be rewritten in term of the normalized orbital and spin form factors f_L and f_S:

$$g(\mathbf{Q}) = 3f_L/f_S \ .$$

The solid line drawn on Fig. 4 gives the prediction of P'_{charge}, (15). The dashed line labelled ($L = 6$, $S = 2$) shows P'_m calculated from (16) and assuming $L = 6$ and $S = 2$. The remaining dashed lines show the calculated values of P'_m assuming either ($L = 0$ and S arbitrary) or (L arbitrary and $S = 0$). The experimental data are consistent with the calculated non-resonant cross sections.

The above results were the first experimental demonstration that non-resonant x-ray scattering is a powerful technique to separate spin and orbital moments in antiferromagnets.

Actinide Compound [16, 17]

From neutron diffraction, at low temperature ($T < 61\,\text{K}$), UAs (which presents the NaCl fcc structure at 300 K) has a **2q** antiferromagnetic structure with a saturated moment of $2.2\,\mu_B$ per uranium atom. The ordered magnetic wave vector is (0 0 1/2) with the net magnetic moment aligned along a $\langle 110 \rangle$ direction. The value of the net moment is below the values expected for either $5f^2$ $(3.28\,\mu_B)$ or $5f^3$ $(3.42\,\mu_B)$ states, in agreement with an hybridization between $5f$ and other electron states. An ambiguity about the electronic configuration was still existing before the achievement of the x-ray magnetic scattering experiments.

Following the first partial measurements of McWhan et al. [16], Langridge et al. [17] measured four magnetic reflections (reflections (1/2 0 2), (0 1/2 2), (1/2 0 4) and (0 1/2 4)) for a photon energy of 8.1 keV (far away from the uranium M and L edges) and for a π-polarized incident beam.

As expected from (9), the only component visible in the (0 1/2 2) and (0 1/2 4) reflections is the rotated one (π-σ) because these reflections sample only components in the scattering plane. The opposite is true for the (1/2 0 2) and the (1/2 0 4) reflections.

The ratio of the intensities measured at the same $\lambda^{-1} \sin\theta$, e.g., $I[\mathbf{Q} = (1/2\ 0\ 4)]_{\pi \to \pi}$ /$I[\mathbf{Q} = (0\ 1/2\ 4)]_{\pi \to \sigma}$ allows the extraction of the ratio $f_L(Q)\,\mu_L/f_S(Q)\,\mu_S$ evaluated at $\mathbf{Q} \neq 0$ (Fig. 5). The solid line is the calculation with the appropriate relativistic radial integrals $\langle j_i \rangle$. The variation with $\lambda^{-1} \sin\theta$, is a consequence of the individual form factors $f_L(Q)$ and $f_S(Q)$ so that the orbital term dominates the scattering at high $\lambda^{-1} \sin\theta$.

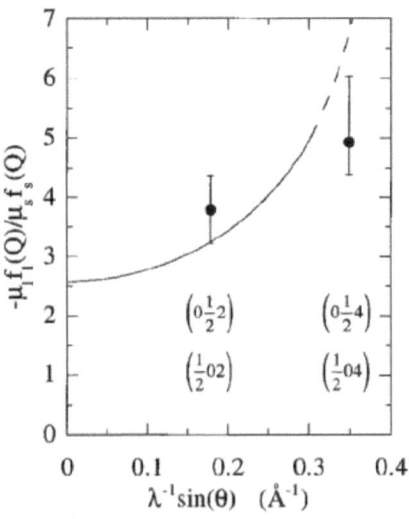

Fig. 5. Predicted variation for the quantity $f_L(Q)\,\mu_L/f_S(Q)\,\mu_S$ in UAs [17] as a function of $\lambda^{-1} \sin\theta$ compared with two experimental determinations, each involving pairs of reflections. The *dashed* section represents the region in which the dipole approximation may no longer be valid. The *solid line* extrapolates to -2.58, the value at $\mathbf{Q} = 0$ for a $5f^3$ configuration

In conclusion, a large orbital moment exists in UAs and the experimental ratio of μ_L/μ_S is close to that projected from a U^{3+} ion. The authors were unable to prove unambiguously whether UAs has the $5f^2$ or $5f^3$ configuration (or something in between). However, considering all the experimental facts, they favour a $5f^3$ configuration.

$3d$ Metals and Oxides [15, 34, 35, 36]

No attempt to extract the L/S ratio in the case of $3d$ transition elements or compounds has been considered until 1998 for two reasons: 1) it is generally assumed that the orbital contribution in these materials is quenched due to the importance of crystal electric fields and 2) the magnetic moments are usually small, which leads to weak x-ray scattered signals.

First experiments were performed on NiO below and well above the K edge of Ni [15]. The ground-state configuration of the Ni^{2+} ion has a $3d^8$ configuration. Below $T_N = 523\,K$, NiO orders in the type-II antiferromagnetic structure where ferromagnetic planes are stacked antiferromagnetically along the [111] axes with their magnetic moments aligned along one of the $[11\bar{2}]$ directions. The ordered phase is orthorhombic, which gives rise to 12 magnetic domains.

The integrated magnetic intensities have been determined for the $(1/2\ 1/2\ 1/2)$, $(3/2\ 3/2\ 3/2)$ and $(5/2\ 5/2\ 5/2)$ reflections. Since it is impossible to control the domain population at every position of the sample, all measured intensities have been averaged over the domain populations by rotating the sample about the relevant threefold $\langle 111 \rangle$ axes. Average intensities allow a straightforward determination of the ratio $L(\mathbf{Q})/2S(\mathbf{Q})$ (Fig. 6 (left)). The results show that a large contribution $(17 \pm 3\%)$ to the magnetization from the orbital moment exists in NiO.

The spin- and orbital-moment form factors have been extracted in absolute numbers. The resulting values are shown in Fig. 6 (right) as a function of the scattering vector. The extrapolated values at zero scattering vector, $S(0) = 0.95 \pm 0.1$ and $L(0) = 0.32 \pm 0.05$, lead to a value of $(2.2 \pm 0.2)\,\mu_B$ for the staggered magnetization at $T = 300\,K$. From extrapolation at $Q = 0$ on Fig. 6 (right), the L/S ratio amounts to 0.34.

The L/S ratio has also been measured in MnO, CoO and CuO [35]. Non zero orbital moments have been found in CoO $(L/S = 0.95)$, and in CuO $(L/S = 0.18)$. Only MnO, with 5 electrons in the d shell, has a vanishing orbital moment. The main conclusion is that the experimental results do not agree with the previous statement predicting that crystal field effects should quench the orbital moment. Neubeck [35] demonstrated that spin-orbit interactions are responsible for the partial reestablishment of the orbital moment (because of the mixing of the t_{2g} states into the e_g states).

Finally, the non-resonant magnetic x-ray diffraction has been used to directly prove that chromium has a zero orbital momentum [36].

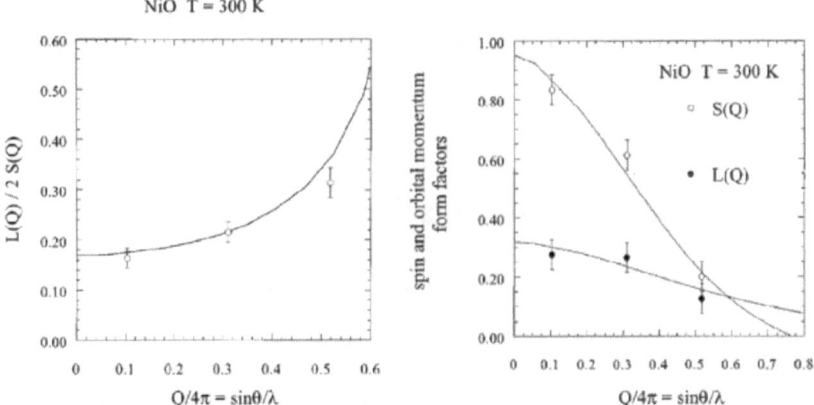

Fig. 6. Measured variation of $L(\mathbf{Q})/2S(\mathbf{Q})$ as a function of $\sin\theta/\lambda = \mathbf{Q}/4\pi$ in NiO [15] (*left*). Spin form factor and orbital-moment form factor in NiO (*right*). The continuous lines are the $\mathbf{Q}$ dependence estimated by Blume [7] adjusted to fit the data with a contraction of the wave function by 17%

3.3 Separation of Contributions from Different Species

Standard scattering techniques do not reveal element-specific properties of materials containing different magnetic elements. The individual behaviours can only be isolated through the analysis of structure factors. The resonant x-ray scattering methods provide direct access to the individual response and the chemical sensitivity is exploited to observe selectively the magnetic response of the different species in alloys, compounds or superlattices.

Actinides Solid Solutions [21]

In pure URu_2Si_2, the U magnetic moment is only $0.02\,\mu_B$; at the other end of the series, $NpRu_2Si_2$ exhibits a sizeable Np moment of $1.7\,\mu_B$. In order to analyze the role of neptunium on the magnetism of uranium atoms, RXMS experiments were performed at the $M_{4,5}$ edges of U and Np [21] in $U_{1-x}Np_xRu_2Si_2$ solid solutions. In pure $NpRu_2Si_2$, the energy dependence of the (0 0 4.808) antiferromagnetic peak shows two well-defined resonances at the M_4 and M_5 edges (top part of Fig. 7) with extremely large enhancements similar to those observed in UAs. The two series of resonances at the U and Np edges can be easily identified in the bottom part of Fig. 7 for $U_{0.5}Np_{0.5}Ru_2Si_2$. By tuning the photon energy to one of the absorption edges, it is possible to select the response of one or the other chemical species and to monitor their evolution with temperature and concentration. It has been shown that the temperature dependence of the uranium and the neptunium moments differ in the mixed $U_{1-x}Np_xRu_2Si_2$ solid solutions.

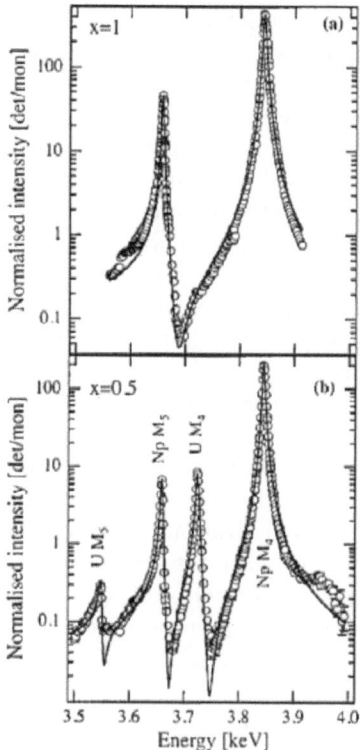

Fig. 7. Resonant scattering from $U_xNp_{1-x}Ru_2Si_2$ at the M_4 and M_5 edges of neptunium and uranium [21]; (**a**) energy dependence of the (0 0 4.808) antiferromagnetic peak in pure $NpRu_2Si_2$. Note the logarithmic scale of the intensity. (**b**) The same energy scan at the same antiferromagnetic peak in $U_{0.5}Np_{0.5}Ru_2Si_2$. The *solid lines* are fits

The observed variation of intensity versus energy has been compared to a calculation for an $x = 0.5$ alloy using a localized model and a coherent superposition of U^{4+} and Np^{3+} ions. The agreement between theory and experiment is reasonable, suggesting a ratio $\mu_U/\mu_{Np} = 0.25$ in this alloy. Since μ_{Np} is known to be $1.5\,\mu_B$ for $0.1 < x < 1$, the uranium moment is $0.4\,\mu_B$. This is much larger than $0.02\,\mu_B$ known to exist in URu_2Si_2 $(x = 0)$. The increase is a consequence of the molecular field of the ordered Np^{3+} moments and is consistent with the crystal-field model proposed for the U^{4+} ground state.

Rare-Earth Superlattices [23, 24]

A pioneering experiment has been performed by Goff et al. [23]: they investigated the magnetic structures of Nd/Pr superlattices using RXMS at both the Nd and Pr L edges. Magnetic dipole resonances were observed at the L_2 ($2p_{1/2} \rightarrow 5d$) edges of both components of the superlattices, proving that a coherent spin-density wave is established in the $5d$ bands of both elements. Multi-$\mathbf{q}$ magnetic structures were induced in the Pr, and the two components of the superlattice were found to adopt identical magnetic modulation vectors. Magnetic superlattice peaks were observed in $\mathbf{Q}$-scans at both edges and, from these, the authors were able to determine the magnetization profiles in each component of the superlattice separately.

From the width of the magnetic peaks, they were able to measure the magnetic coherence length in the Nd layers, and the results demonstrate that the observed spin-density wave in the Pr is responsible for the coupling between successive Nd blocks.

More recently, proximity effects in a Er/Tb superlattice [24] have been shown to lead to the formation of new magnetic phases. Modulated magnetic order, expected for pure Er, and ferromagnetic order, expected for pure Tb, coexist at low temperatures. Employing RXMS, the authors could probe directly the respective spin polarizations of the conduction-band electrons of Er and Tb, providing a mechanism for the interlayer coupling. The different anisotropies of Er and Tb compete with this tendency to long-range magnetic order, leading also to substantial thermal-hysteretic effects.

3.4 Electronic Probe

In the case of incomplete electronic shells, the shell selectivity of the resonant process leads to a separate observation of the magnetic response of electronic shells.

$5d$ Band Polarization in DyFe$_4$Al$_8$ [25]

From neutron diffraction, in the intermetallic system DyFe$_4$Al$_8$, the Fe electrons give rise to the high value of $T_N = 170$ K whereas at lower temperature, below $T_{Dy} = 50$ K, the Dy $4f$ moments order antiferromagnetically with the same periodicity as the Fe moments.

The RXMS data show a remarkable feature (Fig. 8) [25]: the resonant intensity at the Dy L_3 edge falls between 12 K and 30 K, remains approximately constant up to approximately 100 K and then falls to zero near T_N. In contrast, the resonant signal at the Dy L_2 edge initially rises for 12 K $< T < T_{Dy}$ and then falls smoothly to zero at T_N. Consequently, the L_3/L_2 intensity ratio ("branching ratio") is temperature dependent. Such behaviour has never been observed, neither in scattering nor in dichroism experiments. Moreover, let's recall that the initial model of Hannon et al. [11], based on isolated atoms, leads to temperature independent "branching ratio". Langridge et al. introduced solid-state effects, such as rigid-band splittings and antiferromagnetic bands, and reproduced the anomalous behaviour of the L_3/L_2 intensity

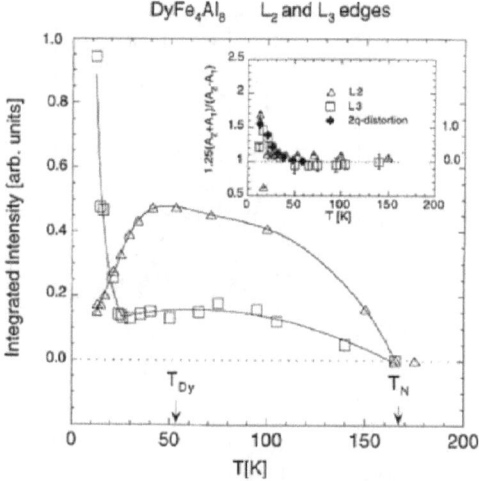

Fig. 8. Temperature dependence of the resonant intensity at the magnetic Bragg peak (4.14 4.14 0) in $DyFe_4Al_8$ as measured in the σ-π channel at the L_2 (*open triangles*) and L_3 (*open squares*) edges of dysprosium [25]. The lines trough the data points are a result of model calculations. The insert gives the temperature variation of fit parameters

ratio: the response observed at the Dy $L_{2,3}$ edges reflects the polarization of the $5d$ bands due to the magnetic ordering of the Fe atoms, independently of the $4f$ shell. This example shows how x-ray resonant scattering provides direct information on the ordering mechanisms in intermetallic systems.

Separate Study of d Orbitals and f Levels in Sm [26]

Stunault et al. [26] have shown, through a detailed spectroscopic study of the resonance at the L_2 and L_3 edges of samarium in a single-crystal epitaxial film, how the exploitation of the atomic effects can lead to a deeper understanding of long-range magnetic order in this material.

At the L_2 and L_3 edges of samarium, both quadrupole $E2$ resonances ($p \rightarrow 4f$) and dipole $E1$ resonances ($p \rightarrow 5d$) have been identified through their polarization and **Q**-dependencies. For the magnetic peaks corresponding to ordering of moments located on the hexagonal sites, energy scans across the Sm L_3 absorption edge (Fig. 9-top) allowed the observation of a quadrupolar contribution (below the absorption edge) and of a split dipolar contribution (above the edge). For the magnetic peaks corresponding to ordering on the cubic sites (Fig. 9-bottom), the quadrupolar contribution is dominant at the L_3 edge. No dipolar part could be extracted, although the long-range RKKY interaction is expected to give rise to a strong polarization of conduction electrons.

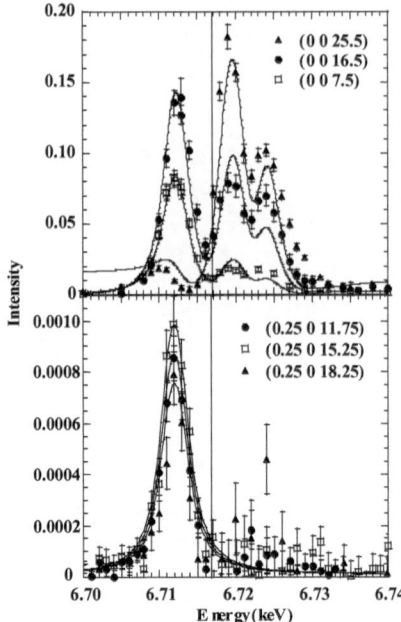

Fig. 9. Energy dependence of the resonance at three different magnetic hexagonal reflections at 50 K (*top*) and at three different magnetic cubic reflections at 8 K (*bottom*) at the Sm L_3 edge for a Sm single-crystal film [26]. The *vertical line* shows the position of the absorption edge. The *solid lines* are the result of a calculation using the model density of states

The quadrupolar contribution (for both hexagonal and cubic reflections) can be fully interpreted using atomic considerations. A systematic study of the intensity of the quadrupolar resonance of several reflections from the hexagonal and the cubic sublattices indicates that the magnitude of the $4f$ moments in samarium is independent of the local environment (cubic or hexagonal).

The energy dependence of the split dipolar resonance has been interpreted in the framework of a model density of $5d$ states (solid lines on Fig. 9).

Finally, the temperature dependences of the $4f$ and $5d$ resonant intensities have been shown to be identical. This result constitutes a direct experimental proof of the RKKY model for the onset of long-range magnetic order in rare earths.

Uranium Compounds [38]

The nature of the RXMS processes at the uranium L_2, L_3, M_4 and M_5 edges of UPd_2Si_2 has been studied. The different polarization dependences of the electric dipole and quadrupole transitions have been used to distinguish

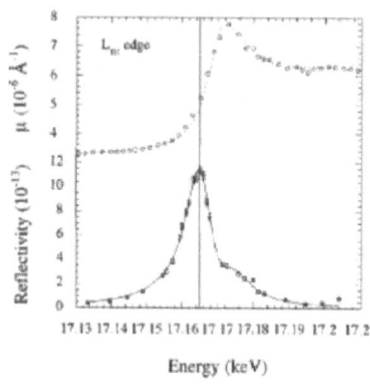 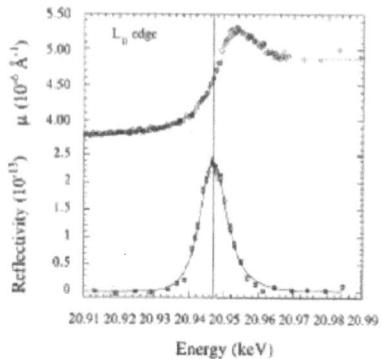

Fig. 10. Integrated reflectivity along the l direction of the magnetic (0 0 13) reflection at $T = 80\,\mathrm{K}$ in the rotated σ-π channel as a function of the incident photon energy around the U L_3 (*left*) and U L_2 (*right*) edge for UPd$_2$Si$_2$ [38]. In the left figure, the solid line is a sum of two resonances centered at different energies. In the right figure, the solid line corresponds to a splitting of 1 eV. For both figures, the upper part shows the absorption coefficient as determined from fluorescence

the two channels. The temperature variation, together with the angular dependence of the magnetic reflections, demonstrates the dipolar origin of the resonance at the uranium L_2 and L_3 edges.

The energy profiles of both L edges are different (Fig. 10). A single resonance is observed at the L_2 edge. The line shape at the L_3 edge shows two distinct contributions, both of dipolar type, which may be due to the influence of the core hole on the density of states of 6d levels. The energy of the maximum of the resonant intensity in both profiles corresponds to the inflection point in the respective absorption coefficient curve. An effort to put the scattering amplitudes on an absolute scale has been made in order to compare with theoretical calculations. In conclusion, RXMS at the L edges explores the spin polarization of the valence 6d band and allows the study of the magnetism of electronic d orbitals separately from f levels probed at the M_4 and M_5 edges in dipole transitions.

3.5 Resonant Enhancement: Small Magnetic Moments and Non Magnetic Ions

The resonant enhancement permits the determination of unresolved structure in compounds with a very small magnetic moment.

Moreover, the nature of induced magnetic moments can be traced by tuning the photon energy to absorption edges of nominally non magnetic but easily polarized species and by studying the polarization dependence of the resonant intensities. It is possible to isolate weak signals arising from these induced moments.

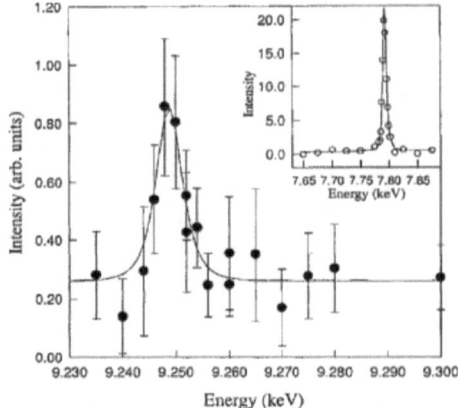

Fig. 11. Q integrated intensity for the $(0\ 0\ 0\ 2+\tau)$ magnetic peak versus energy at the L_3 Lu edge for a $Dy_{0.6}Lu_{0.4}$ alloy [20]. Inset: the same measured through the Dy L_3 edge

NpO_2 [39]

The magnetic structure of NpO_2, unlike the magnetic structure of UO_2, has been unresolved for over 30 years. Specific heat measurements on NpO_2 in 1953 revealed the existence of magnetic order below $25\,K$ as expected for a Kramer's ion with an odd number of $5f$ electrons. The confirmation of the magnetic structure was not forthcoming, possibly due to the small mass of single crystal available. Mannix et al. [39] have reported RXMS measurements on the magnetic structure of NpO_2 with an estimated Np moment of $<0.1\,\mu_B$/atom. Using the polarization and wave-vector dependence of the M_4 edge resonance, an antiferromagnetic arrangement with the long range ordered Np moment parallel to the ordering wave-vector and adopting a triple-**q** structure was proposed. Unfortunately, it is currently not possible to reliably connect the observed RXMS intensity with absolute magnetic moments.

DyLu alloy [20]

In rare earths, it is accepted that magnetic exchange is mediated by the $5d$–$6s$ conduction bands. In systems such as DyLu [20], where the Lu possess a closed $4f$ shell, the observation of a resonant signal at the Lu L_3 edge indicates that the $5d$ band of Lu is polarized by the large $4f$ moments from Dy (Fig. 11). The x-ray results have shown that a spin density wave is formed which involves the Lu $5d$ electrons with the same periodicity and temperature dependencies as the $4f$ Dy moments. The resonance at the Lu L_3 edge has been attributed to the $5d$ conduction electrons because of the filled $4f$ shell of Lu.

Actinide Compounds UGa$_3$ and UAs [40]

Mannix et al. [40] have discovered large (>1000) enhancements of the magnetic scattering intensities at the K edges of nominally non-magnetic anions, e.g. Ga and As in UGa$_3$ and UAs compounds. The width in energy, the position with respect to the white line, and the azimuthal and polarization dependencies permit one to associate the signal with transitions of $E1$ dipole symmetry from $1s$ to $4p$ states. In momentum space, the signal exhibits long-range order at the antiferromagnetic wave vector.

3.6 Multi-q Structure Detection

A single-$\mathbf{q}$ configuration has only one magnetic propagation vector in any given magnetic domain whilst a multi-$\mathbf{q}$ configuration is defined by the *simultaneous* presence of more than one such propagation vector with intersite phase coherence. In general cases, neutron scattering can not distinguish between multi-domain single-$\mathbf{q}$ and multi-$\mathbf{q}$ structures, unless an external symmetry-breaking field is applied to un-balance the domain population. However, the applied field may also modify the phase transition. RXMS can actually detect multi-$\mathbf{q}$ structures, in the absence of external field, owing the presence of higher order terms in the development of the scattering amplitude. In a multi-$\mathbf{q}$ structure, the magnetic moment is:

$$\boldsymbol{\mu}(\mathbf{r}) = \sum_{\mathbf{q}_k} \mu_{\mathbf{q}_k} e^{-i\mathbf{q}_k \cdot \mathbf{r}} \tag{17}$$

where the sum is taken over all the branches of the star of propagation vectors and $\mathbf{r}$ is the position of the magnetic site. The resonant dipole scattering amplitude is:

$$F(\mathbf{Q}) = \sum_n f_{E1}(\mathbf{r}_n) e^{i\mathbf{Q} \cdot \mathbf{r}_n} \tag{18}$$

where f_{E1} is defined in (14). Its development leads to non-zero amplitude at two types of positions in the reciprocal space: i) at $\mathbf{G} \pm \mathbf{q}_k$ due to the first order term in (14); ii) at $\mathbf{G} \pm \mathbf{q}_k \pm \mathbf{q}_j$ due to the second order term in (14). Note that, in single-$\mathbf{q}$ structures ($\mathbf{q}_k = \mathbf{q}_j = \mathbf{q}$), this term leads to magnetic satellites at $\mathbf{G} + 2\mathbf{q}$.

Resonant $\mathbf{G} \pm \mathbf{q}_k \pm \mathbf{q}_j$ satellites with $k \neq j$ have been first observed in UAs at the M_4 edge of uranium, hence confirming the coupling between different $\boldsymbol{\mu}_q$'s [41]. These satellites resonate 10 times weaker than the $\mathbf{G} \pm \mathbf{q}_k$ satellites. Then, the multi-$\mathbf{q}$ nature of the intrinsic magnetic configuration in UAs$_{0.8}$Se$_{0.2}$ has been established through the presence of the same satellites [42]. Their identification by a distinct energy, momentum, polarization and azimuthal dependence yields unique proof of the double-$\mathbf{q}$ and triple-$\mathbf{q}$ states in the unperturbed sample.

Note that the detection of higher order terms in the development of the scattering amplitude is not the only way to evidence multi-q structures using x-ray diffraction. The advantage of the small sampling volume permits to probe a single domain and to avoid confusion between single-q and multi-q magnetic structure (see the example of Nd [43]).

Finally, x-ray scattering permits to observe lattice modulation associated to the multi-q structures [41].

4 Conclusion

Since the pioneering experiment on a NiO crystal [5], and because of the development of synchrotron sources, magnetic x-ray diffraction experiments are currently performed in both the non-resonant and resonant regimes. The studies presented above illustrate the strength of both methods: new features in the physics of magnetic systems have been unravelled and unexpected properties have been evidenced.

In addition to the selected examples presented in this paper, a very large number of materials have been studied including $3d$, $4f$ and $5f$ metals and compounds. Let's underline that, as developed in Sect. 3, a lot of experiments have been performed on systems in which the magnetic periodicity is different from the lattice one: the intensity at the additional reflections is not submerged by the charge one. However, both non-resonant and resonant magnetic scattering have also been measured from ferro- and ferri-magnetic compounds (for reviews see [44] and [45]). In these crystals, interference phenomena between magnetic and charge scattering amplitudes have to be taken into account. In ferromagnets, the main magnetic contribution comes from additional interference term in (5) involving the product of charge and magnetic densities. This interference term changes sign upon magnetization direction reversal. As an example, the reversal of the magnetic field produces scattering intensity variations as large as 10% in CoPt [46]. Moreover, direct measurements of spin/orbital magnetization ratios have been performed for the ferromagnet $HoFe_2$ using a technique based on diffraction of elliptically polarized white synchrotron radiation [47].

The examples reported in this paper were obtained using hard x-rays, in the <20 keV range. In fact, non-resonant magnetic scattering has also been performed using very hard x-ray above 80 keV. The large penetration power makes this radiation a true volume probe just like neutron scattering. The short wavelength and the small cross section eliminate extinction effects, which allows precision measurements of structure factors within the first Born approximation. Moreover, with high energy x-ray diffraction one can determine the spin density distribution independent of the polarization of the incident beam and without analysis of the final polarization after scattering [48]. Due to the higher $\mathbf{Q}$ resolution, magnetoelastic effect accompanying the phase transition can be conveniently investigated by this technique [49].

Recent developments in magnetic x-ray diffraction include experiments under magnetic field and under pressure. Non-resonant magnetic scattering has been performed under magnetic field at 30 keV at the Spring8 synchrotron and allowed the observation of field-induced magnetic and structural transitions in an antiferromagnet MnF_2 [50]. A 10 T magnetic field will be soon available on the ID20 beamline at ESRF for experiments in the 4–20 keV range. Synchrotron radiation experiments and high pressure are two well-suited techniques due to the fact that the synchrotron beam can be focused on a small sample area compatible with sample sizes used in high pressure experiments, including the diamond anvil cell. The scope for magnetic diffraction experiments under pressure is enormous, for the study of many magnetic phenomena beyond the pressure range achievable with neutron scattering techniques (usually 2–3 GPa).

Finally, let's note that resonant x-ray scattering is a powerful technique not only for magnetic systems, as illustrated above, but also for materials that exhibit charge and orbital ordering [51].

In conclusion, magnetic x-ray diffraction methods are still developing and can pick up what neutrons cannot detect. Indeed, the complementarity of the two probes is a real fact [32]. The resonant x-ray methods will mature in the coming years and it is hoped that they will offer new tests for the modelling and understanding of the electronic properties of solids.

References

1. O. Klein and Y. Nishina: Zeit. Phys. **52**, 853 (1929).
2. F.E. Low: Phys. Rev. **96**, 1428 (1954).
3. M. Gell-Mann and M.L. Goldberger: Phys. Rev. **96**, 1433 (1954).
4. P.M. Platzman and N. Tzoar: Phys. Rev. B **2**, 3556 (1970).
5. F. de Bergevin and M. Brunel: Phys. Lett. A **39A**, 141 (1972).
6. F. de Bergevin and M. Brunel: Acta Cryst. **A37**, 314 (1981).
7. M. Blume: J. Appl. Phys. **57**, 3615 (1985).
8. K. Namikawa, M. Ando, T. Nakajima and H. Kawata: J. of the Phys. Soc. of Jap. **54**, 4099 (1985).
9. M. Blume and D. Gibbs: Phys. Rev. B **37**, 1779 (1988).
10. D. Gibbs, D.R. Harshman, E.D. Isaacs, M.B. McWhan, D. Mills and C. Vettier: Phys. Rev. Lett. **61**, 1241 (1988).
11. J.P. Hannon, G.T. Trammell, M. Blume and Doon Gibbs: Phys. Rev. Lett. **61**, 1245 (1988) and Phys. Rev. Lett. **62**, 2644 (1989).
12. E.D. Isaacs, D.B. McWhan, C. Peters, G.E. Ice, D.P. Siddons, J.B. Hastings, C. Vettier and O. Vogt: Phys. Rev. Lett. **62**, 1671 (1989).
13. J.P. Hill and D.F. McMorrow: Acta Cryst. **A52**, 236 (1996).
14. D. Gibbs, D.E. Moncton, K.L. D'Amico, J. Bohr and B.H. Grier: Phys. Rev. Lett. **55**, 234 (1985).
15. V. Fernandez, C. Vettier, F. de Bergevin, C. Giles and W.Neubeck: Phys. Rev. B **57**, 7870 (1998).

272   C. Dufour et al.

16. D.B. McWhan, C. Vettier, E.D. Isaacs, G.E. Ice, D.P. Siddons, J.B. Hastings, C. Peters and O. Vogt: Phys. Rev. B **42**, 6007 (1990).
17. S. Langridge, G.H. Lander, N. Bernhoeft, A. Stunault, C. Vettier, G. Grübel, C. Sutter, W.J. Nuttall, W.G. Stirling, K. Mattenberger and O. Vogt: Phys. Rev. B **55**, 6392 (1997).
18. C. Dufour, K. Dumesnil, S. Soriano, Ph. Mangin, P.J. Brown, A. Stunault and N. Bernhoeft: Phys. Rev. B **66**, 94428 (2002).
19. D.B. Pengra, N.B. Thoft, M. Wulff, R. Feidenhans'l and Bohr: J. Phys.: Condens. Matter **6**, 2409 (1994).
20. B.A. Everitt, M.B. Salamon, B.J. Park, C.P. Flynn, T. Thurston and D. Gibbs: Phys. Rev. Lett. **75**, 3182 (1995).
21. E. Lidström, D. Mannix, A. Hiess, J. Rebizant, F. Wastin, G.H. Lander, I. Marri, P. Carra, C. Vettier and M.J. Longfield: Phys. Rev. B **61**, 1375 (2000).
22. J.P. Hill, D.F. McMorrow, A.T. Boothroyd, A. Stunault, Vettier, L.E. Berman, M.v. Zimmermann and Th. Wolf: Phys. Rev. B **61**, 1251 (2000).
23. J.P. Goff, R.S. Sarthour, D.F. McMorrow, F. Yakhou, A. Stunault, A. Vigilante, R.C.C. Ward and M.R. Wells: J. Phys.: Cond. Matter **11**, L139 (1999).
24. J. Voigt, E. Kentzinger, U. Rücker, D. Wermeille, D. Hupfeld, W. Schweika, W. Schmitt and TH. Brückel: Europhys. Lett. **65**, 560 (2004).
25. S. Langridge, J.A. Paixao, N. Bernhoeft, C. Vettier, G.H. Lander, D. Gibbs, S. Aa. Sørensen, A. Stunault, D. Wermeille and E. Talik: Phys. Rev. Lett. **82**, 2187 (1999).
26. A. Stunault, K. Dumesnil, C. Dufour, C. Vettier and N. Bernhoeft: Phys. Rev. B **65**, 064436 (2002).
27. C. Cohen-Tanoudji, J. Dupont-Roc and G. Grynberg: *Processus d'interaction entre photons et atomes* (Interedition, Paris, 1988), p. 84.
28. L.D. Landau and E.M. Lifshitz: *Quantum Mechanics* (Pergamon Press, Oxford, 1977)
29. M.D. Hamrick (PhD *Thesis,* Rice University, Houston, U.S.A., 1994).
30. C. Vettier: Journal of Electron Spectroscopy and Related Phenomena **117**, 113 (2001).
31. D.F. McMorrow, D. Gibbs and J. Bohr: *Handbook on the Physics and Chemistry of Rare Earths Vol. 26*, Ed. K. Gschneidner and L. Eyring (Elsevier, Amsterdam, 1999) p. 1.
32. J. Bohr, D. Gibbs, D.E. Mocton and K.L. D'Amico: Physica **140A**, 349 (1986).
33. D. Gibbs, G. Grübel, D.R. Harshman, E.D. Isaacs, M.B. McWhan, D. Mills and C. Vettier: Phys. Rev. B **43**, 5663 (1991).
34. W. Neubeck (PhD *Thesis,* University Joseph Fourier, Grenoble I, France, 2000).
35. W. Neubeck, C. Vettier, F. De Bergevin, F. Yakhou, D. Mannix, L. Ranno and T. Chatterji: Journal of Physics and Chemistry of Solids **62**, 2173 (2001).
36. D. Mannix, P.C. de Camargo, C. Giles, A.J.A. de Oliveira, F. Yokaichiya and C. Vettier: Eur. Phys. J. B **20**, 19 (2001).
37. D. Wermeille (PhD *Thesis,* Ecole Polytechnique de Lausanne, Switzerland, 1998).
38. D. Wermeille, C. Vettier, N. Bernhoeft, A. Stunault, S. Langridge, F. de Bergevin, F. Yakhou, E. Lidström, J. Flouquet and P. Lejay: Phys. Rev. B **58**, 9185 (1998).
39. D. Mannix, G.H. Lander, J. Rebizant, R. Caciuffo, N. Berhoeft, E. Lidström and C. Vettier: Phys. Rev. B **60**, 15187 (1999).

40. D. Mannix, A. Stunault, N. Bernhoeft, L. Paolasini, G.H. Lander, C. Vettier, F. de Bergevin, D. Kaczorowski and A. Czopnik: Phys. Rev. Let. **86**, 4128 (2001).
41. C. Vettier, F. de Bergevin, N. Bernhoeft, A. Stunault and D. Wermeille: SPIE Proceedings **3773**, 351 (1999).
42. M.J. Longfield, J.A. Paixao, N. Bernhoeft and G.H. Lander: Phys. Rev. B **66**, 054417 (2002).
43. D. Watson, E.M. Forgan, W.J. Nuttal, W.G. Stirling and D. Fort: Phys. Rev. B **53**, 726 (1996).
44. J.M. Tonnerre: in *Magnetism and Synchrotron Radiation, Mittelwhir 1996*, ed E. Beaurepaire, B. Carrière and J.P. Kappler (Les éditions de Physique, 1997).
45. S.W. Lovesey and S.P. Collins: *X-ray scattering and absorption by magnetic materials* (Oxford Science Publication, Clarendon Press, Oxford 1996).
46. F. de Bergevin, M. Brunel, R.M. Galera, C. Vettier, E. Elkaïm, M. Bessière and S. Lefèbvre: Phys. Rev. B **46**, 10772 (1992).
47. S.P. Collins, D. Laundy and G.Y. Guo: J. Phys.: Condens. Matter **5**, L637 (1993).
48. M. Lippert, T. Brückel, T. Köhler and J.R. Schneider: Europhys. Lett. **27**, 537 (1994).
49. T. Chatterji, K.D. Liss, T. Tschentscher, B. Janossy, J. Stempfer and T. Brückel: Solid State Com. **131**, 713 (2004).
50. K. Katsumata et al.: J. Phys.: Cond. Matter **14**, L619 (2002).
51. C. Vettier: J. Mag. Mag. Mat. **226-230**, 1053 (2001).

Magnetic Imaging

Wolfgang Kuch

Freie Universität Berlin, Institut für Experimentalphysik, Arnimallee 14, 14195
Berlin, Germany
kuch@physik.fu-berlin.de

Abstract. Imaging of magnetic domains has contributed significantly to our
present level of understanding of micromagnetic phenomena. A number of modern techniques is used nowadays routinely for magnetic imaging of magnetic materials and nanostructures. In this chapter, state-of-the art techniques for imaging
magnetic structures are introduced, and the underlying physical phenomena are discussed. The strengths and disadvantages as well as special features of the individual
techniques will be discussed in the light of their applicability and complementarity
for addressing specific questions and problems.

1 Motivation

Magnetic imaging techniques allow the most direct view on magnetic properties on a microscopic scale. Magnetic domains, microscopic regions of identical
magnetization direction, can be viewed in real space, and their arrangement
as a function of geometric and material properties can be studied immediately in a straightforward way. The parallel improvements of magnetic imaging techniques and micromagnetic computer simulations have lead to our
substantial present level of understanding of magnetism on a microscopic
lengthscale.

There are numerous cases that show the importance of magnetic imaging for the investigation of magnetic thin films and nanostructures [1]. One
example that should be mentioned here is the so-called spin reorientation
transitions in ultrathin magnetic films. In such films the energetically preferred axis of magnetization, the easy axis, can be either within the film
plane or perpendicular to it. Transitions between these two cases, the spin
reorientation transitions, may occur when changing the film thickness or the
temperature. Early works on spin reorientation transitions used laterally averaging techniques, and reported a loss of magnetic order within a certain
temperature range right at the spin reorientation transition [2–4]. Later on,
however, it was discovered by magnetic imaging that a strong tendency towards domain formation in films with out-of-plane magnetization close to the
spin reorientation transition explained the observation of zero macroscopic
magnetization [5–7].

Magnetic domains generally occur if a magnetic sample is confined in one
or more dimensions [8]. In this case the dipolar energy includes contributions

W. Kuch: *Magnetic Imaging*, Lect. Notes Phys. **697**, 275–320 (2006)
www.springerlink.com

from the magnetostatic poles at the sample boundaries, the energy of which can be minimized by the formation of magnetic domains. The competition between this energy and the magnetic exchange energy, which tries to align neighboring magnetic spins in parallel, leads to the occurrence of characteristic magnetic domain configurations. Figure 1 gives some explanatory examples. Panels (a) and (b) show square-shaped thin film elements of different sizes. Below a certain size, the exchange energy dominates, and a single domain state is observed. At the boundaries where magnetic flux exits the sample, the right and left edges in Fig. 1 (a), magnetic poles lead to a magnetic field, which inside the sample is opposite to the magnetization. The Zeeman energy of that field, the demagnetizing field, in the sample can be avoided if a magnetic domain configuration as shown in Fig. 1 (b) is assumed. The magnetic flux is now confined inside the sample, however at the expense of exchange energy at the boundaries between domains of different magnetization direction, the domain walls. The competition between these two energy contributions, demagnetizing energy and exchange energy, defines the energetically most favorable domain configuration. In smaller elements generally the exchange energy dominates, leading to a single domain configuration (a), while in larger elements a multi-domain configuration (b) is generally more favorable. Similar flux closure domain patterns are also observed in rectangular elements (c).

As mentioned before, films with an out-of-plane easy axis of magnetization have also a tendency towards domain formation. In an infinitely extended thin film with perpendicular magnetization the magnetic flux exiting the two film surfaces leads to a demagnetizing field inside the film equal in size to the film magnetization, but opposite in sign. This demagnetizing field can be lowered by the formation of stripe domains as shown in Fig. 1 (d). A partial flux closure outside the sample at the domain boundaries between oppositely magnetized domains leads to a reduced demagnetizing energy. Because of the formation of domain walls, this happens at the expense of exchange energy and anisotropy energy. The latter is the spin–orbit related energy difference between a magnetization direction in the film plane and perpendicular to it, the driving energy contribution for the out-of-plane magnetization direction. Since in the domain walls a magnetization component in the film plane is present, its size, together with the exchange energy, defines the total energy balance. The anisotropy energy is reduced close to the spin reorientation transition, thus facilitating the occurrence of stripe domains. An example will be presented in Sect. 3.5.

In the case of small elements, i.e. when the sample extension is limited along the magnetization axis, as well as in ultrathin films with perpendicular magnetization, the magnetic flux closure occurs mainly outside the sample. If the film thickness is increased, the formation of closure domains may become favorable, as in the case of small elements (b), (c). Such a case is schematically depicted in panel (e). Closure domains at the two surfaces of the film with

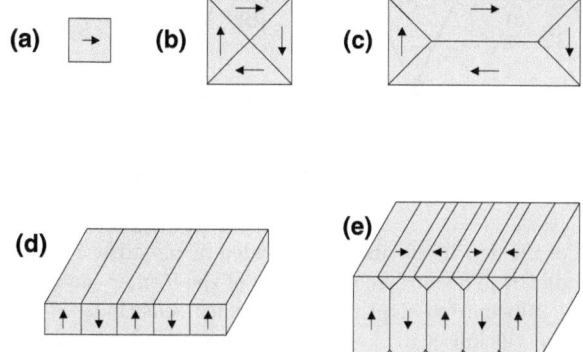

Fig. 1. Schematic explanation of occurrence of magnetic domains in different samples. (**a**): Small square magnetic element in single-domain state. Providing the demagnetization energy due to the poles at the left and right edges where magnetic flux is exiting the element and avoiding domain walls is favorable for elements below a certain size. (**b**), (**c**): In larger elements the energy needed for the creation of domain walls is provided by avoiding demagnetization energy. Closed flux domain patterns as the ones shown (Kittel, Landau pattern) avoid stray fields outside the element. (**d**): Stripe domains in a thin film sample with out-of-plane magnetization. Here the stray field outside the sample at the domain boundaries lowers the demagnetization energy in the film. This energy is balanced by the energy needed for domain wall formation. (**e**) Stripe domains in a thicker film with out-of-plane magnetization. Here the stray field outside the sample is reduced by the creation of closure domains with in-plane magnetization at the two interfaces

magnetization direction in the film plane lower the demagnetizing energy at the cost of exchange and anisotropy energy. Depending on the parameters, this configuration may be more favorable than the situation shown in panel (d). The information depth of the magnetic imaging technique used for the investigation of such samples is thus an important issue to disentangle experimentally the domain structure. The information depth of several commonly used magnetic imaging techniques together with their basic properties will be surveyed in the following section. Examples of each of these techniques will then be presented and shortly discussed along with the advantages and shortcomings in Sect. 3.

2 Overview

This section gives a short overview of the physical interactions underlying some of the most commonly used methods for magnetic characterization, and explains which of them are used for magnetic imaging. A principal distinction of the techniques can be made by the way in which depth sensitivity

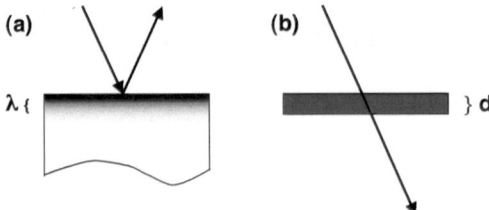

Fig. 2. Sketch of the depth information provided by on-surface (**a**) and transmission imaging methods (**b**). The surface sensitivity of the former can be characterized by the exponential information depth λ, the surface sensitivity of the latter by the maximum sample thickness d. Typically λ amounts to a few nm and d to tens of nm

is achieved. Techniques that acquire the laterally resolved magnetic information from the surface can be characterized by the information depth λ, after which the signal has decayed by a factor of $1/e$. This is schematically shown in Fig. 2 (a). The signal in the final image represents an exponentially weighted average over a certain thickness at the surface. This is represented in Fig. 2 (a) by the gradually changing graytone at the surface. Depending on the relation between sample thickness and probing depth λ, the complete depth of the sample or only the sample surface are probed. We will call techniques that work in such a way "on-surface" techniques. On-surface techniques are well suited for the investigation of very thin samples, and of samples in which the magnetic properties do not change within the sample thickness. An advantage is the insensitivity to anything that is below several times the probing depth, for example supporting layers or substrates.

A second class of magnetic imaging techniques provides information throughout the thickness of the sample, averaged with identical weighting. These are techniques in which magnetic information is gathered by a probe beam that transmits the whole sample, as sketched in Fig. 2 (b). In this case the depth sensitivity can be characterized by the upper limit on the sample thickness d for which the technique provides data within a sensible measuring time. Transmission methods are better suited for the investigation of samples in which the magnetic properties vary over the thickness (in such a case, transmission methods provide a piece of information averaged over the sample thickness in opposition to surface sensitive methods) on, or of samples that exhibit a substantial amount of non-magnetic material at the surface. A disadvantage is that the sample has to be prepared in a free-standing way, i.e., without a substrate or thick supporting layer.

Another distinction can be made by the way in which the lateral information is acquired. As generally in imaging there are two basic ways to do this, namely parallel imaging and scanning. Parallel imaging is the traditional way of image acquisition as it is used for example in optical imaging like photography etc. Parallel imaging techniques have the advantage of rapid image creation, but pose rather high requirements on the optical elements that are used. Parallel imaging therefore often result in different image quality with respect to resolution and distortion in different positions of the image. So for example the resolution is typically higher in the center of an image than at the edges. In scanning techniques a finely focused probe beam is moved across the sample, and the image information is collected by recording a detection signal together with the information about the spatial coordinates of the probe position. The image is then reconstructed in an x-y-oscilloscope or in a computer. The image acquisition time of scanning techniques is longer compared to techniques that use parallel imaging, but the image quality is identical over the entire image, and for several techniques it is much easier to create a finely focused probe beam than to implement parallel imaging.

Magnetic imaging techniques can be further categorized by the physical interaction with the sample magnetization, just like the laterally averaging standard techniques for magnetic measurements. Table 1 lists some of the

Table 1. Commonly used physical interactions to measure integral magnetic properties and their use for magnetic imaging

Physical Interaction	Integral Method	Magnetic Imaging Method
force on electrons by B field of sample	vibrating sample magnetometry (VSM) superconducting quantum interference device (SQUID)	Lorentz microscopy
force on sample by external field	alternating gradient magnetometer (AGM)	magnetic force microscopy (MFM)
precession of magnetic moments	ferromagnetic resonance (FMR)	—
electronic properties	magneto-optical Kerr effect (MOKE) x-ray magnetic circular dichroism (XMCD)	Kerr microscopy XMCD + photoelectron emission microscopy (PEEM) XMCD + transmission x-ray microscopy (M-TXM) scanning electron microscopy with polarization analysis (SEMPA) spin-polarized low energy electron microscopy (SPLEEM) spin-polarized scanning tunneling microscopy (sp-STM)

most commonly used techniques for magnetic characterization along with the underlying physical interaction. This interaction can be the Lorentz force on moving electrons by the B field of the sample, either outside or inside the sample. Standard techniques are vibrating sample magnetometry (VSM), or superconducting quantum interference devices (SQUID). In both techniques the stray field outside the sample is measured. In VSM the sample is moving, and the force on electrons in a coil is measured as induced voltage. In a SQUID also the magnetic flux outside the sample is detected by a ring that contains two Josephson contacts. One can regard this as a force on the electrons that form the supercurrent in that ring. To use the Lorentz force for magnetic imaging, it has to be detected locally at each point of the sample. A straightforward way to achieve this is to use transmission electron microscopy. This is called Lorentz microscopy and utilizes the deflection of electrons that travel through a magnetic sample by the Lorentz force of the sample's B field, inside and outside the sample, to obtain magnetic information with lateral resolution. This allows to obtain magnetic images with high resolution. We will discuss examples of Lorentz microscopy in Sect. 3.2.

A second interaction that is used for magnetic characterization is the force between the magnetic dipole of the sample and an external field. Representative for that interaction is alternating gradient magnetometry (AGM). In AGM an inhomogeneous magnetic field is created by a set of coils, and the force on the sample is measured. To enhance the detection sensitivity, the sign of the field is oscillating, and consequently also the resulting force on the sample. To implement this mechanism into an imaging technique, the force between sample and external field has to be probed locally. This is achieved by magnetic force microscopy (MFM), in which the inhomogeneous external field is generated by a sharp magnetic tip. The tip is brought into close vicinity of the sample and then scanned across the surface, while the force between sample and tip is detected. This has been successfully implemented into atomic force microscopes (AFM), which are designed to detect the small van der Waals forces between the tip and the sample surface. MFM will be discussed in Sect. 3.6.

Excitation of resonant precession of the magnetic moments in an external field by absorption of microwaves is employed during ferromagnetic resonance (FMR) measurements. There is no standard technique for magnetic imaging using local detection of FMR.

The largest number of standard techniques for magnetic imaging belongs to the class of techniques measuring magnetic properties by the corresponding electronic properties. The reason is that it is relatively easy to extend these methods used for the integral characterization of electronic properties like magneto-optical Kerr effect (MOKE) or x-ray magnetic circular dichroism (XMCD) to microscopic lateral resolution. In MOKE measurements the change in polarization of a light beam, typically from a laser, after reflection at the sample surface is detected. It is usually described within the context of macroscopic dielectric theory. The magnetization dependent part of the polarization change after reflection can be extracted when placing the sample into an external magnetic field of varying strength and direction. XMCD measurements rely on the dependence of resonant absorption of circularly polarized soft x-rays at elemental absorption edges on the helicity of the x-rays. Here the magnetic properties of the sample enter from the spin polarization of the unoccupied part states just above the Fermi level. Both, MOKE and XMCD, can be used for magnetic imaging. Laterally resolved MOKE is termed Kerr microscopy, and will be discussed in Sect. 3.1. Lateral detection of XMCD can be achieved in various ways; here we will focus on the two most commonly used ones. Both employ already existing imaging techniques. One is photoelectron emission microscopy (PEEM). PEEM is a parallel imaging technique in which the local intensity of secondary electrons is used to create an image of the sample. After x-ray excitation, the intensity of secondary electrons is proportional to the x-ray absorption cross section, so that PEEM can be used for the laterally resolved detection of the XMCD effect. The other is transmission x-ray microscopy (TXM), in which an x-ray optics is used to obtain an image of the sample from the transmitted x-ray intensity. It is also straightforward to extend TXM for the laterally resolved detection of XMCD. We will present PEEM and magnetic TXM (M-TXM) in Sect. 3.5.

There are more magnetic imaging techniques that have to be counted to the class of techniques employing the laterally resolved detection of electronic properties. All of them have in common that already existing imaging techniques are extended to yield magnetic information. We will discuss three of them, namely scanning electron microscopy with polarization analysis (SEMPA), spin polarized low energy electron microscopy (SPLEEM), and spin-polarized scanning electron microscopy (sp-STM). SEMPA is an extension of scanning electron microscopy (SEM). In SEM a finely focused electron beam is scanned across the sample surface, and the local reflected electron intensity provides the image information. SEMPA takes advantage of the fact that secondary electrons scattered from a magnetic surface are in general spin-polarized because of the spin-polarized electronic states. Inserting a spin detector into a SEM and detecting the spin polarization of the secondary electrons as image information thus allows to obtain magnetic images. SEMPA will be discussed in Sect. 3.3.

A low energy electron microscope (LEEM) comprises an electron optics that uses elastically reflected electrons of low energy to create an image of the sample. In SPLEEM a spin-polarized electron gun is used so that the incoming electrons on the sample are spin-polarized. The reflection coefficient at low electron energies, close to the Fermi level, is strongly spin-dependent due to the spin-polarization of the unoccupied states. The final image in SPLEEM therefore contains magnetic information. SPLEEM will be presented in Sect. 3.4.

Sp-STM is technically different from the other electronic-property-based techniques since it is a scanning probe technique, similar to AFM and MFM. It relies on the dependence of the electron tunnel probability between a magnetic tip and the magnetic sample on the spin-polarized electronic states of both, tip and sample. This can be implemented into a scanning tunneling microscope (STM) by adding a magnetic tip and modulating the tunnel voltage between tip and sample, or the magnetization direction of the tip, resulting in magnetic imaging with highest lateral resolution. This will be discussed in Sect. 3.6.

3 Techniques for Magnetic Domain Imaging

In this section the techniques already listed in the previous section will be shortly presented. The sequence in which they will be presented differs from the list of Table 1 for didactic reasons. A typical example is given for each technique, and the specific features, advantages and shortcomings are discussed. The purpose of this chapter is to give an overview about a number of existing and commonly used techniques for magnetic imaging, and put the reader in a position to judge which of theses techniques is suited best for a certain problem. It is not intended as a comprehensive introduction or textbook for experimentalists who plan to use any of these techniques; for this the reader is referred to the existing specific literature.

3.1 Kerr Microscopy

Kerr microscopy is the laterally resolved measurement of the magneto-optical Kerr effect (MOKE). In macroscopic dielectric theory [9], the magnetization-dependent part of the dielectric tensor leads to a magnetization-dependent change of the light polarization after reflection at a magnetic sample. In a simplified picture this may be considered as the action of the Lorentz force on the oscillating charges in the sample that are induced by the incoming electromagnetic radiation. The emission of an electromagnetic wave by the deflected charges thus exhibits a characteristically different polarization. In general both a rotation of the polarization plane and a change in the ellipticity may occur. In a MOKE experiment this is detected as the intensity change of the reflected beam after passing a polarizing filter as a function of an applied external magnetic field [10]. To employ this effect for magnetic imaging, the imaging process has to be performed either by an optical microscope set-up that provides parallel imaging of the reflected light, or by scanning a focused incoming light spot across the sample. Both methods are applied in laboratory use, where the latter is usually termed scanning Kerr microscopy. For parallel imaging two different set-ups can be employed. For use in an ultrahigh vacuum environment the distance between the sample and the closest optical element has to be typically at least several centimeters.

In this case a set-up can be employed in which the illuminating and imaging paths are completely separated. For the imaging path a light microscope with a corresponding working distance of several centimeters has to be used, which allows to place a polarizing filter as polarization detector between sample and objective lens. Because of the large working distance resolution in this case is limited to one or even two microns. The advantage is the complete polarization analysis because there are no optical elements between sample and the polarizing filters which may influence or diminish the polarization.

To obtain magnetic Kerr microscopy images with a resolution comparable to the best optical microscopes (300–500 nm), the distance between sample and objective lens has to be very small, typically one millimeter or less. This is only possible if the sample is in air. The polarizing filters have then to be positioned behind the objective lens, which has to serve both as a condenser lens for the illuminating light and as objective lens for the reflected light. In this case the polarization may be influenced by the objective lens, which results in a loss of magnetic contrast. To enhance the magnetic contrast and to get rid of other, topographic contrast sources, usually images are digitally compared to a reference image which can be taken at the same spot if the sample is magnetically saturated, for example by a high external field. Compared to other magnetic imaging techniques in this chapter the resolution of Kerr microscopy is quite moderate. In the case of parallel imaging the acquisition time for an image is rather short. Depth information is obtained in an on-surface way, with a probing depth of the order of some tens of nanometers.

An example is shown in Fig. 3. It presents magnetic domains in strained FeCoSiB amorphous ribbons [11]. Depending on the stress, different magnetic domain patterns are observed. One notices that the field of view is several hundreds of microns. This is typical for Kerr microscopy with its moderate lateral resolution, but it may be regarded as a special feature that it is quite easy to zoom out and obtain images with a large field of view. Another advantage is that the method is purely optical, and therefore can be applied in arbitrarily high external magnetic fields.

In layered systems of ultrathin films light reflection is influenced by the interference of reflected waves from the separate interfaces between the different layers. It has been shown recently that this can be used for achieving a depth selectivity by phase adjustment [12]. In this way contributions to the Kerr signal from certain depths in a multilayered sample can be enhanced or cancelled, leading to images that represent the magnetic domain pattern at different depths or of different layers [12]. Since nearly all of the most exciting new discoveries in thin-film magnetism are observed in multilayered structures in which two or more magnetic layers have to be switched separately, a magnetic imaging technique that is able to address the different magnetic layers is a substantial advantage for the microscopic investigation of such structures. With Kerr microscopy this is, within certain limits, possible.

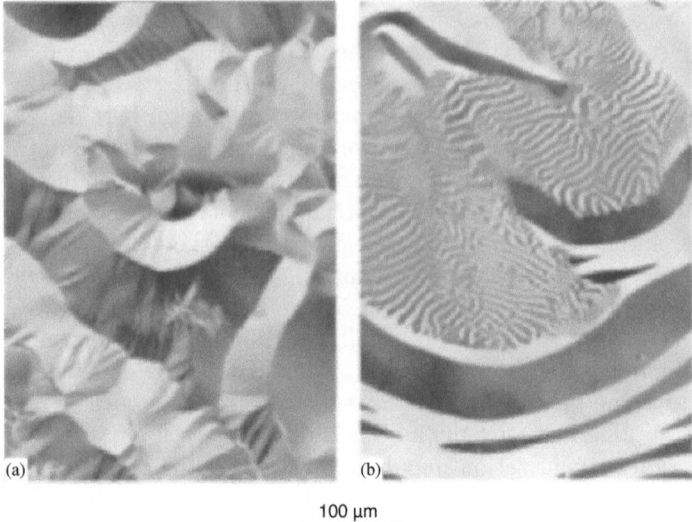

Fig. 3. Example of Kerr microscopy. Shown are magnetic domain images of Fe-CoSiB amorphous ribbons that exhibit a stress-induced change in domain patterns. (Reproduced from [11] with permission, copyright (2000) by Elsevier Science)

The details of how to access the magnetic information of the different magnetic layers, however, depend on the particular sample under study and have to be optimized again for each new structure.

3.2 Lorentz Microscopy

In Lorentz microscopy the force exerted by the magnetic flux density $\boldsymbol{B}$ on moving charges is used for magnetic imaging. It can be implemented without too much effort in commercial transmission electron microscopes [13,14]. The force $\boldsymbol{F}$ on a moving electron with velocity $\boldsymbol{v}$ is given by

$$\boldsymbol{F} = |e|(\boldsymbol{v} \times \boldsymbol{B}) , \tag{1}$$

where e is the electron charge. This force can lead to a deflection in the propagation direction of the electron if the integrated flux density, inside and outside the sample, along the electron path does not vanish. This is depicted schematically in Fig. 4. Panel (a) shows the situation in which stripe domains with in-plane magnetization are separated by domain walls aligned with the magnetization direction. In this case there is no magnetic flux outside the sample. Electrons that are transmitted through the sample experience a Lorentz force according to (1), which at domains of opposite magnetization direction points into opposite directions. The propagation paths of electrons after transmission through the sample are bent into different directions, depending on the local magnetization direction.

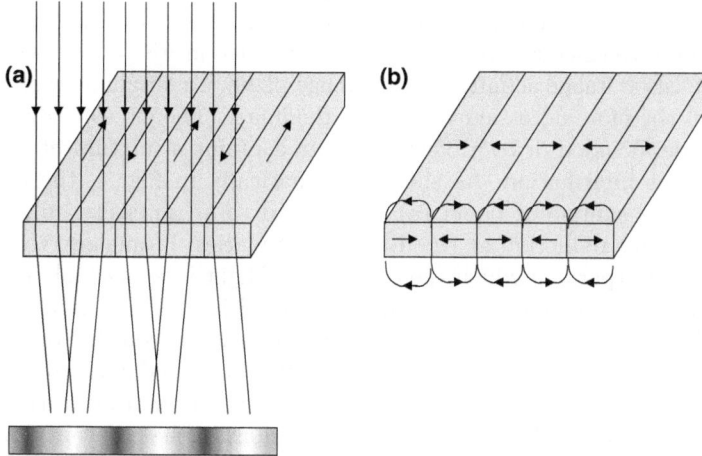

Fig. 4. Schematic explanation of Lorentz microscopy. Electrons that transmit the sample are deflected by the Lorentz force from the **B** field inside and outside the sample. (**a**) Situation in which 180° domains are separated by domain walls running along the magnetization axis. The **B** field is confined to the interior of the sample. The deflection of the electrons after transmission through the sample leads to an intensity modulation in out-of-focus images as shown schematically at the bottom. (**b**) Situation in which 180° domains are separated by domain walls running perpendicular to the magnetization axis. In this case the **B** field lines are closed by the stray field outside the sample. Electrons transmitted through the sample do not experience a net deflection, since the contributions from the interior of the sample and the stray fields in front and behind the sample cancel out

The situation in panel (b) is different. If stripe domains with in-plane magnetization are separated by domain walls running perpendicular to the magnetization direction, as in panel (b), the lines of the magnetic flux density are closed above and below the sample. Electrons that are transmitted through that sample do not experience a net deflection, since the contributions from the Lorentz forces inside and outside (above and below) the sample cancel out.

To implement Lorentz force contrast for magnetic imaging into a transmission electron microscope one has to recall some basic facts of optical imaging. If an ideal lens is used to image an object, all parallel rays emanating from the specimen meet in the focal plane behind the lens, located at the focal length from the optical center of the lens. In other words, all rays that are emitted at the specimen under identical angle pass the focal plane at identical position. Different emission angles, on the other hand, are represented by different positions in the focal plane. A reciprocal space image is thus present at the focal plane. The image plane, on the other hand, is defined as the plane where all rays coming from the same spot of the specimen, irrespective of the emission angle, meet again. This is where a focused real space image of the specimen is present. The consequence for Lorentz magnetic imaging is that a deflection of the electron beam at the sample, like in Fig. 4 (a), is irrelevant for the focused image acquired in the image plane. To obtain a magnetic contrast it is therefore necessary to defocus the image. The intensity distribution of an underfocussed or overfocussed image contains a mixture of real space and angular information. As shown schematically in Fig. 4, the defocused image of the sample of panel (a) would lead to an intensity enhancement or an intensity reduction at the domain boundaries. In analogy to the light optical counterpart, in which the deflection of light in a specimen of varying optical thickness is made visible in a defocused image, this mode of operation of Lorentz microscopy is called Fresnel mode. The contrast of the actual domains in this mode is the same everywhere, but the domain walls appear with a dark or bright contrast, depending on the orientation of the domain wall with respect to the magnetization directions of the adjacent domains.

In another mode of operation part of the beam in the focal plane of the objective lens is blocked or obscured by an aperture. This selectively removes electrons from the final image according to their starting angle at the sample. Domains that lead to a deflection of passing electrons in the direction which is blocked in the focal plane consequently appear darker in the real space image taken at the image plane. By moving the aperture to block different positions according to different directions of electron deflection, the contrast of the various domains can be varied, and statements about the magnetization directions can be made.

Figure 5 shows a typical example of Lorentz microscopy from [15]. Lorentz microscopy images of two different permalloy microstructures of 2.5 μm overall length are presented. The top row, panels (a) and (b), are examples for the Fresnel mode of operation, while panels (c) and (d) show images taken in the Foucault mode of operation. It is nicely seen that in the Fresnel images (a) and (b) black and white lines are present that mark the borders between domains of different magnetization direction. For the interpretation of such images it is sometimes, in particular in the more complicated patterns as in the elements shown in Fig. 5, necessary to simulate the Fresnel image numerically and optimize the input domain configuration until agreement with the experimentally observed image is achieved. The bottom row of images in [15], panels (c) and (d), shows Foucault images of the same elements. Note that, although panels (a) and (c), as well as (b) and (d) show the same microstructures, they were imaged at different external magnetic fields and after different external field history, so that the magnetic configuration is not the same. All images were taken during application of an external field prior to switching. For details see [15]. The difference between the Fresnel and Foucault modes of operation is immediately recognized. In the Foucault images (c) and (d) the different grayscale contrasts can be directly related to the different deflection angles the electrons have acquired by the Lorentz force during their passage through the sample, provided the exact position

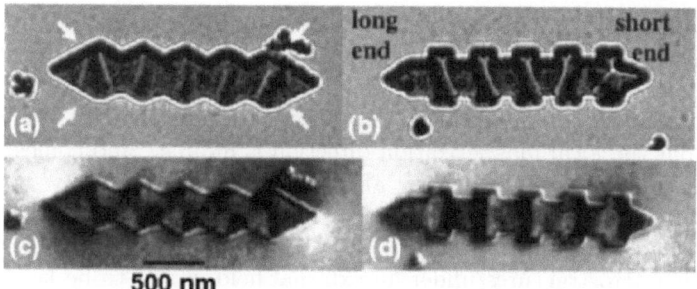

long
end

short
end

(a)

(b)

(c)

(d)

500 nm

Fig. 5. Example of magnetic domain imaging by Lorentz microscopy. Shown are magnetic images of two different permalloy microstructures during reversal by an applied external magnetic field. (**a**), (**b**): Images obtained using the Fresnel mode, (**c**), (**d**): images obtained in Foucault mode (Reproduced from [15] with permission, copyright (2000) by the American Institute of Physics)

of the aperture in the focal plane with respect to the sample orientation is known. However, as already mentioned, also the stray field outside the sample contributes to this deflection. This can be nicely seen in panels (c) and (d) from the dark and white contrast at some of the edges outside the magnetic structure. At these edges stray magnetic flux is coming out of the sample. It may be regarded as an advantage of Lorentz microscopy that the stray field outside magnetic micro- and nanostructures can be imaged. For the design of structures with desired magnetic reversal properties it can be useful to know how much stray field is present at which edge of the sample. To know the stray field outside the sample also helps with the interpretation of the magnetic configuration. It has always to be kept in mind, however, that the stray field above and below the microstructures contributes to the image in an identical way, but is more difficult to separate from the actual domain structure of the element.

Instead of the two modes of operation described above, which can be implemented rather easily into transmission electron microscopes, another way of Foucault imaging is being followed with more specifically dedicated instruments. In a scanning transmission electron microscope four or more electron detectors can be positioned off-axis in the back focal plane of the objective lens in order to detect electrons passing that plane with a certain offset from the optical axis, corresponding to electrons that have experienced a deflection into a certain direction [16,17]. In this scanning mode of operation the difference between pairs of oppositely placed detectors can be used as deflection signal along one certain axis. It is then not necessary to have any detector in the image plane, since the imaging process is realized by scanning the electron beam across the sample. This mode of Lorentz microscopy is referred to as differential phase contrast microscopy.

The main advantage of Lorentz microscopy is the quite high resolution, which can be also judged from the scale of the images of Fig. 5. It can be used to study the details in switching of small magnetic structures. It is only possible to perform Lorentz microscopy in transmission. The information about the local magnetization direction is indirect, and comes from the integrated magnetic flux density along the electron trajectory. If coils are integrated into the sample holder, the technique tolerates some additional in-plane fields. However, when studying the magnetic reversal or switching of magnetic microstructures under an external field, it has to be kept in mind that usually there is a quite high perpendicular field at the sample position from the magnetic lens of the electron microscope. Lorentz microscopy is not element-selective, and there is no means to obtain depth-selective information.

3.3 Scanning Electron Microscopy with Polarization Analysis (SEMPA)

While in the last section a technique for magnetic imaging that used a transmission electron microscope has been presented, this section will deal with a technique using the column of a scanning electron microscope. It takes advantage of the fact that secondary electrons that are emitted after the absorption and inelastic scattering of a primary electron in a magnetic sample are usually spin-polarized along the spin direction of the majority electrons. Scanning the primary electron beam across the sample and detecting the spin of the secondary electrons instead of their bare intensity, as it is done in standard scanning electron microscopy, should thus yield magnetic contrast with the same high resolution as scanning electron microscopy. The difficulty is to obtain reasonable statistics in the final image. The signal of a spin detector is typically orders of magnitude lower compared to the total electron intensity, and the resolution of an electron microscope column scales inversely with the square root of the beam current.

Figure 6 sketches the set-up of scanning electron microscopy with polarization analysis (SEMPA, sometimes also termed spin-SEM). The electron beam from the microscope column hits the sample under a certain angle from the surface normal, and the emitted secondary electrons are collected by an electron optics and transferred to a spin detector [18, 19]. A spin detector usually employs the right–left asymmetry caused by the spin–orbit interaction during the scattering of electrons at high Z materials. In a Mott type detector electrons are scattered at a gold foil at rather high kinetic energies, typically 50–100 keV, while in a LEED type detector Bragg scattering at a certain optimum energy (104.6 eV) at a tungsten single crystal surface is used. The performance of a spin detector is characterized by its figure of merit, a number giving the factor by which the acquisition time has to be extended for spin detection compared to direct electron detection without spin analysis in order to reach the same counting statistics. It is given by the

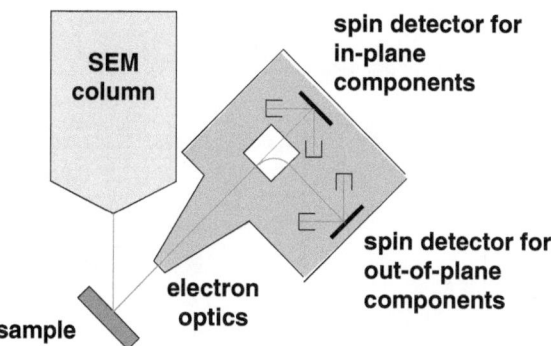

Fig. 6. Schematics of magnetic imaging by scanning electron microscopy with polarization analysis (SEMPA). A finely focussed electron beam from a scanning electron microscope column is scanned across the sample surface, and the emitted secondary electrons are transferred to a spin detector by an electron optics

square of the spin detection efficiency of the spin detector times the reduction in intensity by the spin detection with respect to the incoming signal. Typical numbers are 0.25 for the spin detection efficiency, and $1/1000$ for the intensity attenuation, so that figures of merit are typically of the order of 10^{-4}. This illustrates that it is not straightforward to implement magnetic imaging by SEMPA into a standard commercial scanning electron microscope. There are special requirements to the illuminating column with respect to electron current and resolution, which can be met only with state-of-the-art field emission cathodes. The first publication of magnetic imaging with SEMPA appeared only about 20 years ago [20].

A spin detector is sensitive to the two transverse spin components of the electrons as they enter the detector. In the sketch of Fig. 6 a set-up with two spin detectors is shown. While the spin detector positioned along the surface normal is sensitive to the two in-plane components of magnetization, a second spin detector oriented at 90° delivers information about the out-of-plane component and one in-plane component. Both spin detectors can be alternatively operated by switching on and off an electrostatic electron beam deflector as depicted as a white square. The quasi-simultaneous detection of all three spin components, or in other words of all three magnetization components in the images, is one of the specific strengths of the technique. Since all the imaging process is done by the primary electron beam with a relatively high kinetic energy, all the adjustment of the electron transfer optics and the spin detectors does only influence the contrast, not the position on the image. It is therefore possible to numerically compare images that have been acquired at the same sample position but for different components of magnetization, and calculate, for example, an image of the angle of magnetization. This is one of the specific strengths of SEMPA, which is probably the only method where magnetic images for different magnetization components,

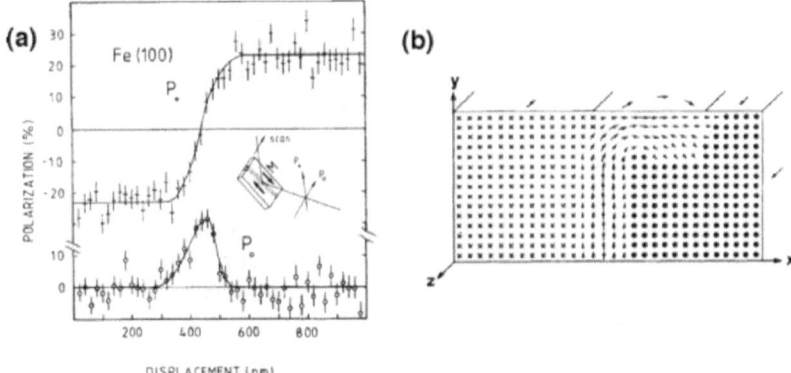

Fig. 7. Example of SEMPA. (**a**): A line scan of the two in-plane components of spin polarization across a Bloch wall of an Fe(001) single crystal reveals that the magnetization direction in the wall at the surface is rotated into the surface plane. (Reproduced from [21] with permission, copyright (1989) by the American Physical Society).) (**b**): Result of micromagnetic simulation demonstrating the magnetization configuration of such a Néel cap of a Bloch wall at the surface. (Reproduced from [22] with permission, copyright (1989) by the American Physical Society))

but with fully identical spatial information can be obtained. SEMPA images in literature are therefore often found color-coded, with colors corresponding to the two-dimensional information about the local magnetization direction, the so-called color wheel.

A famous early example of magnetic imaging by SEMPA, in which the simultaneous sensitivity to different magnetization components was used, is shown in Fig. 7. SEMPA measurements at domain walls of an Fe single crystal revealed that the magnetization direction in the wall at the surface is different from the volume of the sample [21, 22]. Figure 7 (a) shows line profiles of two components of the spin polarization across a domain wall separating two domains of opposite magnetization in Fe(001), as sketched in the inset. The upper curve, labelled P_+, represents the magnetization component along the magnetization of the adjacent domains, and consequently changes sign across the domain boundary. The lower curve, labelled P_0, represents the polarization of the spin component in the surface plane perpendicular to the magnetization of the domains. It shows the presence of a magnetization direction within the film plane inside the domain wall. This means that the magnetization rotates within the surface plane between one domain and the next. This is somewhat surprising if one considers the higher magnetostatic dipolar energy connected to the stray field of such a domain wall structure compared to the energetically more favorable configuration in which the magnetization rotates within the plane of the domain wall, a so-called Bloch wall. Figure 7 (b) shows a micromagnetic simulation. It represents a cross section through the near-surface region of a 180° domain wall. Crosses and full circles correspond to the magnetization directions of the two domains, pointing into and out of the drawing plane, respectively. It is nicely seen how at the surface the spin structure of the domain wall changes significantly. The magnetization direction inside the wall changes from the direction perpendicular to the surface into the surface, and the wall becomes wider. The reason is to avoid magnetic poles at the surface by creating magnetic poles at the edges of the domain wall. The energy balance between these two configurations determines the actual three-dimensional spin structure of the domain wall, as displayed in Fig. 7 (b). In analogy to the so-called Néel wall of ultrathin films such a structure was termed "Néel cap" because of the similar surface spin structure.

This example demonstrates another feature of SEMPA, namely its high surface sensitivity. The attenuation length of secondary electrons is relatively short, about 2 nm; in addition, secondary electrons are created in overlayers, and the spin polarization of passing electrons is influenced by additional scattering in an overlayer. All together the information contained in a SEMPA image stems from the topmost layers in a sample. This means that surface preparation under ultrahigh vacuum conditions is necessary. Often surfaces for SEMPA imaging are prepared by depositing an ultrathin overlayer of a material with a high spin polarization, mostly iron. It is then assumed that, because of the extremely small thickness of this surface layer, the magnetic properties of the sample are not significantly affected.

The high surface sensitivity of SEMPA can be an advantage in certain cases. This is demonstrated in the next example, in which the surface of the layered antiferromagnet $La_{1.4}Sr_{1.6}Mn_2O_7$ is imaged. The left image of Fig. 8 shows the surface topography of the single crystal sample. Steps corresponding to the height of one unit cell, about 1 nm, are recognized as thin white

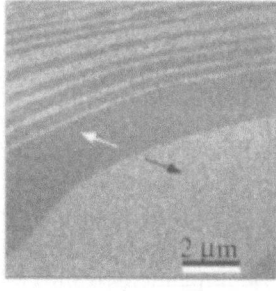

Fig. 8. Example demonstrating the surface sensitivity of SEMPA. *Left*: SEM image of the surface topography of a $La_{1.4}Sr_{1.6}Mn_2O_7$ single crystal. *Right*: Magnetic SEMPA image of the same position of the sample. The alternating grayscale contrast reflects the layered antiferromagnetic spin structure of the sample (Reproduced from [23] with permission, copyright (2004) by the American Physical Society)

lines in the scanning electron microscopy image. The right panel is the magnetic image. It is clearly seen how the spin polarization contrast reverses its sign at every terrace. The different grayscales in this image therefore reflect the opposite spin directions of the topmost layer of the sample, characteristic for such a layered antiferromagnet. This is only possible using an imaging method with a high surface sensitivity. It should be mentioned here that similar studies have been conducted also before by SEMPA on antiferromagnetic Cr films on Fe(001) [24]. Spin-polarized scanning tunneling microscopy, which is presented in Sect. 3.6, is another method that can be used to image the surface spin structure of antiferromagnets [25, 26].

To summarize, SEMPA is a method that can provide real vectorial information about the sample magnetization distribution with high spatial resolution. The resolution is in principle determined only by the electron column, but since there is a certain trade-off between the intensity necessary for spin polarization detection and the resolution, in general the resolution of SEMPA does not exactly match that of scanning electron microscopy. It is a surface sensitive scanning technique. The sample has to be in ultrahigh vacuum, and the sample surface has to be clean and conducting. An advantage is also the variable field of view, which allows to zoom in and out in a quite wide range. SEMPA is not element-specific, and the size of the contrast may be weak for certain elements or surfaces. Since all the imaging is done by the primary electrons of relatively high energy, special set-ups have been developed and can be used to obtain SEMPA images even under moderate magnetic fields [27]. A deflection of the emitted electrons is not critical as long as there can be taken care that the electrons reach the spin detector.

3.4 Spin-Polarized Low-Energy Electron Microscopy (SPLEEM)

Instead of probing the sample with a finely focussed beam of unpolarized high-energy electrons and detecting the spin-polarization of low-energy secondary electrons, as in SEMPA, also the reverse approach is applicable for magnetic imaging. The reflection of spin-polarized low-energy electrons from a magnetic surface can strongly depend on the electron spin direction if, for example, the energy of these electrons matches a gap in the density of states above the vacuum level of either minority or majority electrons, but not of both [28]. The reflection of low energy electrons is enhanced in such gaps [29], which occur at zone boundary crossings in the extended zone scheme.

This can be implemented into a low energy electron microscope (LEEM) in the way shown in Fig. 9. LEEM is a parallel imaging technique, in which the sample is illuminated by a beam of low energy electrons, and the reflected electrons are used to create an image of the sample. As will be discussed below, imaging of emitted electrons is only feasible using immersion electron lenses as objective lens, in which a high voltage between sample and objective accelerates the electrons towards the imaging optics. The only way to get electrons as illumination onto the sample across the electric field in front

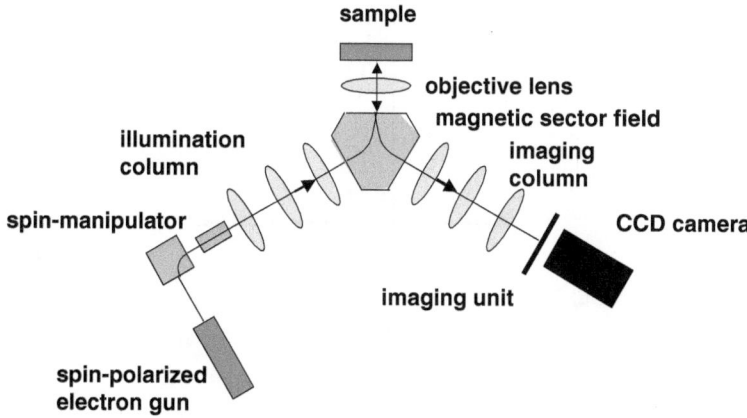

Fig. 9. Schematics of magnetic imaging by spin-polarized low-energy electron emission microscopy. Except for the spin-polarized electron gun and the spin manipulator at the left hand side, all other components are the same as for LEEM. An electron beam is formed in the illumination column, and deflected towards the sample by a magnetic sector field. It reaches the sample surface after passing through the objective lens. Reflected and emitted electrons from the sample are imaged by the objective lens, and deflected into the opposite direction by the magnetic sector field. The image is the projected by the imaging column on the imaging unit and recorded by a CCD camera

of the sample surface is to pass these electrons through the same objective lens. In a LEEM this is usually achieved by a magnetic sector field that acts as a beam separator. The Lorentz force (1) of a magnetic field acting on an electron contains the electron velocity vector, so that the deflection of the illuminating and the reflected electrons has opposite sign. The left part of the sketch of Fig. 9 is the illumination column. The only parts that are added for magnetic imaging are the spin-polarized electron gun and a spin manipulator, which will be described below. The right hand part is the imaging column, in which a parallel image of the sample by the emitted electrons is projected onto an imaging unit, usually a combination of an intensifying multi-channel plate and a fluorescent screen, from which the image can be read by a standard CCD camera.

The LEEM and later also the SPLEEM method were pioneered by E. Bauer and coworkers. Imaging of emitted electrons, though, is much older and dates back to the early work of Brüche in 1933 [30]. In order to obtain a reasonable intensity usually the so-called immersion or cathode lenses are used. Figure 10 shows a sketch of such a cathode lens. It shows an electric tetrode lens, which is called like that because it contains four optical elements. If you are now counting and think that the fourth element is missing in the drawing: It is not, the sample itself is the fourth element. By applying a high voltage between sample and objective lens, electrons are accelerated

294   W. Kuch

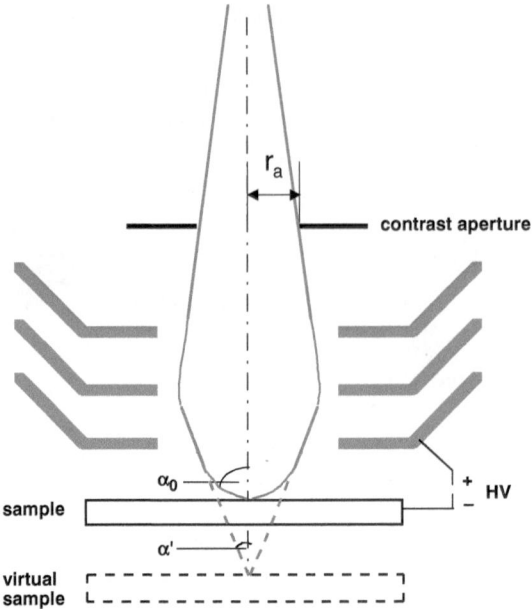

Fig. 10. Sketch of an electrostatic tetrode lens used for electron emission microscopy. Electrons are accelerated towards the entrance of the lens by a high voltage HV. Extrapolating straight rays makes them look like originating from a virtual sample under a virtual starting angle α', which is determined by the size r_a of the contrast aperture, and is much smaller than the real starting angle α_0

towards the entrance of the objective lens. This allows a higher solid angle of emitted electrons to be used for the imaging, and increases the intensity significantly. The electrons thus move on curved trajectories between the sample and the entrance of the objective lens. Optically it is equivalent to assuming that the electrons originate from a virtual sample at a larger distance from the objective lens, and under a virtual starting angle α' that is significantly smaller than the real starting angle α_0. These angles are determined by the contrast aperture that is placed in the back focal plane of the objective lens. By limiting its radius r_a and thus the angle α', spherical and chromatic aberrations can be reduced at the cost of reduced intensity. Magnetic cathode lenses reach theoretical resolutions below 10 nm for monoenergetic electrons and optimum aperture size. In practice, resolutions about a factor of 2–3 worse than the theoretical limit are achieved.

Figure 11 shows a sketch of the spin manipulator. It consists of a combination of electrostatic and magnetostatic electron deflectors, and is used to orient the spin polarization direction of the illuminating electron beam. Electrons that are emitted from the GaAs cathode of a spin-polarized electron gun are spin-polarized in the longitudinal direction, i.e., along the propagation axis. A 90° deflection of the electrons by an electric field preserves the

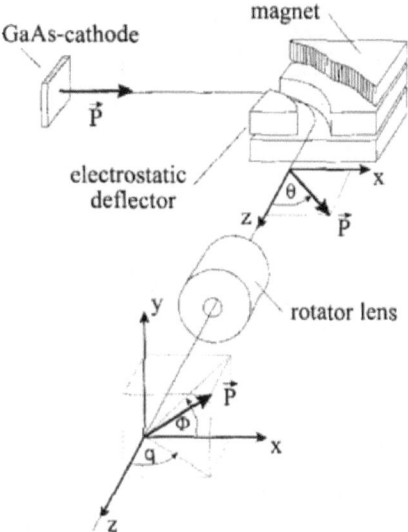

Fig. 11. Working principle of the spin manipulator for SPLEEM. Spin-polarized electrons with polarization vector pointing along x are emitted from a GaAs photocathode. The electron spin direction is preserved after an electrostatic 90° deflection, while it turns into the z direction after a magnetostatic 90° deflection. Any angle Θ between x and z can be selected by a combination of electric and magnetic fields. A subsequent magnetic rotator lens is used to rotate the spin out of the x–z plane by an angle Φ (Reproduced from [28] with permission, copyright (1998) by the World Scientific Publishing Company)

spin direction of the electrons, which are now transversely polarized along the x direction in the coordinate system of Fig. 11. On the other hand, if the 90° deflection is performed by a magnetic field, the electron spin is deflected along with the propagation direction, leading to a longitudinal spin polarization in the z direction. If the 90° deflection is achieved by a combination of electric and magnetic field, the resulting spin polarization can be adjusted at any angle Θ in the x–z plane. A rotation of the spin axis out of this plane is achieved by a second element, a magnetic rotator lens. A magnetic field along the electron propagation direction leaves the electron trajectory unchanged, but rotates the spin into the y direction. By a careful adjustment of all elements of such a spin manipulator the spin of the electrons illuminating the sample can thus be turned into any direction in space. This allows to obtain vectorial information by the SPLEEM technique. It should be mentioned that the spatial images obtained for different spin components are not a priori fully identical, like for SEMPA. While in SEMPA the selection of the spin component is done after the imaging process, in SPLEEM it is done before. Slight changes in the direction and angular spread of the electron beam by the spin manipulator may lead to slightly different spatial images. If the

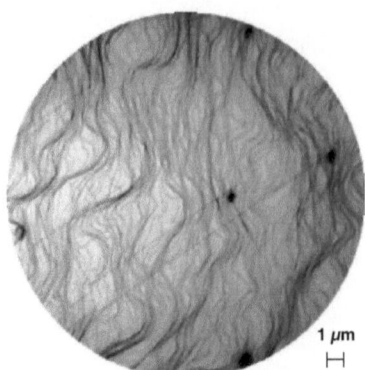

Fig. 12. Example of topographic contrast in LEEM. The image shows the (001) surface of a Cu single crystal. The *dark lines* represent monatomic steps [31]

spin manipulator is carefully adjusted, such deviations can be kept small, and magnetic images obtained for different components of the magnetization may still be compared numerically with each other.

An advantage of SPLEEM is that at the same time with the magnetic image, all the topographic and structural information of LEEM can be obtained. LEEM is a technique that is used by several groups for the study of single crystal surfaces and thin film growth. Figure 12 shows as an example a LEEM image of a Cu(001) surface [31]. The dark lines represent step edges between surface terraces of single atomic height. At some places these step edges are bunched together to form steps of a height of several atoms. The information obtained from LEEM images may be combined with the magnetic images from SPLEEM to identify structure–magnetism relationships on the nanoscopic scale.

The magnetic contrast in SPLEEM is realized by the spin-dependent reflection coefficient for low-energy electrons. The penetration of electrons into the sample is governed by the electronic states, and is smaller in gaps of the unoccupied part of the band structure [29]. Since bands of majority and minority electrons are shifted in energy with respect to each other, such gaps are located at different energies for the two spin directions. The spin-dependence of the reflection coefficient is the biggest at such spin-shifted gaps in the band structure [28]. The conditions for best magnetic contrast consequently change for different samples.

Figure 13 shows an example of a simultaneous structural and magnetic investigation by SPLEEM. Images (a)–(g) show the magnetic domain structure of an iron film at different times during the evaporation on a Cu(001) single crystal at room temperature. The curve in Fig. 13 is the simultaneously determined diffracted electron intensity, and reflects the atomic roughness of the surface [32]. Intensity oscillations are characteristic of a layer-by-layer growth of the Fe film, with maxima in the diffracted intensity corresponding

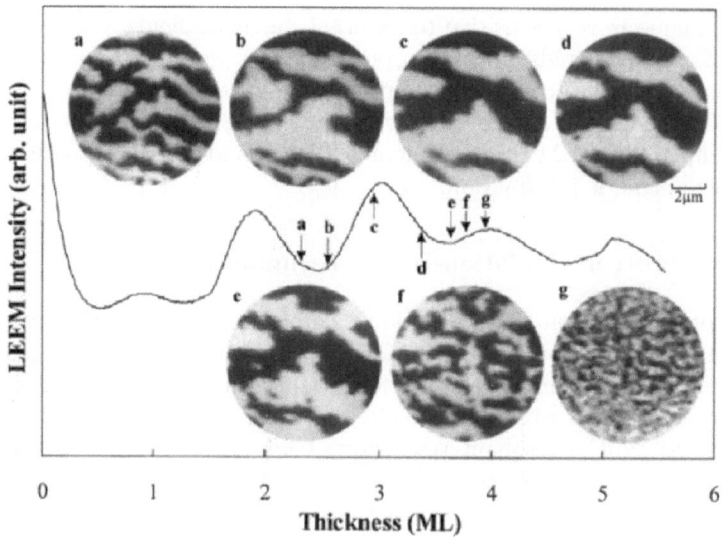

Fig. 13. Example of magnetic imaging by SPLEEM. The curve displays the diffracted electron intensity during the evaporation of Fe on Cu(001). Magnetic domain images (**a**)–(**g**) were taken during film deposition at the thicknesses indicated by arrows on the curve (Reproduced from [32] with permission, copyright (2001) by Elsevier Science)

to smooth surfaces. They appear in layer-by-layer growth at the completion of successive atomic layers, i.e., at film thicknesses of integer numbers of atomic monolayers (ML), and allow a precise thickness calibration of the deposited films. The total acquisition time of each image of Fig. 13 was only 10 s. Considering the low evaporation rate used in [32], this corresponds to the deposition of only 1% of a monolayer during acquisition of one image. Fe/Cu(001) exhibits a perpendicular magnetization in the thickness range of Fig. 13, leading to the occurrence of typical stripe domain patterns as discussed in Sect. 1, see Fig. 1 (d). From the series of images taken during growth of the Fe film the evolution of stripe patterns with different stripe width and shape can be clearly recognized, and correlated to the film thickness with extremely high accuracy.

In summary, SPLEEM is a technique that allows vectorial magnetic imaging at high lateral resolution. It is relatively surface sensitive, and has to be performed under high vacuum conditions. Because of parallel imaging and high electron intensities, it is a very fast technique, which allows real time magnetic imaging of, for example, growth-induced changes of the magnetic properties. A major advantage is that microscopic magnetic and topographic information can be obtained simultaneously. The conditions (electron energy) for best magnetic contrast depend on the sample, and have to be determined beforehand for best results. Since low energy electrons are imaged, the technique is very sensitive to external magnetic fields. Only very small fields of a few Oersteds along the optical axis, i.e., perpendicular to the surface plane, may be tolerated if a special integrated coil design is used [33]. In general, however, it is necessary to image the remanent magnetic state of a sample. SPLEEM is not element-selective, and does not provide depth-selective magnetic information.

3.5 Synchrotron Radiation-Based Techniques

Synchrotron radiation-based techniques employ magnetic dichroisms in x-ray absorption as contrast mechanism for the magnetic imaging. The particular strength of these techniques is the elemental sensitivity of this magnetic contrast mechanism. X-ray magnetic circular dichroism (XMCD) relies on the spin-polarization of resonant transitions between occupied core states and spin-split unoccupied states just above the Fermi edge excited by circularly polarized photons. If the spin-polarization of the transitions, which is determined by the helicity of the exciting radiation, is aligned with the spin polarization of the unoccupied states, the absorption cross section is higher than in the opposite case, when the spin polarization of the transitions is antiparallel to the spin polarization of the unoccupied states. The photon energy at which this effect occurs depends on the binding energy of elemental core levels, and can therefore be used to probe different elements in a sample, or different magnetic layers containing different elements separately.

As in laterally integrating XMCD measurements, there are two basic ways of detecting the x-ray absorption. One can either detect the x-ray absorption by measuring the amount of emitted electrons at the sample surface, or by measuring the transmitted intensity of x-rays after transfer through the sample. In laterally integrating experiments the former is usually termed "total electron yield" detection of x-ray absorption. The latter is a more direct way of measuring the absorption, but requires appropriately thin samples.

In imaging experiments these two approaches are realized by combining resonant excitation by circularly polarized x-rays and existing microscopy techniques. A photoelectron emission microscope can be used to create a magnified image of the sample from the emitted electrons at the surface after photon absorption. In x-ray transmission microscopy an image of the sample is obtained from the locally different transmission of x-rays, which, if circularly polarized and tuned to elemental absorption edges, depends also on the local sample magnetization.

Photoelectron Emission Microscopy (PEEM)

We will first turn to the local electron yield detection of the x-ray absorption. Photoelectron emission microscopy (PEEM) is a relatively old technique, older than LEEM. Schematically it is similar to the imaging column of the

SPLEEM set-up presented in Fig. 9, but with a straight optics, since no deflecting magnetic sector field is necessary. As objective lens also the cathode lens type as the one shown in Fig. 10 or the magnetic counterpart are used. The first experimental realization by Brüche and Pohl dates back to the thirties [30, 34]. After adaption to ultrahigh vacuum environment [35, 36] PEEM was used in a number of studies of surfaces and surface reactions [37–40]. With the availability of synchrotron radiation sources and tunable x-rays the potential of PEEM for the spectromicroscopic detection of x-ray absorption was soon realized [41]; the first demonstration of magnetic contrast using XMCD as contrast mechanism followed a few years later [42]. XMCD-PEEM is routinely used as a technique for magnetic imaging since about 1997 [43–45].

Figure 14 shows an example of how magnetic contrast is obtained using XMCD [31]. Panel (a) shows the raw image of a Co/Ni bilayer on Cu(001), obtained with circularly polarized x-rays of positive helicity tuned to the

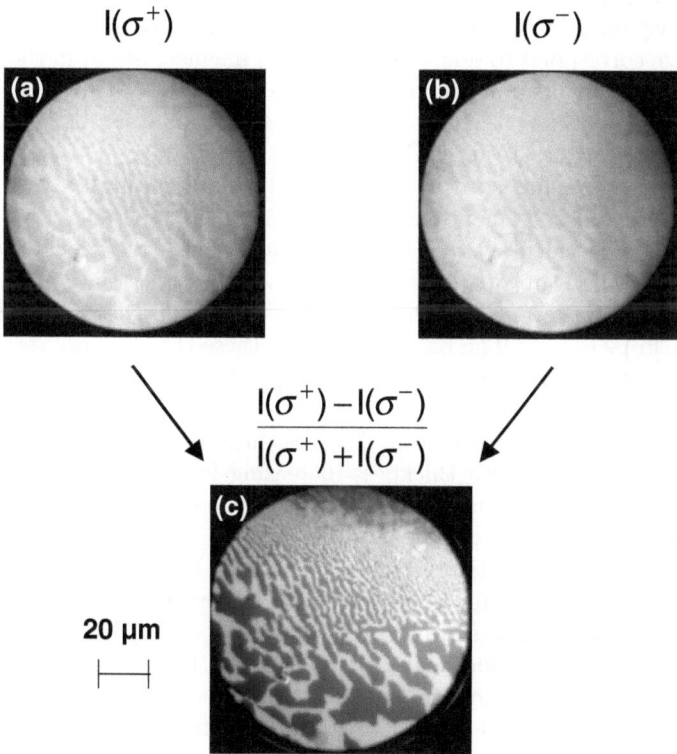

Fig. 14. Demonstration of magnetic contrast in XMCD-PEEM. (**a**): PEEM image of Co/Ni/Cu(001) obtained with circularly polarized x-rays of positive helicity at the Ni L_3 edge, (**b**): same with negative helicity. (**c**): Asymmetry image calculated from the images (a) and (b) showing the magnetic contrast

maximum of the Ni L_3 absorption edge. One can recognize a stripe-like contrast that is not observed in images taken for off-resonant photon energies. This is the magnetic contrast, representing the lateral distribution of XMCD. XMCD leads to a different x-ray absorption of domains with magnetization parallel and antiparallel to the helicity of the incoming x-rays. This is already visualized in the raw image (a), and seen in the live camera image. To prove the magnetic origin, the x-ray helicity can be reversed. Figure 14 (b) shows the same image, but for opposite x-ray helicity. Since all topographic contrast in the image does not depend on the x-ray helicity, all differences between images (a) and (b) are purely magnetic. It is seen that the stripe-like pattern indeed reverses its contrast, as is expected for magnetic contrast if the direction of the helicity vector is reversed. To eliminate all topographic contrast and obtain a magnetic domain image of the sample, the two images obtained for opposite helicity of the circularly polarized x-rays can be subtracted. Usually the result is then divided by the sum of the two images in order to eliminate effects of different illumination across the image. The result, the intensity difference $I(\sigma^+) - I(\sigma^-)$ divided by the sum $I(\sigma^+) + I(\sigma^-)$, is called asymmetry. It is presented in Fig. 14 (c). Dark and bright areas in that image correspond to domains of opposite magnetization, in the example of Fig. 14 to domains pointing into and out of the surface plane, because the thickness combination of Ni and Co in the Co/Ni bilayer was such that an out-of-plane easy axis resulted [31].

The stripe domains seen in Fig. 14 are typical for samples with an out-of-plane easy axis. As mentioned in Sect. 1 when discussing Fig. 1, these samples have a tendency towards domain formation. The demagnetizing energy connected to an out-of-plane magnetization direction can be lowered by formation of alternatingly magnetized domains, on the cost of exchange and anisotropy energy. The balance between these three energy terms, dipolar demagnetizing, exchange, and anisotropy energy, determines the width of the stripe domains. In Fig. 14 the stripe width shrinks across the range of the image from bottom to top. The reason is that the Co overlayer was deposited as a wedge with thickness increasing from bottom to top, which leads to a decrease in anisotropy energy and consequently to a decrease in domain size [31, 46].

In Fig. 14 the magnetization direction in the domains was known from the anisotropy of the sample, and confirmed by qualitative arguments about the shape of the domains. In general one does not know the magnetization direction from a single XMCD-PEEM image, but only the projection of the magnetization direction on the light helicity vector. Vectorial magnetic imaging as in SEMPA or SPLEEM would require to change the direction of the incoming synchrotron radiation with respect to the microscope. This would be evidently only possible by rotating or moving the vacuum chamber containing the PEEM; it has however not been realized to-date. Some instruments allow instead the azimuthal rotation of the sample, so that different

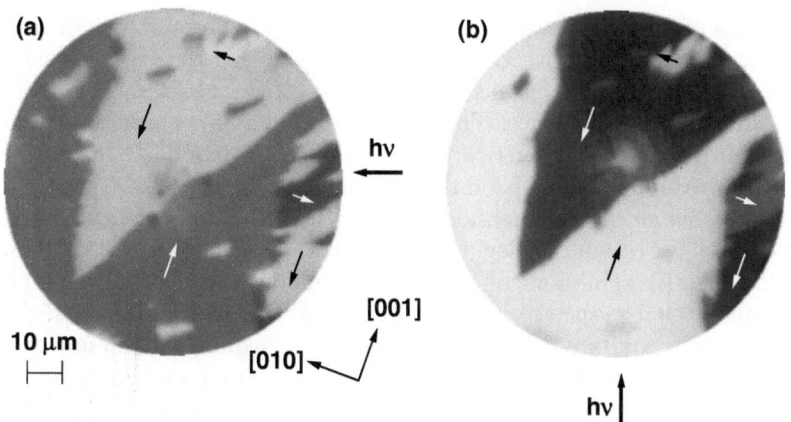

Fig. 15. Example of vectorial magnetic domain imaging using XMCD-PEEM. Magnetic domains in Fe/W(001) display different contrast depending on the azimuthal direction of the incoming x-rays, from the *right* in (**a**), from the *bottom* in (**b**). The local magnetization direction is indicated by arrows in some domains

light incidence directions with respect to the domain structure of the sample are possible. The imaging conditions in this case are not fully preserved after the rotation of the sample, since a tiny sample tilt or a minute asymmetric distortion of the electric field between sample and objective lens can cause slightly different image proportions or focusing conditions. Since the camera is not rotated together with the sample, it is also necessary to rotate the resulting image numerically. It is therefore normally not possible to compare images acquired for different sample azimuth numerically with each other, as it can be done in SEMPA or under certain conditions also in SPLEEM. What is possible, though, is a manual comparison of images for different light incidence in order to determine the magnetization vector in space. If the absolute value of the magnetization vector is known or can be assumed to be constant in different domains, two measurements, i.e., two images for different light incidence directions, are sufficient to determine the remaining two degrees of freedom of the magnetization vector in space. An example is given in Fig. 15. The two images (a) and (b) show magnetic domains of an Fe film on W(001) at about the same position of the sample [47]. The difference between the two images is the light incidence direction, which is indicated by arrows labelled "$h\nu$". Between the acquisition of the two images the sample was rotated by 90°, and afterwards the images were turned in the computer to display identical sample alignment. If one tries to overlay the images (a) and (b) of Fig. 15 it becomes evident that the domain boundaries do not match perfectly, and that the resolution is slightly better in panel (a) than in panel (b). From the manual quantitative comparison of the dichroic contrast in each of the domains, however, the magnetization direction can be determined. It is indicated by arrows in the larger domains. Four different directions of magnetization are observed, corresponding to the four in-plane ⟨100⟩ crystallographic directions. It is nicely seen how the grayscale contrast varies with the different projections of these directions on the two different directions of light incidence. This works also for more complicated configurations in which the magnetization direction is not confined to the sample plane but exhibits in-plane and out-of-plane components [48, 49].

XMCD as an integral method is widely used in connection with a set of sum-rules [50, 51], which allow to extract quantitative magnetic information like the effective spin moment or the orbital moment per atom from a pair of x-ray absorption spectra for opposite helicity. If the photon energy is scanned in small steps and a stack of PEEM images is acquired for many different energies around the absorption edges for both helicities, a quantitative analysis of this microspectroscopic data set is possible. The result are microscopic images of the spin and orbital moments [52, 53].

As already mentioned in the introduction to Sect. 3.5, one of the main advantages of techniques using XMCD as contrast mechanism is the element-selectivity of the magnetic information. In samples that contain different magnetic layers of different elements this can be used for the layer-selective magnetic imaging. Nearly all of the most exciting new discoveries in thin film magnetism are observed in multilayered structures in which two or more magnetic layers have to be switched separately. To address the different layers separately and to obtain microscopic magnetic information about the different magnetic layers in a simple way is thus crucial for the microscopic investigation of such structures. With XMCD-based techniques this is possible by setting the x-ray monochromator of a soft x-ray beamline at a synchrotron radiation facility to the tabulated value of the photon energy corresponding to the elemental core–valence excitations. If the elemental composition of the multilayered structure is known, it is therefore straightforward to take images of the different layers. Figure 16 shows an example of layer-resolved magnetic images, taken at an FeNi/Cu/Co trilayer deposited on FeMn/Cu(001) [47]. It shows the layer-resolved magnetic domain structure after application of an external magnetic field of 340 Oe along the direction indicated by "H". Image (a) was obtained by tuning the photon energy to the energy of the Fe L_3 absorption edge, and thus represents the domain image of the FeNi top ferromagnetic layer, image (b) was acquired with the photon energy tuned to the Co L_3 absorption edge and shows the domain image of the Co bottom ferromagnetic layer. The domains in the Co layer are mainly oriented along two opposite ⟨110⟩ directions as indicated by arrows, namely along [1$\bar{1}$0] and [$\bar{1}$10], corresponding to dark gray and lighter gray contrast, respectively. Besides, there are also some small domains with brighter and darker contrast; they exhibit magnetization directions along [110] and [$\bar{1}\bar{1}$0], respectively. The observed domain pattern in the Co layer was not changed by the application of the 340 Oe external field.

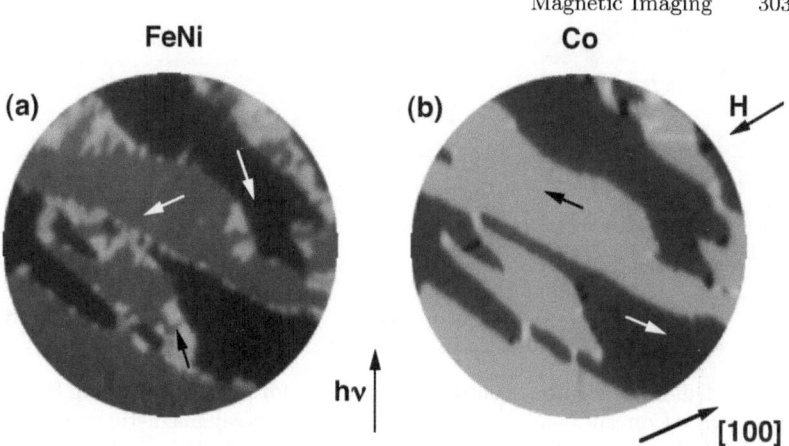

Fig. 16. Example of layer-resolved magnetic domain imaging by XMCD-PEEM. (a) and (b) show the magnetic domain images of the FeNi and the Co layer, respectively, of an FeNi/Cu/Co trilayer on FeMn/Cu(001) after application of an external magnetic field of 340 Oe in the direction indicated by "H". Domain walls in the Co layer are found to be decorated by small domains with bright contrast in the FeNi layer, pointing towards a local interlayer coupling mechanism mediated by the magnetostatic stray fields of domain walls

The domain image of the FeNi layer (a) shows a qualitatively similar pattern, but with different contrast. An analysis reveals that here the magnetization directions in the different domains are along the $\langle 100 \rangle$ in-plane directions, as indicated by arrows in some domains. The comparison with the domain image of the Co layer (b) shows that everywhere the magnetization directions of the two magnetic layers include an angle of 45°. This is explained by a different anisotropy energy in the two layers, and a parallel interlayer coupling [47]. Because of the direction of the applied field H, the magnetization in the FeNi layer points mainly along $[\bar{1}00]$ and $[0\bar{1}0]$. The interesting point in these images is that the positions of the domain boundaries of the Co layer are decorated in the FeNi layer by small domains with a brighter contrast, corresponding mainly to magnetization along $[010]$. This is an indication of a locally enhanced interlayer coupling at the domain walls of the Co layer. Such a local interlayer coupling has also been observed previously by XMCD-PEEM in Co/Cu/Ni trilayers, and is attributed to the interaction between the layers by the magnetostatic stray field emanating from the domain walls [54]. The identification and investigation of this coupling mechanism, which is highly relevant for the magnetization reversal and control of magnetic trilayers in devices, requires a technique that can provide microscopic lateral resolution and layer-resolved magnetic information at the same time. XMCD based methods do exactly fulfill such requirements.

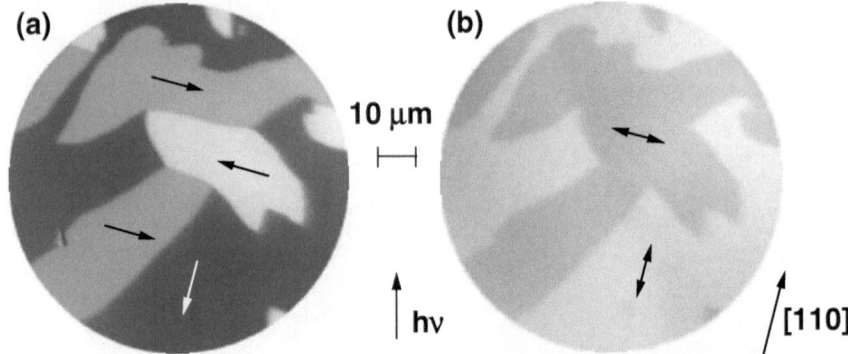

Fig. 17. Different magnetic contrast mechanisms for magnetic imaging by XMCD-PEEM. (**a**): "Conventional" magnetic imaging using x-ray magnetic circular dichroism (XMCD) as contrast mechanism. The graytone in the domain image is proportional to the projection of local magnetization direction on the incoming x-ray direction "$h\nu$". (**b**): Image at the same sample position obtained by x-ray magnetic linear dichroism (XMLD) as contrast mechanism. The graytone is proportional to the cosine square of the angle between the linear polarization axis of the incoming x-ray and the local magnetization direction

Although XMCD is the most widely used contrast mechanism for magnetic imaging with synchrotron radiation, also linearly polarized x-rays can be used to obtain magnetic contrast. The absorption of linearly polarized x-rays depends on the orientation of the electric field vector with respect to a uniaxial deviation from cubic symmetry of the electronic states involved in the transitions. Such a deviation can be caused by the presence of magnetic order. In that case absorption spectra for parallel and perpendicular alignment of polarization and magnetization axes show a typical plus/minus shaped difference at the absorption edges. This linear magnetic dichroism can be sizeable in oxides [55], but is much smaller in metals [56,57], typically more than a factor of 10 smaller than the circular dichroism. After a careful adjustment of the experiment and using an appropriate exposure time it still can be used for magnetic imaging. Figure 17 shows the comparison between magnetic domain images of a metal Co film on Cu(001) obtained by XMCD (a) and XMLD (b) [58]. Note that the span of the grayscale is not the same in (a) and (b), but has been adjusted to clearly show the magnetic domains. While the helicity vector of circularly polarized photons defines an orientation, the linear polarization vector defines only an axis. For symmetry reasons domains with opposite magnetization direction can therefore not be distinguished by linear magnetic dichroism, where the contrast follows a cosine-square behavior, and is maximum between domains with a 90° difference in magnetization direction. This is very nicely seen in Fig. 17, where in panel (b) the contrast between the bright and medium gray domains of panel (a) has vanished.

This property of XMLD, the insensitivity to the magnetization direction, allows to image domains in collinear antiferromagnets with different spin axes. A combination of XMCD- and XMLD-PEEM has indeed been successfully used to image antiferromagnetic domains in NiO [59] and LaFeO$_3$ samples [60,61], as well as their interaction with ferromagnetic domains of an adjacent ferromagnetic layer.

To be complete it should also be mentioned that dichroisms in photo-electron spectroscopy can be used for magnetic imaging, provided the instrument is equipped with an imaging energy filter to suppress the low energy secondary electrons and use only primary photoelectrons for the imaging process. In this case circular and linear magnetic dichroisms in photoemission lead to a magnetic contrast, the different angular behavior of which can deliver complementary information to XMCD-based magnetic images [62].

In summary, magnetic imaging by XMCD-PEEM provides element-selective magnetic domain images. The moderate surface sensitivity of the secondary electron yield detection of the x-ray absorption with an information depth of about 2 nm [63] allows to image also buried layers and interfaces close to the surface. PEEM has a good resolution, however using it in connection with synchrotron radiation increases the chromatic aberrations so that the attainable resolution is somewhat worse compared to LEEM. It is a parallel imaging method that has to be performed in vacuum, although not necessarily in ultrahigh vacuum. Vectorial information can be obtained manually by analyzing images obtained for different sample azimuth angle. Since low energy electrons are imaged, it is very sensitive to external magnetic fields, and allows only images of the remanent magnetic state of a sample. In connection with microspectroscopic data the quantitative magnetic information inherent to XMCD spectra can be combined with lateral resolution. Synchrotron radiation is needed.

Magnetic X-ray Transmission Microscopy (M-XTM)

Magnetic x-ray transmission microscopy is the transmission counterpart to XMCD-PEEM for the laterally resolved detection of XMCD. It is performed either by parallel imaging, or by scanning. X-ray transmission microscopes are mostly used to image biological systems with high resolution using synchrotron radiation x-rays. X-ray microscopes work with zone plates as x-ray lenses. Zone plates are tiny lithographically fabricated concentric ring structures in which the width of the alternating open and closed circles varies as a function of the radius. These rings give rise to a diffraction of the x-rays. In order to focus the first order diffraction of a parallel x-ray beam into a common focal point, the diffraction angle has to be larger for rays that are further away from the optical axis. This is achieved if the ring width is inversely proportional to the radial position from the center. Because the diffraction angle depends on the wavelength, the focal length is proportional to the x-ray photon energy. The attainable resolution of a zone plate is determined by the width of the outermost ring. With state-of-the-art lithography a resolution of 20 nm and below can be achieved.

A sketch of a set-up of a transmission x-ray microscope for parallel imaging is shown in Fig. 18 [64]. A condenser zone plate focusses the illuminating x-rays onto the sample. A microscopic image of the transmitted intensity is created by an objective zone plate ("micro zone plate" in Fig. 18) behind the sample, where it is read by an imaging unit, for example a CCD chip. In a scanning transmission x-ray microscope a focussed x-ray beam is created by a zone plate optics in front of the sample, the sample is moved by a piezo scanner, and the transmitted x-ray intensity behind the sample is recorded by an integral x-ray detector as a function of sample position. The distance between the sample and zone plate is quite small, and usually transmission x-ray microscopy is not performed under ultrahigh vacuum conditions.

In the set-up shown in Fig. 18, where the x-rays reach the sample under normal incidence, the magnetic contrast in the resulting image represents only the out-of-plane component of magnetization. Sensitivity to in-plane components can be achieved by a modification of the XTM set-up so that the sample can be tilted with respect to the optical axis [65]. Since the focal length of a zone plate depends on the wavelength of the x-rays, the distances between zone plates, sample, and detector have to be re-adjusted if a different photon energy is to be used.

Compared to PEEM, the detection of XMCD by TXM differs in several ways: The resolution is usually higher in TXM, but the field of view can not be easily changed. Since TXM is a photon-in/photon-out method, it is not sensitive to external magnetic fields. It is thus ideally suited for the imaging of magnetization reversal processes, for example in small magnetic structures. The main difference between M-TXM and XMCD-PEEM, however, is the way the magnetic information is obtained. XMCD-PEEM is an on-surface

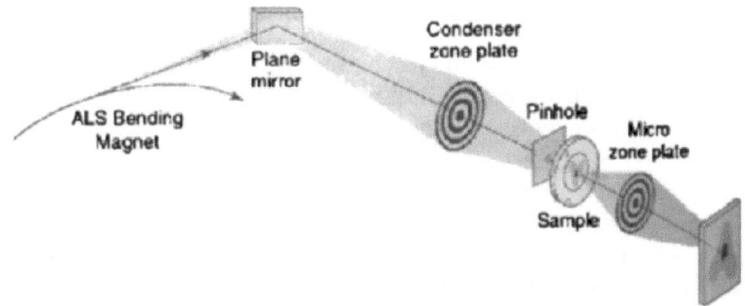

Fig. 18. Schematics of a transmission x-ray microscope. The x-ray beam from a synchrotron radiation source is focussed onto the sample by the condenser zone plate. A magnified image of the transmitted intensity is projected onto a detector by the micro zone plate (Reproduced from [64] with permission, copyright (2001) by IEEE)

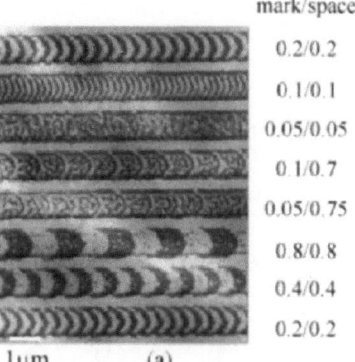

mark/space

	0.2/0.2
	0.1/0.1
	0.05/0.05
	0.1/0.7
	0.05/0.75
	0.8/0.8
	0.4/0.4
	0.2/0.2

1 μm   (a)

Fig. 19. Example of magnetic domain imaging using a transmission x-ray micro-scope showing a test pattern in a $Tb_{25}(Fe_{75}Co_{25})_{75}$ film for magneto-optical data storage. The mark/space assignments of the different stripe patterns are given in μm (Reproduced from [66] with permission, copyright (2001) by the American Institute of Physics)

technique with a probing depth of about 2 nm, while TXM is a transmission method. PEEM is thus more sensitive for signals of ultrathin films or inter-faces, while information from more deeply buried layers (below $\approx$ 10 nm) is only obtained by M-TXM. Substrates for M-TXM have to be suitably chosen membranes.

Figure 19 shows an example of magnetic imaging by magnetic x-ray trans-mission microscopy. Displayed is a magnetic domain image of a 50 nm thick $Tb_{25}(Fe_{75}Co_{25})_{75}$ magneto-optical storage media with out-of-plane easy axis, in which test patterns with different bit sizes and periods have been written. It shows that the thermally assisted write process works well for periods of 200 nm and higher, while it fails for smaller bits or smaller bit spacings [66].

3.6 Scanning Probe Techniques

In scanning probe techniques a sharp tip is laterally moved above the sample surface. The size of an interaction between tip and sample is measured, and usually a feedback loop is used to adjust the tip height in such a way to keep the strength of this tip-sample interaction constant. Depending on the type of tip-sample interaction several different scanning probe techniques can be dis-tinguished. Prominent examples are scanning tunneling microscopy (STM), in which the size of the tunnel current between the tip and a conducting sample represents the tip-sample interaction, and atomic force microscopy, in which either the repulsive binding force upon close approach between tip and sample, or van der Waals forces at larger distances are used to regu-late the tip-sample distance. In the following two scanning probe techniques for magnetic imaging that are based on these two non-magnetic imaging

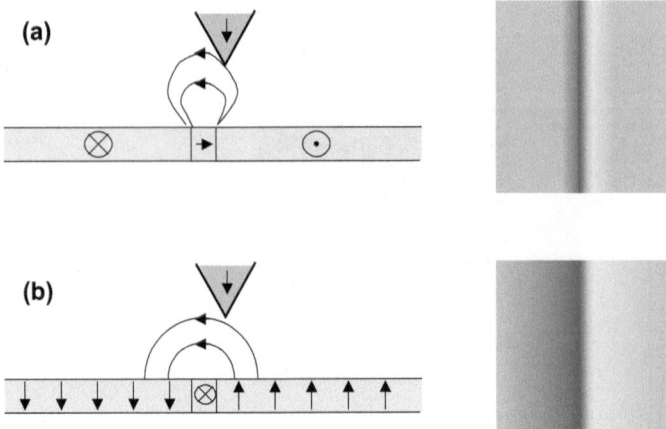

Fig. 20. Schematic explanation of the contrast in magnetic force microscopy. (**a**): Thin film sample with in-plane magnetization. The stray field originating from a Néel wall between two domains with opposite magnetization direction leads to a black/white contrast in the MFM image (*right*). (**b**): Thin film sample with out-of-plane magnetization. The stray field originating from the edges of the domains leads to a contrast as schematically displayed at the right

techniques are presented. As magnetic tip-sample interaction serve the force between the stray field of the sample and a magnetic tip in one case, and the spin-dependence of the tunnel current in the other.

Magnetic Force Microscopy (MFM)

Magnetic force microscopy (MFM) is one of the most widely used techniques for magnetic imaging. This is due to the easy implementation of MFM into the existing scanning probe technology, the high resolution, and the insensitivity to the surrounding which makes MFM applicable also under ambient pressure. In MFM the force exerted on a magnetic tip by the stray field of the sample above the surface is measured. State-of-the-art atomic force microscopes are able to detect very small forces between sample and tip by a change in resonance frequency of a cantilever on which the tip is mounted. This can be used also for the detection of the small magnetic forces that occur when the tip is positioned above the sample surface. MFM is an indirect imaging technique, since not the magnetic domains, but the stray fields above the surface of the magnetic domain pattern are imaged.

Figure 20 shows schematically the working principle of MFM. In the upper panel (a) a situation is considered in which two oppositely magnetized domains in a thin film are separated by a Néel wall. Magnetic poles at both ends of the domain wall lead to a stray field outside the sample as indicated by arrows. If a magnetic tip with magnetization direction as sketched in Fig. 20

(a) is moved across the domain wall, it will experience a repulsive force on the right side of the domain wall, and an attractive force on the left side. An image of the cantilever oscillation frequency would qualitatively look like the one shown on the right hand side, where a brighter contrast means a higher oscillation frequency, corresponding to a repulsive force.

The case of a sample with out-of-plane magnetization is sketched in Fig. 20 (b). An infinitely extended sample with a single domain with perpendicular magnetization does not exhibit a stray field outside the sample, since the magnetic flux density is completely closed inside the film by the demagnetizing field, as discussed in Sect. 1. However, at the domain boundaries part of the flux is also closed outside the sample, as sketched in Fig. 20 (b). This part of the flux is actually exactly the one responsible for the reduction of the demagnetizing energy, which leads to the occurrence of stripe domains in films with perpendicular anisotropy. For MFM this means that close to the domain walls there is a magnetic force acting on the tip. As in the case of the Néel wall in panel (a), the perpendicular component of this force changes sign at the center of the domain wall, resulting in an image as depicted at the right. If the width of the stripe domains is similar to the tip-sample distance, typically about 50 nm or more, the domains appear with a constant alternating contrast in the final image. In this case it looks like an image of the sample magnetization, but one should keep in mind that magnetic images obtained by MFM do not directly represent the magnetization distribution of the sample, but contain only indirect information about it. The actual domain pattern has to be reconstructed from the force images. In the case of narrow out-of-plane domains this is quite easy, but can be more complicated for in-plane domains with different magnetization angles. The long range nature of dipolar interactions is an additional obstacle. In addition, the interaction of the stray field of the tip may influence the sample magnetization in soft magnetic samples. An example is shown in Fig. 21. It shows in panels (b) and (c) MFM images of a $2 \times 2 \ \mu m^2$ permalloy microstructure of 16 nm thickness, taken for opposite tip magnetization direction. The arrows in panel (a) show the result of a micromagnetic simulation for this structure, while the grayscale represents the simulated MFM contrast resulting from this calculated magnetization distribution. As schematically shown in Fig. 1 (b), the energetically most favorable state in square microstructures of a certain size is the so-called Landau pattern, in which no flux exits at the sample's outer edges. At the 90° domain boundaries, however, some flux is present above and below the structure. This is imaged by the MFM, where it leads to the typical black/white contrast at these lines. The sequence of black and white is reversed if the tip magnetization is reversed, which is clearly recognized in Fig. 21. Compared to the simulated pattern (a), however, the experimentally observed domain patterns (b) and (c) exhibit a more curved appearance of the domain walls. Furthermore, the curvature is different in (b) and (c). It is

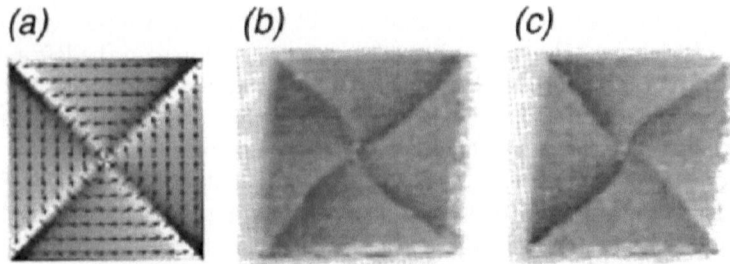

Fig. 21. Example of magnetic imaging using MFM. (**a**) Micromagnetic simulation of a 2×2 μm^2 permalloy element of 16 nm thickness. (**b**), (**c**): MFM images of this element obtained with opposite tip magnetization directions. From [67]

attributed to the perturbation of the sample magnetization distribution by the stray field of the tip [67].

Although the information about the magnetization distribution in a sample is only indirect, MFM is nevertheless probably the most widely used technique for magnetic imaging. Reasons are the high resolution, which until shortly made MFM a standard technique in development and production of magnetic storage devices, the easy use under environmental conditions (no vacuum necessary), and the commercial availability of instruments. Another advantage of MFM is that within the same instrument also topographic information can be obtained from the atomic force mode of operation. If the magnetization reversal of the tip is taken into account, MFM tolerates quite high external fields, and allows the study of magnetization reversal processes.

The lateral resolution depends on the distance between tip and sample, and on the shape of the tip. Routinely about 50 nm are achieved. To improve the resolution, both the design of the tip and the method of tip-sample distance control have to be worked on [68]. Although MFM does not directly provide quantitative information, several attempts for quantitative calibration of MFM have been performed [69–72]. This is not straightforward since also the shape and the magnetization distribution of the tip enter in the equations for the force. A calibration of the tip properties with a known sample is therefore indispensable for obtaining quantitative information from MFM. MFM is not element-selective. Due to the long-range nature of the dipolar interaction also information from deeper layers of a sample is obtained, but it may be sometimes difficult to single out the contributions to the force coming from certain depths of a sample.

Spin-Polarized Scanning Tunnelling Microscopy (Sp-STM)

The other tip-sample interaction that is being used for magnetic imaging is the size of the electron current tunnelling through the vacuum gap between a metallic tip and the sample. The tunnel current between two magnetic electrodes depends in general on the relative orientation of the two magnetization directions. This is explained by Fig. 22. It gives a schematic representation of the involved spin-split densities of states. The left graph shows the spin-resolved density of states of the sample. The horizontal line is the Fermi edge. Majority and minority states are shown left and right of the vertical energy axis, respectively. The middle graph shows the density of states of a tip that is magnetized parallel to the sample. The applied bias voltage U_{bias} between sample and tip appears as an energy difference between the Fermi levels of sample and tip. Since the electron spin is preserved in the tunnel process, tunnelling occurs between majority states of the tip and majority states of the sample, as well as between minority states of the tip and minority states of the sample. For a positive bias voltage as drawn in Fig. 22, electrons tunnel from occupied states of the tip into unoccupied states of the sample. The graph on the right hand side of Fig. 22 shows the density of states of the tip if its magnetization direction is reversed, i.e., if majority and minority states are exchanged. In this case, when tip and sample magnetizations are antiparallel, minority electrons from the tip can tunnel into majority states of the sample, and vice versa. The tunnel current in these two cases, tip magnetization parallel or antiparallel to the sample magnetization, will generally be different. This is called tunnel magnetoresistance [73], and can be used to obtain information about the sample magnetization in a scanning tunnelling microscopy experiment.

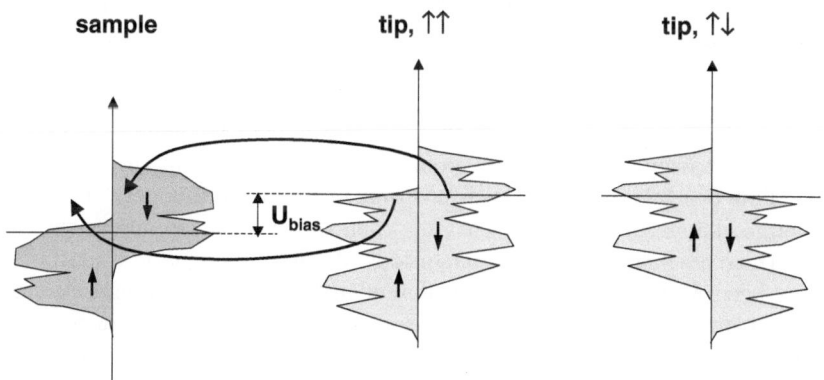

Fig. 22. Schematic explanation of tunnelling magnetoresistance between a magnetic tip and the sample in scanning tunnelling microscopy. *Left*: spin-split density of states of the sample, *middle*: spin-split density of states of a tip with magnetization parallel to that of the sample, *right*: spin-split density of states of a tip with magnetization antiparallel to that of the sample. Electron tunnelling occurs between states of tip and sample of the same spin direction. Separation between magnetic and topographic contrast can be obtained by using a lock-in technique and modulating either the tip magnetization direction or the bias voltage between tip and sample

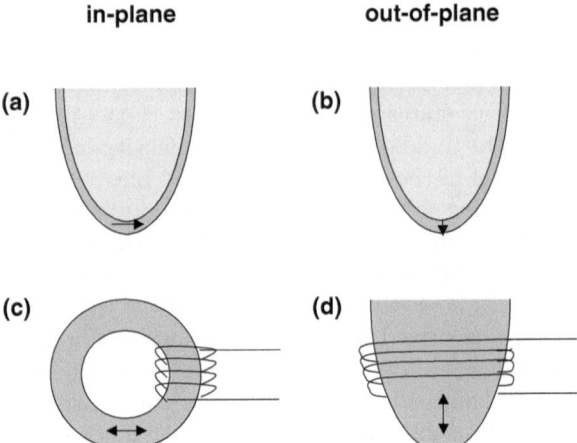

Fig. 23. Schematics of different ways of experimental realization of spin-polarized scanning tunneling microscopy. (**a**), (**b**): Tip covered by a magnetic thin film, used for sp-STM by modulating the bias voltage between tip and sample. (**c**), (**d**): Bulk tips used for sp-STM by reversing the tip magnetization. (a) and (c) show geometries with sensitivity for in-plane magnetization, (b) and (d) geometries with sensitivity for out-of-plane magnetization

The great difficulty is that besides this dependence on the relative orientation of sample and tip magnetization the tunnel current also depends exponentially on the distance between tip and sample. This is the normal contrast mechanism used in STM to obtain topographic contrast. If one wants to use the tunnel magnetoresistance as contrast mechanism for magnetic imaging, the task is to separate magnetic from topographic signal in the tunnel current. This has to be done at each position in an image independently, since the large change in tunnelling current when moving the tip across the sample surface does not allow any conclusions about the magnetic contribution. It has been accomplished by two different approaches. The magnetic signal can be identified by modulating a parameter and using a lock-in technique to separate the much smaller magnetic signal from the topographic signal. In the two different approaches the parameter to be modulated was in one case the bias voltage, in the other the direction of tip magnetization. These two ways of realizing sp-STM are shown in Fig. 23. The top row shows a tip covered with a thin magnetic film. Depending on the magnetic anisotropy of the film, the magnetization at the tip apex will be either perpendicular to the tip axis (a) or along the tip axis (b), probing thus samples with in-plane or out-of-plane magnetization, respectively. Using such a tip, the bias voltage U_{bias} has to be modulated. The lock-in signal will then yield the energy derivative of the tunnel current, which, like the tunnel current itself, will in general depend on the relative orientation of tip and sample magnetization. It is then possible to use the magnitude of the tunnel current to control the tip-sample distance, and the lock-in signal to obtain information about the sample magnetization. This spectroscopic approach has been pioneered by Bode, Wiesendanger, and coworkers [25, 74]. The problem is that the magnetic signal varies in size as a function of bias voltage, and that different elements will also lead to a different dI/dU_{bias}. The advantage is that once a tip is covered with a magnetic film it is easily implemented into any existing STM. It is also possible to use antiferromagnetic tips in order to make the technique more insensitive to the application of external fields [75].

The other approach, modulating the tip magnetization, is shown in Fig. 23 (c) and (d). It was pioneered by Wulfhekel, Kirschner, and coworkers [76,77]. Here the tip has to be fully made from a magnetic material. The tip magnetization is then periodically reversed by the current through a coil. For a normal tip (d) this leads to a sensitivity to the out-of-plane component of magnetization. Note that according to the scale the tips are sketched in Fig. 23, the coil would be way up and much larger. To detect in-plane components, a special ring-like design of the tip (c) has to be used in order to reverse the magnetization direction [78]. Here the electrons tunnel between the sample and the bottommost atoms of a tiny washer-like ring. Modulating the sample magnetization yields directly the tunnel magnetoresistance from the lock-in output. The signal in this method is present and can be used for magnetic imaging at any bias voltage, which is an advantage in particular if the spectroscopic properties of a sample are not known beforehand. Technical problems are the choice of an appropriate tip material, which has to be magnetically soft and at the same time exhibit very low magnetostriction.

An example of magnetic contrast by sp-STM is shown in Fig. 24. Like for SEMPA, the high surface sensitivity of Sp-STM allows the investigation of layered antiferromagnetic samples. This has been successfully used for the imaging of Cr(001) [25] and Mn/Fe(001) [26, 78]. Figure 24 gives an example from [78]. Panel (a) shows the topographic STM image of 7 ML Mn on an Fe(001) single crystal. Black, gray, and white contrast corresponds to increasing terrace height in steps of one atom. Figure 24 (b) shows the simultaneously acquired magnetic signal, using the technique as sketched in Fig. 23 (c). Now the different grayscales correspond to different local magnetoresistance between tip and sample. Comparing Figs. 24 (a) and (b), the layered antiferromagnetic spin structure of the Mn film is nicely recognized, similar to the example shown in Fig. 8.

Sp-STM is at present the magnetic imaging technique with the highest resolution, less than 1 nm. It is very surface sensitive, and has to be performed under ultrahigh vacuum conditions. Topographic information as in STM is available from the same measurement. The high surface sensitivity implies that the signal from buried layers is not accessible by Sp-STM.

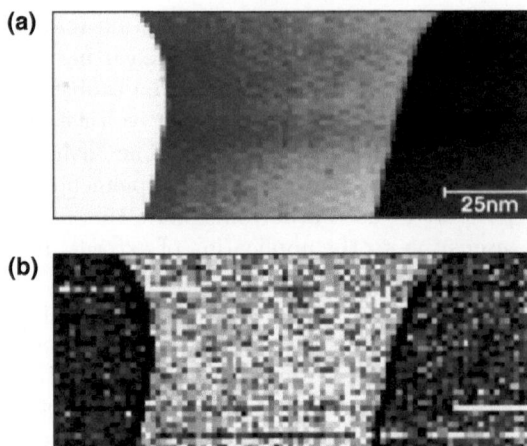

Fig. 24. Example for magnetic imaging by sp-STM. (**a**): Topographic STM image of a Mn film of 7 ML thickness on Fe(001). (**b**): Magnetic sp-STM image at the same position of the sample, acquired simultaneously with the topographic information (Reproduced from [78] with permission, copyright (2003) by the American Institute of Physics)

4 Summary and Outlook

The features of the various techniques for magnetic imaging that have been presented in the previous chapters are summarized by pictograms in Table 2. The first column, headed by a magnifying glass, refers to the resolution. The resolution of the listed techniques ranges between that of Kerr microscopy (500 nm) and sp-STM (< 1 nm), the latter being characterized by a double smiley. Due to the chromatic aberration, the resolution of XMCD-PEEM is somewhat worse compared to SPLEEM despite the identical electron imaging technique.

The second column, headed by an hourglass, gives an overview of the speed of image acquisition. Although the speed depends on image size, resolution, and the actual system under study, in general the scanning probe techniques and SEMPA as scanning technique with the intensity loss in the spin detection process have to be considered as relatively slow. The type of image acquisition, scanning or parallel imaging, is listed in the third column, titled with an eye symbol. The magnifying glass pictograms indicate parallel imaging, while the page with lines pictogram refers to scanning imaging.

An important point is the sensitivity to external magnetic fields. This is summarized in the fourth column, headed by a horseshoe magnet symbol. Here the techniques that use low energy secondary electrons for the imaging as SPLEEM and XMCD-PEEM are marked by bomb symbols, indicating a high sensitivity to external magnetic fields. In SEMPA also slow electrons are detected, but they do not form the image, which is determined by the

Table 2. Summary of the features of the discussed magnetic imaging techniques. Columns with pictograms refer (from left to right) to resolution, image acquisition speed, type of imaging (parallel imaging or scanning), sensitivity to applied magnetic fields, type of depth information (surface based or transmission), information depth (path length for exponential weighting for surface based techniques, maximum sample thickness for transmission techniques), possibility to obtain depth selective information

	(resolution)	(speed)	(imaging type)	(sensitivity)	(depth type)	(info depth)	(depth selective)
Kerr microscopy	☹	☺	☗ ▤	☺	surface	< 20 nm	😐
Lorentz microscopy	☺	😐	☗ ▤	😐	transmission	< 100 nm	☹
SEMPA	☺	☹	▤	☹	surface	< 0.5 nm	💣
SPLEEM	☺	☺	☗	💣	surface	< 1 nm	💣
XMCD-PEEM	😐	☺	☗	💣	surface	< 5 nm	☺
M-TXM	☺	☺	☗ ▤	☺	transmission	< 200 nm	☺
MFM	☺	☹	▤	☺	surface	< 2 μm	☹
sp-STM	☺ ☺	☹	▤	☺ ☹	surface	< 0.2 nm	💣

high energy primary electrons, so that moderate magnetic fields can be tolerated in certain dedicated set-ups. Two different symbols are shown in the field for sp-STM. They refer to the two different experimental approaches: If the magnetization of a soft magnetic tip is being modulated, sensitivity to external magnetic fields is higher than in the approach where the sample bias voltage is modulated.

The fifth column, with a rectangular box as title, displays the way the magnetic information is obtained. The different pictograms found in that column indicate on-surface acquisition and transmission measurements, respectively. Modified on-surface symbols are being used for the scanning probe techniques MFM and sp-STM, to indicate that the surface is not, as in the other techniques, probed by an incident electron or photon beam. The approximate corresponding length scales, information depth in the case of on-surface measurements, and maximum sample thickness in the case of transmission measurements, are listed in the sixth column, titled with a distance pictogram.

The last column summarizes another important aspect of magnetic imaging techniques, namely the possibility to obtain depth-resolved information, for example in multilayered magnetic samples. This is most easily achieved by the XMCD-based techniques, XMCD-PEEM and M-TXM, which due to their element-selectivity can address different magnetic layers at different depths separately. A depth sensitivity can also be achieved in Kerr microscopy by a careful adjustment of the phase. Lorentz microscopy as a transmission technique provides information from all depths of a sample; with some additional information at hand, features in the final image may be attributed to different layers at different depths. The same holds for MFM, which due to the long-range dipolar interaction provides also information about magnetic domain walls in more deeply buried layers. No depth-resolved information is available from images of the more surface-sensitive techniques like SEMPA, SPLEEM, and sp-STM.

A trend that has already started is to furnish magnetic imaging with time resolution to obtain laterally resolved images of the dynamic magnetic response of the sample. Especially with respect to the increased data rate in magnetic data storage the dynamic behavior of small magnetic structures is becoming more and more important. Time-resolution has been successfully integrated into scanning Kerr microscopy in a stroboscopic pump–probe approach [79–81]. Since a laser beam is used for the illumination of the sample, it is quite straightforward to include the time structure of a pulsed laser source for the probe in a pump–probe scheme. The pump has to be an ultrashort synchronized magnetic field pulse, which can be obtained by hitting a photoconductive switch with a laser pulse, and passing the generated current pulse through a stripline underneath the sample.

Similar time-resolved pump–probe experiments have also been already implemented in synchrotron radiation-based magnetic imaging techniques. Here the pulsed time structure of the synchrotron radiation provides the probe. Compared to a laser pulse the typical width of a synchrotron radiation pulse is much larger, typically about 50 ps, but still smaller than typical rise times in stripline-generated magnetic field pulses. Stroboscopic time-resolved magnetic imaging has been demonstrated in XMCD-PEEM by several groups [82–85], and also in M-XTM [86]. The element-selectivity of XMCD is an advantage also for the time-resolved measurements, since magnetization reversal dynamics of the different magnetic layers in a magnetic trilayer can be studied separately [87].

The electron imaging techniques like SPLEEM and XMCD-PEEM, but also Lorentz microscopy, may benefit in the near future from novel developments for aberration correction. In light optical lens systems aberration correction is accomplished by suitable combinations of convex and concave lenses. Since there are no concave lenses for electron imaging, aberration correction can not readily be performed in electron optics. The first suggestion for an aberration-corrected electron optics by non-rotational-symmetric lenses has been made by Scherzer more than half a century ago [88]. 43 years later Rose took up the idea [89], which lead to the construction of an aberration corrected transmission electron microscope [90]. An aberration correction scheme for LEEM and PEEM based on the reflection of electrons in a tetrode electron mirror was also suggested [91], and currently two experimental realizations of aberration-corrected instruments are under development at BESSY in Berlin [92] and at the ALS in Berkeley. The theoretical resolution is expected to be better than 1 nm; in addition, a huge increase in intensity is expected compared to non-corrected instruments for identical resolution settings because a bigger contrast aperture can be selected in the corrected instrument, which leads to a larger electron acceptance angle (cf. Fig. 10).

New developments in the so-called "lensless imaging" should be also mentioned here, although strictly speaking this is not an imaging technique but belongs to the class of diffraction experiments. A collaboration of researchers from BESSY and the ALS/Stanford University achieved the holographic reconstruction of a magnetic domain image from a resonant x-ray scattering experiment using coherent x-ray radiation in combination with the XMCD effect [93, 94]. In their set-up a coherent x-ray beam, tuned to an elemental absorption edge of the sample, was transmitted through a thin magnetic film with out-of-plane magnetization. The diffraction pattern of a small reference hole next to the sample was superimposed for phase retrieval. The resulting coherent scattering speckle pattern was then recorded. This allows the reconstruction of the magnetic real-space image [93, 94]. It is expected that this way of magnetic imaging will greatly benefit from the availability of free electron based sources of extremely brilliant and ultrashort x-ray pulses.

The new developments mentioned in this section have just begun. Together with other improvements on the existing magnetic imaging methods they will push the whole field forward, and open new ways for the investigation of magnetic properties for both fundamental and applied research. This is necessary since the shrinking dimensions of structures involved in magnetic recording and data storage technology pose a serious challenge to the resolution and the speed of magnetic imaging techniques. Magnetic imaging can only be competitive if the resolution satisfies the needs. On the other hand this is a big chance, since it increases the quest for microscopic information to understand the fundamentals of dynamic micromagnetism in confined structures.

Acknowledgement

My thanks go to L.I. Chelaru, K. Fukumoto, F. Offi, J. Wang, M. Kotsugi, C. Quitmann, F. Nolting, T. Ramsvik, and J. Kirschner for the collaboration on results presented in Sect. 3.5 of this chapter. This research has been partly financed by the BMBF (grant no. 05 KS1EFA6). I am grateful to W. Wulfhekel, U. Schlickum, and H.P. Oepen for critical reading of parts of the manuscript. Finally I also want to thank the organizers of the IVth International School on Magnetism and Synchrotron Radiation in Mittelwihr (France) for the invitation to present the above lecture.

318   W. Kuch

References

1. H. Hopster, H.P. Oepen (Eds.): *Magnetic Microscopy of Nanostructures*, (Springer, Berlin Heidelberg New York 2004)
2. D.P. Pappas, K.-P. Kämper, H. Hopster: Phys. Rev. Lett. **64**, 3179 (1990)
3. D.P. Pappas, C.R. Brundle, H. Hopster: Phys. Rev. B **45**, 8169 (1992)
4. Z.Q. Qiu, J. Pearson, S.D. Bader: Phys. Rev. Lett. **70**, 1006 (1993)
5. R. Allenspach, A. Bischof: Phys. Rev. Lett. **69**, 3385 (1992)
6. M. Speckmann, H.P. Oepen, H. Ibach: Phys. Rev. Lett. **75**, 2035 (1995)
7. K. Fukumoto, H. Daimon, L.I. Chelaru et al.: Surf. Sci. **514**, 151 (2002)
8. A. Hubert, R. Schäfer: *Magnetic Domains*, (Springer, Berlin Heidelberg New York 1998)
9. L.D. Landau, E.M. Lifshitz: *Electrodynamics of Continuous Media*, (Pergamon, London 1960)
10. S.D. Bader: J. Magn. Magn. Mater. **100**, 440 (1991)
11. R. Schäfer: J. Magn. Magn. Mater. **215-216**, 652 (2000)
12. R. Schäfer, R. Urban, D. Ullmann et al.: Phys. Rev. B **65**, 144405 (2002)
13. M. de Graef: Lorentz Microscopy: Theoretical basis and image simulations. In: *Magnetic imaging and its applications to materials*, ed by M. de Graef, Y. Zhu (Academic Press, San Diego 2001) pp 27–67
14. A.K. Petford-Long, J.N. Chapman: Lorentz Microscopy. In: *Magnetic Microscopy of Nanostructures*, ed by H. Hopster, H.P. Oepen, (Springer, Berlin Heidelberg New York 2004) pp 67–86
15. M. Herrmann, S. McVitie, J.N. Chapman: J. Appl. Phys. **87**, 2994 (2000)
16. G.R. Morrison, J.N. Chapman: Optik **64**, 1 (1983)
17. G.R. Morrison, H. Gong, J.N. Chapman et al.: J. Appl. Phys. **64**, 1338 (1988)
18. J. Unguris: Scanning electron microscopy with polarization analysis (SEMPA) and its applications. In: *Magnetic imaging and its applications to materials*, ed by M. de Graef, Y. Zhu (Academic Press, San Diego 2001) pp 167–193
19. H.P. Oepen, H. Hopster: SEMPA Studies of Thin Films, Structures, and Exchange Coupled Layers. In: *Magnetic Microscopy of Nanostructures*, ed by H. Hopster, H.P. Oepen, (Springer, Berlin Heidelberg New York 2004) pp 137–167
20. K. Koike, H. Hasegawa: Appl. Phys. Lett. **45**, 585 (1985)
21. H.P. Oepen, J. Kirschner: Phys. Rev. Lett. **62**, 819 (1989)
22. M.R. Scheinfein, J. Unguris, R.J. Celotta et al.: Phys. Rev. Lett. **63**, 668 (1989)
23. M. Konoto, T. Kohashi, K. Koike et al.: Phys. Rev. Lett. **93**, 107201 (2004)
24. J. Unguris, R.J. Celotta, D.T. Pierce: Phys. Rev. Lett. **69**, 1125 (1992)
25. M. Kleiber, M. Bode, R. Ravlić et al.: Phys. Rev. Lett. **85**, 4606 (2000)
26. U. Schlickum, N. Janke-Gilman, W. Wulfhekel et al.: Phys. Rev. Lett. **92**, 107203 (2004)
27. G. Steierl, G. Liu, D. Iorgov et al.: Rev. Sci. Instrum. **73**, 4264 (2002)
28. T. Duden, E. Bauer: Surf. Rev. Lett. **5**, 1213 (1998)
29. V.N. Strocov, H.I. Starnberg, P.O. Nilsson: Phys. Rev. B **56**, 1717 (1997)
30. E. Brüche: Z. Phys. **86**, 448 (1933)
31. W. Kuch, K. Fukumoto, J. Wang, C. Quitmann, F. Nolting, T. Ramsvik, unpublished result.
32. K.L. Man, M.S. Altman, H. Poppa: Surf. Sci. **480**, 163 (2001)
33. H. Poppa, E.D. Tober, A.K. Schmid: J. Appl. Phys. **91**, 6932 (2002).

34. J. Pohl: Zeitschr. f. techn. Physik **12**, 579 (1934)
35. H. Bethke, M. Klaua: Ultramicroscopy **11**, 207 (1983)
36. W. Engel, M.E. Kordesch, H.H. Rotermund et al.: Ultramicroscopy **36**, 148 (1991)
37. M.E. Kordesch, W. Engel, G.J. Lapeyre et al.: Appl. Phys. A **49**, 399 (1989)
38. M. Mundschau, M.E. Kordesch, B. Rausenberger et al.: Surf. Sci. **227**, 246 (1990)
39. H.H. Rotermund, S. Nettesheim, A. von Oertzen et al.: Surf. Sci. **275**, L645 (1992)
40. S. Nettesheim, A. von Oertzen, H.H. Rotermund et al.: J. Chem. Phys. **98**, 9977 (1993)
41. B.P. Tonner, G.R. Harp: Rev. Sci. Instrum. **59**, 853 (1988)
42. J. Stöhr, Y. Wu, M.G. Samant et al.: Science **259**, 658 (1993).
43. W. Swiech, G.H. Fecher, C. Ziethen et al.: J. Electron Spectrosc. Relat. Phenom. **84**, 171 (1997)
44. W. Kuch, R. Frömter, J. Gilles et al.: Surf. Rev. Lett. **5**, 1241 (1998)
45. S. Anders, H.A. Padmore, R.M. Duarte et al.: Rev. Sci. Instrum. **70**, 3973 (1999)
46. W. Kuch, J. Gilles, S.S. Kang et al.: Phys. Rev. B **62**, 3824 (2000)
47. L.I. Chelaru: Microscopic studies of interlayer magnetic coupling across non-magnetic and antiferromagnetic spacer layers. PhD thesis, Martin-Luther-Universität Halle–Wittenberg, Halle (2003) (http://sundoc.bibliothek.uni-halle.de/diss-online/03/04H051)
48. W. Kuch, J. Gilles, X. Gao et al.: J. Magn. Magn. Mater. **242-245**, 1246 (2002)
49. W. Kuch, X. Gao, J. Kirschner: Phys. Rev. B **65**, 064406 (2002)
50. B.T. Thole, P. Carra, F. Sette et al.: Phys. Rev. Lett. **68**, 1943 (1992)
51. P. Carra, B.T. Thole, M. Altarelli et al.: Phys. Rev. Lett. **70**, 694 (1993)
52. W. Kuch: Imaging magnetic microspectroscopy. In: *Magnetic Microscopy of Nanostructures*, ed by H. Hopster, H.P. Oepen, (Springer, Berlin Heidelberg New York 2004) pp 1–28
53. W. Kuch, J. Gilles, F. Offi et al.: Surf. Sci. **480**, 153 (2001)
54. W. Kuch, L.I. Chelaru, K. Fukumoto et al.: Phys. Rev. B **67**, 214403 (2003)
55. D. Alders, L.H. Tjeng, F.C. Voogt et al.: Phys. Rev. B **57**, 11623 (1998)
56. S.S. Dhesi, G. van der Laan, E. Dudzik: Appl. Phys. Lett. **80**, 1613 (2002)
57. W. Kuch, L.I. Chelaru, F. Offi et al.: Phys. Rev. Lett. **92**, 017201 (2004)
58. W. Kuch, F. Offi, L.I. Chelaru, J. Wang, M. Kotsugi, and K. Fukumoto: unpublished result.
59. H. Ohldag, A. Scholl, F. Nolting et al.: Phys. Rev. Lett. **86**, 2878 (2001)
60. A. Scholl, J. Stöhr, J. Lüning et al.: Science **287**, 1014 (2000)
61. F. Nolting, A. Scholl, J. Stöhr et al.: Nature **405**, 767 (2000)
62. W. Kuch, L.I. Chelaru, F. Offi et al.: J. Vac. Sci. Technol. B **20**, 2543 (2002)
63. R. Nakajima, J. Stöhr, Y.U. Idzerda: Phys. Rev. B **59**, 6421 (1999)
64. G. Denbeaux, P. Fischer, G. Kusinski et al.: IEEE Trans. Magn. **37**, 2764 (2001)
65. P. Fischer, T. Eimüller, G. Schütz et al.: J. Appl. Phys. **89**, 7159 (2001)
66. P. Fischer, T. Eimüller, G. Schütz et al.: Rev. Sci. Instrum. **72**, 2322 (2001)
67. A. Thiaville, J. Miltat, J.M. García: Magnetic Force Microscopy: Images of Nanostructures and Contrast Modeling. In: *Magnetic Microscopy of Nanostructures*, ed by H. Hopster, H.P. Oepen, (Springer, Berlin Heidelberg New York 2004) pp 225–251

68. L. Abelmann, A. van den Bos, C. Lodder: Magnetic Force Microscopy –
    Towards Higher Resolution. In: *Magnetic Microscopy of Nanostructures*, ed by
    H. Hopster, H.P. Oepen, (Springer, Berlin Heidelberg New York 2004) pp 253–
    283
69. H.J. Hug, B. Stiefel, P.J.A. Van Schendel et al.: J. Appl. Phys. **83**, 5609 (1998)
70. J. Lohau, S. Kirsch, A. Carl et al.: J. Appl. Phys. **86**, 3410 (1999)
71. J. Lohau, S. Kirsch, A. Carl et al.: Appl. Phys. Lett. **76**, 3094 (2000)
72. P.J.A. Van Schendel, H.J. Hug, B. Stiefel et al.: J. Appl. Phys. **88**, 435 (2000)
73. M. Jullière: Phys. Lett. **54A**, 225 (1975)
74. M. Bode, M. Getzlaff, R. Wiesendanger: Phys. Rev. Lett. **81**, 4256 (1998)
75. A. Kubetzka, M. Bode, O. Pietzsch et al.: Phys. Rev. Lett. **88**, 057201 (2002)
76. W. Wulfhekel, J. Kirschner: Appl. Phys. Lett. **75**, 1944 (1999)
77. H.F. Ding, W. Wulfhekel, J. Kirschner: Europhys. Lett. **57**, 100 (2002)
78. U. Schlickum, W. Wulfhekel, J. Kirschner: Appl. Phys. Lett. **83**, 2016 (2003)
79. W.K. Hiebert, A. Stankiewicz, M.R. Freeman: Phys. Rev. Lett. **79**, 1134 (1997)
80. B.C. Choi, M. Belov, W.K. Hiebert et al.: Phys. Rev. Lett. **86**, 728 (2001)
81. A. Barman, V.V. Kruglyak, R.J. Hicken et al.: Appl. Phys. Lett. **82**, 3065
    (2003)
82. J. Vogel, W. Kuch, M. Bonfim et al.: Appl. Phys. Lett. **82**, 2299 (2003)
83. A. Krasyuk, A. Oelsner, S.A. Nepijko et al.: App. Phys. A **76**, 863 (2003)
84. C.M. Schneider, A. Kuksov, A. Krasyuk et al.: Appl. Phys. Lett. **85**, 2562
    (2004)
85. S.-B. Choe, Y. Acreman, A. Scholl et al.: Science **304**, 420 (2004)
86. H. Stoll, A. Puzic, B. van Waeyenberge et al.: Appl. Phys. Lett. **84**, 3328 (2004)
87. W. Kuch, J. Vogel, J. Camarero et al.: Appl. Phys. Lett. **85**, 440 (2004)
88. O. Scherzer, Optik **2**, 114 (1947)
89. H. Rose, Optik **85**, 19 (1990)
90. M. Haider, S. Uhlemann, E. Schwan et al.: Nature **392**, 768 (1998)
91. H. Rose, D. Preikszas: Nucl. Instrum. and Meth. A **363**, 301 (1995)
92. R. Fink, M.R. Weiss, E. Umbach et al.: J. Electron Spectrosc. Relat. Phenom.
    84, 231 (1997)
93. S. Eisebitt, M. Lörgen, W. Eberhardt et al.: Appl. Phys. Lett. **84**, 3373 (2004)
94. S. Eisebitt, J. Lüning, W.F. Schlotter et al.: Nature **432**, (2004)

Dynamic Aspects of Magnetism

Christian H. Back[1], Korbinian Perzlmaier[1], and Matthias Buess[1,2]

[1] Institut für Experimentelle und Angewandte Physik, Universität Regensburg, Universitätsstr. 31, 93040 Regensburg, Germany
christian.back@physik.uni-regensburg.de
[2] Laboratorium für Festkörperphysik, Eidgenössische Technische Hochschule Zürich, 8093 Zürich, Switzerland
buess@phys.ethz.ch

Abstract. This chapter touches some of the current interests in magnetization dynamics in the precessional regime. In particular, experiments with high spatio-temporal resolution are described. Here, we focus on experiments aimed at identifying the modal structure of inhomogeneously magnetized ferromagnetic elements. Recent developments at synchrotron sources are discussed.

1 Introduction

In general, experiments aimed at exploring magnetization or spin dynamics in the time domain can be divided into two groups. The magnetization may be disturbed by an intense ultra short laser pulse [1–8], or by a short magnetic-field pulse that tips the magnetization out of its equilibrium position [9–27]. In special cases a tipping field pulse may also be produced by the ultra fast change of an internal field component due to irradiation with an ultra short laser pulse [28–30]. In this chapter we will discuss some of the recent developments in magnetic-field-pulse induced magnetization dynamics. Experiments with high resolution in both, space and time, have become increasingly important in the field of magnetization dynamics in recent years [9–13, 21–27, 31, 32]. Beside the fundamental interest in magnetization dynamics occurring in nano- or micron-sized confined magnetic structures also technological interest drives this development. The temporal response of recording heads in modern hard drives, to name just one example, is routinely studied with high resolution time-resolved Kerr microscopy [31–33]. More fundamental experiments aim at the detailed understanding of the modal structure and – intimately connected – the boundary conditions of the small angle excitations of ferromagnetic elements [13, 22–24, 34–36]. For large angle excitations, the ultimate goal is the control and reliable realization of precessional magnetization reversal [16, 18–20, 37, 38].

C.H. Back et al.: *Dynamic Aspects of Magnetism*, Lect. Notes Phys. **697**, 321–343 (2006)
www.springerlink.com

2 Introduction to Ferromagneto-Dynamics

2.1 Precession

The equation of motion for the spin is the Landau-Lifshitz equation. Let us first consider a single electron in an external magnetic field, and derive the equations of motion for the spin. The time evolution of an observable is given by its commutator with the Hamilton operator. For the spin operator this means:

$$i\hbar \frac{d}{dt}\langle S \rangle = \langle [S, \mathcal{H}] \rangle . \tag{1}$$

In our case, the Hamilton operator only consists of the Zeeman term

$$\mathcal{H} = -\frac{g\mu_B}{\hbar} S \cdot B \tag{2}$$

where we define $\mu_B < 0$. We now look for example at the z-component and obtain:

$$\begin{aligned}
[S_z, \mathcal{H}] &= -\frac{g\mu_B}{\hbar}[S_z, S_x B_x + S_y B_y + S_z B_z] \\
&= -\frac{g\mu_B}{\hbar}(B_x[S_z, S_x] + B_y[S_z, S_y]) \\
&= \frac{g\mu_B}{\hbar} i\hbar(S_x B_y - S_y B_x)
\end{aligned}$$

whereby the last step makes use of the commutation rules for the spin operator: $[S_i, S_j] = i\hbar\epsilon_{ijk}S_k$. All together we obtain:

$$\frac{d}{dt}\langle S_z \rangle = \frac{1}{i\hbar}\langle [S_z, \mathcal{H}] \rangle = \frac{g\mu_B}{\hbar}(\langle S \rangle \times B)_z$$

The same holds for the other components of S thus yielding:

$$\frac{d}{dt}\langle S \rangle = \frac{g\mu_B}{\hbar}(\langle S \rangle \times B) . \tag{3}$$

We now want to extend this equation to the magnetization. In the *macrospin* model the magnetization is considered to be uniform and is given as the average of the spin magnetic moments:

$$M = \frac{g\mu_B}{\hbar}\langle S \rangle .$$

From here we arrive at the analog relation to (3) for the magnetization:

$$\frac{d}{dt}M = \gamma M \times B$$

with $\gamma = \frac{g\mu_B}{\hbar}$, the gyromagnetic ratio and $g = 2.0023$ the gyromagnetic splitting factor for a free electron. Introducing the H-field with $B = \mu_0 H$ this becomes

$$\frac{d}{dt}\boldsymbol{M} = \mu_0 \gamma \boldsymbol{M} \times \boldsymbol{H}$$

which is usually written as:

$$\frac{d}{dt}\boldsymbol{M} = -\gamma_0 \boldsymbol{M} \times \boldsymbol{H} \tag{4}$$

with $\gamma_0 = -\mu_0 \gamma > 0$. This is the first part of the *Landau-Lifshitz* equation. Apart from the factor γ_0, it has exactly the same form as the classical equation known from mechanics dealing with precession of rotating bodies. We note here, that due to the summation over the spins making up the total magnetization, it is not clear anymore if the length of the magnetization vector needs to be preserved! Let us imagine that one spin flips at any given location. This will lead to a reduction of the length of the vector $\boldsymbol{M}$. However, once the magnetization has been found and it obeys (4), then both, the length of $\boldsymbol{M}$ and its precession angle are preserved. The frequency of revolution for this precessional motion is directly related to the magnetic field:

$$\omega_0 = \gamma_0 H \ . \tag{5}$$

Thus, the total magnetic field and the gyromagnetic splitting factor hidden in the gyromagnetic ratio set the timescale for the precessional motion of the magnetization of $\approx 17.6\,\mathrm{MHz/Oe}$ for $3d$ transition metals with a g-factor close to the one of the free electron.

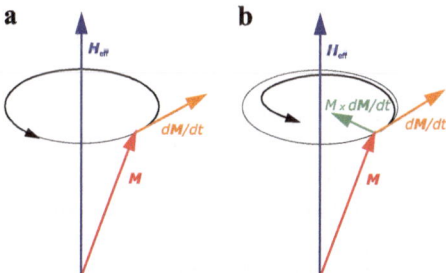

Fig. 1. (a) Illustrates the Landau-Lifschitz equation. The magnetization (red) precesses around the local effective field (blue) in analogy to a spinning top. (b) The Gilbert damping term leads to an additional component (green) moving $\boldsymbol{M}$ towards the low energy equilibrium state of the system

2.2 Damping

One may imagine that in nature this precessional motion does not hold up forever, otherwise one would not be able to use simple magnetic devices such as a compass. Thus, in the second part of the equation of motion a simple expression for energy dissipation must be introduced. This phenomenological *damping term* describes how the magnetization reaches its energetic minimum while performing a precessional motion. The energy is dissipated and the magnetization gradually becomes aligned with the effective field. This damping may be described by the Gilbert term:

$$\frac{\alpha}{M}\left(M \times \frac{d}{dt}M\right) \tag{6}$$

with α being a dimension-less constant. Now with the two parts (4) and (6) the Landau-Lifshitz equation reads:

$$\frac{d}{dt}M = -\gamma_0 M \times H + \frac{\alpha}{M}\left(M \times \frac{d}{dt}M\right). \tag{7}$$

In this form the Landau-Lifshitz equation is known as the Landau-Lifshitz-Gilbert equation. For 3d transition metals the intrinsic damping parameter α is as small as 0.003 and thus even for small angle excitations the magnetization vector performs many revolutions before it reaches its energetic minimum. Equation (7) describes the motion of the magnetization in an external magnetic field. In its simplest application one can consider the motion of the magnetization of a complex material as a macrospin subjected to an externally applied magnetic field. This of course is not the full story, but for simple experiments this approach may prove to give valuable insight.

3 The Energy Landscape

However, before examining simple experiments in the macrospin model we have to perform one refinement. So far we have considered the magnetization to be subjected to an external field only. Let us think back to our introductory physics course and imagine two magnetic needles (dipoles) coupled by their own stray fields. If one aligns them with their long axis perpendicular to each other (a high energy state) and subsequently lets them go, a complicated motion develops and after quite some time they will align with their long axis aligned (low energy state). The same holds on a microscopic level and one immediately realizes that the Ansatz of an external field acting on the magnetization alone is not an adequate one. In general, one needs to take all internal and external fields into account. The most important field terms will be summarized in the following section.

For this purpose we will use the continuum representation of the magnetization and describe all magnetic quantities on length scales comparable to the exchange length (typically some 1–5 nm). This is justified if all relevant vector quantities vary slowly in space. For most magnetic objects this assumption holds, but one has to be careful when applying micromagnetics to objects with length scales comparable to the exchange length.

In this section only the four most relevant energy terms (Zeeman energy, stray field energy, exchange energy and anisotropy energy) will be discussed. Other magnetic energies, such as for example the magneto-elastic energy, are omitted in the following, as they are irrelevant for the material used within this chapter.

The basic variable is the magnetization $M(r) = M_s m(r)$, its norm being considered constant, only the direction of the magnetization $m(r)$ changing with the constraint $m^2 = 1$, and M_s being the constant saturation magnetization.

3.1 Zeeman Energy

As shown above, the Zeeman energy only appears upon applying an external magnetic field. It is given by

$$E_H = -\mu_0 M_s \int_{\text{sample}} H_{ex} \cdot m \, dV \tag{8}$$

and is important to account for the effects of the external magnetic excitation.

3.2 Exchange Energy

The tendency of ferromagnetic spins to align parallel is described by the exchange energy, representing an energy penalty for non-parallel alignment of neighboring spins. This can be written in the so-called "stiffness" expression[1]

$$E_x = A \int_{\text{sample}} (\text{grad } m)^2 \, dV \tag{9}$$

with the exchange constant A being a material constant.

3.3 Anisotropy Energy

There is a wide variety of possible anisotropy energies that are characteristic for different ferromagnetic materials. In this chapter we will consider poly-cristalline Co with a sizeable uniaxial anisotropy field of $168 \frac{\text{kA}}{\text{m}}$ in the plane of the film. A uniaxial anisotropy term can be accounted for in the following way:

$$E_{Ku} = \int_{\text{sample}} \varepsilon_{Ku} \, dV \ , \quad \varepsilon_{Ku} = K_u \sin^2 \vartheta \tag{10}$$

where K_u describes the strength of the uniaxial anisotropy and ϑ being the angle between the magnetization direction and the easy axis, the easy axis being aligned along the x-axis in our case.

---

[1] This formula can be derived from the standard spin-spin interaction energy $E \sim \sum_{i,j} S_i \cdot S_j$.

3.4 Stray Field Energy

The stray field energy is caused by the interaction of the magnetization with the magnetic field of the sample itself. Starting from Maxwell's law div $\boldsymbol{B} = \mu_0 \operatorname{div}(\boldsymbol{H} + \boldsymbol{M}) = 0$, the stray field $\boldsymbol{H}_d$ can be defined as

$$\operatorname{div} \boldsymbol{H}_d = -\operatorname{div}(\boldsymbol{M}) . \tag{11}$$

Accordingly, the stray field energy can be written as

$$E_d = -\frac{\mu_0}{2} \int_{\text{sample}} \boldsymbol{H}_d \cdot \boldsymbol{M} \, dV . \tag{12}$$

Analogous to electrostatics, volume and surface "charges"

$$\rho = -\operatorname{div} \boldsymbol{m} \text{ and } \sigma_s = \boldsymbol{m} \cdot \boldsymbol{n} \tag{13}$$

can be defined, $\boldsymbol{n}$ being the (outward directed) surface normal. With this definition and again analogous to electrostatics, the potential of the stray field is given as:

$$\Phi_d(\boldsymbol{r}) = \frac{M_s}{4\pi} \left[\int_{\text{sample}} \frac{\rho(\boldsymbol{r}')}{|\boldsymbol{r} - \boldsymbol{r}'|} dV' + \int_{\text{surface}} \frac{\sigma_s(\boldsymbol{r}')}{|\boldsymbol{r} - \boldsymbol{r}'|} dS' \right] . \tag{14}$$

If the explicit form of the stray field was required, this could be derived by $\boldsymbol{H}_d(\boldsymbol{r}) = -\operatorname{grad}\Phi_d(\boldsymbol{r})$. The stray field energy can be derived by integration, yielding

$$E_d = M_s \left[\int_{\text{sample}} \rho(\boldsymbol{r}) \Phi_d(\boldsymbol{r}) \, dV + \int_{\text{surface}} \sigma_s(\boldsymbol{r}) \Phi_d(\boldsymbol{r}) \, dS \right] . \tag{15}$$

Note that, in contrast to the exchange energy, the stray field energy cannot be evaluated locally but rather has to be calculated by performing an integration over the complete sample for every point $\boldsymbol{r}$. This is the main reason why micromagnetic simulations take so much computing time.

In contrast to the exchange energy, the stray field energy is reduced when the mean value of the magnetic moment of a sample disappears due to changes of its magnetic configuration. This is the reason why closure domain states are energetically favored compared to single domain states for samples in the mesoscopic size range.

For homogeneously magnetized ellipsoids the stray field energy takes a particularly simple form. In this case, the demagnetizing tensor has only diagonal elements N_x, N_y and N_z and the stray field $\boldsymbol{H}_d$ may be written as:

$$\boldsymbol{H}_d = -M_s(N_x m_x \boldsymbol{e}_x + N_y m_y \boldsymbol{e}_y + N_z m_z \boldsymbol{e}_z) \tag{16}$$

For an infinite thin plate lying in the x-y-plane: $N_x = N_y = 0, N_z = 1$.

What remains left to do is the calculation of the effective field entering the Landau-Lifshitz equation, thus one needs to calculate the derivative of the total energy density with respect to m. This procedure results in:

$$H_{eff} = -\frac{1}{\mu_0 M_s}\frac{\delta \varepsilon_{tot}}{\delta m} \tag{17}$$

$$H_{eff} = \left[\frac{2A}{\mu_0 M_s}\nabla^2 m - \frac{1}{\mu_0 M_s}\mathrm{grad}_m\varepsilon_{an}(m)\right] + H_{ext} + H_d. \tag{18}$$

This is the relevant field that allows to describe the motion of the magnetization according to the Landau-Lifshitz equation in a complex magnetic sample. The complications that arise are clear. As soon as the magnetization vector in any place in a given sample starts to deviate from its ground state, it influences the total magnetic field acting on the magnetization in other locations within the sample. Thus, the energy functional and the effective field are dynamic ones. They have to be recalculated for every time step of the motion of the magnetization. However, for simple models even a simplified macrospin model with no interaction between different magnetic cells within the sample may explain the basic underlying physics. This will be the theme for the following experimental chapters.

3.5 Boundary Conditions

Just a short comment on the boundary conditions. In general the boundary conditions of the Maxwell equations for B and H at the boundaries of a ferromagnetic sample have to be fulfilled. The dynamical boundary conditions using the Landau-Lifshitz equation impose additional conditions. A thorough discussion of some resulting implications may be found in [39]. For the case of surface anisotropy and exchange interaction the boundary conditions proposed by Rado and Wertmann [40] are often used.

$$\frac{\partial m_i}{\partial \xi} + Dm_i = 0. \tag{19}$$

In this case, pinning of the dynamic magnetization is described by the parameter $D = \frac{K_s}{A}$ that balances the strength of surface anisotropy K_s with the exchange stiffness constant A. It describes the behavior of the dynamical magnetization component m_i across the surface normal ξ at the boundaries. Without anisotropy, it reduces to $\frac{\partial m_i}{\partial \xi} = 0$. When we include the long-ranged dipolar interaction, we obtain different boundary conditions for samples of different sizes and geometries, depending on the class of magnetic excitations like exchange or dipolar dominated spin waves [41].

4 The SLAC Experiment: A Simple Experiment Explained by the Macrospin Model

Let us now consider a very simple experiment. A ferromagnetic thin film with an easy magnetization axis perpendicular to the film plane will be described in the macrospin model, see Fig. 2a. We neglect the exchange interaction and consider each cell of the film as a macrospin that interacts with an external field pulse via the Zeeman term. The dipolar interaction will be taken into account in a "global" way only. The thin film represents an ellipsoid with $N_x = N_y = 0$, $N_z = 1$. As the easy anisotropy axis is pointing perpendicular to the film plane, its anisotropy field and the effective stray field are pointing along the same axis with the same angular dependence $\propto cos(\vartheta)$, where ϑ is the angle between the magnetization and the film normal. Thus, the total magnetic field $\boldsymbol{H}_{eff}$ acting on the individual macrospins is the sum of the effective anisotropy field and the external field. Let us now assume we apply a short, but strong field pulse along the y-axis in the plane of the film. Due to the rotational symmetry of the problem the field pulse can in fact be applied along any direction in the plane of the film. The total field vector is now pointing at some angle θ with respect to the film normal and the magnetization will begin to precess about this field direction. During the precession process the total internal field decreases as $cos(\vartheta)$. If the externally applied field pulse is strong enough, θ may overcome 45 degrees and the trajectory of the precessional motion may cross the plane of the film. If the field pulse is turned off just when the magnetization has crossed the plane, the magnetization will subsequently relax into the nearest energetic minimum, in this case the opposite magnetization direction, see the trajectories in Fig. 3a.

It is easily understood that for precessional switching to occur (in the limit of low damping) two requirements have to be met: (i) the field pulse must be strong enough to overcome the energy barrier ΔE imposed by the

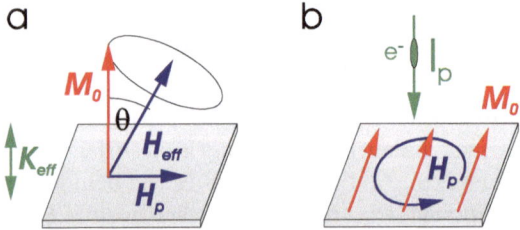

Fig. 2. (a) A magnetic thin film with strong perpendicular uniaxial anisotropy is subjected to an ultra short in-plane field pulse. While the field pulse is present, the effective total field $\boldsymbol{H}_{eff}$ is tilted away from the film normal and $\boldsymbol{M}$ starts to precess. In this geometry the initial field pulse at $t = 0$ is always perpendicular to $\boldsymbol{M}$. (b) When the uniaxial magnetization lies in the plane of the film, $\boldsymbol{M}$ and the pulse field $\boldsymbol{H}_p$ encompass all possible angles. Thus the torque acting on $\boldsymbol{M}$ can take any value between zero and the maximum torque

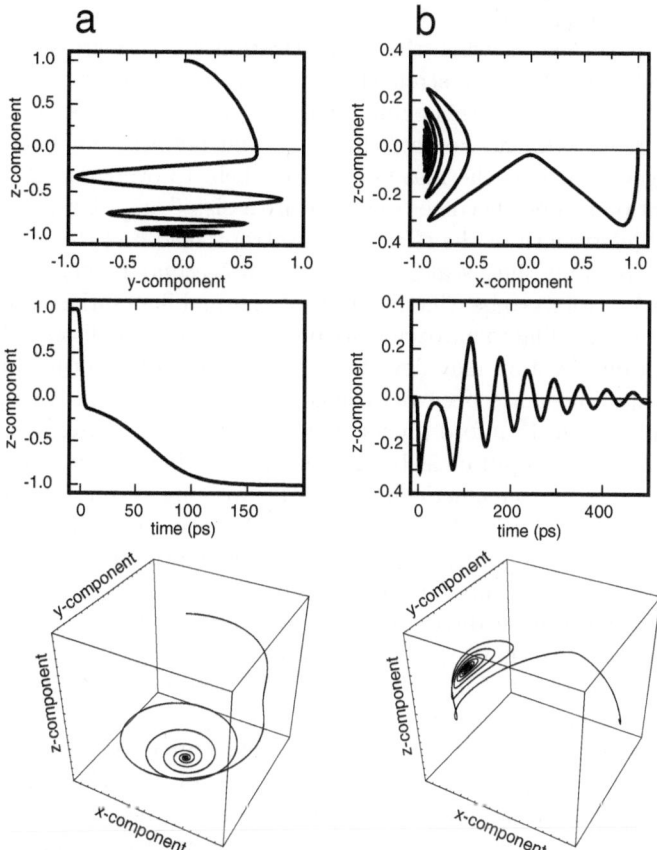

Fig. 3. (a) Trajectories of M for a magnetic film with perpendicular easy magnetization direction when subjected to a Gaussian field pulse in time ($\sigma_t = 2\,\mathrm{ps}$). The strength of the field pulse is adjusted so that the magnetization has just crossed the plane of the film after the field pulse has passed. In the upper panel a "side-view" of the precessing magnetization is shown. In the middle panel we show the evolution of the perpendicular component of M as a function of time. It can be seen that after about 5 ps the magnetization has reached the point of no return, it subsequently relaxes into the new equilibrium. The lower panel shows the full trajectory in space. (b) Trajectories of M for a magnetic film with in-plane easy magnetization direction when subjected to a Gaussian field pulse in time ($\sigma_t = 2\,\mathrm{ps}$). The strength of the field pulse is adjusted so that switching will occur. In the upper panel a "side-view" of the precessing magnetization is shown. In the middle panel we show the evolution of the perpendicular component of M as a function of time. It can be seen that after about 50 ps the magnetization has reached the point of no return, it subsequently relaxes into the new equilibrium. The lower panel shows the full trajectory in space

effective anisotropy field and (ii) the length of the field pulse must be adjusted so that it is turned off when the magnetization has just crossed the plane. The product of field pulse strength and duration is the crucial parameter. We would like to note at this point, that this experiment is essentially a zero temperature experiment. For switching experiments performed at timescales much longer than 1 ns thermal fluctuations help to overcome ΔE and thus magnetization reversal becomes temperature assisted [37]. Here, the timescale is set by the strength of the effective anisotropy field and the gyromagnetic ratio, see (5), and time scales are in the picosecond range, too short for thermal fluctuations to be effective. In fact, thermal fluctuations are frozen at this time scale leading to a frozen distribution of initial conditions. The initial magnetization direction may deviate slightly from its perfect alignment.

This experiment has been performed at the Final Focus Test Beam section (FFTB) at the Stanford Linear Accelerator Center (SLAC). A short but strong magnetic field pulse has been applied in the hard plane of perpendicularly magnetized ferromagnetic films and the evolving magnetization patterns have been inspected using Kerr microscopy. As seen above, in this particular geometry the resulting initial reversal occurs while the field pulse is present and the magnetization subsequently, on a much longer time scale, relaxes into the easy magnetization direction given by the magnetic anisotropy. In this experiment we have changed the product of field strength and field duration by using a field pulse of fixed length, but with varying amplitude. The experimental set-up is sketched in Fig. 2. The finely focused 46.6 GeV electron beam in the FFTB is used as a magnetic field source for pulses as short as 2–6 ps and field amplitudes up to several Tesla. In each electron bunch the number of electrons was $(9.6 \pm 0.2) \cdot 10^9$. These bunches were focussed to a small spot. The Gaussian half widths of the electron beam in the x and y directions were determined to be $\sigma_x = (3.8 \pm 0.4)\,\mu m$ and $\sigma_y = (0.8 \pm 0.2)\,\mu m$ respectively. The temporal pulse lengths in the experiments could be tuned to 2, 3 and 4.4 ps. The magnetic field produced by the electron beam is computed from the current density $j(x, y, t) = n(x, y, t) \cdot e \cdot c$ by simply applying Ampere's law. Here $n(x, y, t)$ is the number of electrons at a given position and time, determined by the three dimensional Gaussian beam. Large fields up to 20 T have been reached close to the surface of the beam.

The result for a single shot experiment is shown in Fig. 4a. The domain image has been recorded in a Kerr microscope weeks after the exposure. Polycrystalline thin film multilayer samples of $10 \times [\text{Co x Å/Pt 12 Å}]$ grown at $\cong 150°\text{C}$ onto (111) textured, 200 Å thick Pt buffer layers deposited at $400°\text{C}$ onto SiN_x coated Si(100) substrates have been exposed to one electron bunch. By tuning the Co-layer thickness, the effective perpendicular anisotropy field could be tuned over a wide range and for the sample shown in Fig. 4a it was $\text{H}_{Keff} = 2552\,\text{kA/m}$ leading to an effective demagnetizing field of $\mu_0 M_s = 1.6\,\text{T}$. From the Kerr contrast we immediately see, that in the black areas the magnetization has switched its direction and the

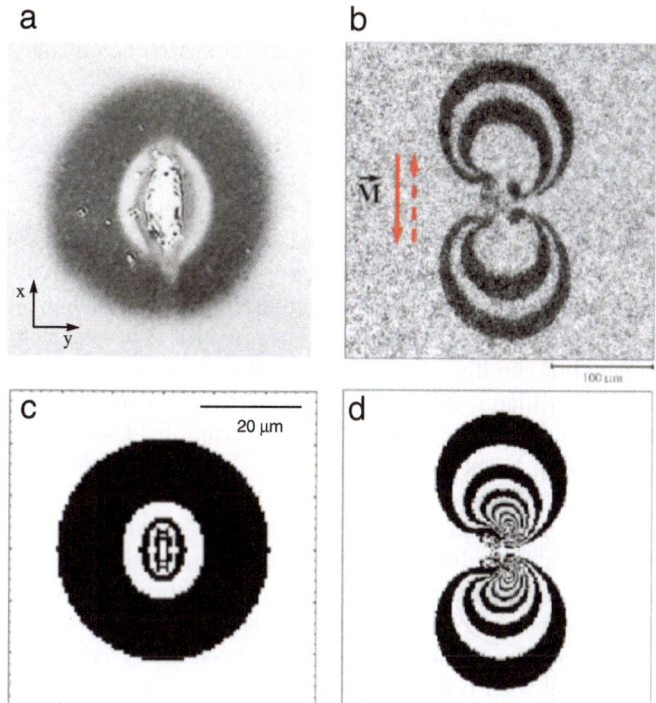

Fig. 4. (a) Kerr microscopy image of a perpendicularly magnetized magnetic thin film after exposure to one electron bunch in the FFTB. The original magnetization direction shows white contrast, the black areas are areas where the magnetization has switched to the opposite direction. The center of the sample has been destroyed by the high energy electron beam. (b) SEMPA image of an in-plane magnetized magnetic thin film after exposure to one electron bunch in the FFTB. The original magnetization direction shows white contrast and is pointing along the −x-direction, the black areas are areas where the magnetization has switched to the opposite direction. Several rings corresponding to one, two, three and four switches can be observed. (c) Macrospin simulation of the experiment of (a), excellent agreement can be found. (d) Macrospin simulation of (b). In this case only the gross features are reproduced. This is a direct consequence of the fact that for this geometry the time needed to reach the "point of no return" is much longer than in (a). Consequently, deviations from the macrospin model are more prominent

outermost contour line represents the line were the product of a field pulse with 2 ps Gaussian width and the minimum field strength required for precessional switching is reached (in this case 2300 kA/m for a Gaussian pulse shape in time). When we compare this switching pattern to the macrospin model introduced above we find excellent agreement, see Fig. 4c).

This experiment is the simplest precessional switching experiment possible. For the perpendicular geometry switching is governed by the precessional motion only, as only half a precessional period is required for switching to occur. It is obvious, that damping does not affect switching on this short time scale. Let us now use a more complicated geometry.

Imagine the magnetization is lying in the plane of the film and is pointing along the $-$x-axis due to a uniaxial anisotropy, see Fig. 2b. The stray field can again be expressed via the demagnetizing factors $N_x = N_y = 0$, $N_z = 1$. Now, the anisotropy forces the easy magnetization to point along the x-axis. The field pulse is still applied along the y-axis. As soon as the field pulse is turned on the magnetization starts to precess out of the plane of the film, thus "turning on" the demagnetizing field which would like to pull the magnetization back into the plane. In virtue of this field pointing along the normal of the film, the precessional motion picks up a component parallel to the film plane. M starts to swing towards the +x-direction. As soon as M picks up a y-component, the anisotropy field sets in and tries to pull M back towards the x-axis. A complicated motion develops, see Fig. 3b. If the initial field pulse was strong enough, the magnetization swings over to the new energetic minimum and relaxes into the +x-direction in a damped precessional motion. The driving field for this complicated motion is the demagnetizing field which helps to overcome the energy barrier, in this case given by the in-plane anisotropy. Figure 4b shows the results of this experiment. In this case the domain pattern has been analyzed using scanning electron microscopy with polarization analysis [38]. Again, white areas show regions where the magnetization is pointing along its initial $-$x-direction, while black contrast demonstrates that the magnetization has switched into the opposite +x-direction. Along a line along the x-axis and through the center of the domain pattern eight transitions can be observed corresponding to one, two, three and four reversals on either side of the location of electron beam impact. This is a consequence of the increasing field pulse strength when approaching the center. When comparing the experimental data to a macrospin simulation (Fig. 4d) using the experimentally obtained sample parameters ($\mu_0 M_s = 1.7\,\text{T}$, $H_k = 168\,\text{kA/m}$) one immediately recognizes a striking similarity. The gross features of the experiment are well reproduced within this simple model. However, two problems arise. First the effective damping parameter α introduced in the Landau-Lifshitz equation has to be adjusted to 0.02 to accommodate for the size of the pattern. This value is much too large compared to FMR values for small angle excitations. Second, the inner structure of the domain pattern is not reproduced by the simulation. Clearly, the model is oversimplified and one needs to take exchange interaction and true dipolar interactions into account. However, given the large size of the structure it is not feasible using micromagnetic simulations. It is clear that spatially resolved experiments that allow measuring the trajectory of the magnetization during the reversal process are needed to gain further insight into precessional switching phenomena [20].

5 Time-Resolved Kerr Microscopy

One of the most successful experimental approaches has been time resolved Kerr microscopy. It offers the advantage of fairly high optical resolution of 300 nm in combination with high temporal resolution limited in principle only by the laser pulse width of modern femtosecond laser systems, typically 150 fs. The two essential elements of a stroboscopic temporally and spatially resolved pump-probe experiment aimed at mapping the local time evolution of the magnetization vector are: (i) the generation of ultra-fast magnetic field pulses used to tip the magnetization away from its equilibrium position. This pulse sets the initial conditions under which M evolves. (ii) The evolution of the precessional motion of M is monitored by a probe pulse aimed at measuring one or more components of M after some variable time delay Δt. Obviously, pump and probe pulses must be synchronized to allow stroboscopic experiments. A single laser pulse is not sufficient to get a high enough signal to noise ratio. Therefore, a stroboscopic technique is needed to accumulate the signal of many pulses. At this point it is important to note that in stroboscopic experiments one is limited – apart from very rare cases [42] – to measure the repetitive part of the magnetic signal. A typical set-up is described in Fig. 5. In our setup, the pulsed light source is a commercial Ti:Sapphire laser system. By frequency doubling using a LBO crystal, we obtain laser pulses shorter than 150 fs at a wavelength of 400 nm. This beam is focused onto the sample using a polarization conserving optical microscope. To pump the sample, a photoconductive switch is used to generate the magnetic field pulse. The 800 nm dump beam of the frequency doubler is focused onto the switch. To control the delay between the pump- and the probe beam, the pump beam is delayed by using a motorized stage as an optical delay line.

The sample is mounted on a piezoelectric table which allows to move the sample under the focused beam and to obtain images of the local magnetization by scanning at a fixed time Δt after the excitation. The spatial resolution is given by the quality of the beam and the numerical aperture of the objective lens. In our case, we reach ≈ 300 nm for the polar Kerr component. We use time resolved Kerr microscopy. When linearly polarized light is reflected from a ferromagnetic surface, both its intensity and its polarization are changed by the magnetization of the sample: this is known as the Kerr effect. The Kerr effect offers the possibility to measure the magnetization in an external field. As the penetration depth of the light is of the order of 10–20 nm in metals, it is possible to investigate buried magnetically active films capped by a protection layer. To measure the perpendicular component of the magnetization one uses the polar Kerr effect. In this geometry, the laser beam impinges perpendicularly onto the surface of the sample. Using a polarization sensitive detector, the polarization of the reflected light is detected. For symmetry reasons, the rotation of the polarization only depends on the out-of-plane component of the magnetization. In our setup, we use

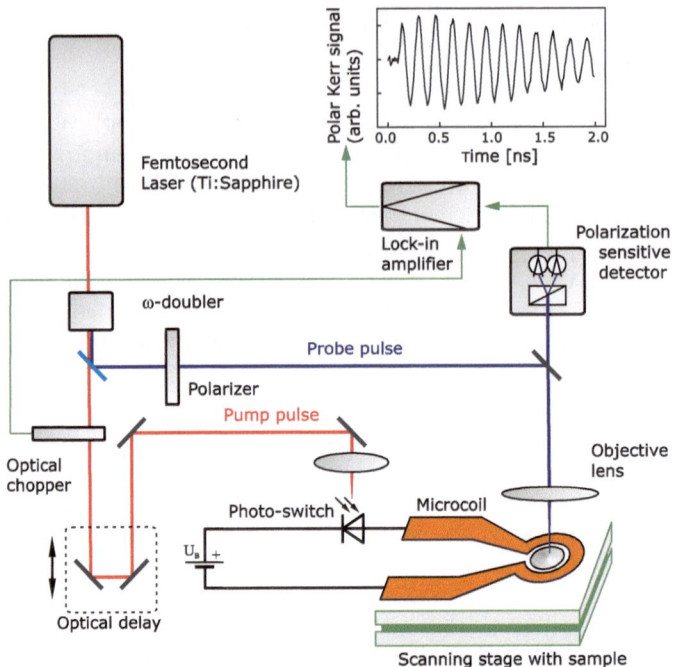

Fig. 5. Diagram of the experimental set-up. Using a photo switch, the probe pulse triggers an electric pulse that produces a magnetic field pulse in the microcoil where the sample is located. Positioning using a piezo stage in combination with the microscope objective lens provides spatial resolution of 300 nm

a Wollaston prism which splits the reflected beam into its two polarization components. The intensity of the two components is measured using balanced photodiodes. The difference of the two intensities is proportional to the Kerr rotation.

To suppress changes of the reflectivity which are not caused by the magnetization, the excitation is gated with a frequency of about 2 kHz and a lock-in amplifier picks up the signal from the photodiodes. Thus, we measure the difference of the Kerr signal with and without field pulse $(M_z(\Delta t) - M_z(\Delta t = 0))$. The detected Kerr signal is proportional to the change of the magnetization due to the excitation.

One possible way of generating a fast, local magnetic field pulse is to launch a current pulse into a microcoil or into microstriplines. Figure 6 shows a lithographically defined single turn coil surrounding a ferromagnetic disk. The microcoil is connected to a microstripline which is successively connected to the current source. The current source may be a fast electrical pulse generator or an optically controlled switch [9, 13, 14, 20]. In the latter – the so called Auston switch – a laser pulse is directed onto a GaAs crystal. The photo-generated carriers cause a current pulse to flow between two contacts evaporated onto the crystal. In this case current rise times can be as short as 1 ps, but the fall time is dominated by the recombination time of the carriers in the GaAs substrate. Recently, Geritts et al. reported of pulse shaping using a superposition of current pulses generated in two optical switches connected to the same microstripline [20]. In the case of an electronic pulse generator the shape and amplitude of the current pulses can be varied. The delay between the pump pulse and the optical probe pulse is controlled electronically, and can be as long as 1 μs with an electronic delay. The drawback of electrical pulse generators is that they have a slower rise time than Auston switches: State of the art pulse generators provide rise and fall times of the order of 50 ps, while Auston switches have a rise time of a few ps. Notice that in both cases the current pulse has to be brought from the current source to the location of the magnetic element through electrical waveguides, a process which is typically accompanied by distortions, damping and unwanted reflections. Although the photoconductive switch can be placed close to the micro-coil, pulse shape and amplitude are often unknown.

5.1 Pulsed Precessional Motion

In this section we will describe a time resolved Kerr microscopy experiment and will explain the magnetic response to a tipping field pulse of a particular magnetic configuration: the magnetic vortex state. Apart from uniformly magnetized ellipsoids this is one of the simplest magnetic configurations with inhomogeneous internal spin structure. A typical element is sketched in Fig. 6. In the ground state the magnetic element exhibits a flux closure vortex configuration[2]. It is thus natural to describe the magnetization in cylindrical coordinates $M(r,t) = M(\rho, \varphi, z, t)$. This state is now disturbed by a short, but weak – typical field strengths do not exceed 50 Oe – magnetic

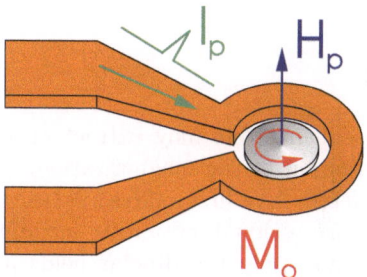

Fig. 6. Sketch of the experimental geometry. A short perpendicular tipping field pulse excites the magnetization of the 6 micron diameter Co disk which in its ground state is in a flux closure state that can be approximated by a vortex state

---

[2] In reality the Co disk exhibits a flux-closure multi-domain state as observed using SEMPA. However, as a first approach we assume a perfect vortex state.

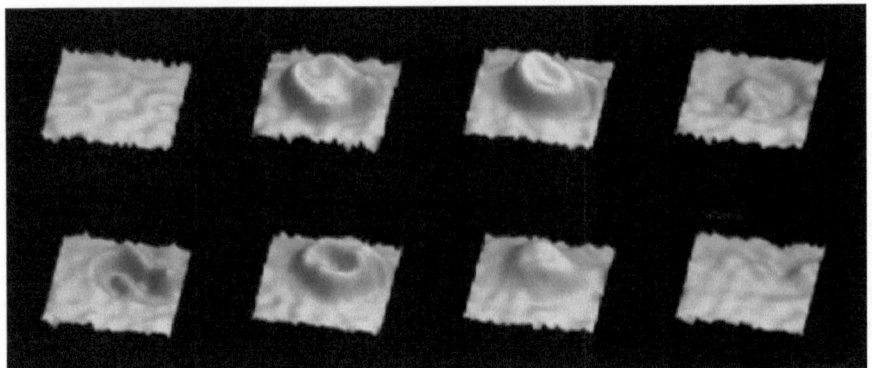

Fig. 7. Selection of polar Kerr effect images of the 6 micron Co disk recorded at a variable time delay Δt after application of the tipping field pulse. For the case of the polar component the images correspond to the dynamic perpendicular component of the magnetization as in the ground state the magnetization lies in the plane of the disk. Images are taken at $\Delta t = 20$, 60, 100, 140, 180, 220, 260 and 300 ps

tipping pulse $\boldsymbol{H}_p$ pointing along the z-direction. The precessional motion of the spins in this highly symmetric problem is excited and one is left to observe the response on a macroscopic length scale given by our optical experiment with 300 nm spatial resolution [13]. Polar Kerr images are obtained as a function of time delay between application of the tipping pulse and the probe pulse in a stroboscopic experiment. In the linear limit, it is sufficient to consider only one component in order to have the full knowledge of the dynamical response. We have plotted here only the component that could be detected with the greatest precision, namely the one perpendicular to the disk (M_z). A movie of the z-component of the magnetization is recorded with a 10 ps frame rate. Figure 7 shows a selection of images for a 6 micron diameter Co-disk.

5.2 Uniform Model

From such an image sequence one may extract an overall periodicity corresponding to the precession of the magnetization. A value of the order of $f = 6$ GHz can be extracted. As a first analysis we use this frequency to investigate the origin of the precessional motion. In a simple approach we neglect exchange interactions and apply the dipolar fields in a local "mean field" approach. We can write down the most simple form of the Landau-Lifshitz equation by considering only the local demagnetizing factors N_r and N_z in the linearized model: $\dot{M}_r = \gamma_o N_z M_z M_\varphi$, $\dot{M}_z = -\gamma_o N_r M_r M_\varphi$, $\dot{M}_\varphi = 0$. For a disk with the geometrical dimensions of the one in Fig. 7, $N_r = 0.17$ and $N_z = (4\pi - 0.17)$ [43]. By using an oscillatory Ansatz one indeed arrives at a frequency of about 6 GHz.

We have used only the dipolar interaction for a sample with our aspect ratio and can determine the frequency within experimental errors. We conclude that dipolar interactions are responsible for the observed precessional motion. The spin motion, however, is not uniform. By analyzing the time evolution at different positions on the sample one can extract a position dependence of the apparent oscillation frequency [44]. It is clear that the simple mean field model cannot explain the data in detail. We assume in the following that dipolar interactions are responsible for the observed motion of M, but a superposition of various dipolar modes has to be taken into account.

5.3 Dipolar Model

The idealized magnetic element is a thin circular platelet with small aspect ratio d/R. Its ground state is the closed flux vortex state which can be best described in circular coordinates: $M_o(r, \varphi, z) = M_s \cdot (0, 1, 0)$.

We consider the Landau-Lifshitz Equation (4) of motion where the functional derivative of the energy density represents the effective field $H_{eff} = H_{ext} + H_d$ acting onto M. In a "back of the envelope" model we consider exclusively the external pulse field H_{ext} along z direction and the dipolar interaction H_d. For the circularly symmetric sample in the flux closure state, no anisotropies are introduced. Generally, exchange interaction tends to align neighboring spins and establishes long range ferromagnetic order. However, the circulating flux configuration is the ground state for the sample and can be explained by dipolar interaction. As seen in the section above, dipolar interactions seem to be responsible for the observed precessional frequency. In addition, micromagnetic simulations have been conducted to explore the influence of exchange interactions and show that they will become relevant for smaller samples and high mode numbers. In the size regime we examined, no deviations of the modal frequencies from the dipolar dominated model are observed. Therefore, exchange interaction can be neglected[3] for the size range of our platelets, which is much bigger than the exchange length and the optical resolution of the experiment.

Again, the formulation of virtual magnetic charges is used to describe the total dipolar energy inside the sample. The magnetostatic equations specify that both the rotation of the magnetic field H_d and the divergence of the magnetic flux density B must vanish. Thus,

$$\nabla \cdot B = \mu_0 \nabla \cdot (H_d + M) = 0 . \tag{20}$$

Despite the difference in the vector properties of magnetic and electric fields, a "magnetic poisson equation" can be formulated as

---

[3] For disks smaller than one micron and higher mode numbers, the stronger effect of exchange interaction leads to deviations between theory and micromagnetic simulation.

$$\nabla^2 \Phi_d = -\rho \tag{21}$$

with the magnetic charge $\rho = \nabla \cdot \boldsymbol{m}$ and a potential $\Phi_d(\boldsymbol{r}) = -\frac{M_s}{4\pi} \int \frac{\rho(\boldsymbol{r}')}{|\boldsymbol{r}-\boldsymbol{r}'|} \, d\boldsymbol{r}'$. This formulation is analogous to electro-statics and is consistent with (20). It can be used to calculate the stray-field

$$\boldsymbol{H}_d = -\nabla \Phi_d \,. \tag{22}$$

This field depends on $\boldsymbol{M}$ and therefore changes in time during the motion. It is evaluated by using the gradient in cylindrical coordinates. The relevant component is the radial field $H_r[M_r]$,

$$H_r[M_r] = \frac{d}{4\pi} \frac{\partial}{\partial r} \int_{disk} \frac{1}{|\boldsymbol{r}-\boldsymbol{r}'|} \frac{1}{r'} \frac{\partial}{\partial r'} (r' M_{r'}) \, dr' \tag{23}$$

which is a linear functional of M_r. We must add a perpendicular component to the dipolar field, the standard demagnetizing field $-M_z$ pointing in the direction opposite to M_z. This is the field of a perpendicularly magnetized plate with infinite radius. With all relevant fields specified, we obtain the system of coupled linear equations

$$\frac{\partial M_r}{\partial t} = -\gamma_o M_s (H_{ext} - M_z) \tag{24}$$

$$\frac{\partial M_\varphi}{\partial t} = 0 \tag{25}$$

$$\frac{\partial M_z}{\partial t} = +\gamma_o M_s H_r[M_r] \,. \tag{26}$$

Equation (25) states that we are considering only small deviations from the ground state[4]. Generally, the Landau-Lifschitz (LL) equations couple different components. By taking the second derivative of (24) and inserting in (26), the linearized LL equations can be decoupled to

$$\frac{\partial^2 M_r}{\partial t^2} = -\gamma_o M_s \frac{\partial H_{ext}}{\partial t} + \gamma_o^2 (M_s)^2 H_r[M_r] \tag{27}$$

$$M_z(r,t) = \frac{1}{\gamma_o M_s} \left[\sum_i \dot{c}_i(t) M_r^i(r) + \gamma_o M_s H_{ext}(\boldsymbol{r},t) \right] \,. \tag{28}$$

Equation (27) shows explicitly that the problem cannot be reduced to a standard wave equation. Equation (28) allows M_z to be straightforwardly calculated once M_r is known. We seek a solution of (27) with the separation Ansatz $M_r(r,t) = \sum_i c_i(t) M_r^i(r)$. The radial functions $M_r^i(r)$ are the solutions of the eigenvalue equation

---

[4] We estimate the deviation from the ground-state configuration to be smaller than 5%, so that the component of $\boldsymbol{M}$ along φ is assumed to remain equal to the ground-state value $M_\varphi = M_{o\,\varphi} = M_s$ for all times and the derivatives along the φ direction are neglected to lowest order.

$$H_r[M_r^i] = -N_r^i M_r^i \, . \tag{29}$$

Inserting the separation Ansatz in (27) leads to a set of decoupled ordinary differential equations for the coefficient c_i

$$\ddot{c}_i + \omega_i^2 c_i = -\gamma_o M_s \Big(H_{ext}(r), M_r^i(r) \Big) \dot{H}_{ext}(t) \tag{30}$$

where the eigenfrequencies ω_i are related to the sought-for eigenvalues N_r^i by the relation

$$\omega_i^2 = N_r^i \gamma_o^2 M_s^2 \, . \tag{31}$$

We use the scalar product $(a, b) = \int_0^R a(r) \, b(r) 2\pi r dr$ and the external field pulse is written as $H_{ext}(t) H_{ext}(r)$. Both, t- and r-dependence are known approximately. As (30) is the equation of motion of a classical undamped forced harmonic oscillator, it can be solved exactly, provided eigenmodes and eigenvalues are known, so that the coefficients c_i and thus $M_z(r, t)$ can be calculated analytically:

$$c_i = -\gamma_o M_s \Big(H_{ext}(r), M_r^i(r) \Big) \int_{-\infty}^{\infty} H_{ext}(\tau) \, G(t, \tau) \, d\tau \tag{32}$$

with the Greens function $G(t, \tau) = \frac{1}{\omega_i} \sin(\omega_i(t - \tau))$.

The ideal case of $\boldsymbol{H}_{ext}(\boldsymbol{r}, t) = \boldsymbol{e}_z H_o \, \delta(t)$ can be solved at a glance and produces a solution equivalent to imposing a finite value to M_r at $t = 0$. This is the "initial condition" scenario referred to as "pulsed precessional motion". When we consider only this most simple case, we can insert (29) and $H_{ext}(t > 0) = 0$ into the LL equations and search for oscillating solutions $M_r, M_z \sim e^{i\omega t}$. By multiplying (24) by (26) we directly find (31).

The key elements of this problem are the eigenmodes $M_r^i(r)$ which are the solutions of the eigenvalue equation. To determine them we first notice that the integral in (23) diverges for $r = 0$ unless $M_r |_{r=0}$ is zero. This establishes the first boundary condition: the radial component of the magnetization must vanish in the center of the disk where in reality the magnetization is pinned by the vortex. Next, we notice that the exact solution of the eigenvalue equation for a disk with infinite radius is

$$N_r = \frac{1}{2} \, d \, k_r \tag{33}$$

and $M_r \propto J_1(k_r r)$, J_1 being the 1-st order Bessel function and k_r being the in-plane radial wave vector labelling the low energy excitations with frequency $\omega \propto \sqrt{k_r}$. When the disk has a finite radius, $H_r[M_r]$ contains a contribution arising from the magnetic charge ρ building up at $r = R$ in virtue of abrupt change of M_r from $M_r(R)$ to zero. This contribution diverges unless $M_r |_{R=0} = 0$. This establishes the second boundary condition.[5] Thus, within

---

[5] Of course, magnetic fields are, strictly speaking, only singular at boundaries in the two-dimensional limit which we have adopted here. In practice, bound-

our two dimensional model, the appearance, during the motion, of a finite radial component M_r at the center of the disk or at the boundary $r = R$ is associated with an infinite magnetostatic energy, so that pinning $M_r \mid_{0,R} = 0$ must be introduced to avoid this divergence. By virtue of the vanishing of M_r at $r = R$, the operator $H_r[M_r]$ defined on a disk with finite radius becomes an hermitic one. In the spirit of the Ritz variational principle, $J_1(k_r)$ are "good" eigenfunctions for finite R as well, provided k_r is chosen to fulfill the boundary condition $J_1(k_r) \mid_{r=R} = 0$. This produces a discrete set of eigenvalues N_r^i and a complete orthonormal basis set M_r^i on the disk.

For given time and space dependence of the external field pulse the solution of the problem becomes clear. What is left to do is to map the field pulse onto the set of eigenfunctions to find the correct time dependent expansion coefficients c_i by using (32). Once these are found, the response of the magnetization can be found by superimposing the weighted eigenfunctions. The result of this procedure is shown in Fig. 8. We have calculated the external field pulse $H_{ext}(t)H_{ext}(r)$ from the geometry of our coil and from an estimate of fall and rise times (12 ps rise time and 150 ps fall time). Figure 8 shows the time evolution of M_z as a function of position x. The top image shows the data and the bottom shows the calculated values using 10 eigenfunctions. Notice, that the picture obtained after superposition of the first six modes is essentially the same. As the field pulse has a finite band width and the higher modes have a smaller projection onto the quite uniform external field pulse, their contribution to the precessional motion of M is negligible. To check for the accuracy of the variational eigenvalues, the eigenvalue problem was solved numerically within a trial space consisting of 15 basis functions, see [24].

6 Connection to Synchrotron Based Experiments

Modern synchrotron sources may bring some advanced to the field of spatio-temporal magnetization dynamics. For synchrotron methods such as transmission X-ray microscopy (TXM) or x-ray photo electron emission microscopy (XPEEM) to be successful two criteria must be met: (i) the time resolution must be around 10 ps and (ii) the spatial resolution must by far surpass the spatial resolution of 300 nm of Ti:Sapphire laser experiments to justify experimental complications. Modern sources fulfill the latter easily. The resolution of state of the art instruments lies in the range of 20–50 nm. The first criteria is not met in most synchrotron sources today. The typical pulse width of the x-ray flashes produced in third generation sources is of the order of 50–70 ps. However, special filling patterns may allow for shorter

---

aries have a finite thickness and although the fields might become large they remain finite at two-dimensional surfaces. In the present case, $H_\rho \mid_{z=0,\rho \to R^-} \to -2\pi M_\rho(R))$

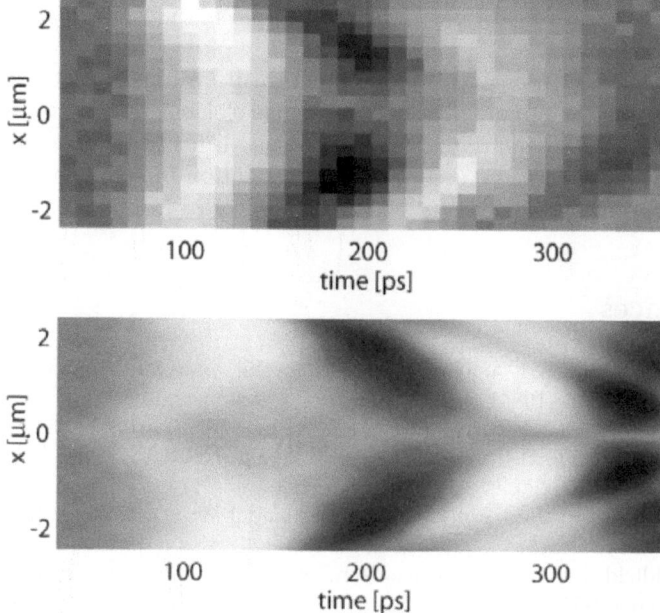

Fig. 8. *Top*: Plot of the perpendicular component of M along the x-direction crossing the center of the disk. The line scan is plotted as a function of time elapsed after application of the tipping field pulse. White corresponds to maximum positive, black to maximum negative excursion of M. *Bottom*: Calculated response of the magnetization to the tipping field pulse. The result has been obtained by superimposing 10 eigenmodes and using an external field pulse with 12 ps rise time and 150 ps decay time. Figure adapted from [24]

pulses at the expense of the average current in the ring. In the so called low alpha mode at BESSY II pulse widths of the order of 10 ps have been achieved. The near future will show if laser based slicing techniques may become relevant for magnetization dynamics. In any case, synchrotron based techniques offer one bonus: using the element specificity of x-ray absorption it is possible to distinguish magnetization dynamics in for example trilayer systems.

Pioneering experiments have been performed by Sirotti and co-workers at LURE based on photoemission techniques [45]. Due to the time resolution of the storage ring the experiments were limited to the nanosecond range. At the ESRF first experiment using the element specificity of x-ray magnetic dichroism (XMCD) techniques have been performed in spin valve systems [46]. Stoll et al. [27] and Choe et al. [26] reported of the first high resolution experiments in the precessional regime using XMCD based techniques. While Stoll et al. observed the dynamics in the walls of micron sized permalloy squares after a perpendicular field pulse excitation, Choe et al. used an in-plane excitation to investigated the gyrotropic motion of the vortex core in similar samples.

In the near future several experiments will be finalized at various light sources giving the opportunity to investigate element specific magnetization dynamics with high spatial resolution. This will for example allow to investigate the coupling of the precessing magnetization in spin valve stacks or in high resolution experiments to observe the effect of the boundaries directly.

References

1. M. B. Agranat, S. I. Ashikov, A. B. Gronovskii, and G. I. Rukmann: Sov. Phys. JETP, **59**, 804 (1984)
2. A. Vaterlaus, T. Beutler, and F. Meier: Phys. Rev. Lett, **67**, 3314 (1991)
3. A. Vaterlaus, T. Beutler, D. Guarisco, M. Lutz, and F. Meier: Phys. Rev. B, **46**, 5280 (1992)
4. E. Beaurepaire, J.-C. Merle, A. Daunois, and J.-Y. Bigot: Phys. Rev. Lett, **76**, 4250 (1996)
5. J. Hohlfeld, E. Matthias, R. Knorren, and K. H. Bennemann: Phys. Rev. Lett., **78**, 4861 (1997)
6. A. Scholl, L. Baumgarten, R. Jacquemin, and W. Eberhardt: Phys. Rev. Lett, **79**, 5146 (1997)
7. B. Koopmanns, M. van Kampen, J. T. Kohlhepp, and W. J. M. de Jonge: Phys. Rev. Lett, **85**, 844 (2000)
8. T. Kampfrath, R. G. Ulbrich, F. Leuenberger, M. Münzenberg, B. Sass, and W. Felsch: Phys. Rev. B, **65**, 104429 (2004)
9. W. K. Hiebert, A. Stankiewicz, and M. R. Freeman: Phys. Rev. Lett., **79**, 1134 (1997)
10. M. R. Freeman, W. K. Hiebert, and A. Stankiewicz: J. Appl. Phys., **83**, 6217 (1998)
11. B. C. Choi, M. Belov, W. K. Hiebert, G. E. Ballentine, and M. R. Freeman: Phys. Rev. Lett., **86**, 708 (2001)
12. M. Belov, Z. Liu, R. D. Sydora, and M. R. Freeman: Phys. Rev. B, **69**, 094414 (2004)
13. Y. Acremann, C. H. Back, M. Buess, O. Portmann, A. Vaterlaus, D. Pescia, and H. Melchior: Science, **290**, 492 (2000)
14. Y. Acremann, M. Buess, C. H. Back, M. Dumm, G. Bayreuther, and D. Pescia: Nature, **414**, 51 (2001)
15. T. J. Silva, C. S. Lee, T. M. Crawford, and C. T. Rogers: J. Appl. Phys., **85**, 7849 (1999)
16. S. Kaka and S. E. Russek: J. Appl. Phys., **87**, 6391 (2000)
17. R. H. Koch, D. W. Abraham, P. L. Trouilloud, R. A. Altman, Yu Lu, W. J. Gallagher, R. E. Scheuerlin, K. P. Roche, and S. S. P. Parkin: Phys. Rev. Lett., **81**, 4512 (1998)
18. H. W. Schumacher, C. Chappert, P. Crozat, R. C. Sousa, P. P. Freitas, J. Miltat, J. Fassbender, and B. Hillebrands: Phys. Rev. Lett., **90**, 017201 (2003)
19. H. W. Schumacher, C. Chappert, R. C. Sousa, P. P. Freitas, and J. Miltat: Phys. Rev. Let, **90**, 017204 (2003)

20. Th. Gerrits, H. A. M. van den Berg, J. Hohlfeld, L. Br, and Th. Rasing: Nature, **418**, 509 (2002)
21. R. J. Hicken, A. Barman, V. V. Kruglyak, and S. Ladak: J. Phys. D: Appl. Phys., **36**, 2183 (2003)
22. J. P. Park, P. Eames, D. M. Engebretson, J. Berezovsky, and P. A. Crowell: Phys. Rev. Lett., **89**, 277201 (2002)
23. J. P. Park, P. Eames, D. M. Engebretson, J. Berezovsky, and P. A. Crowell: Phys. Rev. B, **67**:020403(R) (2003)
24. M. Buess, Y. Acremann, A. Kashuba, C. H. Back, and D. Pescia: J. Phys.: Condens. Matter, **15**, R1093 (2003)
25. M. Buess, T. P. J. Knowles, U. Ramsperger, D. Pescia, and C. H. Back: Phys. Rev. B, **69**, 174422 (2004)
26. S.-B. Choe, Y. Acremann, A. Scholl, A. Bauer, A. Doran, J. Stöhr, and H. A. Padmore: Science, **304**, 420 (2004)
27. H. Stoll, A. Puzic, B. van Waeyenberge, P. Fischer, J. Raabe, M. Buess, T. Haug, R. Höllinger, C. H. Back, D. Weiss, and G. Denbeaux: Appl. Phys. Lett., **84**, 3328 (2004)
28. Ganping Ju, A. Vertikov, A. V. Nurmikko, C. Canady, Gang Xiao, R. F. C. Farrow, and A. Cebollada: Phys. Rev. B, **57**:R700 (1998)
29. Ganping Ju, A. V. Nurmikko, R. F. C. Farrow, R.F. Marks, M. J. Carey, and B. A. Gurney: Phys. Rev. B, **62**, 1171(2000)
30. M. van Kampen, C. Jozsa, J. T. Kohlhepp, P. LeClair, L. Lagae, W. J. M. de Jonge, and B. Koopmans: Phys. Rev. Lett., **88**, 227201 (2002)
31. M. R. Freeman and J. F. Smyth: J. Appl. Phys., **79**, 5898 (1996)
32. C. H. Back, A. Taratorin, and J. Heidmann. Appl. Phys., **86**, 3377 (1999)
33. C. H. Back, J. Heidmann, and J. McCord: IEEE Trans. Magn., **35**, 637(1999)
34. edited by B. Hillebrands and K. Ounadjela: Spin Dynamics in Confined Magnetic Structures I. Springer Verlag, Berlin (2002)
35. K. Yu. Guslienko, S. O. Demokritov, B. Hillebrands, and A. N. Slavin: Phys. Rev. B, **66**, 132402 (2002)
36. C. Bayer, J. P. Park, H. Wang, M. Yan, C. E. Campbell, and P. A. Crowell: Phys. Rev. B, **69**, 134401 (2004)
37. C. H. Back, D. Weller, J. Heidmann, D. Mauri, D. Guarisco, E. L. Garwin, and H. C. Siegmann: Phys. Rev. Lett., **81**, 3251 (1998)
38. C. H. Back, R. Allenspach, W. Weber, S. S. P. Parkin, D. Weller, E. L. Garwin, and H. C. Siegmann: Science, **285**, 864 (1999)
39. J. F. Cochran, B. Heinrich, and A. S. Arrott: Phys. Rev. B, **34**, 7788 (1986)
40. G. T. Rado and J. R. Wertmann: J. Phys. Chem. Solids, **11**, 315 (1959)
41. S. O. Democritov: J. Phys.: Condens. Matter, **15**, 2575 (2003)
42. M. R. Freeman, R. W. Hunt, and G. M. Steeves: Appl. Phys. Lett., **77**, 717 (2000)
43. R. B. Goldfarb D.-X. Chen, J. A. Brug: IEEE Trans.Magn., **27**, 3601 (1991)
44. Y. Acremann, A. Kashuba, M. Buess, D. Pescia, and C. H. Back: J. Mag. Magn. Mat., **239**, 346 (2002)
45. F. Sirotti, R. Bosshard, P. Prieto, G. Panaccione, L. Floreano, A. Jucha, J.D. Bellier, and G. Rossi: J. Appl. Phys., **83**, 1563 (1998)
46. M. Bonfim, G. Ghiringhelli, F. Montaigne, S. Pizzini, N.B. Brookes, F. Petroff, J. Vogel, J. Camarero, and A. Fontaine: Phys. Rev. Lett., **86**, 3646 (2001)

Molecular Magnetism

Stephen J. Blundell

Oxford University Department of Physics, Clarendon Laboratory, Parks Road, Oxford OX1 3PU, United Kingdom
s.blundell@physics.ox.ac.uk

Abstract. Most materials in magnetic applications are based on *inorganic* materials. Recently, however, *organic* and *molecular* materials have begun to show increasing promise. Purely organic ferromagnets, based upon nitronyl nitroxide radicals, show long range magnetic order at very low temperatures in the region of 1 K, while sulphur-based radicals show weak ferromagnetism at temperatures of up to 36 K. It is also possible to prepare molecule–based magnets in which transition-metal ions are used to provide the magnetic moment, but organic groups mediate the interactions. This strategy has produced magnetic materials with a large variety of structures, including chains, layered systems, and three-dimensional networks, some of which show ordering at room temperature and some of which have very high coercivity. Even if long range magnetic order is not achieved, interesting materials displaying the spin crossover effect may be prepared and these can have useful applications. Further magnetic materials may be obtained by constructing charge-transfer salts, which can produce metallic molecular magnets. A very exciting recent development is the preparation of single molecule magnets, which are small magnetic clusters. These materials can show macroscopic quantum tunnelling of the magnetization and may have uses as memory devices or in quantum computation applications. These systems can be powerfully studied using various experimental methods, including magnetometry, neutron scattering, muon-spin rotation and synchrotron radiation techniques.

1 Introduction

Solid-state physics, and the study of ferromagnetism in particular, has been traditionally concerned almost exclusively with the study of *inorganic* elements (e.g. Fe, Co, Ni), alloys (e.g. permalloy), and simple compounds (e.g. transition-metal oxides). However, in the last few years there has been increasing interest in various new organic materials, the building blocks of which are not atoms but molecules. The shape of these molecules and their electronic structure play a crucial rôle in determining the resulting crystallographic structure, and hence the observed physical properties. Intermolecular forces are typically weaker than interatomic forces and also rather short-range. They differ markedly therefore from the type of strong, long-range Coulombic forces found in ionic crystals. Molecular crystals are consequently rather soft and and hence their properties can be extremely tunable using

S.J. Blundell: *Molecular Magnetism*, Lect. Notes Phys. **697**, 345–373 (2006)
www.springerlink.com   © Springer-Verlag Berlin Heidelberg 2006

applied pressure. Organic materials can often possess interesting optical properties, thus allowing the possibility of devices which implement new functionality. This chapter gives a short introduction to this field of *molecular magnetism* (for more detailed accounts, see [1, 2]) and the approach that I will take is to illustrate the field by discussing various example systems.

2 Magnets based on Stable Free Radicals

Neutral organic radicals possess an odd number of electrons (and hence one unpaired electron) and are often highly chemically reactive. They can be made more stable by adding aromatic rings to delocalize the unpaired electron or by introducing bulky substituents. Crystals of neutral organic radicals exhibit paramagnetism at high temperatures, provided dimerization can be prevented. They then usually show a very small negative Weiss constant, indicating weak antiferromagnetic interactions. This is to be expected, since the intermolecular overlaps of the singly occupied molecular orbitals (SOMO) will tend to lead to a state with the character of a bonding orbital, i.e. one in which one SOMO is spin-up and the other is spin-down.

Prospects for ferromagnetic interaction between organic molecules therefore seemed limited, but in 1969 ferromagnetic interactions were identified in crystals of galvinoxyl [4]. Magnetic susceptibility measurements gave a positive Weiss constant (19 K), but the crystal undergoes a phase transition at 85 K to a low temperature phase with strong antiferromagnetic interactions [5, 6] ($J/k_B = -260$ K). In galvinoxyl, the almost planar radicals are arranged in stacks along the c-axis, and in fact the high temperature susceptibility is well reproduced by a one-dimensional ferromagnetic Heisenberg model with $J/k_B = (13 \pm 1)$ K [3] (see Fig. 1). The first-order phase transition is hysteretic (width 5 K) and the low temperature susceptibility is consistent with a singlet-triplet model with $J/k_B = (-230 \pm 20)$ K [3]. The magnetic moment on the galvinoxyl radical is due to the SOMOs. Galvinoxyl has a large spin polarization because the large intramolecular exchange means that the next-highest occupied molecular orbital (NHOMO) for ↓-spin (often denoted β) is higher in energy than the SOMO for ↑-spin (often denoted α), thus stabilizing the latter state. It is found that the intermolecular SOMO–SOMO overlap is very small (this is because the π-orbitals are spread out over the molecule, and the SOMO–SOMO overlap is positive in some regions and negative in others, so averages to a small value overall). Hence the SOMOs on neighbouring molecules are almost orthogonal.

In many organic radicals, a singlet state (i.e. antiferromagnetic alignment) is stabilized by resonance with an excited charge-transfer configuration involving intermolecular SOMO–SOMO overlap. Because this overlap is small in galvinoxyl, the triplet state (i.e. ferromagnetic alignment) is stabilized by resonance with excited charge-transfer configurations involving intermolecular interactions between the SOMO on one molecule and the fully occupied

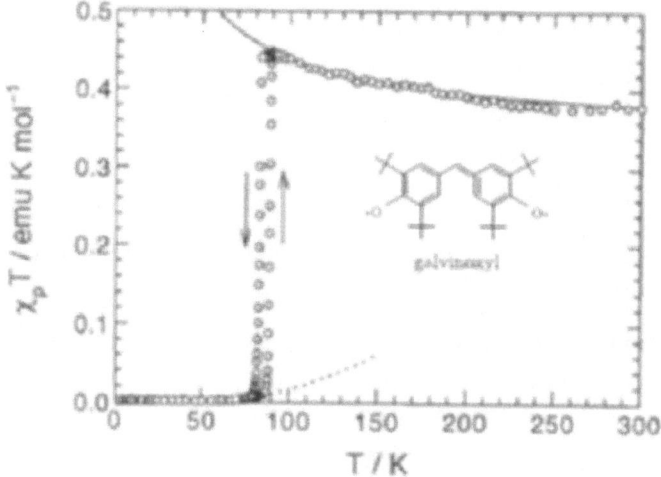

Fig. 1. Magnetic susceptibility of galvinoxyl [3]. Ferromagnetic interactions dominate above the phase transition at 85 K. The high temperature behaviour fits to a
one-dimensional ferromagnetic Heisenberg model and the low temperature behaviour fits to a singlet-triplet model

molecular orbital on the other [7]. Large spin polarization can be achieved
by promoting intramolecular exchange, and this can be helped by using elec-
tronegative atoms such as oxygen or nitrogen, which lead to a large probability of occupation by an unpaired π-electron. An extended π-system also
helps ferromagnetic interactions by promoting the SOMO–NHOMO interaction [8]. The crystal structure is also a vital ingredient; this must be such as
to minimize the intermolecular SOMO–SOMO overlaps. The overlap between
the SOMO on one molecule and the NHOMO/NLUMO (NLUMO = next-
lowest unoccupied molecular orbital) on the other can be such that regions
of positive spin density on one molecule overlap with regions of negative spin
density on the other molecule, thus promoting ferromagnetic interactions [9].
Galvinoxyl therefore fulfilled many of the conditions necessary to be an organic ferromagnet, but unfortunately only above the 85 K phase transition.

Another early candidate organic ferromagnet was tanol suberate, which
is a biradical with formula $(C_{13}H_{23}O_2NO)_2$ (for molecular structure, see
Fig. 2c). The spin density is found to be located on the NO group and almost equally shared between the oxygen and the nitrogen atoms [10]. The
magnetic susceptibility measured down to liquid helium temperatures follows
a Curie-Weiss law with a positive Curie temperature (+0.7 K) and it therefore looked like a good candidate for an organic ferromagnet. The specific
heat exhibits a λ anomaly [11] at 0.38 K, but tanol suberate was actually
found to be an antiferromagnet with a metamagnetic transition in a field of
6 mT, resulting in ferromagnetic spin alignment [12,13]. In tanol suberate the

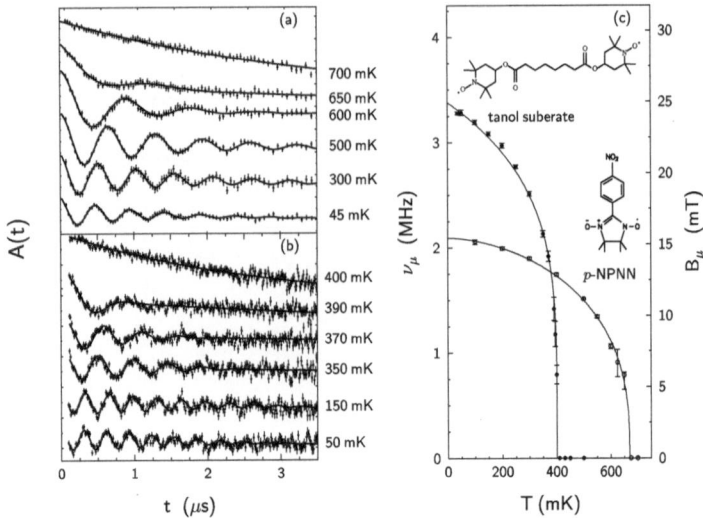

Fig. 2. Zero-field muon spin rotation frequency in (**a**) the organic ferromagnet *p*-NPNN (after [18]) and (**b**) the organic antiferromagnet tanol suberate (after [15]). In both cases the data for different temperatures are offset vertically for clarity. (**c**) Temperature dependence of the zero-field muon spin rotation frequency in *p*-NPNN and tanol suberate

molecules are arranged in sheet-like layers in which the magnetic moments are almost localized [14]. Muon-spin rotation (μSR) experiments [15] yield clear spin precession oscillations (see Fig. 2b). The temperature dependence of the precession frequency (Fig. 2c) fits to $\nu_\mu(T) = \nu_\mu(0)(1 - T/T_C)^\beta$ where $\beta = 0.22$, consistent with a two-dimensional XY magnet [16] and also with the temperature dependence of the magnetic susceptibility [17], suggesting that the ordered state is dominated by two-dimensional interactions.

Following these early attempts, organic ferromagnetism was first achieved using a member of a family of organic radicals called nitronyl nitroxides. The unpaired electron in nitronyl nitroxides is mainly distributed over the two NO moieties, although some unpaired spin density is also distributed over the rest of the molecule. The central carbon atom of the O–N–C–N–O moiety is a node of the SOMO. Nitronyl nitroxides are chemically stable but the vast majority of them do not show long range ferromagnetic order. Therefore, the discovery of long-range ferromagnetism in one of the crystal phases (the β phase) of *para*-nitrophenyl nitronyl nitroxide ($C_{13}H_{16}N_3O_4$, abbreviated to *p*-NPNN) was particularly exciting, even though the transition temperature was a disappointingly low 0.65 K [19] and only present in one of its crystal phases.

The transition to ferromagnetic order was indicated by a λ-type peak in the heat capacity and a divergence in the *ac* susceptibility at the critical temperature [19, 20]. The magnetization below T_C saturates in a very small field, demonstrating that p-NPNN is a very soft ferromagnet. μSR experiments on p-NPNN (Fig. 2a) show the development of coherent spin precession oscillations below T_C [18, 21]. The temperature dependence of the precession frequency and the corresponding local field at the muon-site is shown in Fig. 2c. This is fitted to a phenomenological functional form $\nu_\mu(T) = \nu_\mu(0)(1 - (T/T_C)^\alpha)^\beta$ yielding $\alpha = 1.7 \pm 0.4$ and $\beta = 0.36 \pm 0.05$. In this expression, the parameter β controls the behaviour near T_C, while the parameter α determines the power law at $T \ll T_C$. This is consistent with three-dimensional long range magnetic order [18, 21]. Near T_C the critical exponent is as expected for a three-dimensional Heisenberg model (one expects $\beta \approx 0.36$ in this case). At low temperatures the reduction in local field with increasing temperature is consistent with a Bloch-$T^{3/2}$ law ($\alpha = 1.5$), indicative of three dimensional spin waves.

The overlaps in all these materials which favour ferromagnetism appear to agree with the McConnell mechanism [9]: as a result of spin polarization effects, positive and negative spin-density may exist on different parts of each molecule; intermolecular exchange interactions tend to be antiferromagnetic, so if the dominant overlaps are between positive (majority) spin-density on one molecule and negative (minority) spin density on another molecule, the overall intermolecular interaction may be ferromagnetic. Though the mechanism for ferromagnetism is electronic, the low values of T_C imply that the magnetic dipolar interactions will play a rôle in contributing to the precise value of T_C. Dipolar interactions are also particularly important in determining the easy magnetization axis [18, 22]. This too depends on the crystal structure, which in turn depends on the molecular shape.

The spin density in nitronyl nitroxides can be studied using neutron scattering. Neutrons interact both with nuclei and with magnetic moments in a sample. These interactions can be separated by the use of spin-polarized neutrons. The scattered intensity $I(\mathbf{k})$ for scattering vector $\mathbf{k}$ is given by

$$I(\mathbf{k}) = |F_N(\mathbf{k})|^2 + \mathbf{P} \cdot (\mathbf{F}_M^\perp(\mathbf{k})F_N^*(\mathbf{k}) + \mathbf{F}_M^{\perp*}(\mathbf{k})F_N(\mathbf{k})) + |\mathbf{F}_M^\perp(\mathbf{k})|^2 \quad (1)$$

where $F_N(\mathbf{k}) = \sum_j b_j e^{i\mathbf{k}\cdot\mathbf{r}_j}e^{-W_j}$ is the nuclear structure factor, $\mathbf{F}_M^\perp(\mathbf{k})$ is the projection of the magnetic structure factor $\mathbf{F}_M(\mathbf{k}) = \int_{\text{cell}} \mathbf{M}(\mathbf{r})e^{i\mathbf{k}\cdot\mathbf{r}} \, d^3r$ on to the plane perpendicular to $\mathbf{k}$, $\mathbf{P}$ is the polarization of the beam, and W_j is a Debye-Waller factor for the jth atom in the unit cell at position r_j with scattering length b_j. The sample is magnetized by an external field and the intensity distribution of scattered polarized neutrons is measured. By measuring the scattered intensity with the polarization of the beam being either parallel or antiparallel to the magnetic field, the sign of the interference term in (1) can be measured, yielding an enormous sensitivity in determination of the magnetic structure factor. This can allow the mapping of the spin density distribution inside the unit cell, either by direct Fourier inversion or by maximum entropy reconstruction [24].

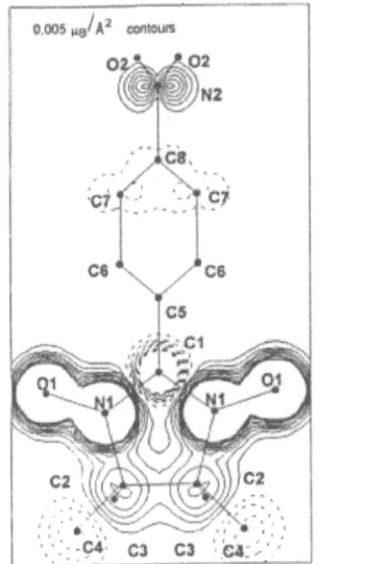

Fig. 3. Spin density map obtained from polarized neutron diffraction experiments (after [23]) for p-NPNN (molecular structure shown to the right), contours at $0.005\,\mu_B/\text{Å}^2$

Figure 3 shows the spin density map of p-NPNN [23], which reveals a strong region of positive spin density which is delocalized over the nitronyl nitroxide group, as expected, but also a region of positive spin density on the nitrogen of the nitro group (labelled N2) corresponding to a p-orbital perpendicular to the plane of the NO$_2$ group. The orthogonality of the p-orbitals of N2 and O1 favours a ferromagnetic coupling between molecules. Each molecule of p-NPNN has a magnetic moment because of the radical (unpaired) electron which is delocalized over the O–N–C–N–O moiety (as shown in Fig. 3).

Various chemical modifications can be made to p-NPNN in order to change the properties, particularly substituting different groups for the para-nitrophenyl group. However, such small changes can lead to significant modifications of the crystal structure, thereby altering the intermolecular overlaps and thus the magnetic interactions between unpaired spins on neighbouring molecules.

Forming polyradical molecules with ferromagnetic intramolecular coupling leads to high-spin molecules with potential for improved transition temperatures. A biradical molecule currently holds the transition temperature record for a fully organic ferromagnet, which was measured at $1.48\,\text{K}$ in the compound diazaadamantane dinitroxide (adamantane for short) [25] (see Fig. 4 for molecular structure). Asymmetric molecular structure in a triradical can lead to a ferrimagnetic molecular moment. An example of such

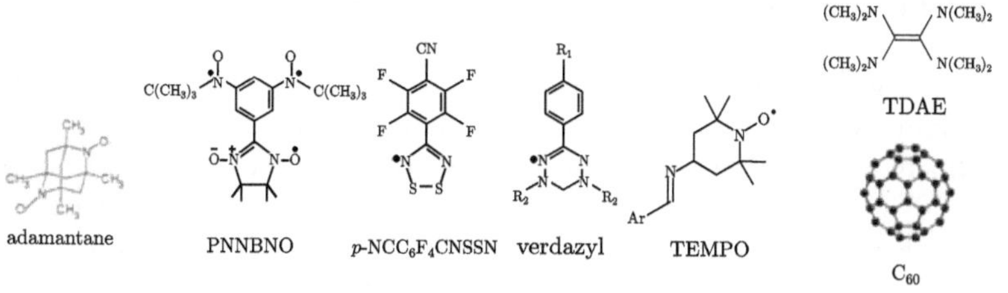

Fig. 4. Molecular structure of several molecules which are used in molecular magnets

a nitroxide based organic ferrimagnet, PNNBNO (see Fig. 4 for molecular structure), has recently been prepared [26] with a three-dimensional phase transition at 0.28 K. The molecular triradical has two spins ferromagnetic coupled to give an $S = 1$ unit and a separate $S = \frac{1}{2}$ unit within a single molecule. The two units are connected by an antiferromagnetic intramolecular exchange. Coupling between molecules is dominated by antiferromagnetic interaction between the different types of spin, producing a bulk ferrimagnet.

Other stable nitroxide organic radicals showing ferromagnetic interactions have been found, including those based on TEMPO, verdazyl, and thiazyl radicals (see Fig. 4 for examples). A notable example has been the recent discovery of spontaneous magnetization below 35 K, albeit associated with noncollinear antiferromagnetism, in the β crystal phase of the dithia-diazolyl molecular radical p-NC(C_6F_4)(CNSSN) [27] (see Fig. 4). The transition temperature for this material could be raised still further to 65 K by applying a hydrostatic pressure of 16 kbar [28]. Fullerenes can also be employed in making purely organic magnets, and much interest followed the discovery of ferromagnetism in TDAE-C_{60} (see Fig. 4) at the relatively high temperature of 16 K [29].

Room temperature ferromagnetism in samples of C_{60} polymerized at high temperature and pressure has been reported [30–32], though the magnetization is very small and far from being uniformly distributed throughout the sample. Different mechanisms for the origin of this effect have been postulated [33] and magnetic effects have been introduced into carbon nanostructures by ion irradiation [34]. Whatever the actual mechanism underlying these intriguing results, the discovery that such a strong magnetic coupling might be possible in a system containing only s and p electrons motivates further study in this area. Carbon nanotubes have been thought as promising candidates for unusual magnetic properties, but a very recent exciting development has been room temperature ferromagnetism in nanotubes, not of carbon, but of vanadium oxide (VO_x), grown by self-assembly, which can be controlled by electron or hole doping [35].

3 Prussian Blues

Prussian blue, $Fe_4^{III}[Fe^{II}(CN)_6]_3 \cdot 14H_2O$, was one of the first, modern, artificially manufactured pigments, and was made by the colourmaker Diesbach of Berlin in about 1704. It is a deep blue colour due to an intense absorption band in which an electron is transferred from an Fe^{II} site to an Fe^{III} site. The Fe^{III} ($3d^5$) ions carry a spin, but the Fe^{II} ($3d^6$) ions are low-spin $S = 0$. However, because of strong spin delocalization from the Fe^{III} ions onto neighbouring atoms, there is significant magnetic coupling through the Fe^{III}–C–N–Fe^{II}–N–C–Fe^{III} linkages and the magnetic transition temperature is $T_C = 5.6\,K$ [36, 37]. Replacing the two metal ions (Fe^{II} and Fe^{III}) by other magnetic ions can be expected to increase the transition temperature and in fact high transition temperatures in compounds with the general formula $A_k[B(CN)_6]_l \cdot nH_2O$ have been found [38, 39] (see Fig. 5). In this formula, A is a high-spin ion and B is a low-spin ion (the environment of B, surrounded as it is by six cyano groups, leads to a large electron pairing energy, forcing B to be low-spin). Varying the metals A and B leads to a variety of behaviours, and depending on the charges of A and B, structural defects (such as missing $B(CN)_6$ groups) can be introduced which can be filled with water [40]. One of the most spectacular examples of one of these compounds is that of $V[Cr(CN)_6]_{0.86} \cdot 2.8H_2O$ which has $T_C = 315\,K$ [41]. The vanadium is believed to be in a $V_{0.42}^{II}V_{0.58}^{III}$ state and the moments on both V^{II} and V^{III} ions are antiparallel to those on the Cr^{II} ions so that the system is a ferrimagnet. The saturation magnetization is around $0.15\mu_B$ per formula unit, while the coercive field is a rather low $1\,mT$ [41]. Prussian blue compounds with even higher transition temperatures have now been found [42, 43]. X-ray magnetic circular dichroism (XMCD) can be used to probe directly the spin–state configuration of each transition metal ion in a Prussian blue and to extract the relative spin and orbital contributions to the total magnetic moment [44].

4 Spin Crossover

Even if there is no transition to long range magnetic order, a molecular magnetic material may yet exhibit bistability and hence be useful in memory applications. A spectacular example of this bistability is the spin crossover effect, sometimes known as a spin transition. A transition metal ion of configuration $3d^n$ ($n = 4$ to 7) in octahedral surroundings can have a low spin (LS) or high spin (HS) ground state [45], depending on the magnitude of the energy gap Δ between e_g and t_{2g} orbitals compared to the mean spin pairing energy P (Hund's rule coupling). When Δ and P are of comparable magnitude, the energy difference between the lowest vibronic levels of the potential wells of the two states may be sufficiently small that a change in spin state may occur due to the application of a relatively minor external perturbation [46]. A spin crossover between the LS and HS states is associated

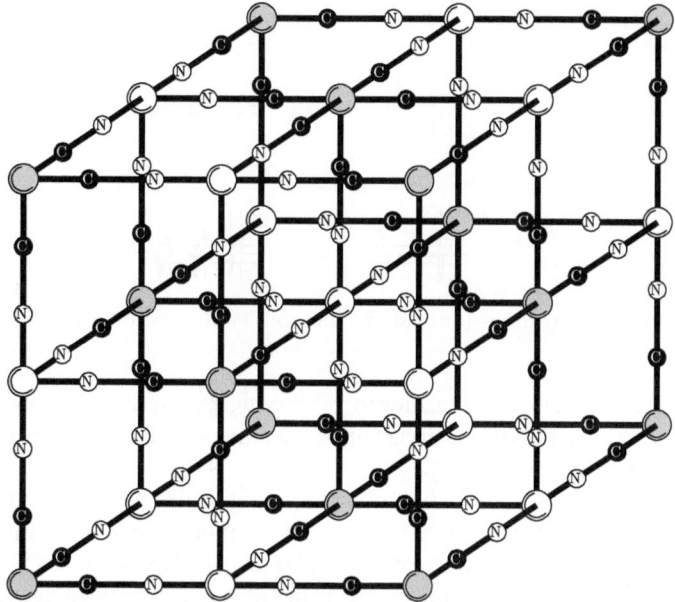

Fig. 5. The crystal structure of generalized Prussian blue compounds $A_k[B(CN)_6]_l \cdot nH_2O$ (not showing the water of crystallization). The shaded ions are A (high-spin) and the unshaded atoms are B (low-spin) (after [2])

with a change of magnetic properties and often by a change of colour. Iron (II) systems ($3d^6$) can show a crossover between LS ($S = 0$) and HS ($S = 2$) (see Fig. 6) induced by changes in temperature or by light irradiation [47]. In the HS and LS states there are different orbital occupancies and hence the metal–donor atom distance in complexes is very sensitive to spin state. For example, Fe–N bond lengths are often about 0.2 Å longer in the HS state due to the occupancy of the antibonding e_g^* orbitals. This effect has important implications in biological systems: for example, a spin transition occurs in haemoglobin and the accompanying change in Fe–N bond length allows it to absorb or release oxygen [48].

The crossover as a function of temperature is relatively smooth and gradual in solution, and is driven only by the entropy gain for LS → HS, due partly to a magnetic contribution $R[\ln(2S+1)_{HS} - \ln(2S+1)_{LS}]$, but mainly to a vibrational contribution. However, the crossover is very sharp in the solid state, showing that cooperativity plays an important rôle. The change in metal–donor atom distance accompanying the spin crossover sets up an internal pressure which is communicated to the surrounding molecules via phonons. As the transition progresses through the sample, those atoms which have switched spin state produce an increasing internal pressure which accelerated the transition on the other un-switched metal centres [49]. The sharpness of the transition therefore can be increased by enhancing the cooperativity of

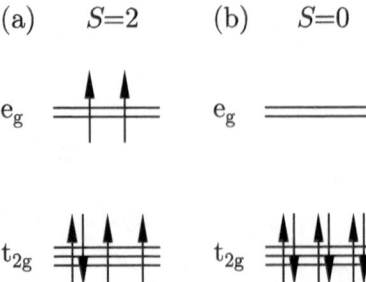

Fig. 6. *Left*: The (a) high-spin and (b) low-spin configuration for Fe^{2+} ($3d^6$) ions

the transition. This can be achieved by linking the active sites by chemical bridges, designing polymeric structures, stacking aromatic groups and introducing hydrogen bonding [50].

The spin crossover can be followed using bulk measurements of the magnetic susceptibility χ as a function of temperature T (see Fig. 7). For Fe^{2+} ions, the molar fraction of molecules in the HS state, x, is given by $x = \chi T/(\chi T)_{HS}$ where $(\chi T)_{HS}$ is the value of χT when all molecules are in the HS state. Usually $x = 0$ at low temperatures and $x = 1$ at high temperatures, reflecting the fact that the HS state maximizes entropy while the LS state minimizes energy. However, the energy difference is often very small and often the spin crossover is incomplete at low temperatures so that $x \neq 0$ as $T \to 0$. The spin crossover can be hysteretic so $T_c(\uparrow) \neq T_c(\downarrow)$ where $T_c(\uparrow)$ and $T_c(\downarrow)$ are the transition temperatures for warming and cooling respectively.

Subtle changes between the low-spin and high-spin states are also visible in the XANES spectra (Fig. 8), illustrating that the Fe–N distances are shortened during the transition from HS to LS, enhancing the hybridization and mixing the N (2p) ligands with the metal Fe (4s,4p) [53].

Mössbauer spectroscopy is also highly valuable for the study of the spin crossover in Fe(II) complexes. The quadrupole splittings and isomer shifts are much higher in HS Fe(II) than LS Fe(II), allowing the spin state to be easily followed [46]. The switching process in dinuclear Fe(II) compounds can also be measured in this way, and since Mössbauer spectroscopy gives local information, the number of LS–LS, LS–HS and HS–HS pairs can be extracted. The cooperative process can then be followed by measuring the probability of the second spin in a dinuclear pair undergoing a transition if the first one has already done so [54]. EXAFS and nuclear forward scattering can also be used to study the spin transition [55].

The spin crossover can in fact also be induced by light, and at low temperature it is possible to convert a LS state into a metastable HS state using light of a particular wavelength. This is known as the light-induced excited spin state trapping (LIESST) effect [47, 56]. The reverse process is also possible using light of different wavelength. The metastable state needs to be prepared

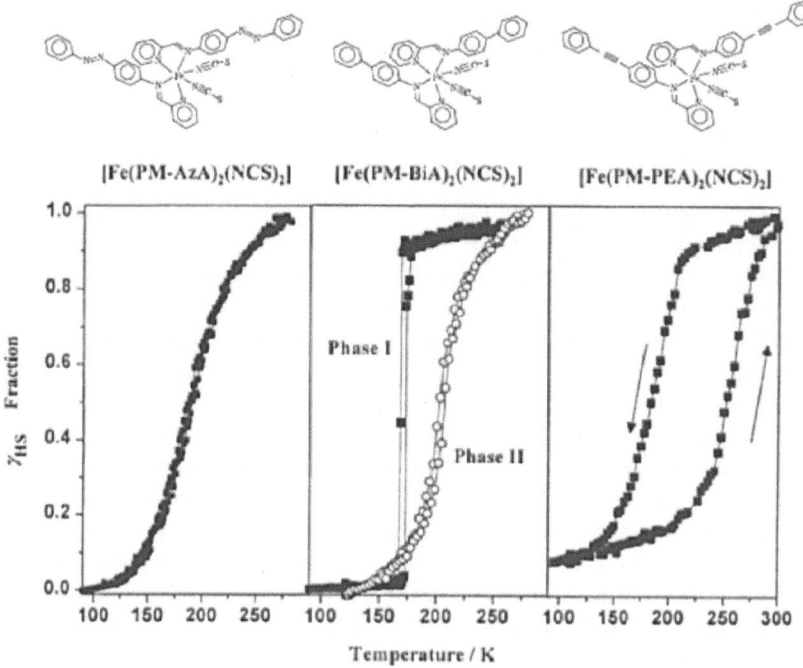

Fig. 7. Measured susceptibility for the spin crossover compounds Fe(PM-AzA)$_2$ (NCS)$_2$ Fe(PM-BiA)$_2$(NCS)$_2$ and Fe(PM-PEA)$_2$(NCS)$_2$ [51, 52]

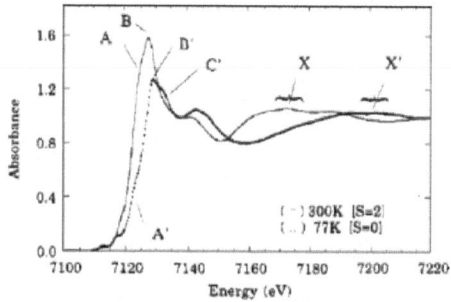

Fig. 8. The XANES spectra for the spin transition in Fe(phen)$_2$(NCS)$_2$ complex at the iron K edge. The *solid line* shows the high-spin isomer recorded at 300 K. The *dotted line* shows the low-spin isomer recorded at 77 K (after [53])

at low temperatures otherwise its lifetime becomes very small, and vanishes at a temperature T_c(LIESST). The photoswitching process can exhibit non-linear characteristics [57] and evidence is emerging that the photoinduced HS phase is structurally different from the thermally induced HS phase [58]. The spin crossover effect has also recently been used to produce a nanoporous metal-organic framework material which displays reversible uptake and release of guest molecules as the crossover proceeds [59].

5 TCNE Salts

Magnetic ordering has been found in various metallocene TCNE based magnets, specifically in [FeCp$_2^*$]TCNE [60] (sometimes known as (DMeFc)TCNE) where Cp* = C$_5$Me$_5$ = pentamethylcyclopentadienide and TCNE is an acceptor (see Fig. 9). This charge-transfer salt contains parallel chains of alternating [FeCp$_2^*$]$^+$ cations and TCNE$^-$ anions and undergoes a phase transition to a ferromagnetically ordered state at $T_C = 4.8$ K [60]. There is a sharp anomaly in the specific heat at T_C and a broad maximum at 15 K, corresponding to an exchange interaction of $J \approx 35$ K along the chain axis [61]. The coercive field is large, $\sim$0.1 T at 2 K. The transition temperature increases with pressure, reaching 7.8 K at 14 kbar [62]. Single crystal polarized neutron diffraction studies [63] confirm that the spin density of TCNE$^-$ is delocalized over all atoms. Muons have been used to measure the development of short-range spin correlations above this temperature which are very slow in this quasi-1D material and follow an activated temperature dependence [64], as expected in a spin chain with Ising character. The extracted activation energy E_a is 57.8 K, which is consistent with $E_a = 2J$ where $J/k_B = 28$ K was estimated from specific heat measurements.

Various similar salts were subsequently prepared and [MnCp$_2^*$]TCNE [65] has a transition temperature of 8.8 K, which is consistent with a mean-field treatment [66] which takes account of the larger moment on the Mn site than on the Fe site. Since [V^I(C$_6$H$_6$)$_2$]$^+$ is an $S = 1$ cation like [Mn^{III}Cp$_2^*$]$^+$, this was mixed with TCNE in dichloromethane in the hope of making a new molecular magnet. The resulting black precipitate had formula [V(TCNE)$_x$]·yCH$_2$Cl$_2$ with $x = 2$ and $y = \frac{1}{2}$ and showed ferromagnetism

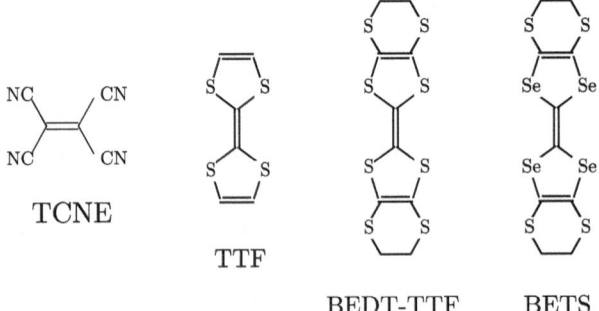

Fig. 9. Molecular structure of TCNE, TTF, BEDT-TTF (also known as ET) and BEDT-TSF (also known as BETS)

at room temperature [67]. The critical temperature exceeds 350 K, the thermal decomposition temperature of the compound. The coercive field at room temperature is quite small ($\sim$6 mT). It is thought that $S = \frac{3}{2}$ V^{II} and the two $S = \frac{1}{2}$ TCNE ligands are antiferromagnetically coupled, so that the system is probably a ferrimagnet. Substituting V by Fe in this material reduces the ferrimagnetic transition to 100 K [68]. Although the transition temperatures are impressively high for these materials, their structures are rather disordered and not yet properly determined. This structural disorder may be responsible for some glassy aspects of the magnetic behaviour. For a more detailed review of TCNE magnets, see [69].

6 BEDT-TTF and BEDT-TSF Salts

The molecule TTF (tetrathiafulvalene) is an electronic donor and can be used to make a number of interesting magnetic salts. Various modifications of the TTF molecule can be made (see Fig. 9) and one of the most important is the molecule BEDT-TTF (short for bis(ethylenedithio)tetrathiafulvalene, sometimes also known as "ET"). This molecule turns out to be a very versatile donor and a number of charge-transfer salts derived from BEDT-TTF have been found to be very good organic metals and sometimes superconductors. The BEDT-TTF molecules contain sulphur atoms on the side, and intermolecular S–S overlaps are the most important in these salts.

The basic structure of BEDT-TTF salts is illustrated by the example shown in Fig. 10. These materials are naturally layered, with alternating layers of the BEDT-TTF molecules, stacked side-to-side so that the molecular orbitals overlap, and layers of anions. In this way, the charge-transfer salts are

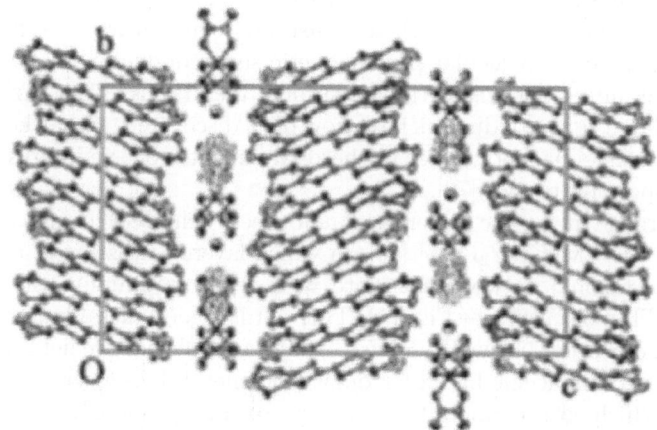

Fig. 10. Monoclinic crystal structure of β''-(BEDT-TTF)$_4$[(H$_3$O)M(C$_2$O$_4$)$_3$]$\cdot$ C$_5$H$_5$N projected along the a axis [70]

"organic-inorganic molecular composites" or "chemically constructed multi-layers" [71]. Within the BEDT-TTF layers, the molecules are in close proximity to each other, allowing substantial overlap of the molecular orbitals. Usually two (or sometimes three) BEDT-TTF molecules will jointly donate an electron to the anion, and the charge transfer leaves behind a hole on the BEDT-TTF molecules. This means that the bands formed by the overlap of the BEDT-TTF molecular orbitals will be partially filled, leading to the possibility of metallic behaviour. The transfer integrals which parameterize the ease of hopping of electrons between BEDT-TTF molecules, will be relatively large within the BEDT-TTF planes. Conversely, in the direction perpendicular to the BEDT-TTF planes, the BEDT-TTF molecules are well separated from each other; the transfer integrals will be much smaller in this direction. This results in electronic properties which for many purposes can be considered to be two dimensional [72, 73]. In the Mott insulator regime of these metals, the π-system can be regarded as having a localized $S = \frac{1}{2}$ on each BEDT-TTF molecule, and magnetic interactions between BEDT-TTF dimers are important for understanding the system [74, 75] (see also [76]).

Introduction of magnetic ions into the anion layer can lead to some magnetically interesting compounds. The compound $(BEDT-TTF)_3CuCl_4\cdot H_2O$ was an early example of this idea, combining a localized moment on the Cu ion and delocalized electrons in the BEDT-TTF layer. The salt remains metallic down to $0.4\,K$ while distinct magnetic resonances from the localized spins and conduction electrons can be resolved [77]. The interest here is that the electrons close the Fermi surface are largely confined to the frontier orbitals of the BEDT-TTF, while the magnetic moments are localized on the anions, providing a degree of spatial separation between the two. Very often, the introduction of magnetic anions produces insulating salts (e.g. [78]), though this is true of non-magnetic anions as well. Therefore, an important breakthrough was the discovery of $(BEDT-TTF)_2H_2OFe(C_2O_4)_3C_6H_5CN$, which is a superconductor with a transition temperature of 7 K and was the first discovered molecular paramagnetic superconductor (i.e. a superconductor containing paramagnetic transitional metal ions) [79]. Various other closely related superconductors have now been discovered, and although there is no long range magnetic order in these systems, they have interesting magnetotransport properties [70].

Long-range magnetic order has now been achieved in the compound $(BEDT-TTF)_3[MnCr(C_2O_4)_3]$ [80], in which layers consisting of an oxalate-bridged hexagonal network alternate with layers containing the β-packed BEDT-TTF molecules (see Fig. 11). This salt is a good metal but shows weak ferromagnetic order below $T_C = 5.5\,K$. It has a small coercive field ($0.5–1\,mT$) and is the first molecule-based ferromagnetic metal. It is essentially a hybrid material which owes its metallic properties to the organic BEDT-TTF layers and its magnetic properties to the inorganic oxalate-bridged layers.

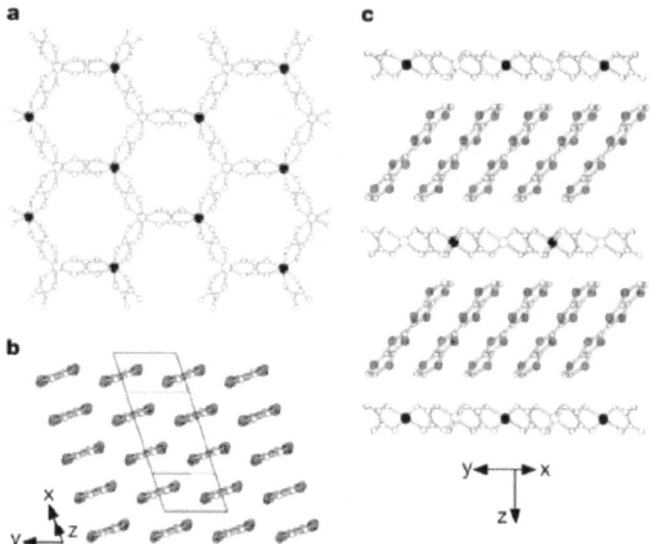

Fig. 11. Structure of $(BEDT\text{-}TTF)_3[MnCr(C_2O_4)_3]$ [80]. View of (**a**) oxalate layers, (**b**) BEDT-TTF layers (illustrating the β-packing, (**c**) the hybrid structure (showing the alternating organic/inorganic layers). After [80]

An entirely different family of organic metals can be obtained when the innermost four sulphur atoms of BEDT-TTF are replaced by selenium. The resulting molecule is BEDT-TSF (short for bis(ethylenedithio)tetraselena-fulvalene), which is usually known by the name BETS (see Fig. 9 for molecular structure). The selenium atoms are larger than the sulphur atoms and tend to broaden the electronic bands. Several salts based on the BETS molecular donor have been found to be metallic and superconducting, with several salts of the form $BETS_2X$, where X is a magnetic anion, maintaining good metallic and superconducting properties in the presence of magnetism.

A fascinating example is $\lambda\text{-}BETS_2GaCl_4$ (here λ refers to a particular crystallographic packing type) which was found to be metallic with a super-conducting transition at $\sim$6 K [81,82]. The isostructural salt $\lambda\text{-}BETS_2FeCl_4$ shows a metal-insulator transition at around 8 K, which is associated with antiferromagnetic order of the Fe^{3+} ($S = \frac{5}{2}$) spins [81,83,84]. However, the antiferromagnetic insulating phase can be destabilized by the application of a magnetic field of $\sim$10 T which stabilizes a paramagnetic metallic phase due to the gain of Zeeman energy of the Fe^{3+} moments. Below 1 K, and with a field applied parallel to the layers, superconductivity is induced above 17 T (and then destroyed above 42 T) [85,86]. The field-induced superconductivity appears to involve a Jaccarino-Peter [87] mechanism in which the exchange field is cancelled by the applied field. Precise orientation of the field is important to avoid orbital dissipation in the superconducting phase. Alloying

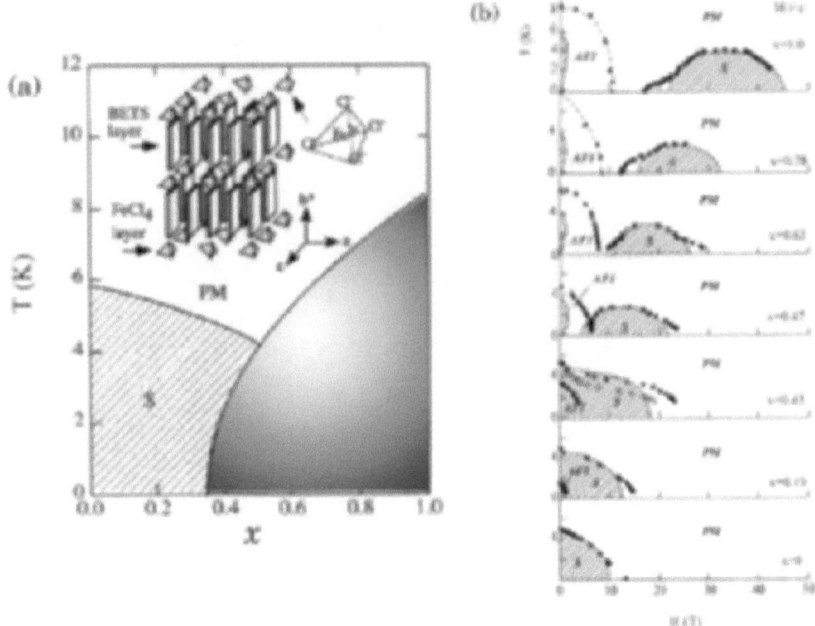

Fig. 12. Phase diagram of in λ-BETS$_2$Fe$_x$Ga$_{1-x}$Cl$_4$ (**a**) in zero magnetic field as a function of x, and (**b**) as a function of magnetic field for different values of x (after [88])

of the Fe and Ga salts produces an extremely rich phase diagram (shown in Fig. 12) and allows the field-induced superconductivity phase to be brought down to lower magnetic fields [88]. This phase diagram can be understood in terms of the Jaccarino-Peter compensation mechanism.

The interplay between coexisting magnetism and superconductivity can be studied in the salts κ-BETS$_2$FeBr$_4$ and κ-BETS$_2$FeCl$_4$. The κ-crystal phase is an orthorhombic layered structure in which the BETS molecules are arranged in 2D sheets of interacting dimers in the ac plane. These sheets alternate with layers of magnetic anions as one goes along the b axis. The spatial separation of the highly conducting molecular layers and the strongly magnetic layers is a key feature of this structure. Magnetic and transport measurements on the FeBr$_4$ salt indicate that the Fe^{3+} is in a high spin state ($S = \frac{5}{2}$) with an antiferromagnetic transition at $T_N = 2.5\,\mathrm{K}$ and a superconducting transition taking place at $T_c = 1\,\mathrm{K}$ [89].

Zero-field muon spin relaxation measurements on the organic metal κ-BETS$_2$FeCl$_4$ clearly show the formation of an antiferromagnetically ordered state below $T_N = 0.45\,\mathrm{K}$. The magnetic order remains unperturbed on cooling through the superconducting transition $T_c \sim 0.17\,\mathrm{K}$, providing unambiguous evidence for the coexistence of antiferromagnetic order and superconductivity

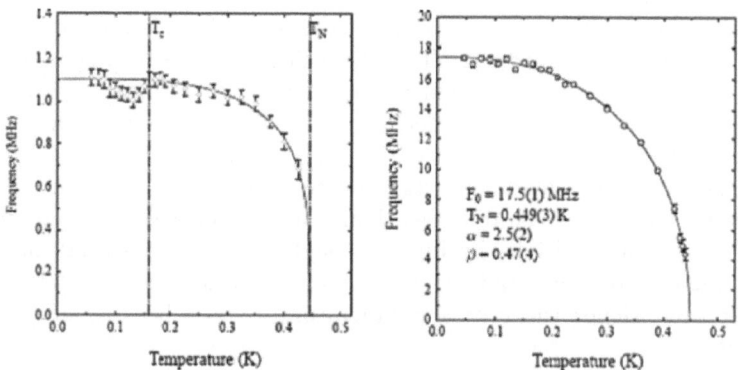

Fig. 13. Temperature dependence of the two muon precession frequencies measured in κ-BETS$_2$FeCl$_4$ (after [90])

in this system. The internal field seen at a muon site depends on its position with respect to the magnetic structure and hence a number of different frequencies are to be expected from the large low-symmetry unit cell of this material. The precession amplitude observed for each site depends both on the probability of occupancy of the site and on the alignment of the internal field at the site with respect to the polarization of the incoming muon. Two clear precession frequencies are observed (see Fig. 13, [90]). The antiferromagnetic nature of the Fe ordering means that high local fields such as reflected by the higher frequency can only be present close to the anion plane. For the lower frequency, sites arranged between FeCl$_4$ along the c-axis are a possibility, however the local field in the region round the centre of the BETS sheets is also consistent. Superconductivity is expected to modify the Spin Density Wave (SDW), so the drop in frequency seen on entering the superconducting state may provide evidence for the presence in the normal state of a weak SDW within the BETS layers, as previous calculations have suggested [91].

7 One-dimensional Systems

A large number of magnetic spin chains can be produced using organic ligands and metal ions. As an example, a family of magnetically ordered materials can be prepared by combining hexafluoroacetylacetonate (hfac) hfac, nitronyl nitroxides and transition metal or lanthanide ions. Each nitronyl nitroxide radical can bind to two different ions leading to compounds with a chain-like structure [92]. One such example is Mn(hfac)$_2$NITPhOMe [93] where NITPhOMe is a nitronyl nitroxide with methoxyphenyl attached to the carbon of the O–N–C–N–O moiety. The structure consists of helices in which Mn(hfac)$_2$ moieties are bridged by the radicals coordinated through two equivalent oxygen atoms [93]. It orders ferrimagnetically below 4.8 K with a moment corresponding to $S = 2$, as expected for antiparallel alignment of $S = \frac{5}{2}$ (Mn^{2+}) and $S = \frac{1}{2}$ (nitronyl nitroxide radical). The helical structure provides natural optical activity which may lead to interesting magneto-optical properties. (For an X-ray magnetic circular dichroism study on a similar chain, see [94].)

Because the interchain contacts are very weak, and there are no superexchange pathways between different chains, it was believed that the ordering in these systems is driven by dipolar coupling [93]. The strong intrachain metal–radical coupling leads to long spin correlation lengths $\xi(T)$ at low temperature. Hence the critical temperature T_C can be estimated by the relation

$$k_B T_C = \xi(T_C)|E_{dip}| , \qquad (2)$$

where E_{dip} is the dipolar energy in the preferred spin orientation [95]. However, the single-ion anisotropy is also believed to play a role in determining the three-dimensional ordering [96].

It is possible to avoid a three-dimensional magnetically ordered ground state if the magnetoelastic coupling in a spin-chain is large. In this case, a spin-Peierls transition may occur. This is an intrinsic lattice instability in spin-$\frac{1}{2}$ antiferromagnetic Heisenberg chains. Above the transition temperature T_{SP}, there is a uniform antiferromagnetic next-neighbour exchange in each chain; below T_{SP} there is an elastic distortion resulting in dimerization and hence two, unequal, alternating exchange constants (see Fig. 14(a,b)). The alternating chain possesses an energy gap between the singlet ground state and the lowest lying band of triplet excited states which closes up above T_{SP}. The transition temperature may be related to the relevant coupling constants; whereas the conventional Peierls distortion is expected at a

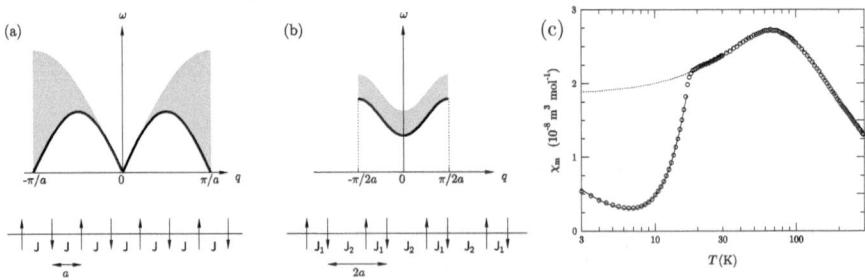

Fig. 14. Schematic representation of the elementary excitations in (a) a uniform Heisenberg antiferromagnetic chain and (b) an alternating chain (for which the ground state is a singlet state at $q = 0$), and for which the unit cell is doubled (after [45]). (c) Bulk magnetic susceptibility for $MEM(TCNQ)_2$. The *dotted line* represents a fit to a Bonner-Fisher expression [97] at high temperature, which yields an exchange constant of 46 K. Note the knee at 18 K, indicative of a spin-Peierls transition. The low temperature fit (*solid line*) is to a combination of Curie impurity and spin gap terms (after [98])

temperature $T_P \sim (E_F/k_B)\exp(-1/\lambda)$, where E_F is the Fermi energy of the system and λ is the electron-phonon coupling constant, the spin-Peierls transition is expected at $T_{sP} \sim (J/k_B)\exp(-1/\lambda)$. Since $J \ll E_F$, this implies that $T_{sP} \ll T_P$ [99].

One example is MEM(TCNQ)$_2$, which consists of one-dimensional stacks of planar TCNQ molecules, each of which has a charge of $-\frac{1}{2}e$ associated with it. Adjacent stacks are separated by arrangements of MEM molecules, each of which possess a localized charge of $+e$. It undergoes two structural distortions. The first, which occurs at 335 K, is a conventional Peierls transition in which the TCNQ chains dimerize. This results in a change from metallic to insulating behaviour as a single electronic charge becomes localized on each TCNQ dimer; the single spin on each dimer couples antiferromagnetically to its neighbours. This phase persists down to the spin-Peierls transition at 18 K, where a further dimerization of the TCNQ stacks takes place (this is a tetramerization of the original chain).

Magnetic susceptibility data for MEM(TCNQ)$_2$ are shown in Fig. 14(c) (first measured by [100], but the data shown are from [98]). The high temperature dependence is well fitted by a Bonner-Fisher [97] expression for a uniform Heisenberg antiferromagnet, with a J of 46 K. There is a rapid drop in the susceptibility at 18 K, which is indicative of the opening of a gap in the magnetic excitation spectrum. The size of the BCS pair excitation gap, $2\delta(0)$, at absolute zero can be estimated by using the result of Bulaevskii [101] who calculated the temperature dependence of χ in the dimerized state; this yields a value of 56 K. It is necessary to include a contribution from paramagnetic defect-spins at the lowest temperatures. These may arise, for instance, from chain ends where there are some spins which do not go into a dimerized configuration.

8 Single Molecule Magnets

The magnetic properties of small magnetic particles have been studied for a long time [102, 103], originally in the context of the switching of minute magnetic grains in rocks in the Earth's magnetic field (which is a function of geological time). A small magnetic particle with uniaxial anisotropy will have its magnetization parallel to an easy axis. Switching between one easy axis direction and another will take place via thermal activation in a time τ given by

$$\tau = \tau_0 \exp\left(\frac{U}{k_B T}\right), \tag{3}$$

where U is the energy barrier, which is equal to KV where K is the anisotropy energy density and V is the volume. The prefactor τ_0 is usually in the range 10^{-9}–10^{-11} s. When the temperature T is much larger than U/k_B, the magnetization can fluctuate readily, but when T is much smaller than U/k_B the

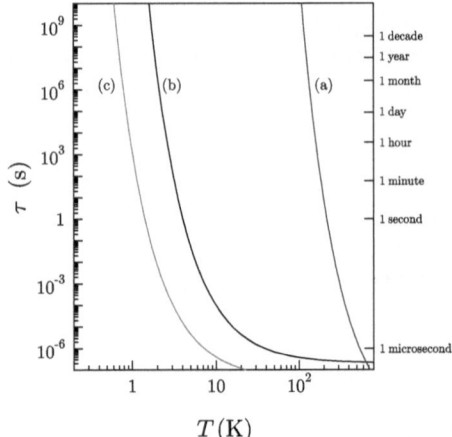

Fig. 15. The correlation time of thermally activated fluctuations following (3) for (**a**) a small magnetic particle with energy barrier $U = 5000\,\mathrm{K}$, (**b**) Mn_{12} and (**c**) Fe_8

fluctuations become "blocked". The blocking occurs quite suddenly as the particles are cooled (because of the exponential in (3)), and sets in at a temperature determined mainly by $U \propto V$ but also by the timescale of the experiment (albeit only logarithmically). This behaviour is shown in Fig. 15.

Thermal activation is not the only way to cross the barrier; another possibility is quantum tunnelling, in an analogous manner to the tunnelling of an alpha particle from a radioactive nucleus [104]. The exchange interaction constrains all the electronic spins inside the particle to remain parallel with one another, but the "macro-spin" can coherently tunnel from one state to the other. This coherent process of a macroscopic system undergoing a tunnelling process is known as macroscopic quantum tunnelling (MQT) [105,106] or quantum tunnelling of magnetization (QTM).

Conventional methods of preparing small magnetic particles suffer from a rather broad range of particle sizes which leads to a broad distribution of intrinsic switching rates, complicating the interpretation. On the other hand, the chemical synthesis of polynuclear cage compounds is an attractive route to prepare identical molecular clusters [107]. The resulting molecular clusters turn out to be appealing systems in which to study these effects since the molecular approach provides ensembles of completely identical, iso-oriented nanomagnets [108].

These materials have a large spin S which results from the intramolecular exchange between the transition metal ions in the cluster. In zero-applied field, to a first approximation the Hamiltonian of this system can be written as

$$\mathcal{H} = -DS_z^2 \,, \tag{4}$$

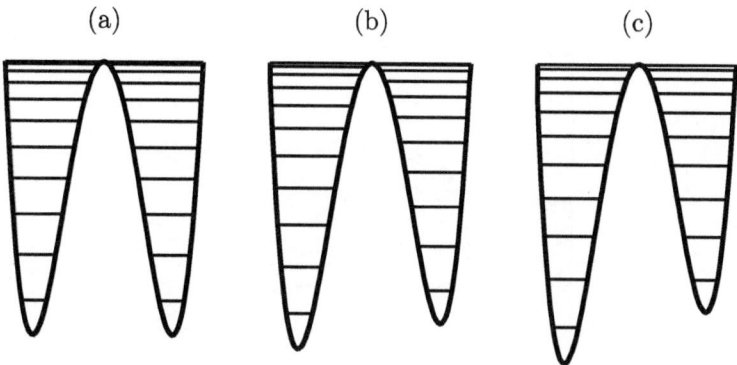

Fig. 16. The energy levels in a single molecule magnet with Hamiltonian given by $\mathcal{H} = -DS_z^2 + g\mu_{\mathrm{B}}BS_z$ with $D > 0$ for (**a**) $B = 0$, (**b**) $B = D/2g\mu_{\mathrm{B}}$ and (**c**) $B = D/g\mu_{\mathrm{B}}$

so that for $D > 0$ the lowest energy states have $M = \pm S$. At low temperatures ($k_{\mathrm{B}}T \ll D$), only the states with $M = \pm S$ are thermally accessible. Moreover, if a molecule has $M = -S$, there is an activation energy barrier equal to $|D|S^2$ if S is an integer (or $|D|(S^2 - \frac{1}{4})$ if S is half-integer) which separates the $M = +S$ state from the $M = -S$ state (see Fig. 16a). Molecules with $D > 0$ are known as single-molecule magnets (SMMs).

If a field is applied in the z-direction to a SMM, the Hamiltonian becomes

$$\mathcal{H} = -DS_z^2 + g\mu_{\mathrm{B}}S_z B ,\qquad(5)$$

which causes the levels with $\pm M$ to no longer be degenerate (see Fig. 16b). In fact, the $M = -S$ and the $M = S - n$ levels have the same energy when the resonant field B_n given by

$$B_n = \frac{nD}{g\mu_{\mathrm{B}}}\qquad(6)$$

is applied (see Fig. 16c for the case of $n = 1$). In real single molecule magnets, there are usually additional small terms in the Hamiltonian, so that it should be written in general as

$$\mathcal{H} = -DS_z^2 + g\mu_{\mathrm{B}}\mathbf{S}\cdot\mathbf{B} + \mathcal{H}' ,\qquad(7)$$

where $\mathcal{H}'$ is an anisotropy term such as $E(S_x^2 - S_y^2)$. These additional terms can be (as is the case for $E(S_x^2 - S_y^2)$) symmetry-violating, and hence induce tunneling between states on opposite sides of the barrier. Such tunneling becomes enhanced when resonant magnetic fields described by (6) are reached [109].

Terms such as $E(S_x^2 - S_y^2)$ act like a transverse field in the Hamiltonian in that they mix states with different M. The field-dependence of the energy for

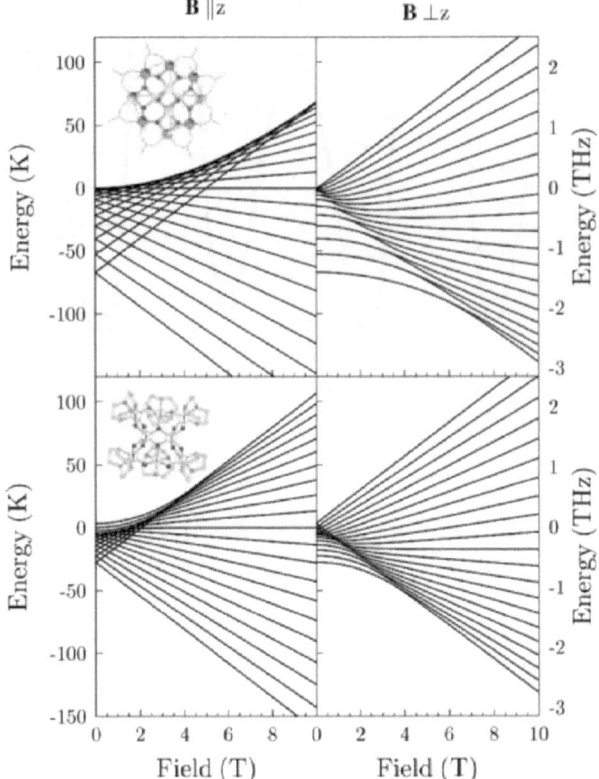

Fig. 17. Calculated energy spectra for Mn$_{12}$ (*top*) and Fe$_8$ (*bottom*) as a function of field applied parallel (*left*) or perpendicular (*right*) to the easy-axis

two molecular magnets, Mn$_{12}$ and Fe$_8$, is shown in Fig. 17 for two orientations of the magnetic field B, parallel and perpendicular to the easy-axis which is assumed to lie along z. Both these systems have an $S = 10$ ground state, but different values of D and different forms of $\mathcal{H}'$. When B is parallel to z, the states can be labelled by M, but when B is parallel to x, significant mixing of states occurs, except at high field.

As an example of real SMM behaviour, consider the mixed-valent manganese cage Mn$_9$ which possesses an $S = 17/2$ ground state as a result of an antiferromagnetic interaction between three ferromagnetically coupled Mn^{4+} ions and a wheel of four Mn^{3+} and two Mn^{2+} ions. Data for this compound are shown in Fig. 18 and demonstrate the Arrhenius behaviour of the relaxation time as a function of $1/T$. The hysteresis loops are not smooth, but show steps in the magnetization at regular intervals of field, indicative of resonant QTM. As the applied field changes, the magnetization does not respond unless the field passes through a resonant field B_n. These steps in the magnetization show up more when the field sweep rate is faster, so that

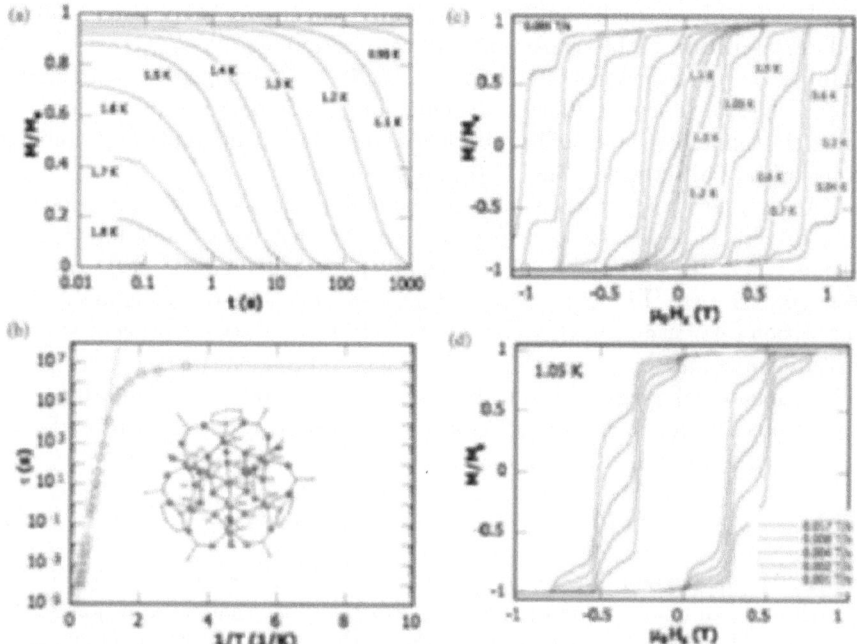

Fig. 18. SMM behaviour in "Mn9", i.e. Mn9O7(OAc)11(py)3(H2O)2 (molecular structure shown in inset to (b)). (a) Relaxation data, plotted as fraction of maximum value M_s versus time. (b) Arrhenius plot using ac data and dc decay data on a single crystal. The *solid line* is the fit of the data in the thermally activated region and the *dashed line* the fit of the temperature-independent data. (c,d) Magnetization versus applied magnetic field. The resulting hysteresis loops are shown at (c) different temperatures and (d) different field sweep rates. (After [110])

there is less time for all the tunneling to take place at zero applied field and therefore a step becomes resolved. The field separation between the zero-field resonance and the next gives a value of D which is in good agreement with that obtained from the bulk DC measurements [110].

The cluster $[Fe_8O_2(OH)_{12}(tacn)_6]^-$ (abbreviated to Fe8) has an $S = 10$ ground state due to interactions between the $\frac{5}{2}$ spins of the eight Fe(III) centres. It has no axial symmetry and hence its spin Hamiltonian contains a term $E(S_x^2 - S_y^2)$. The ESR data were satisfactorily simulated with $D = 0.27$ K and $E = -0.046$ K [111]. Below 0.4 K, temperature dependent tunneling through the barrier (of magnitude $DS^2 \approx 27$ K) is observed, corresponding to a quantum tunneling of the magnetization [112]. This is a more clear-cut case than Mn12 in which temperature-dependent (i.e. non-quantum) relaxation is observed even down to 0.06 K [113].

When two levels begin to cross, there is actually an avoided level crossing characterized by an energy gap called a tunnel splitting. This tunnel splitting

can be tuned by a small transverse field. In fact, the tunnel splitting is an os-
cillatory function of the transverse field with period $2k_B[2E(E+|D|)]^{1/2}/g\mu_B$
[114], as has been demonstrated experimentally in Fe_8 using an array of micro-
SQUIDS [115]. These oscillations result from a Berry phase [116] which enters
the spin tunnelling probability and leads to radically different behaviour for
integer and half-integer spins [117] (see also [118]).

At very low temperature, phonon-mediated relaxation can be neglected
and in zero field the $M = \pm 10$ states are coupled by a tunnelling matrix
element $\Delta \sim 10^{-7}$ K. The local dipolar magnetic fields induce an energy bias
$\xi = g\mu_B SB \sim 10^{-1}$ K which forces almost all molecules off resonance. Magnetization relaxation can therefore only occur if there is some process, such as
fast dynamic nuclear fluctuations, which acts to broaden the resonance [119].
This scenario can be described using a "spin-bath" model [120], and this leads
to the prediction of a relaxation time with a "square–root time" dependence,
so that at short times t, the magnetization $M(t)$ is given by

$$M(t) = M(0) + [M_{eq} - M(0)]\sqrt{\Gamma t} , \qquad (8)$$

where M_{eq} is the equilibrium magnetization, $\Gamma \sim \Delta^2 P(\xi)/\hbar$ and $P(\xi)$ is the
distribution of energy bias in the sample [121]. Note that in equation 8, the
quantities M_{eq}, $M(t)$ and Γ are functions of the applied field. This behaviour
has been observed in crystals of Fe_8 [122], and the distribution $P(\xi)$ has hence
been studied as a function of temperature [121].

Moreover, $P(\xi)$ is found to evolve during relaxation in a non-trivial way,
illustrating how the tunnelling is proceeding in the sample. This gives rise
to the so-called "hole-digging" method, in which a hole in $P(\xi)$ is "dug" by
depleting the population of certain spins at a given applied field [121] by
letting the relaxation occur. The hole line width is found to be independent
of the initial value of the magnetization and is related to the inhomogeneous
level broadening arising from the nuclear spins [119, 121]. This interpretation
has been confirmed by studying samples with ^{56}Fe replaced by ^{57}Fe, and also
those in which ^{1}H is partially replaced by ^{2}H. In each case, the width of
the hole in $P(\xi)$ agrees with numerical simulations of the level broadening
arising from the altered hyperfine couplings [123]. The hole-digging method
has also now been performed for Mn_{12} [124], with similar results obtained
at low temperature. At much higher temperatures, the effect of the nuclear
spins on the relaxation is much less important, and the spin-phonon effect is
expected to be dominant.

Although the intermolecular interactions between single molecule magnets can usually be ignored, this is certainly not the case in a supramole-
cular dimer. In one such system, known as $[Mn4]_2$, two Mn_4 cubane–based
molecules, each with ground state $S = \frac{9}{2}$, are coupled antiferromagnetically
($J = 0.1$ K) resulting in an $S = 0$ ground state [125]. The coupling results
in an exchange bias of the tunneling transitions [125] and also in additional
quantum resonances associated with spin-spin cross relaxation [126], allowing

the effective exchange coupling to be deduced [127]. ESR transitions involve well-defined superposition states on both molecules [128] and the measured hysteresis loops demonstrate quantum tunnelling of the magnetization via entangled states of the dimer [129].

Many new clusters are currently being prepared and are studied using micro-SQUID measurements, ESR, NMR, μSR, neutron scattering and many other techniques. X-ray magnetic circular dichroism is increasingly being employed, confirming for example that Mn_{12}-acetate has a ferrimagnetic structure due to individual Mn(III) and Mn(IV) centres and showing that both have negligible orbital magnetic moments [130,131].

9 Conclusion

Molecular magnetism is emerging as a highly complex and endlessly surprising field. The flexibility of carbon chemistry means that it is hard to predict what new families of materials may be possible in the future, but further applications in both technology and in further fundamental physics experiments seem highly likely.

References

1. O. Kahn: *Molecular Magnetism* (Wiley-VCH, Weinheim, 1993); J. S. Miller and M. Drillon (eds) *Magnetism: Molecules to Materials* volumes 1–4 (Wiley-VCH, Weinheim, 2001); P. M. Lahti (ed) *Magnetic Properties of Organic Materials* (Dekker, 1999); K. Itoh and M. Kinoshita (eds) *Molecular Magnetism, New Magnetic Materials* (Gordon and Breach, Amsterdam, 2000).
2. S. J. Blundell, F. L. Pratt: J. Phys.: Condens. Matter **16**, R771, (2004)
3. T. Sugano: Polyhedron **20**, 1285 (2001)
4. K. Mukai: Bull. Chem. Soc. Jpn. **42**, 40 (1969)
5. K. Awaga, T. Sugano, M. Kinoshita: Chem. Phys. Lett. **128**, 587 (1986)
6. K. Awaga, T. Sugano, M. Kinoshita: J. Chem. Phys. **85**, 1211 (1986)
7. H. M. McConnell: Proc. R. A. Welch Found. Chem. Res. **11**, 144 (1967)
8. K. Awaga, T. Sugano, M. Kinoshita: Chem. Phys. Lett. **141**, 540 (1987)
9. H. M. McConnell: J. Chem. Phys. **39**, 1910 (1963)
10. P. J. Brown, A. Capiomont, B. Gillon, J. Schweizer: J. Mag. Magn. Mat. **14**, 289 (1979)
11. M. Saint-Paul, C. Veyret: Phys. Lett. **45A**, 362 (1973)
12. A. Benoit, J. Flouquet, B. Gillon, J. Schweizer: J. Magn. Magn. Mater. **31–34**, 1155 (1983)
13. G. Chouteau, C. Veyret-Jeandey: J. de Phys. **42**, 1441 (1981)
14. A. Capiomont: Acta Cryst. B **28**, 2298 (1972)
15. S. J. Blundell, A. Husmann, T. Jestadt, F. L. Pratt, I. M. Marshall, B. W. Lovett, M. Kurmoo, T. Sugano, W. Hayes: Physica B **289**, 115 (2000)
16. S. T. Bramwell, P. C. W. Holdsworth: J. Phys.: Condens. Matter **5**, L53 (1993)

17. T. Sugano, S. J. Blundell, F. L. Pratt, T. Jestadt, B. W. Lovett, W. Hayes, P. Day: Mol. Cryst. Liq. Cryst. **334**, 477 (1999)
18. S. J. Blundell, P. A. Pattenden, F. L. Pratt, R. M. Valladares, T. Sugano, W. Hayes: Europhys. Lett. **31**, 573 (1995)
19. M. Tamura, Y. Nakazawa, D. Shiomi, K. Nozawa, Y. Hosokoshi, M. Ishikawa, M. Takahashi, M. Kinoshita: Chem. Phys. Lett. **186**, 401 (1991)
20. Y. Nakazawa, M. Tamura, N. Shirakawa, D. Shiomi, M. Takahashi, Mand. Ishikawa. M. Kinoshita: Phys. Rev. B **46**, 8906 (1992)
21. L. P. Le, A. Keren, G. M. Luke, W. D. Wu, Y. J. Uemura, M. Tamura, M. Ishikawa, M. Kinoshita: Chem. Phys. Lett. **206**, 405 (1993)
22. M. Kinoshita: Phil. Trans. R. Soc. Lond. A **357**, 2855 (1999)
23. A. Zheludev, E. Ressouche, J. Schweitzer, M. Wan, H. Wang: J. Mag. Magn. Mat. **135**, 147 (1994)
24. E. Ressouche: J. Phys. Chem. Sol. **62**, 2161 (2001)
25. R. Chiarrelli, M. A. Novak, A. Rassat, J. L. Tholence: Nature **363**, 147 (1993)
26. Y. Hosokoshi, K. Katoh, Y. Nakazawa, H. Nakano, K. Inoue: J. Am. Chem. Soc. **123**, 7921 (2001)
27. F. Palacio, G. Antorrena, M. Castro, R. Burriel, J. Rawson, J. N. B. Smith, N. Bricklebank, J. Novoa, C. Ritter: Phys. Rev. Lett. **79**, 2336 (1997)
28. M. Mito, T. Kawae, K. Takeda, S. Takagi, Y. Matsushita, H. Deguchi, J. Rawson, F. Palacio: Polyhedron **20**, 1509 (2001)
29. P. M. Allemand, K. C. Khemani, A. Koch, F. Wudl, K. Holczer, S. Donovan, G. Gruner, J. D. Thompson: Science **253**, 301 (1991)
30. T. L. Makarova, B. Sundqvist, R. Hohne, P. Esquinazi, Y. Kopelevich, P. Scharff, V. A. Davydov, L. S. Kashevarova, Rakhmanina: Nature **413**, 716 (2001)
31. R. A. Wood, M. H. Lewis, M. R. Lees, S. M. Bennington, M. G. Cain, N. Kitamura: J. Phys.: Condens. Matter **14**, L385 (2002)
32. V. N. Narozhnyi, K–M Müller, D. Eckert, A. Teresiak, L. Dunsch, V. A. Davydov, L. S. Kashevarova, A. V. Rakhmanina *Physica B* **329–333**, 1217 (2003)
33. M. Coey and S. Sanvito: Physics World **17**, 33 (2004)
34. S. Talapatra, P. G. Ganesan, T. Kim, R. Vajtai, M. Huang, M. Shima, G. Ramanath, D. Srivastava, S. C. Deevi and P. M. Ajayan: Phys. Rev. Lett. **95**, 097201 (2005)
35. L. Krusin-Elbaum, D. M. Newns, H. Zeng, V. Derycke, J. Z. Sun, R. Sandstrom: Nature **431**, 672 (2004)
36. A. N. Hoden, B. T. Matthias, P. W. Anderson, H. W. Lewis: Phys. Rev. **102**, 1463 (1956)
37. R. M. Bozorth, H. J. William, D. E. Walsh: Phys. Rev. **103**, 572 (1956)
38. T. Mallah, S. Thiebalt, M. Verdaguer, P. Veillet: Science **262**, 1554 (1993)
39. W. R. Entley, G. S. Girolami: Science **268**, 397 (1995)
40. J. S. Miller: MRS Bulletin (November) 33 (2000)
41. S. Ferlay, T. Mallah, L. Ouhahes, P. Veillet, M. Verdaguer: Nature **378**, 701 (1995).
42. Ø. Hatlevik, W. E. Buschmann, J. Zhang, J. L. Manson, J. S. Miller Adv. Mater. **11**, 914 (1999)
43. S. D. Holmes, G. S. Girolami: J. Am. Chem. Soc. **121**, 5593 (1999)
44. M.-A. Arrio, Ph. Sainctavit, Ch. Cartier dit Moulin, Ch. Brouder, F. M. F. de Groot, T. Mallah, M. Verdaguer: J. Phys. Chem. **100**, 4679 (1996)

45. S. Blundell: *Magnetism in Condensed Matter* OUP (2001).
46. P. Gütlich, Y. Garcia, H. A. Goodwin: Chem. Soc. Rev. **29**, 419 (2000)
47. S. Decurtins, P. Gütlich, C. P. Köhler, H. Spiering, A. Hauser: Chem. Phys. Lett. **105**, 1 (1984).
48. W. R. Scheidt, C. A. Reed: Chem. Soc. Rev. **81**, 543 (1981)
49. N. Willembacher, H. Spiering: J. Phys. C **21**, 1423 (1988)
50. O. Kahn O, C. J. Martinez: *Science* **279**, 44 (1988).
51. J.-F. Létard, P. Guionneau, E. Codjovi, O. Lavastre, G. Bravic, D. Chasseau, O. Kahn: J. Am. Chem. Soc. **119**, 10861 (1997)
52. H. Daubric, C. Cantin, C. Thomas, J. Kliava, J.-F. Létard, O. Kahn: Chem. Phys. Lett. **244**, 75 (1999)
53. V. Briois, Ch. Cartier dit Moulin, Ph. Sainctavit, C. Brouder, A-M. Flank: J. Am. Chem. Soc. **117**, 1019 (1995)
54. V. Ksenofontov, H. Spiering, S. Reiman, Y. Garcia, A. B. Gaspar, N. Moliner, J. A. Real, P. Gütlich: Chem. Phys. Lett. **348** 381 (2001)
55. H. Paulsen, H. Grünsteudel, W. Meyer-Klaucke, M. Gerdan, H. F. Grünsteudel, A. I. Chumakov, R. Rüffer,H. Winkler, H. Toftlund, A. X. Trautwein: Eur. Phys. J. B **23**, 463 (2001)
56. A. Hauser: Chem. Phys. Lett. **124**, 543 (1986)
57. Y. Ogawa, S. Koshihara, K. Koshino, T. Ogawa, C. Urano, H. Takagi: Phys. Rev. Lett. **84**, 3181 (2000)
58. T. Tayagaki, K. Tanaka: Phys. Rev. Lett. **86**, 2886 (2001)
59. G. J. Halder, C. J. Kepert, B. Moubaraki, K. S. Murray, J. D. Cashion: Science **298**, 1762 (2002)
60. S. Chittipeddi, K. R. Cromack, J. S. Miller, A. J. Epstein: Phys. Rev. Lett. **58**, 2695 (1987)
61. A. Chakraborty, A. J. Epstein, W. N. Lawless, J. S. Miller: Phys. Rev. B **40**, 11422 (1989)
62. Z. J. Huang, F. Chen, Y. T. Ren, Y. Y. Xue, C. W. Chu, J. S. Miller: J. Appl. Phys. **73**, 6563 (1993)
63. A. Zheludev, A. Grand, E. Ressouche, J. Schweizer, B. Morin, A. J. Epstein, D. A. Dixon, J. S. Miller: J. Am. Chem. Soc. **116**, 7243 (1994)
64. Y. J. Uemura, A. Keren, L. P. Le, G. M. Luke, B. J. Sternlieb, W. D. Wu: Hyp. Int. **85**, 133 (1994)
65. G. T. Yee, J. M. Manriquez, D. A. Dixon, R. S. McLean, D. M. Groski, R. B. Flippen, K. S. Narayan, A. J. Epstein, J. S. Miller: Adv. Mater. **3**, 309 (1991)
66. J. S. Miller, A. J. Epstein: Angew. Chem. Int. Ed. Engl. **33**, 385 (1994)
67. J. M. Manriquez, G. T. Yee, R. S. McLean, A. J. Epstein, J. S. Miller: Science **252**, 1415 (1991)
68. K. I. Pokhodnya, N. Peterson, J. S. Miller: Inorg. Chem. **41**, 1996 (2002)
69. J. S. Miller: Inorg. Chem. **39**, 4392 (2000)
70. A. I. Coldea, A. F. Bangura, J. Singleton, Ardavan. A. A., A. Akutsu-Sato, H. Akutsu, S. S. Turner, P. Day: Phys. Rev. B **69**, 165103 (2004)
71. P. Day: Phys. Scr. **T49**, 726 (1993)
72. J. Singleton: Rep. Prog. Phys. **63**, 1111 (2000)
73. J. Wosnitza: *Fermi surfaces of low-dimensional organic metals and supercon-ductors*by (Springer-Verlag, Berlin 1996).
74. T. Mori: J. Solid State Chem. **168**, 433 (2002)
75. T. Kawakami, T. Taniguchi, S. Nakano, Y. Kitagawa, K. Yamaguchi: Polyhe-dron **22**, 2051 (2003)

76. T. Enoki, T. Umeyama, M. Enomoto, J. Yamaura, K. Yamaguchi, A. Miyazaki, E. Ogura, Y. Kuwatani, M. Iyoda, K. Kikuchi: Synth. Met. **103**, 2275 (1999)

77. P. Day, M. Kurmoo, T. Mallah, I. R. Marsden, R. H. Friend, F. L. Pratt, W. Hayes, D. Chasseau, J. Gaultier, G. Bravic, L. Ducasse: J. Am. Chem. Soc. **114**, 10722 (1992)

78. M. Kurmoo, P. Day, P. Guionneau, G. Bravic, D. Chasseau, L. Ducasse, M. L. Allan, I. D. Marsden, R. H. Friend: Inorg. Chem. **35**, 4719 (1996)

79. M. Kurmoo, A. W. Graham, P. Day, S. J. Coles, M. B. Hursthouse, J. L. Caulfield, J. Singleton, F. L. Pratt, W. Hayes, L. Ducasse, P. Guionneau: J. Am. Chem. Soc. **117**, 12209 (1995)

80. E. Coronado, J. R. Galán-Mascarós, C. J. andLaukhin. V. Gómez-García: Nature **408**, 447 (2000)

81. A. Kobayashi, T. Udagawa, H. Tomita, T. Naito, H. Kobayashi: Chem. Lett. **2179**, (1993)

82. H. Tanaka, A. Kobayashi, A. Sato, H. Akutsu, H. Kobayashi: J. Am. Chem. Soc. **121**, 760 (1999)

83. H. Kobayashi, H. Tomita, T. Naito, A. Kobayashi, F. Sakai, T. Watanabe, P. Cassoux: J. Am. Chem. Soc. **118**, 368 (1996)

84. O. Cepas, R. H. McKenzie, J. Merino: Phys. Rev. B **65**, 100502 (2002)

85. S. Uji, H. Shinagawa, T. Terashima, T. Yakabe, Y. Terai, M. Tokumoto, A. Kobayashi, H. Tanaka, H. Kobayashi: Nature **410**, 908 (2001)

86. L. B. Balicas, J. S. Brooks, K. Storr, S. Uji, M. Tokumoto, H. Tanaka, H. Kobayashi, A. Kobayashi, V. Barzykin, L. P. Gor'kov: Phys. Rev. Lett. **87**, 067002 (2001)

87. V. Jaccarino, M. Peter: Phys. Rev. Lett. **9**, 260 (1962)

88. S. Uji, T. Terashima, C. Terakura, T. Yakabe, Y. Terai, S. Yasuzuka, Y. Imanaka, T. Takamasu, M. Tokumoto, F. Sakai, A. Kobayashi, H. Tanaka, H. Kobayashi, L. Balicas, J. S. Brooks: Synth. Met. **137**, 1183 (2003)

89. H. Kobayashi, A. Kobayashi, P. Cassoux: Chem. Soc. Rev. **29**, 325 (2000)

90. F. L. Pratt, S. L. Lee, S. J. Blundell, I. M. Marshall, H. Uozaki, N. Toyota: Synth. Met. **133**, 489 (2003)

91. C. Hotta, H. Fukuyama: J. Phys. Soc. Jpn. **69**, 2577 (2000)

92. A. Caneschi, D. Gatteschi, R. Sessoli: Inorg. Chem. **32**, 4612 (1993)

93. A. Caneschi, D. Gatteschi, P. Rey, R. Sessoli: Inorg. Chem. **30**, 3936 (1991)

94. G. Champion, N. Lalioti, V. Tangoulis, M. A. Arrio, P. Sainctavit, F. Villain, A. Caneschi, D. Gatteschi, C. Giorgetti, F. Baudelet, M. Verdaguer, C. Cartier dit Moulin: J. Am. Chem. Soc. **125**, 8371, (2003)

95. J. Villain, J. M. Loveluck: J. Phys. Lett. (Paris) **38**, L77 (1977)

96. A. Caneschi, D. Gatteschi, A. Lelirzin: J. Mater. Chem. **4319**, (1994)

97. J. C. Bonner, M. E. Fisher: Phys. Rev. **135** A640 (1964)

98. B. W. Lovett, S. J. Blundell, F. L. Pratt, T. Jestadt, W. Hayes, S. Tagaki, M. Kurmoo: Phys. Rev. B **61**, 12241 (2000)

99. J. W. Bray, L. V. Interrante, I. S. Jacobs, J. C. Bonner: *Extended Linear Chain Compounds* **3** 353 (Plenum, New York, 1983)

100. S. Huizinga, J. Kommandeur, G. A. Sawatzky, B. T. Thole, K. Kopinga, de Jonge. W. J. M., J. Roos: Phys. Rev. B **19**, 4723 (1979)

101. L. N. Bulaevskii: Sov. Phys. Solid State **11**, 921 (1969)

102. L. Néel: C.R. Acad. Sci. **34**, 3397 (1949)

103. A. H. Morrish, S. P. Yu: Phys. Rev. **102**, 670 (1956)

104. C. P. Bean, J. D. Livingstone: J. Appl. Phys. **30**, 120S (1959)

105. A. O. Caldeira, A. J. Leggett: Phys. Rev. Lett. **46**, 211 (1981)

106. A. J. Leggett, S. Chakravarty, A. T. Dorsey, M. P. A. Fisher, A. Garg, W. Zwerger: Rev. Mod. Phys. **59**, 1 (1987)

107. R. E. P. Winpenny: J. Chem. Soc. Dalton Trans. 1 (2002)

108. D. Gatteschi, A. Caneschi, L. Pardi, R. Sessoli: Science **265**, 1054 (1994)

109. J. Tejada: Polyhedron **20**, 1751 (2001)

110. E. K. Brechin, M. Soler, G. Christou, J. Davidson, D. N. Hendrickson, S. Parsons, W. Wernsdorfer: Polyhedron **22**, 1771 (2003)

111. A. L. Barra, P. Debrunner, D. Gatteschi, C. E. Schulz, R. Sessoli: Europhys. Lett. **35**, 133 (1996)

112. C. Sangregorio, T. Ohm, C. Paulsen, R. Sessoli, D. Gatteschi: Phys. Rev. Lett. **78**, 4645 (1997)

113. J. A. A. J. Perenboom, J. S. Brooks, S. Hill, T. Hathaway, N. S. Dalal: Phys. Rev. B **58**, 330 (1998)

114. A. Garg: Europhys. Lett. **22**, 205 (1993)

115. W. Wernsdorfer, R. Sessoli: Science **284**, 133 (1999)

116. M. V. Berry: Proc. Roy. Soc. A **392**, 45 (1984)

117. D. Loss, D. P. Di Vincenzo, G. Grinstein: Phys. Rev. Lett. **69**, 3232 (1992)

118. S. E. Barnes: Phys. Rev. Lett. **87**, 167201 (2001)

119. N. V. Prokof'ev, P. C. E. Stamp: Phys. Rev. Lett. **80**, 5794 (1998)

120. N. V. Prokof'ev, P. C. E. Stamp: Rep. Prog. Phys. **63**, 669 (2000)

121. W. Wernsdorfer, T. Ohm, C. Sangregorio, R. Sessoli, D. Mailly, C. Paulsen: Phys. Rev. Lett. **82**, 3903 (1999)

122. T. Ohm, C. Sangregorio, C. Paulsen: Euro Phys. J. B **6**, 195 (1998)

123. W. Wernsdorfer, A. Caneschi, R. Sessoli, D. Gatteschi, A. Cornia, V. Villar, C. Paulsen: Phys. Rev. Lett. **842965**, (2000)

124. W. Wernsdorfer, R. Sessoli, D. Gatteschi: Europhys. Lett. **47**, 254 (1999)

125. W. Wernsdorfer, N. Allaga-Alcalde, D. N. Hendrickson, G. Christou: Nature **416**, 406 (2002)

126. W. Wernsdorfer, S. Hendrickson. D. N. Bhaduri, G. Christou: Phys. Rev. Lett. **89**, 197201 (2002)

127. R. Tiron, W. Wernsdorfer, N. Aliaga-Alcalde, G. Christou: Phys. Rev. B **68**, 140407 (2003)

128. S. Hill, R. S. Edwards, N. Aliaga-Alcalde, G. Christou: Science **302**, 1015 (2003)

129. R. Tiron, W. Wernsdorfer, D. Foguet-Albiol, N. Aliaga-Alcalde, G. Christou: Phys. Rev. Lett. **91**, 227203 (2003)

130. P. Ghigna, A. Campana, A. Lascialfari, A. Caneschi, D. Gatteschi, A. Tagliaferri, F. Borgatti: Phys. Rev. B **65**, 109903 (2001)

131. R. Moroni, C. Cartier dit Moulin, G. Champion, M. A. Arrio, P. Sainctavit, M. Verdaguer, D. Gatteschi: Phys. Rev. B **68**, 064407 (2003)

Magnetism under Pressure
with Synchrotron Radiation

Matteo d'Astuto[1], Alessendro Barla[2,3,4], Nolwell Kernavanois[5], Jean-Pascal Rueff[6], Francois Baudelet[7], Rudolf Rüffer[4], Luigi Paolasini[4], and Bernard Couzinet[1]

[1] Institut de Minéralogie et de Physique des Milieux Condensés – CNRS URM7590 Université Pierre et Marie Curie – B 77, 4 Place Jussieu, 75252 Paris Cedex 05, France
`dastuto@ccr.jussieu.fr`
[2] DRFMC/SPSMS, CEA Grenoble, 17 rue des Martyrs, 38054 Grenoble Cedex 9, France
[3] Institut de Physique et Chimie des Matériaux de Strasbourg, CNRS UMR7504, 23 rue du Loess BP 43, 67034 Strasbourg Cedex 2, France
[4] European Synchrotron Radiation Facility, BP 220, 38043 Grenoble Cedex, France
[5] Institut Laue-Langevin, BP 156, 38042 Grenoble Cedex 9, France
[6] Laboratoire de Chimie Physique – CNRS UMR7614, 11 rue Pierre et Marie Curie, 75231 Paris Cedex 05, France
[7] SOLEIL – UR1 L'Orme des Merisiers, Saint-Aubin BP 48, 91192 Gif surYvette Cedex, France

Abstract. The magnetic properties of materials depend strongly on the distance between ions and on the volume available for each magnetic ion. In this article, we introduce a bird's eye view of the various possible effects of pressure on the magnetic state and the consequences of these effects on the fundamental understanding of the magnetism in condensed matter and on the physics and chemistry of minerals in the interior of the Earth and planets. We will give basic information about the experimental possibilities offered by synchrotron-radiation-based techniques in order to explore the magnetic state of matter at high pressure, their interest and feasibility. Finally, we will present some experimental highlights where x-ray techniques using synchrotron radiation have been successfully applied to show and explain the magnetic state of matter at high density.

1 Introduction

The magnetic properties of materials depend strongly both on the distance between ions and on the volume available for each magnetic ion. The most direct way to change these parameters in a controlled way is to apply external pressure. However, common magnetic experiments are often very difficult to perform with the heavy environment necessary to apply external pressure. Synchrotron radiation in the x-ray range allows us to investigate the magnetic properties of materials, while at the same time one can directly access atomic properties, necessary for a modelling of the materials studied. In this

M. d'Astuto et al.: *Magnetism under Pressure with Synchrotron Radiation*, Lect. Notes Phys. **697**, 375–399 (2006)
`www.springerlink.com`

course, we will introduce in Sect. 2 a bird's eye view of the various possible effects of pressure on the magnetic state, with no pretension of completeness. Subsequently, we will discuss the consequences of these effects on the fundamental understanding of the magnetism in condensed matter and on the physics and chemistry of minerals in the interior of the Earth and planets. Then, in Sect. 3 we will give the basic information of the experimental possibilities offered by synchrotron radiation based techniques in order to explore the magnetic state of matter at high pressure, their interest and feasibility. In Sect. 4 we will present some experimental highlights, where x-ray techniques using synchrotron radiation have been successfully used to show and explain the magnetic state of matter at high density. Finally, some general conclusion will be drawn in Sect. 5.

2 Magnetic State and Interactions in Dense Systems

2.1 Exchange Interaction with One-Band, Direct Exchange Case

In 1936 Néel [1, 2] calculated the interaction energy between two localised moments on two neighbouring atoms with overlapping orbitals. This energy is related to the ordering temperature, which can be taken as a measure of the strength of the magnetic interactions. Consider a simple system: two neighbouring atoms having their n valence-shell electrons with parallel spins, like the hydrogen molecule (Fig. 1).

Indeed, if the overlap is large, i.e. the two atoms are very close (Fig. 1), the probability to find both electrons in the same volume δV is high, hence, in order to respect Pauli's principle, the electrons tend to have opposite spins, i.e. to form a singlet state. The exchange integral $\mathcal{J}$, i.e. the crossed term in the Hartree integral, will take a negative value, which favours an anti-parallel alignment of the spins[1]. In atomic physics this corresponds to a spin-singlet state while in solid state it corresponds to a state that can be realised by an antiferromagnetic long range order, or any other state with zero net total magnetic moment. Conversely, going towards larger distances, the system will have little chance to see both electrons in the same volume δV, and a configuration which minimises the electrostatic repulsion will be preferred; therefore a spatially antisymmetric wave-function, with minimum overlap will be chosen as well as a symmetric spin configuration (spin-triplet, see Fig. 1 lower panel) to conserve a total antisymmetric wave-function. In a solid, one will observe a ferromagnetic state.

On the basis of these simple ideas and using an approach developed by Slater [2, 5], Néel showed a great variation of the energy $\mathcal{J}$ with d-δ where d is the inter-atomic distance and δ the radius of the magnetic shell. This is illustrated in Fig. 2 where the molecular field ($\frac{w}{N_a\mu^2} \times 10^{-2}$) is plotted

---

[1] See [3] and references therein for a complete formulation.

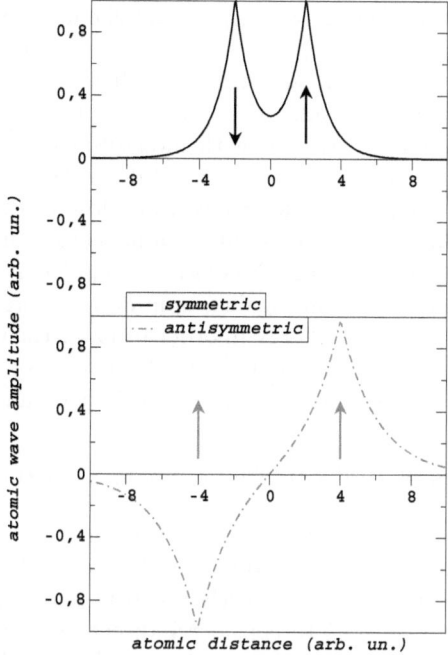

Fig. 1. Molecular electron wave

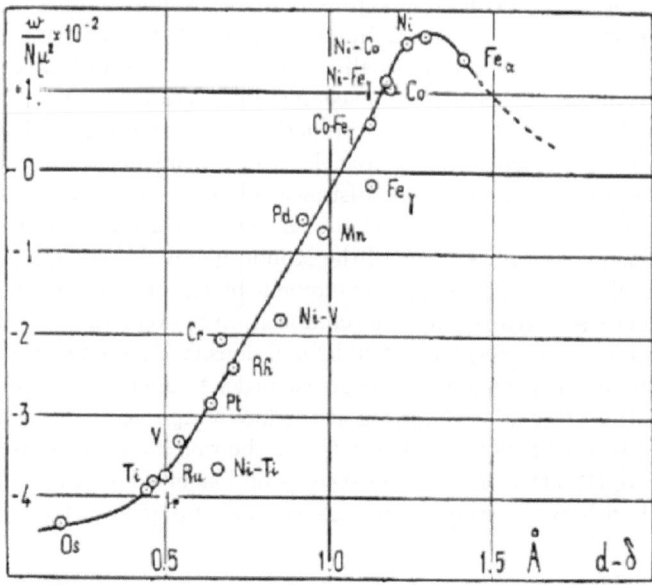

Fig. 2. Molecular field versus inter-atomic distance (after [4])

as a function of d-δ, where $\omega = \mathcal{J}$, μ is the chemical potential and N_a is Avogadro's number.

This heuristic view in real space has a well known analogue in reciprocal space, where similar considerations can be applied to a simple metallic system with one band of width W with inter-site electrostatic repulsion U. Then a system with large W/U (large orbital overlap) will favour an equal population of spin-up and spin-down electrons close to the Fermi level (nonmagnetic state or, more precisely, Pauli paramagnetic state), whilst a system with small value of W/U (large electrostatic repulsion) will prefer an unbalanced population of spin-up against spin-down electrons (ferromagnetic state), with a gain in kinetic energy in order to lower the Coulomb repulsion. On this simple idea is based the well known Stoner criterion (see the chapter *Introduction to magnetism* by C. Lacroix in this book and [6]).

2.2 Simple Cases of Indirect Exchange in Two-Band Models

The previous example was about a one-band system (one-orbital from a molecular point of view) and is called direct exchange as the electrons carrying the magnetic (spin) moment are directly involved in the exchange process. A further step in complexity is to consider indirect exchange where the magnetic electrons interact *via* an intermediate exchange with a non-magnetic band (orbital). The simplest case in the metallic systems is perhaps the one where very strongly localised electrons (i.e. with small bandwidth, typically a $4f$ rare-earth band) carrying a magnetic moment interact with delocalised (large bandwidth) electrons, typically of a lighter metal. In that case we can have two types of effects: RKKY exchange interaction and Kondo magnetic screening. The RKKY exchange interaction is the indirect interaction between localised moments *via* the free electrons of the metal band. It changes as a function of the inter-atomic distance following the Friedel oscillations, and generally increases as a function of density for a typical pressure range of 1–10 GPa[2]. A simplified view of the Kondo magnetic screening is a renormalisation of the localised magnetic moment by the spins of the free-electron cloud. It increases with density. The competition between the two effects gives rise in some special cases, for a well defined density, to a magnetic state with a large magnetic exchange but a magnetic order temperature (Néel or Curie) $T_c = 0$. In this state a large portion of the low-temperature phase diagram is dominated by quantum fluctuations: this is the well known Quantum Critical Point, where the electronic ground state is no longer a Fermi liquid but often associated with exotic states as unconventional superconductivity [7–9].

---

[2] The RKKY interaction is developped in the chapter of C. Lacroix (see (3)), where a prefactor proportional to k_F is omitted, so that the interaction between two magnetic moments separated by a distance r increases with $1/r$.

2.3 Mott-Hubbard and Charge Transfer Three-Band Systems

Indirect exchange in systems of wide interest as perovskites and perovskite-related oxides typically involves 3 bands and is connected with a large spectrum of phenomena involving transport, magnetic, structural and dynamical properties, like the Jahn-Teller effect, the Mott metal-insulator transition, all with large volume changes. Spectacular effects, as high-temperature super-conductivity and giant magneto-resistance have been observed in this class of materials [10]. The origin of these effects is still under debate. An investigation under pressure can give some additional hints or evidence, as it allows one to change the electronic structure in a controlled way. Magnetic perovskite oxides of iron are one of the main components of the earth's mantle (see e.g. [11,12]). For all these reasons, they constitute one of the most interesting subjects in the field of magnetism under pressure. Due to the lack of space for such a wide subject we refer the interested reader to the previous Mittelwihr lecture on the subject [7] and to the introductory chapter in this book by C. Lacroix and the references therein.

2.4 Magnetic Moment of Ions under Pressure

Until now we have referred only to the effect of pressure on the inter-atomic distance and, as a consequence, on the magnetic exchange between ions in a solid. However, also the magnetic state of an ion can be affected by a volume change. This magnetic state is normally determined by the well known Hund's rules (see the introductory chapter by C. Lacroix in this book and [6]). But the energy of an atomic state depends on the spatial extension of its wave-function, and a change of its volume can lead to a different electronic population of its different orbitals and, therefore, to a violation of Hund's rules with a consequent different magnetic state. A typical case is the one of spin-crossover phenomena in Fe(II) complexes, where two electronic configurations, called high-spin and low-spin states, are possible (see for example the chapter by S. J. Blundell in this book as well as the review of [13]). These ideas have often been used even in simple systems (see e.g. [11,12]). They are also applied in metallic systems as for example iron and invar alloys but care must be taken as these systems are very far from a localised ionic state; the transfer of the local orbital pictures to such delocalised bands is only possible as a very simple heuristic picture.

3 Synchrotron Radiation Methods for Magnetism under High Pressure

In order to densify a system we need to embed the sample under investigation in an environment, typically some kind of press, which will surround it in all directions, especially if one needs to apply pressure in a way as close as possible to hydrostatic conditions. As a consequence, this will practically limit synchrotron light experiments to photon-in photon-out hard x-ray techniques. The definition of hard x-ray will depend on the materials and thickness of the pressure device (see Fig. 4 in Sect. 3.1) but as an order of magnitude one can establish a limit at a minimal energy of 1 keV. These conditions limit the possible synchrotron radiation methods for magnetism under high pressure to the following ones:

1. Dichroism in the hard x-ray range.
2. Inelastic x-ray scattering, in particular for x-ray emission spectroscopy (XES).
3. X-ray resonant magnetic scattering in the hard x-ray range;
4. Nuclear resonance scattering (NRS).

The techniques of points 1–3 are widely described in the present book and the previous Mittelwihr course, therefore we will only shortly describe their advantage for the investigations of magnetic states under pressure and their feasibility (Sects. 3.2 to 3.4), while more place will be dedicated to nuclear resonance scattering and nuclear forward scattering (Sect. 3.5). Section 3.1 will briefly introduce the most used device for high pressure in x-ray experiments, the diamond anvil cell (DAC).

3.1 Diamond Anvil Cells and Related Devices for High Pressure

The working principle of a membrane diamond Anvil cell (MDAC), first described by Letoullec et al. [14], is illustrated in Fig. 3 centre, where we present a special design made for x-ray circular dichroism (XMCD) experiments and developed by J.C. Chervin (PMD-IMPMC, CNRS UMR7590 Université Pierre et Marie Curie, Paris). As we see in Fig. 3 left, two diamonds (or other hard materials as SiC moissanite) are used as opposite anvils.

Between these, a metal gasket encloses a sample which is embedded in a suitable pressure medium. The pressure is generated by an axial force (metallic inflatable membrane) on the anvils and is measured by a ruby-chip fluorescent pressure gauge [15]. Photons can reach and leave the sample either by travelling through the diamond or through the metal gasket. In the latter case the only practical choice for the metal is beryllium. It is also possible to mix the two ways, e.g. accepting incoming photons through the diamond and detecting photons coming out from the gasket at ~90°. The choice will depend on the specific needs of the experiment in terms of pressure range, angular span and energy of the photons, taking into account the difference in absorption length between diamond and beryllium. The optimum choice is still a matter of debate in some cases. One can argue that a path through the beryllium gasket is always a better configuration taking into account its larger penetration depth for x-rays compared to diamond (see Fig. 4) and the fact that it allows a 180° angular span around the sample. However this solution has several deficiencies.

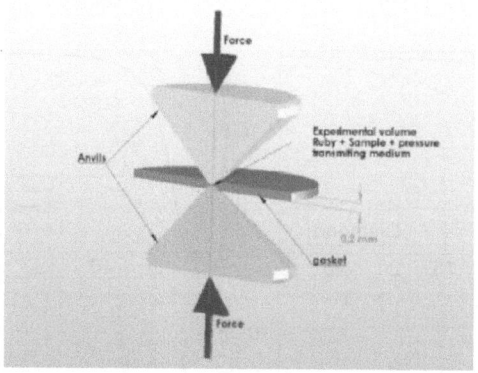

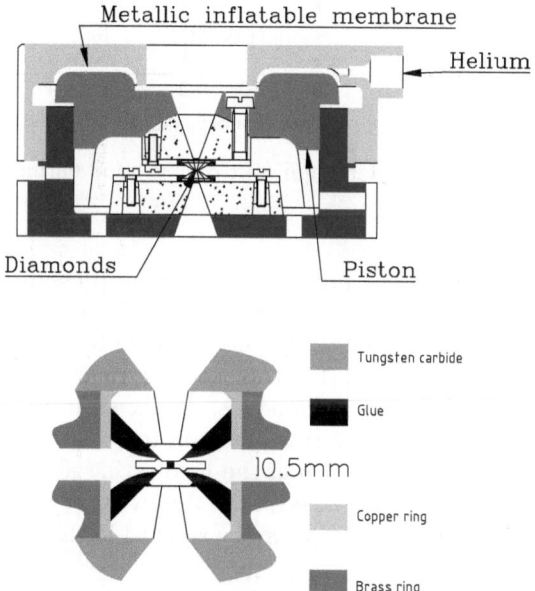

Fig. 3. DAC principle (*top*), the dedicated MDAC for XMCD made of a copper beryllium alloy (*centre*), the drilled diamonds mounting system (*bottom*). Courtesy J.C. Chervin (PMD-IMPMC, CNRS UMR7590 Université Pierre et Marie Curie, Paris)

First of all, Be is brittle, a severe handicap for use as a gasket, especially taking into account that Be powder easily reacts with air producing highly toxic beryllium oxide. For the same reason Be is difficult to machine. Moreover, detecting the photons going out from the gasket implies a signal coming from the border of the sample, especially taking into account self-absorption effects as for experiments close to an absorption edge. This may induce problems

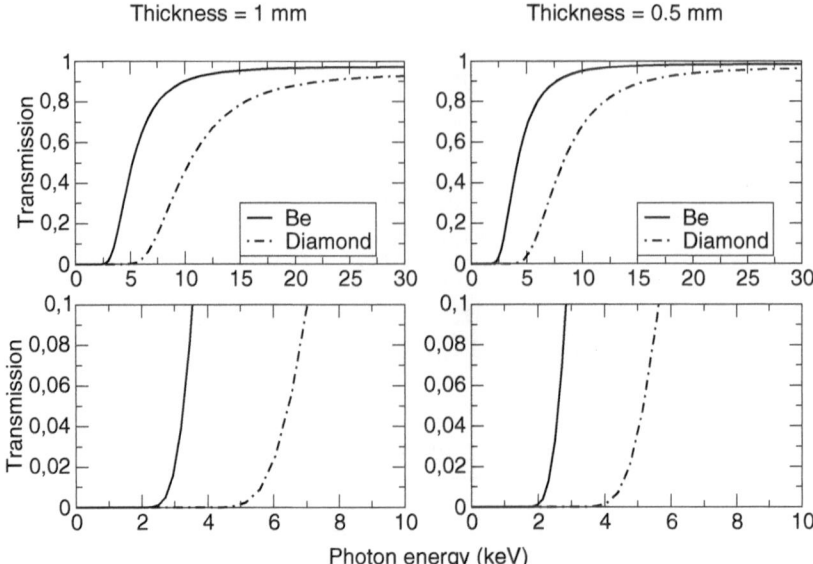

Fig. 4. Transmission factor for 1 mm (*left column*) and 0.5 mm (*right column*) of Be and diamond

in the determination of the pressure as a strong pressure gradient is often present at the borders. All these difficulties can be overcome using thin diamonds and a cell with a wide acceptance angle in front of the diamonds. As introduced by Dadashev et al. [16], transparency to low energy x-Rays can be further improved using drilled diamonds. As shown in Fig. 3 (right), a very thin diamond anvil (e.g. 500 μm) is glued on top of a drilled diamond. This solution limits the angular opening, the hole acting as a collimator, as well as the pressure range because of both the thin window and the drilled stone. However, pressures up to 100 GPa can be reached in this way and 30 GPa are routinely achieved.

3.2 Dichroism

Dichroism and, in particular, x-ray magnetic circular dichroism (XMCD) is well described in the present book (see the chapters by A. Yaresko and G. van der Laan) as well as in the previous Mittelwihr proceedings (see various contributions in [17] and in particular the one by J.-M. Mariot and Ch. Brouder [18]. We will not insist here further on the technique, but just note some peculiarities.

First of all, as stressed above in this section, only hard x-ray dichroism in photon detection (emission or absorption) is possible [7], therefore it is practically limited to some atomic x-ray absorption edges (K of transition metals, L of rare earths and M of actinides). The meaning of "hard x-ray" depends on the experimental conditions as explained above. Secondly, most if not all examples in the literature have been performed using a particular setup, the dispersive absorption one, as described in the previous Mittelwihr course by F. Baudelet on High pressure Magnetism and Magnetic Circular Dichroism [7], and more recently, in a review article by O. Mathon [19].

Last but not least, only circular polarisation experiments have been performed so far to our knowledge while linear dichroism [20,21] has never been used to study magnetism under (high) pressure, e.g. to study the magnetocristalline anisotropy against atomic distances.

3.3 Inelastic X-ray Scattering

Inelastic x-ray scattering is introduced in the chapter by M. Altarelli in this book, and well developed by the contributions of J.-M. Mariot and Ch. Brouder [18] and C. F. Hague [22] in the previous Mittelwihr course. For magnetism under pressure, it allows in particular x-ray emission Spectroscopy 4.1.

3.4 X-ray Resonant Magnetic Scattering

The importance and feasibility of magnetic diffraction experiments in the high-energy and high-angle regime is discussed in this book by the lectures of C. Dufour and M. Altarelli. A specific device for high pressure at low temperature has been recently developed at ESRF on the x-ray resonant magnetic scattering beam-line ID20. More details can be found in [23].

3.5 Nuclear Forward Scattering

Nuclear Resonance Scattering

Nuclear resonance scattering (NRS) is a resonant x-ray scattering technique based on the Mössbauer effect. While conventional Mössbauer spectroscopy is an absorption technique normally carried out in the energy domain, NRS experiments are performed in the time domain. Both processes are connected by the Heisenberg principle. In the case of Mössbauer spectroscopy a radioactive source provides γ quanta with a sharp energy distribution (neV – μeV); by varying the energy of the emitted γ quanta by Doppler effect one measures the absorption spectrum of the nuclei in the sample. In NRS an x-ray pulse from a synchrotron radiation source, which is very sharp in time ($\sim$100 ps) and very broad in energy (typically at least $\sim$meV) excites the Mössbauer level (usually the first excited nuclear state) in the sample. The deexcitation of this level gives rise to an exponential decay of the scattered intensity as a function of time. This exponential function corresponds to the Fourier transform of the Lorentzian absorption line in Mössbauer spectroscopy.

Nuclear Forward Scattering

There are several techniques based on NRS but here we focus on one technique used for magnetic studies under pressure: Nuclear forward scattering (NFS). NFS is the closest synchrotron-based analogue to Mössbauer spectroscopy, and like the latter it allows one to study the electronic and magnetic properties of solids through the hyperfine interactions that split the nuclear energy levels: the magnetic hyperfine field B_{hf} and the electric quadrupole interaction ΔE_Q.

However, while Mössbauer is an absorption spectroscopy (one γ quantum is absorbed by one nucleus), NFS is a scattering spectroscopy. The "white" synchrotron radiation, whose energy bandwidth is orders of magnitude larger than the splitting of nuclear levels by hyperfine interactions, excites all sublevels of the nuclear excited state and creates a coherent collective nuclear state. This collective nuclear state will decay giving rise to an excess of intensity emitted in the forward direction. The time scale of the nuclear deexcitation is determined by the lifetime τ_0 of the involved nuclear level, which typically lies in the range of nanoseconds to microseconds. The nuclear scattering is therefore much slower than the electronic scattering, so that a time discrimination allows one to separate the intense "prompt" electronic scattering from the much weaker "delayed" nuclear scattering. When the Mössbauer level is not split by hyperfine interactions, an exponential decay of the forward scattered intensity is observed. In the case of a Mössbauer level split by hyperfine interactions an interference pattern, the so called quantum beat structure, is superimposed to the exponential decay and contains the information about the hyperfine interaction (see next paragraph on quantum beats). Multiple scattering may as well influence the measured time decay, giving rise to dynamical beats (for further details see [24]).

Quantum Beats

The nuclear levels are characterised by their spin I (e.g. $I_g = 1/2$ for the ground state of ^{57}Fe and $I_e = 3/2$ for its first excited state, which is its most commonly used Mössbauer level) and its projection along the quantisation axis m_I (the magnetic quantum number), taking all values between $-I$ and I separated by unity. Each state has a degeneracy of $2I + 1$ in the absence of hyperfine interactions. However, this degeneracy may be partially or completely lifted by hyperfine interactions. For example, in the case of electric quadrupolar interactions acting on ^{57}Fe, the excited nuclear state splits into a $m_I = \pm 3/2$ and a $m_I = \pm 1/2$ sub-levels while the ground state remains fully degenerate. Two transition channels are therefore allowed between ground and excited states, separated by an energy ΔE typically of the order of nanoelectronvolts (see Fig. 5). In Mössbauer spectroscopy the presence of two absorption lines separated by ΔE reveals this splitting whereas in NRS an interference pattern, the quantum beats, appears with a single frequency $\Omega \propto \Delta E$ (Fig. 5).

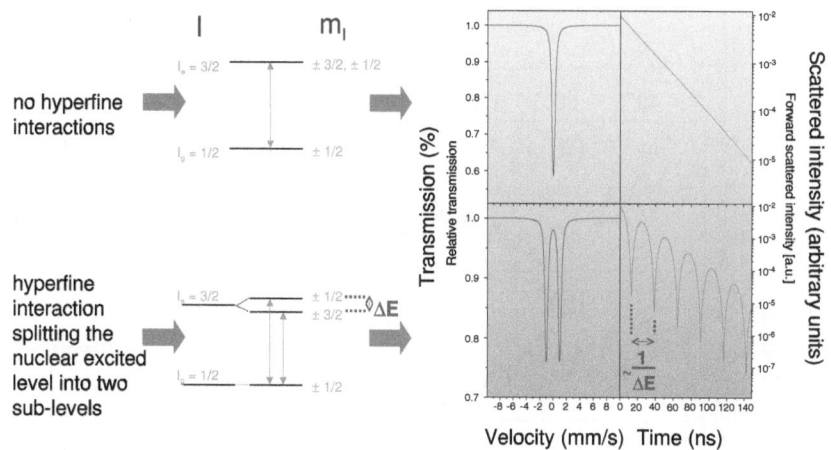

Fig. 5. *Left panel*: energy level scheme for ^{57}Fe nuclei in the absence of hyperfine interactions (*top*) and in the presence of an electric quadrupole interaction (*bottom*). *Right panel*: corresponding Mössbauer and NFS spectra

In the presence of a magnetic hyperfine field a full splitting of the nuclear levels occurs giving rise to six allowed nuclear transitions (those with $\Delta m_I = 0, \pm 1$), with six absorption lines in the Mössbauer spectrum and a complicated quantum beat pattern in the NFS spectrum. Contrary to Mössbauer spectroscopy where the photons from the radioactive source are normally unpolarized, the photons from the synchrotron radiation source are linearly polarized. This polarization can affect strongly the time spectra, as shown in Fig. 6, and can be used to excite selectively only some transitions according to the relative directions of a magnetic field **H**, the wave vector **k** and the polarization vector **E** of the incident x-rays.

4 Application of Synchrotron Radiation Techniques to Typical Case Studies of Magnetic Properties under Pressure

In this section we will present some results obtained with the techniques described above used to investigate magnetism under pressure. The first example concerns the $\alpha \rightarrow \epsilon$ phase transition in iron. This example is didactically very interesting as it has been studied with several of the above mentioned techniques (XMCD, XES, NFS), therefore giving a comparative idea of the different approaches. Then we give an example of a valence and magnetic transition under pressure studied by NFS. All these techniques will not give any direct information on the inter-site magnetic correlations, which can only be obtained in diffraction experiments. Therefore, a last example will concern the recent developments of RXMS under pressure, the only technique which

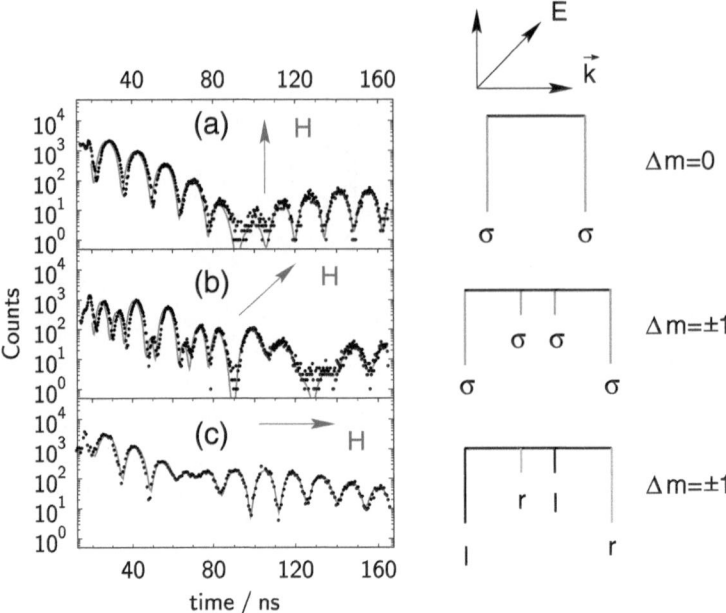

Fig. 6. *Left panel*: measured nuclear forward scattering time spectra (*circles*) of iron in case of a magnetic hyperfine interaction for different alignments of a magnetic field $\boldsymbol{H}$ with respect to the wave vector $\boldsymbol{k}$ and the polarisation vector $\boldsymbol{E}$ of the incident x-rays. The *solid lines* are fits according to the theory of NFS. *Right panel*: corresponding nuclear transition lines with their polarisation state (σ-σ linearly polarised, l and r – *left* and *right* polarised, respectively). Δm is the change of the magnetic quantum number for the different transitions (after [24, 25])

allows one to obtain this type of information but for which experiments under pressure have been developed only recently.

4.1 Dynamics of the Magnetic and Structural $\alpha \rightarrow \epsilon$ Phase Transition in Iron

The iron phase diagram has attracted considerable interest for a long time. At the beginning the motivation was its central role in the behaviour of alloys and steel [26]. Later, its geophysical importance was underlined because of its predominant abundance in the Earth's core [27, 28]. The phase diagram of iron at extreme pressure and temperature conditions is still far from being established. Under the application of external pressure iron undergoes a transition at 13 GPa from the bcc α-phase to the hcp ϵ-phase structure [29], with the loss of its ferromagnetic long-range order [30]. In the literature the transition extends over 8 GPa at ambient temperature within which the bcc and hcp phases coexist [31]. In the pure hcp ϵ-phase superconductivity appears close to the transition between 15 and 30 GPa [32, 33].

The evolution with pressure of the magnetic and structural states across this transition is still a subject of active theoretical and experimental research. Considerable ab initio theoretical studies of this transition have been carried out [26, 27, 33–38]. The first-order nature of the structural and magnetic transitions is well established as well as the non-ferromagnetic state of the hcp phase. It is also clear that the ferromagnetism of iron has a fundamental role for the ground state structure in the stabilization of the bcc phase with respect to hcp or fcc. The driving role of magnetism in the $\alpha \to \epsilon$ phase transition was predicted for the first time in [34]. More recently, [35] presents a complete map of the transition path between bcc and hcp phase using a spin-polarized full-potential total energy calculation within the generalized-gradient method (GGA). It is found that the effect of pressure is essentially to broaden the d bands and to decrease the density of states at the Fermi level below the stability limit for ferromagnetism given by the Stoner criterion. Therefore, they conclude that the $\alpha \to \epsilon$ phase transition is not primarily due to phonon softening but to the effect of pressure on the magnetism of iron. Recent calculations based on a multi-scale model containing a quantum-mechanics-based multi-well energy function propose an initiation of the bcc-hcp phase transition primarily due to shear [39]. During the transition, a low-spin state is also suggested. At higher densities, an incommensurate spin-density-wave ordered state is predicted for a small pressure range starting with the onset of the hcp phase, as for superconductivity [40].

Structural investigations have been reported using x-ray diffraction [41] and more recently EXAFS [31]. The iron bcc-hcp phase transition is described as a martensitic transition with a slow variation of the relative bcc and hcp phase abundance. Distorted bcc and hcp phases with anomalously large lattice constant (bcc) and c/a ratio (hcp) when the relative amounts become small are reported [31]. These anomalies are attributed to interfacial strain between the bcc and hcp phases. A transition model based on lattice shearing movements is proposed with a possible intermediate fcc structure.

Magnetic measurements are traditionally based on Mössbauer techniques [30, 42–44] and, recently, on inelastic x-ray scattering [45]. The authors of [45] report on modifications of a satellite in the Fe-K_β fluorescence line, which show that the magnetic moment of iron decreases over the same pressure range as the bcc to hcp transition i.e. 8 GPa. These data agree with Mössbauer results [44] which indicate that the main part of the magnetic moment decrease occurs in the same pressure domain. Nevertheless the lack of reproducibility of high-pressure conditions does not allow a precise correlation between the structural and the magnetic transition when measured separately. A recent study reports measurements of phonon dispersion curves by inelastic neutron scattering of bcc iron under pressure up to 10 GPa over the entire stability range of the α phase [46]. A surprisingly uniform pressure dependence of the α phase fraction and a lack of any significant pre-transitional change close to the bcc-hcp transition was found, and it was concluded that this behaviour confirms the driving role of magnetism of iron in the transition.

More recently O. Mathon et al. [47] have obtained a direct correlation at room temperature between magnetic and structural data using polarised x-ray absorption spectroscopy as detailed in the next paragraph.

Absorption Spectroscopy Results

Figure 7 compares the normalized Fe K-edge XAS (left panel) and XMCD (right panel) data of pure Fe foil at ambient conditions (bottom curves), to examples of spectra recorded within the DAC at different pressure values, between (10.0 ± 0.1) GPa and (22.4 ± 0.1) GPa, during a pressure increase ramp. The energy range diffracted by the polychromator is limited to the XANES region and the first oscillations of the EXAFS domain. The XAS data shows very clearly that both the electronic structure and the local structure around Fe are drastically modified above 14 GPa: the pre-peak at 7116 eV becomes more pronounced, the maximum of the absorption drifts towards higher

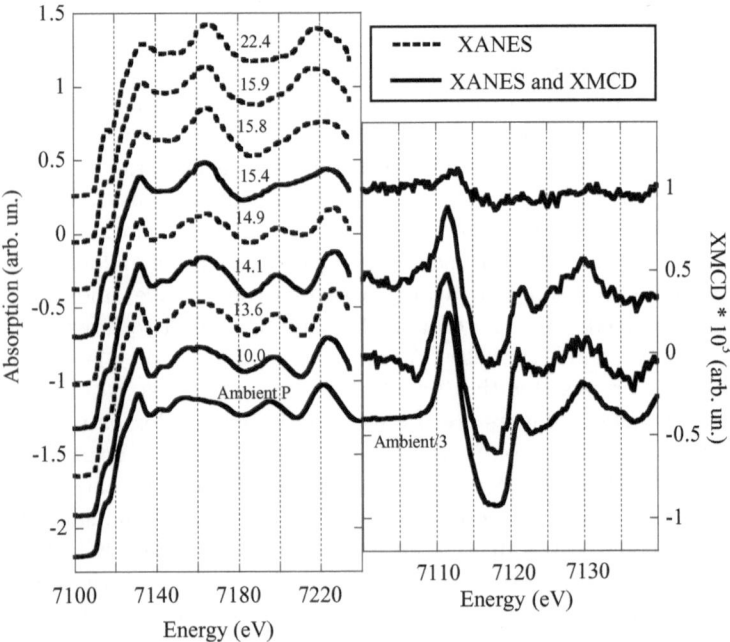

Fig. 7. Fe K-edge XAS (*left panel*) and some examples of XMCD (*right panel*) as a function of pressure between the ambient pressure bcc phase and the high pressure hcp phase. Continuous lines correspond to data measured simultaneously. The small glitches in the XAS at 7158 and 7171 eV in the high-pressure data are artifacts due to small defects of the polychromator crystal (after [47])

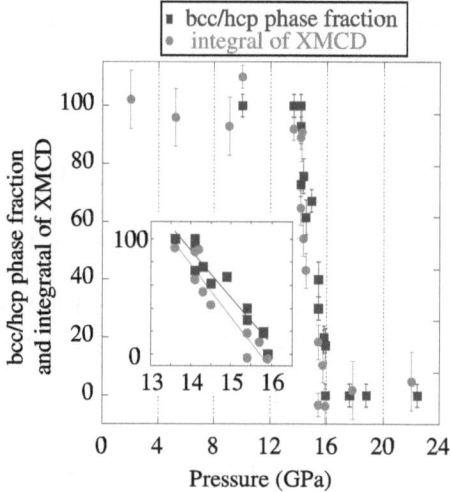

Fig. 8. Evolution of the amplitude of the derivative of the XAS (*full squares*) compared to the reduction of the amplitude of the XMCD signals (*full circles*). The inset is a zoom of the transition region, the two lines are a linear fit of the XAS and XMCD data points in this pressure region (after [47])

energies (from 7132 to 7134 eV) and becomes less intense, and the frequency of the EXAFS oscillations is drastically reduced. The EXAFS signature of the different steps of the Fe bcc-hcp transition was already clearly identified by Wang and Ingalls [31] and the data presented here are consistent with the structure reported. The XMCD curves in the right correspond to the simultaneously obtained XAS spectra shown as continuous lines in the left. The amplitude of the ambient pressure XMCD signal obtained with the sample in the DAC (not shown) is equal to approximately 30% of its value measured outside the DAC due to the smaller applied magnetic field which cannot overcome the demagnetization factor of the probed iron foil. In order to better identify the onset and the evolution with pressure of this phase transition, each pair of XAS and XMCD spectra are carefully analyzed as follows: the first derivative of the XAS spectra in the pressure range close to the transition region was calculated. The amplitude of the derivative signal changes drastically between the bcc phase and the hcp phase at E = 7137 eV, E = 7205 eV and E = 7220 eV. At these energy points, one has the highest sensitivity to the bcc/hcp phase fraction (note that the phase fraction calculated at the three different energies is identical within the error bar). To quantify the amplitude of the XMCD signals, the absolute value of the background subtracted data are integrated in the energy range 7100–7122 eV. The magnetic/non-magnetic phase fraction is then determined.

In Fig. 8 we plot (full squares) the evolution of the bcc/hcp phase fraction and of the magnetic/non-magnetic phase fraction (full circles) as a function of pressure. The onset of the structural transition occurs at about (13.8 ± 0.1) GPa and terminates at (16.2 ± 0.1) GPa. In [31], the transition started at a lower pressure and occurred in a larger pressure range with respect to that observed in the present work. This could be attributed to the different conditions of non-hydrostaticity within the cell, and underlines the difficulty of reproducing the same thermodynamic conditions in different experiments. The magnetic transition is indeed quite abrupt occurring within a pressure range of (2.2 ± 0.2) GPa, with an onset at (13.5 ± 0.1) GPa.

These data show that between 13.5 and (15.7 ± 0.1) GPa the XMCD is suddenly reduced to zero, within the error bars, indicating the disappearence of the room-temperature ferromagnetic order of iron. The abrupt drop to zero of the iron magnetic moment when the bcc to hcp phase transition occurs proves the first-order nature of the pressure-induced magnetic transition, as predicted by Ekman et al. [35]. Note that the observed absence of macroscopic magnetization above 15.7 GPa is in good agreement with the recent prediction of antiferromagnetic fluctuations for a small pressure range starting with the onset of the hcp phase [40], which is suggested as a possible origin of superconductivity. The narrower pressure transition domain for both structural and the magnetic phase transitions found in the data presented here from [47] with respect to previous experiments [31,44,45] could be attributed to a lower pressure gradient within the probed region, directly correlated to the smaller sample volume probed by the x-ray beam [48]. The influence of different experimental conditions on the pressure value at the transition onset and on the pressure range of phase coexistence is well known and frequently addressed in literature (see for example [31]). For this reason, the simultaneous measurement of element-specific magnetic properties using XMCD and of element-specific local structural and electronic properties using XAS, that overcomes all uncertainty related to different hydrostatic conditions within the pressure cell and different probed volumes, is fundamental to address issues concerning correlations between magnetic, electronic and structural degrees of freedom of a system during phase transitions. These results can be compared with previous measurements obtained from NFS [49] and Inelastic x-ray scattering (IXS) [45] experiments detailed below.

Nuclear Forward Scattering

The NFS experiments on iron under pressure were performed at the early stage of the technique as a feasibility test for high-pressure measurements, thanks to the possibility to compare the results with the large literature available on the subject and based on Mössbauer spectroscopy studies [49]. In Fig. 9 one can see the quantum beats originating from the magnetic hyperfine field on the ^{57}Fe nuclei in the magnetic state of α-iron at 3 GPa. The beat contrast is reduced just below the transition (14 GPa) and persists just above it (18 GPa), disappearing completely only at a considerably higher pressure (21 GPa). One may ask oneself if the residual signal seen at 18 GPa (Fig. 9)

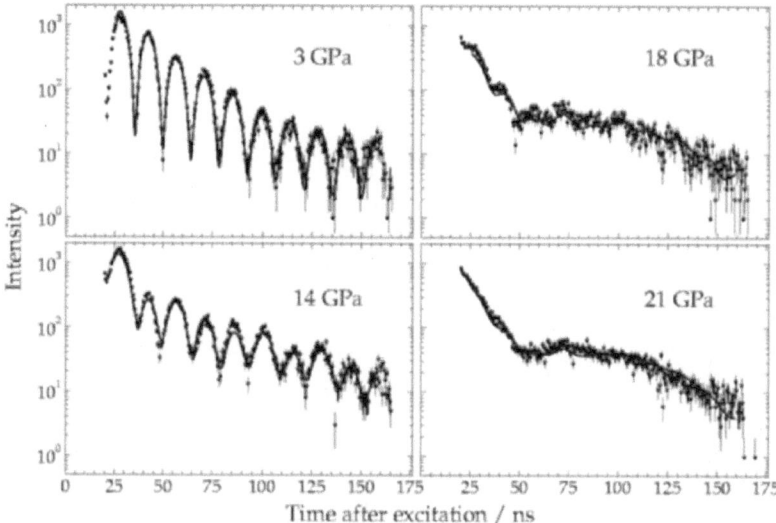

Fig. 9. *Left panel*: measured nuclear forward scattering time spectra of iron (*circles*) for different pressures below (3 and 14 GPa) and above (18 and 21 GPa) the $\alpha \rightarrow \epsilon$ phase transition, at 300 K (after [49])

could be related to a local moment with no ferromagnetic order. Such an explanation would be in contradiction with Mössbauer results which do not detect any signal and set a higher-bound for the magnetic moment of ϵ-iron to $0.05\,\mu_{\mathrm{B}}$ [50]. This will set the time scale for the first quantum beat well outside the window allowed by the NFS technique. Therefore we suppose that the residual signal observed originates from a pressure gradient in the DAC cell.

Inelastic X-ray Scattering

Figure 10 (left panel), shows the XES spectra measured at the K_β emission line in pure Fe under pressure. The emission process results from the decay of a $3p$ electron to a $1s$ vacant state, following primary excitation by incident photons of energy chosen above the Fe-K absorption edge. Only the two extreme pressure points measured on both sides of the α-ϵ transition are represented. The K_β line at ambient pressure can be decomposed into a main line of high intensity and a low-energy satellite, sensibly broader. Schematically, the splitting between the two features arises from the exchange interaction in the XES final state between the $3p$ hole state and the $3d$ magnetic shell. The intensity and position of the satellite is sensitive to the $3d$ local spin magnetic moment through this final-state effect, and changes in the emission line shape are expected as the Fe-ion spin-state is reduced. Such changes are clearly visible in the high pressure XES spectrum, which shows a sizable decrease of the satellite intensity with respect to the main peak.

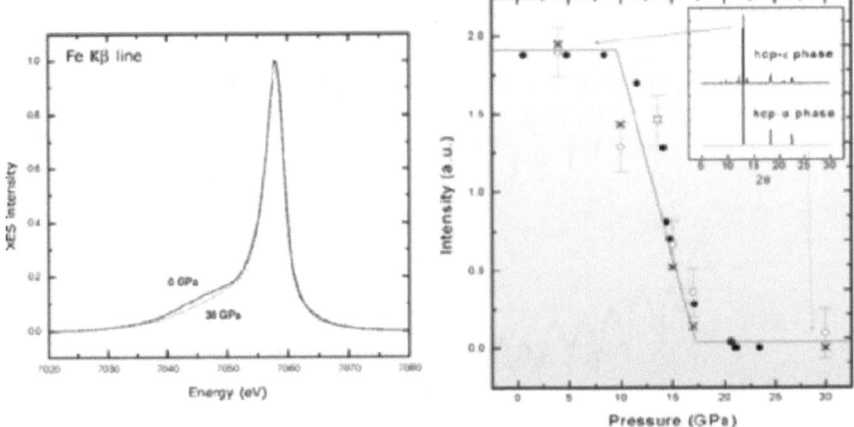

Fig. 10. *Left panel*: XES spectra measured at 0 and 30 GPa in pure Fe. The main peak is normalized to unity and aligned to 7058 eV. At low pressure (*solid line*), the satellite at 7045.5 eV is characteristic of the magnetic state. Its disappearance at higher pressure (*dotted lines*) denotes the transition to the non-magnetic state. *Right panel*: integrated intensity of the satellite of the Fe K_β line (*open circles*) compared to the α-phase fraction determined by Mössbauer spectroscopy (Taylor et al., 1991) (*solid circles*) upon pressure increase. The latter curve has been scaled to the low-pressure XES satellite intensity. The *solid line* is a guide to the eyes. The intensity of the XES satellite was calculated from the difference spectra obtained by subtracting the 23 GPa spectrum from each scan. The plateau at low (high) pressure corresponds to the magnetic (non-magnetic) phase. The diffraction pattern measured in both phases, shown in the inset, confirms the structural change (after [45])

The variation of the emission line shape, thus of the Fe magnetic moment, as a function of pressure is more precisely indicated in Fig. 10 (right) for the complete set of data. The quantification of the spectral change is derived from a sum-rule analysis, which relates the absolute value of the integrated difference signal (between two pressures) to the change of the spin-magnetic moment. The magnetic state of Fe suddenly collapses above 10 GPa, denoting the onset of the α-ϵ transition, which is fully terminated around 15 GPa. The results agree well with Mössbauer data, also shown in Fig. 10. The spread of the transient state, where both α and ϵ phases coexist, might result from a pressure gradient in the pressure cell, or a quantum superposition of the magnetic and non-magnetic states in the vicinity of the transition. The XMCD twin experiment discussed earlier gives evidence in favour of the former argument.

XES provides complementary information compared to other high-pressure magnetic probes. It is a high-energy probe fully compatible with high-pressure sample environments, which can be easily applied to the whole transition-metal series without worrying about isotopic substitution. As a pure local probe, XES does not require the sample to bear a macroscopic magnetic moment, and is especially well suited to study antiferromagnetic and paramagnetic phases. Finally, the XES spectra can be simulated within the full multiplet approach. The choice of parameters, adjusted in the calculations by comparison with the experiment, will in turn provide information about fundamental quantities in the material, such as electron correlation strength, crystal-field splitting or charge-transfer energy, and their evolution at high pressure.

4.2 Pressure-Induced Magnetic Order in "Golden" SmS

At ambient pressure SmS is a non-magnetic semiconductor (black phase), with divalent samarium ions ($4f^6$: $^7F_0$). At $\sim$0.65 GPa and room temperature (RT) it undergoes a pressure-induced isostructural (NaCl-type) first-order transition towards a metallic phase (termed "gold phase") with a large volume collapse ($\sim$15%). Gold SmS is in an intermediate valence (IV) state with a Sm valence of about 2.7 just above the transition [51]. The electrical resistivity shows however that the ground state of the gold phase is a semiconductor with a gap that decreases with increasing pressure. The gap closes completely at $p_\Delta \approx 2$ GPa [52,53], above which a metallic ground-state is observed. The interesting question concerns the possible crossover of the IV state of SmS to the trivalent ($4f^5$: $^6H_{5/2}$) state, which is magnetic, at a critical pressure p_c. Measurements at the Sm L_3 absorption edge show that the Sm valence increases non-linearly above the transition pressure from the black (B) to the gold (G) phase $p_{B-G} \sim 0.65$ GPa and the trivalent state is reached only at $p \sim 13$–15 GPa [54,55].

Using ^{149}Sm high pressure nuclear forward scattering (NFS) of synchrotron radiation it has been clearly shown that above p_Δ the collapse of the semiconducting gap coincides with the appearance of magnetic ordering [56]. From these measurements one obtains information about the hyperfine interactions (the magnetic hyperfine field B_{hf} and the electric quadrupole splitting ΔE_Q) at the ^{149}Sm nuclei as a function of pressure and temperature. Figure 11(left) shows some selected ^{149}Sm NFS spectra collected for SmS up to 19 GPa and at 3 K. As shown in the figure, at 1.73 GPa one observes a spectrum characteristic of unsplit nuclear levels, i.e. quadrupole or magnetic interactions are absent. The Sm ions are thus in a non-magnetic state and, as expected for a NaCl-type structure, in a cubic symmetry. At 2.35 GPa, the spectral shape changes significantly and shows quantum beats indicating that the nuclear levels are now split by hyperfine interactions.

The best fit to the data is obtained by assuming a superposition of non-magnetic and magnetic components with relative weights of about 28% and 72%, respectively (see Fig. 12). The magnetic component in the NFS spectra grows with pressure at the expense of the non-magnetic component and at 5.1 GPa and 3 K all the Sm ions feel combined magnetic and quadrupole interactions. Above 5.1 GPa and up to 19 GPa one observes only a slight increase

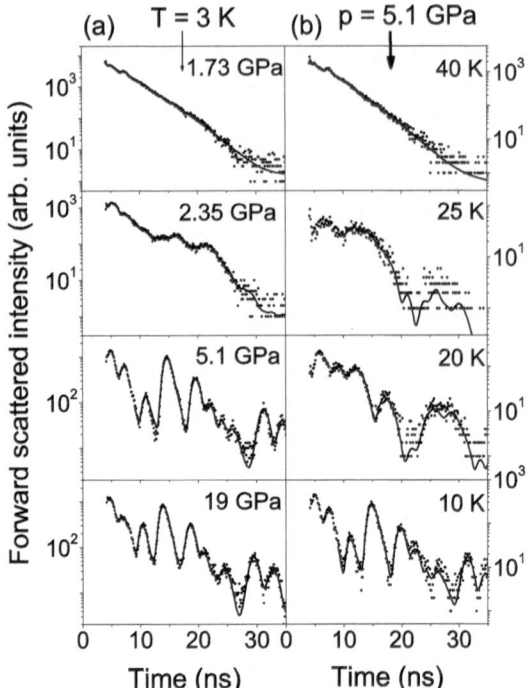

Fig. 11. ^{149}Sm NFS spectra of SmS at $T = 3\,\mathrm{K}$ (*left*) for some selected pressures and at $p = 5.1\,\mathrm{GPa}$ and different temperatures (*right*). The dots represent experimental data points while the lines are fits

of the hyperfine parameters. The pressure dependences of B_{hf}, ΔE_Q and of the magnetic component fraction at $3\,\mathrm{K}$ (see Fig. 12) evidence a pressure-induced phase transition at $p_{\mathrm{c}} \approx 2$ GPa from a non-magnetic state to a magnetically ordered state. The steep variation of both B_{hf} and ΔE_Q as well as the coexistence of the low- and high-pressure phases in the vicinity of p_{c} point towards a first-order transition.

The long-range character of the magnetic order observed by NFS at low temperatures above $\sim$2 GPa is confirmed by specific-heat measurements [57] that show the appearance of a phase transition for $p \geq 2\,\mathrm{GPa}$, at temperatures $T_{\mathrm{m,sh}}$ increasing with pressure (see Fig. 13). The temperature dependence of the hyperfine parameters as measured by NFS, however, does not reveal any sharp transition at $T_{\mathrm{m,sh}}$, but rather a continuous decrease with increasing temperature up to $T_{\mathrm{m,NFS}} \sim 2T_{\mathrm{m,sh}}$ (see Fig. 13). This can be attributed to the persistence of short-range magnetic correlations in the paramagnetic phase.

These ^{149}Sm NFS experiments show the power of this method in detecting the magnetic phase transition induced by pressure in SmS, which other techniques had failed to detect. The results show that at $p_{\mathrm{c}} \approx 2\,\mathrm{GPa}$ SmS

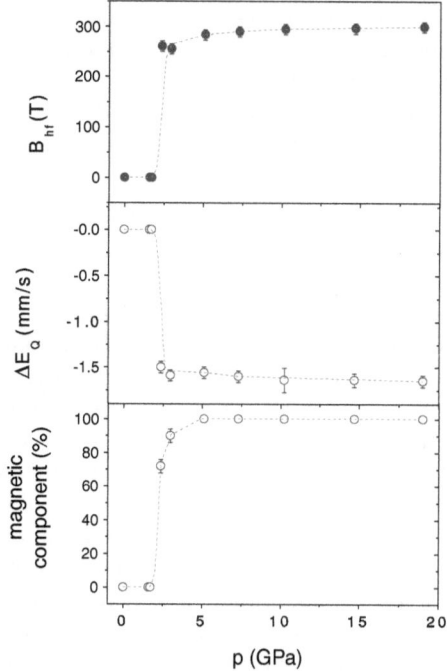

Fig. 12. Pressure dependences of the magnetic hyperfine field (B_{hf}) (*bottom*), the quadrupole splitting (ΔE_Q) (*middle*) and the magnetic component fraction at $T = 3\,\mathrm{K}$ (*top*). The *dashed lines* through the data points are guides to the eye (after [56])

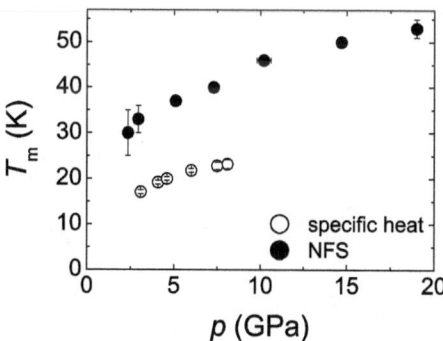

Fig. 13. Pressure dependence of the magnetic ordering temperature T_m of SmS as obtained from high pressure ^{149}Sm NFS measurements (*full circles*) in comparison with the values inferred from high pressure specific heat (*empty circles*) measurements (after [56])

undergoes a transition from a non-magnetic (semiconducting) state to a magnetically ordered (metallic) state with $\mu_{Sm} \approx 0.5~\mu_B$. The transition has a first-order character and the high-pressure magnetically-ordered state can be interpreted in the frame of trivalent Sm ions with a Γ_8 quartet crystal field ground state, although the Sm ions are in an IV state up to about 15 GPa. For further details see [55–58].

4.3 High Pressure Resonant Magnetic X-ray Diffraction and Experiments in $Ce(Co_{0.07}Fe_{0.93})_2$

$CeFe_2$ under pressure presents an antiferromagnetic phase. Co doping acts as a positive chemical pressure. A first Resonant x-ray magnetic scattering (RXMS) feasibility experiment under pressure has been performed on the 7% Co-doped $CeFe_2$ compound, at the Ce-L_3 absorption edge (5.720 keV). The 7% cobalt content has been chosen in order to start from an antiferromagnetic phase which can be studied at ambient pressure conditions. The rocking curves of the structural charge peak (222) at 300 K and of the magnetic peak $(\frac{5}{2}, \frac{5}{2}, \frac{5}{2})$ at 10 K do not show any significant change of width with pressure, indicating that the applied pressure is quite hydrostatic.

Integrated intensity variations of the $(\frac{5}{2}, \frac{5}{2}, \frac{5}{2})$ magnetic peak as a function of both x-ray energy and temperature are presented in Fig. 14 at three different applied pressures (0, 0.5 and 9.5 kbar). The energy scans of magnetic reflections (see inset of Fig. 14) do not present any change in this pressure range. On the other hand, temperature dependences show the increase of the

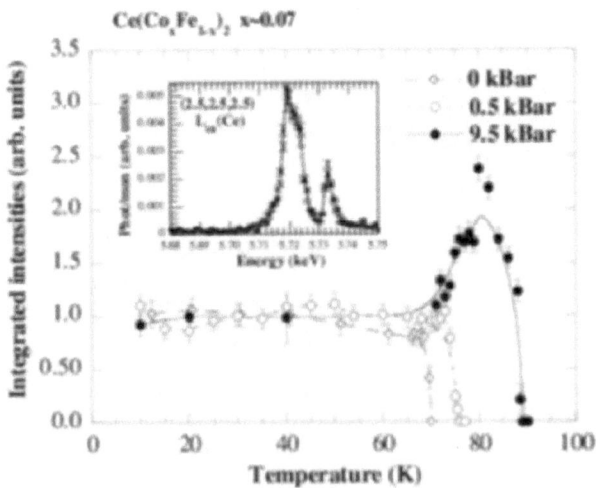

Fig. 14. Integrated intensity at the Ce-L_3 absorption edge of the $(\frac{3}{2}, \frac{3}{2}, \frac{3}{2})$ magnetic peak as a function of temperature at 0, 0.5 and 0.95 GPa. The inset shows the energy scans measured in the σ-π channel at 10 K under 0.5 and 0.95 GPa (after [23, 60])

Néel temperature, T_N, confirming the stabilization of the antiferromagnetic state with increasing pressure, as found in magnetization measurement under pressure [59]. Details of this study will be published elsewhere [60]. With this preliminary measurement, the feasibility and the reliability of RXMS under pressure at low temperature has been evidenced from both a technical and a physical point of view.

5 Conclusion

The importance of studying magnetism under pressure follows from the fact that the magnetic properties of materials depend on both the distance between ions and the volume available for each magnetic ion. To these investigations, as for all the studies under pressure, synchrotron light gives a major impulse thanks to the possibility to perform direct measurements of *atomic* properties, necessary for a modelling of the materials studied under extreme conditions. In the present work we indicated how to proceed in order to perform such kind of experiments. Moreover, the examples given in the present course show some highlights where x-ray techniques using synchrotron radiation have been successfully used to evidence and explain the magnetic state of matter at high density.

Acknowledgement

We acknowledge Ch. Brouder for useful discussions and critical reading. The authors are grateful to J.C. Chervin for technical help.

References

1. L. Néel: Oeuvres scientifiques (CNRS 1978), A **23** (1978) p 141
2. J.C. Slater: Phys. Rev. **36**, 57 (1930)
3. Ch. Cartier dit Moulin: Molecular Magnetism. In: *Magnetism and Synchrotron Radiation*, (Lecture Notes in Physics vol 565), ed. by E. Beaurepaire, F. Scheurer, G. Krill, J.-P. Kappler (Springer, Berlin, Heidelberg, New York 2001) pp 209–224
4. L. Néel: Ann. de Phys. **5**, 232 (1936)
5. L. Néel: C.R.Ac.Sc. **198**, 1311 (1934)
6. D. Givord: Introduction to Magnetism and Magnetic Materials. In: *Magnetism and Synchrotron Radiation*, (Lecture Notes in Physics vol 565), ed. by E. Beaurepaire, F. Scheurer, G. Krill, J.-P. Kappler (Springer, Berlin Heidelberg New York 2001) pp 3–23

7. F. Baudelet: High-Pressure Magnetism and Magnetic Circular Dichroism. In: *Magnetism and Synchrotron Radiation*, (Lecture Notes in Physics vol 565), ed. by E. Beaurepaire, F. Scheurer, G. Krill, J.-P. Kappler (Springer, Berlin Heidelberg New York 2001) pp 254–272
8. J.A. Hertz: Phys. Rev. B **14**, 1165 (1976)
9. G.R. Stewart: Rev. Mod. Phys. **56**, 000755 (1984)
10. M. Iamada, A. Fujimori, Y. Tokura: Rev. Mod. Phys. **70**, 1039 (1998)
11. J. Badro, G. Fiquet, F. Guyot et al: Science **300**, 789 (2003)
12. J. Badro, J.P. Rueff, G. Vanko et al: Science **305**, 383 (2004)
13. P. Gutlich, Y. Garcia, H.A. Goodwin: Chem. Soc. Rev. **29**, 419 (2000)
14. R. Letoullec, J.P. Pinceaux, P. Loubeyre: High Pressure Research **1**, 77 (1988)
15. J.C. Chervin, B. Canny, M. Mancinelli: High Press. Res. **21**, 305 (2001)
16. A. Dadashev, M.P. Pasternak, G.Kh. Rozenberg, R.D. Taylor: Rev. Sci. Instrum. **72**, 2633 (2001)
17. *Magnetism and Synchrotron Radiation*, (Lecture Notes in Physics vol 565), ed. by E. Beaurepaire, F. Scheurer, G. Krill, J.-P. Kappler (Springer, Berlin Heidelberg New York 2001)
18. J.-M. Mariot, Ch. Brouder: Spectroscopy and Magnetism: An Introduction. In: *Magnetism and Synchrotron Radiation*, (Lecture Notes in Physics vol 565), ed. by E. Beaurepaire, F. Scheurer, G. Krill, J.-P. Kappler (Springer, Berlin Heidelberg New York 2001) pp 24–59
19. O. Mathon, F. Baudelet, J.-P. Itié et al.: J. Synchr. Rad. **11**, 423 (2004)
20. G. van der Laan: Relation Between X-ray Magnetic Linear Dichroism and Magnetocristalline Anisotropy. In: *Magnetism and Synchrotron Radiation*, (Lecture Notes in Physics vol 565), ed. by E. Beaurepaire, F. Scheurer, G. Krill, J.-P. Kappler (Springer, Berlin Heidelberg New York 2001) pp 339–342
21. J. Goulon et al.: X-ray Gyrotropy Related Spectroscopies. In: *Magnetism and Synchrotron Radiation*, (Lecture Notes in Physics vol 565), ed. by E. Beaurepaire, F. Scheurer, G. Krill, J.-P. Kappler (Springer, Berlin Heidelberg New York 2001) pp 374–381
22. C.F. Hague: Resonant Inelastic X-ray Scattering. In: *Magnetism and Synchrotron Radiation*, (Lecture Notes in Physics vol 565), ed. by E. Beaurepaire, F. Scheurer, G. Krill, J.-P. Kappler (Springer, Berlin Heidelberg New York 2001) pp 273–290
23. N. Kernavanois, P.P. Deen, D. Braithwaite, L. Paolasini: Rev. Scient. Instr. *in press* (2005)
24. R. Rüeffer, J. Teillet: Diffusion nucléaire résonnante avec le rayonnement synchrotron In: *Techniques de l'Ingénieur, traité Analyse et Caractérisation* (Techniques de l'Ingénieur), P2605 pp 1–12
25. O. Leupold, K. Rupprecht, G. Wortmann: Structural Chemistry **14**, 97 (2003)
26. H. Hasegawa and D.G. Pettifor Phys. Rev. Lett. **50**, 130 (1983)
27. L. Stixrude, R.E. Cohen and D.J. Singh: Phys. Rev. B **50**, 6442 (1994)
28. R. Jeanloz, Ann. Rev. Earth Planet. Sci. **18**, 357 (1990)
29. D. Bancroft, E.L. Peterson and S. Minshall: J. Appl. Phys. **27**, 291 (1956)
30. M. Nicol and G. Jura: Science **141**, 1035 (1963)
31. F.M. Wang and R. Ingalls: Phys. Rev. B **57**, 5647 (1998)
32. K. Shimizu, T. Kimura, S. Furomoto et al.: Nature **412**, 316 (2001)
33. S.K. Bose, O.V. Dolgov, J. Kortus et al.: Phys. Rev. B **67**, 214518 (2003)
34. T. Asada and K. Terakura: Phys. Rev. B **46**, 13599 (1992)

35. M. Ekman, B. Sadigh, K. Einarsdotter and P. Blaha: Phys. Rev. B **58**, 5296 (1998)
36. H.C. Herper, E. Hoffmann and P. Entel: Phys. Rev. B **60**, 3839 (1999)
37. P. Söderlind and J.A. Moriarty, Phys. Rev. B **53**, 14063 (1996)
38. W. Pepperhoff and M. Acet, In *Constitution and magnetism of Iron and its alloys* (Springer-Verlag, Berlin Heidelberg, 2001)
39. K.J. Caspersen, A. Lew, M. Ortiz and E.A. Carter: Phys. Rev. Lett. **93**, 115501 (2004)
40. V. Thakor, J.B. Stauton, J. Poulter et al.: Phys. Rev. B **67**, R180405 (2003)
41. E. Huang, W.A. Basset and P. Tao: J. Geophys Res. **92**, 8129 (1987)
42. G. Cort, R.D. Taylor and J. Willis: J. Appl. Phys. **53**, 2064 (1982)
43. R.D. Taylor, G. Cort and J. Willis: J. Appl. Phys. **53**, 8199 (1982)
44. R.D. Taylor, M.P. Pasternak and R. Jeanloz: J. Appl. Phys. **69**, 6126 (1991)
45. J.P. Rueff, M. Krisch, Y.Q. Cai et al.: Phys. Rev. B **60**, 14510 (1999)
46. S. Klotz and M. Braden: Phys. Rev. Lett. **85**, 3209 (2000)
47. O. Mathon, F. Baudelet, J.P. Itié et al.: Phys. Rev. Lett. **93**, 255503 (2004)
48. S. Pascarelli, O. Mathon and G. Aquilanti: J. of Alloys and Compounds **362**, 33 (2004)
49. H.F. Grünsteudel: *Der α-ε-Übergang in Eisen als Beispiel für nukleare Vorwärtsstreuung von Synchrotronstrahlung.* Ph.D. Thesis, Universität Paderborn, Paderborn (1997)
50. S. Nasu, T. Sasaki, T. Kawakami et al.: J. Phys.: Condens. Matter **14**, 11167 (2002)
51. P. Wachter: Handbook on the Physics and Chemistry of Rare Earths vol 19 ed. by K.A. Gschneidner et al. (Amsterdam: North-Holland 1994) p 383
52. F. Holtzberg and J. Wittig: Solid State Coomun. **40**, 315 (1981)
53. F. Lapierre, M. Ribault, F. Holtzberg, J. Flouquet: Solid State Commun. **40**, 347 (1981)
54. J. Röhler: Handbook on the Physics and Chemistry of Rare Earths vol 10 ed. by K.A. Gschneidner et al. (Amsterdam: Elsevier Science 1987) p 453
55. C. Dallera, E. Annese, J.-P. Rueff et al.: J. Phys.: Condens. Matter **17**, S849 (2005)
56. A. Barla, J.P. Sanchez, Y. Haga et al.: Phys. Rev. Lett. **92**, 066401 (2004)
57. Y. Haga, J. Derr, A. Barla et al.: Phys. Rev. B **70**, 220406 (2004)
58. A. Barla, J.P. Sanchez, J. Derr et al.: J. Phys.: Condens. Matter **17**, S837 (2005)
59. K. Koyama, K. Fukushima, M. Yamada et al.: Physica B **346–347**, 187 (2004)
60. P.P. Deen, L. Paolasini, N. Kernavanois, et al.: Submitted to J. Phys.: Condens. Matter (2005); D. Braithwaite, L. Paolasini, P. Deen et al.: submitted to Physica B (2005)

X-ray Spectroscopy and Magnetism in Mineralogy

Philippe Sainctavit, Sandrine Brice-Profeta, Emilie Gaudry, Isabelle Letard, and Marie-Anne Arrio

Institut de Minénalogie et de Physique des Milieux Condensés
Université Pierre et Marie Curie, CNRS UMR7590, 4 place Jussieu, 75252 Paris cedex 05, France
Philippe.Sainctavit@impmc.jussieu.fr

Abstract. The objective of this paper is to present the kind of information that can be gained in the field of mineralogy from the use of x-ray magnetic spectroscopies. We review some of the questions that are unsettled and that could benefit from an interdisciplinary approach where magnetism, spectroscopy and mineralogy could be mixed. Most of the attention is focused on iron and some other $3d$ transition elements. The mineralogy of planetary cores and its relation with known meteorites are exemplified. The various oxide phases in the mantle and the nature of iron in these phases is also underlined. The presence of transition elements in insulating minerals and its relation with macroscopic properties such as the color of gemstones are reviewed. Finally an introduction to paleomagnetism is given with a special attention to nanomaghemites.

1 Introduction

Mineralogy is the study of the natural mineral phases that exist on Earth or on other planetary objects. The large majority of the mineral phases are diamagnetic and for these very little is expected to be gained from the application of x-ray magnetic spectroscopy. Nevertheless the large abundance of iron on Earth and also on most other planets and meteorites, explains that "magnetic" phases (i.e. ferromagnetic, ferrimagnetic, antiferromagnetic or even paramagnetic phases) are not so seldom. This article is not intended to cover the complete field of Earth magnetism and in the following we shall concentrate on some questions in mineralogy that could be addressed by synchrotron spectrocopists. We shall discuss three types of phases:

- The first group refers to iron containing minerals that are intermetallic phases and where the other atoms are metallic atoms such as nickel. These phases are present in the core of planets or in metallic meteorites. Depending on the ratio of iron and nickel a wide variety of phases exists.
- The second group of minerals are insulating phases where iron or $3d$ transition elements are present in various concentrations. In most cases, the $3d$ transition elements are paramagnetic ions with $3d$ electrons sitting in the gap of the insulator. The interplay between the electronic structure of the

P. Sainctavit et al.: *X-ray Spectroscopy and Magnetism in Mineralogy*, Lect. Notes Phys. **697**, 401–427 (2006)
www.springerlink.com

$3d$ ions and the other atoms (anions and cations) of the insulating phase yields complicated electronic behaviors: the color of gemstones is such an example.
- The third class of minerals concerns the "real" magnetic phases in the sense that these phases are either ferromagnetic or ferrimagnetic phases. In many cases, the phases bear a measurable remanent magnetization that is used as an indication of the Earth's magnetic field at ancient times. Again most of the phases are iron oxides, sometimes iron sulfides.

The Earth can be described as a succession of various shells.

The inner core extends from the center to 5100 km deep and is made of solid iron and nickel. Temperatures and pressures are extremely high: pressure can be as large as 365 GPa and temperature be above 4000 K. It is believed that lighter elements such as oxygen, sulfur or silicon might be present as well. At depth between 5100 and 2900 km, one finds the liquid outer core with a composition similar to that of the inner core. In the outer core, pressures vary between 365 GPa and 135 GPa and temperatures are not below 4000 K. Temperatures and pressures of the core, either solid or liquid, are so high that one does not expect any magnetic properties of core phases.

Though the core phases are difficult to measure at high pressure and high temperature, there are related phases found in iron meteorites that are much easier to study. The chemical composition of these meteorites is similar to the one of the planetary cores. Such meteorites are supposed to have been formed at the commencement of the Solar system and to be representative of the formation conditions of the planets.

Surrounding the core, the mantel is the largest shell. It is commonly divided into lower mantle (from 2900 km to 660 km) and upper mantle (660 km to around a few tens of kilometers). In the lower mantle two major phases are present: magnesiowüstite (Mg, Fe)O and iron containing perovskite (Mg, Fe)SiO_3, at pressures varying between 23 and 135 GPa and temperatures between 1900 K and 3000 K. In the upper mantle many phases are present and in this paper we shall skip this complicated region.

The outer envelop of the Earth is the crust. It is made of the various minerals that can be found on the continents (down to $\approx$50 km) as well as at the bottom of the oceans (down to $\approx$6 km below oceanic floor). The crust is essentially made of oxygen (47%), silicon (28%), aluminum (8%) and iron (5%). Among the four more abundant atoms, iron is the only paramagnetic one [64]. The next most abundant paramagnetic ions are titanium (0.4%) and manganese (0.09%). Others such as vanadium, chromium, nickel or cerium have even lower concentrations (less than 0.01%). From what precedes one clearly understands the importance of iron for the mineralogist. Despite the low overall abundance of paramagnetic ions such as vanadium or chromium, their potential industrial applications put them at the center of various investigations. For instance, chromium is a powerful coloring agent in various wide-gap insulators, leading to nicely colored gemstones such as ruby or emerald.

2 Metallic Phases of the Planetary Cores

The core mainly consists of a mixture of iron and nickel with other light elements. Depending on the depth, the core can be solid (from the center to 5100 km deep) or liquid (from 2900 to 5100 km deep). The understanding of the various phases present in the core can be addressed by high-pressure measurements [1, 2]. Diffraction techniques at high pressure and high temperature can be quite useful to determine the crystallography. Very little is known concerning the exact magnetic state of iron in the core phases. A related question concerns the iron status in iron meteorites that contain iron and nickel. These meteorites are a good image of the primordial matter in the Solar system before the formation of the planets. The meteorites are the result of very special synthesis routes. From a molten meteorite with temperature above several thousand kelvin, the meteorites are cooled down at an extremely low cooling rate: around 1 kelvin per million years. In the meteorite group of octahedrites, one observes very special patterns, called Widmanstätten patterns (Fig. 1).

The electronic structure of the iron meteorites is still far from being known. There are two major crystallographic phases: a nickel-poor kamacite phase and and a nickel-rich taenite phase. From Mössbauer spectroscopy, the taenite phases would appear under various electronic and magnetic versions: pure taenite, tetrataenite and antitaenite [3]. These phases derived from

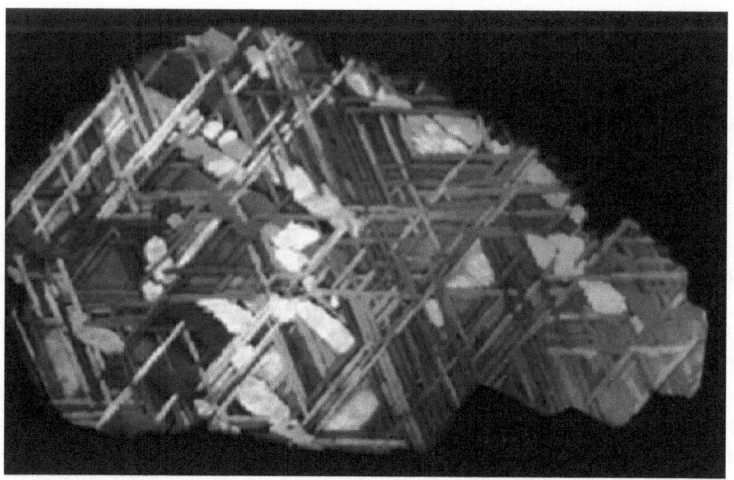

Fig. 1. Widmanstätten patterns of meteorites

taenite have in common a more or less fixed iron/nickel ratio. Tetrataenite is a well defined phase that is different from taenite. On the contrary, taenite and antitaenite would be quite similar from the crystallographic point of view but for the magnetization of iron. In taenite the magnetic moment per iron atom would be large ($\approx 2.8\,\mu_B$), while it would only be a few tenths of a Bohr magneton for antitaenite. In taenite the down-spin $3d$-band is almost empty. In antitaenite, there would be an almost equal filling of the up- and down-spin $3d$-bands [4]. The question of the magnetic state of iron in such intermetallic compounds is closely related to the well-known case of Invar alloys. In such compounds the Invar effect has been attributed to a high-spin to low-spin transition of iron, though the question is still actively debated [5]. In ataxite meteorites, that are iron meteorites containing large amount of nickel, it has been found that antitaenite and tetrataenites were stabilized at the nanoscale in a honeycomb matrix [4]. Most of the magnetic information has been gained from Mössbauer measurements that tracked the magnetization of iron. It is quite clear that magnetic spectroscopic measurements such as XMCD or XMLD are very likely to bring some light onto the subject. Previous XMCD studies have been exploring the field of Invar alloys and observed a high-spin to low-spin conversion with pressure. The measurements were sometimes indirect since iron was not directly probed: in a Fe/Pt alloy the $L_{2,3}$ edges of Pt were measured [5]. When iron has been probed, this has been done at the iron K-edge where the interpretation of the XMCD data is quite difficult [6], since the relation between K-edges XMCD signal and $3d$ magnetic polarization is all but straightforward. The local sensitivity of XMCD, its direct interpretation when it is performed at iron $L_{2,3}$ edges and its ability to be coupled to x-ray microscopy should help in understanding the various phases causing the complicated Widmanstätten patterns.

3 Localized Magnetism

Except for the quite special case of iron/nickel metallic phases for which band magnetism is at stake, Earth sciences are mainly dealing with magnetic questions that can be understood in the framework of localized magnetism. Iron is the most abundant magnetic ion, so that most of the magnetic questions concern the magnetic state of iron. Some unanswered questions are nevertheless related to other paramagnetic ions:

- $3d$ transition elements other than iron can be at the origin of the color of minerals,
- The various paramagnetic ground states of the $3d$ impurities are responsible for the dielectric properties of the minerals,
- Uranium and to a lesser extent the other actinides can also have various magnetic states that can govern the chemical stability of the mineral phases in which they are incorporated. The quite large amount of uranium and its

environmental impact related to nuclear wastes have promoted the field of actinide spectroscopy: see for instance [7, 8].

In the following we shall stick to the case of $3d$ transition elements and we shall illustrate how x-ray magnetic spectroscopies can determine their ground state.

3.1 Iron Ground State in the Lower Mantle of the Earth

The lower mantle of the Earth is the region that extends between 600 km and 2900 km. From various geophysical and geochemical mesurements, one knows that the lower mantle is mainly made of two iron containing phases. A mineral with perovskite structure (Mg, Fe)SiO_3 and an other one with wüstite structure (Mg, Fe)O. The mineral MgO is periclase and the mineral FeO is wüstite so that depending on the authors and on the Mg/Fe ratio the mineral (Mg, Fe)O can be called magnesiowüstite or ferropericlase. The questions concerning the status of iron in this region can be extended to the deep interior of other planets of the Solar system. For instance, it is believed that a "thin" shell between 1800 and 2000 km deep on Mars would have the same mineralogical composition as the one of the Earth lower mantle. Both MgSiO$_3$ (perovskite) and MgO (periclase) are wide gap insultors. The presence of a paramagnetic ion in these wide gap insulators is very likely to change the dielectric response of the pure phases. This is expected to yield dramatic change in the heat transfer between the various regions of the Earth interior. The determination of the dielectric tensor is a first step towards the constitution of a general model of the Earth. Moreover, in the region at the boundary between the outer core (liquid core) and the lower mantle, the seismic measurements have found some anomaly, labelled the D'' anomaly. The origin of this anomaly is believed to be related to the magnetic ground state of iron. From the knowledge of the chemical properties of the perovskite and periclase minerals, one knows that the magnetic ground state of iron will have a strong incidence on the partioning of iron between the two phases. Though this general picture is quite accepted, it rested until recently on very few measurements, performed at temperatures and pressures not prevailing in the lower mantle.

To answer the previous questions, the team of Guillaume Fiquet has applied the Resonant Inelastic X-ray scattering (RIXS) methods (see chapter by M. Altarelli in this book) to periclase and perovskite samples at high pressure [9, 10]. One knows from the theoretical work by de Groot that the K_β fluorescence line of paramagnetic ferrous iron gives strong indications on the ratio between electrons with up-spins and electrons with down-spins [11]. In the iron K_β fluorescence spectrum, the presence of a bump at 7040 eV is the indication that there are more up-spins than down-spins. For ferrous iron, this corresponds to a high-spin ground state with total spin S = 2. When the bump at 7040 eV is absent, ferrous iron is in the low-spin configuration with an equal number of up- and down-spin (i.e. S = 0).

These RIXS experiments can be understood as a first step towards the determination of the iron status in the lower mantle. The concentration of iron between perovskite and periclase is not known and the authors had to start from a composition believed to be relevant to what is present in the lower mantle. A sample with composition $(Mg_{0.83}Fe_{0.17})O$ was synthesized and mounted into an anvil cell. The K_β fluorescence line was measured as a function of pressure and in the region around 60 to 70 GPa, the authors observed a high-spin to low-spin conversion of iron. The spin conversion is found to be reversible upon decompression. It is also found that the transition pressure increases with the iron content. One of the important outcome from this study is that the iron partitionning between the periclase phase and the perovskite phase is expected to depend very much on the high-spin to low-spin conversion in periclase. Indeed, at pressures where iron would be in a low-spin configuration in periclase, iron would still be in high-spin configuration in perovskite. Then in the two-phase system, the iron spin conversion would be accompanied with a strong modification of iron *repartion* between the two phases.

To keep on exploring the phase diagram of the lower mantle minerals, the same group measured also the K_β fluorescence line of iron in the perovskite $(Mg_{0.9}Fe_{0.1})SiO_3$ [10]. The chemistry of this phase is more complicated than for the iron bearing periclase. Indeed in (Mg, Fe)O iron is known to be in the ferrous oxidation state and sitting at the octahedral magnesium site. On the contrary in the case of the iron bearing perovskite, iron can be sitting either at the dodecahedral sites or at the octahedral sites. Moreover iron can be in the ferrous or the ferric configurations. From the RIXS measurements, the authors observed two phase transitions related to iron spin conversion. The first spin conversion appears around 70 GPa. After correcting for the temperature effect, this would correspond to a depth of 1700 km. A second transition is observed at 120 GPa. This pressure corresponds to a depth of 2600 km, that is the region where the D'' anomaly occurs. One then sees that by the use of a spectroscopy sensitive to the total spin momentum of iron, it is possible to attribute a given anomaly in the seismic data to some iron spin conversion.

It has been shown that XAS at the iron K-edge was also sensitive to the high-spin to low-spin conversion, through the large modification of iron-ligand bond distance that accompanies the spin conversion [12]. The development of high-pressure and high-temperature XAS measurement would strongly help the determination of the iron state in the case of perovskite where one is dealing with two types of sites and two ionization states. Since the target of such studies is to understand the lower mantle phases one needs to stick as much as possible to what is known on these phases: Aluminum is thought to be present in non-negligible quantities in the perovskite. This would likely change the pressure transitions.

3.2 Color of Minerals

A well-known cause for colors in insulating minerals is related to transition-metal ions present as impurities. As an example, Cr^{3+} is responsible for the red color of ruby (α-Al_2O_3:Cr^{3+}) and the green color of eskolaite (α-Cr_2O_3). Though the role of chromium as a coloring agent has long been known, the relation between color and the electronic and crystallographic structure around chromium is far from being settled. This undetermination stems from the inherent complexity of the understanding of the origin of color. In wide gap insulators, the color caused by $3d$ paramagnetic ions is interpreted as multielectronic transitions within the $3d$ levels. The observed optical transitions can be assigned to multielectronic transitions following the crystal-field theory developed among others by Tanabe and Sugano [13]. The position of the optical transitions can be understood as resulting from crystal-field splittings coming from the neighboring ligands. The main drawback of such a description is that it is essentially unable to predict the intensities or the shapes of the transitions. Moreover, crystal-field parameters are radial integrals that are considered as free adjustable parameters. Since the crystal-field parameters are empirical parameters their exact physical meaning might be not so obvious. This is clearly exemplified in the work by Sugano and Peter on Cr^{3+} [14]. The energy of the first optical transition ($\approx 2\,eV$) in Cr^{3+} is commonly said to equal the value of $10Dq$. The $10Dq$ parameter represents the energy difference between monoelectronic e_g and t_{2g} levels of Cr^{3+} $3d$ electrons in octahedral symmetry. In a pure ionic picture, $10Dq$ is also equal to the transition between the $^4A_{1g}$ multielectronic ground state and the $^4T_{2g}$ multielectronic excited state for Cr^{3+} in a pure octahedron. Since the crystal-field parameters such as $10Dq$ are radial integrals, one often tries to relate the parameters extracted from UV–V is spectroscopy to the local environment of the paramagnetic ion. For instance, in the point charge model $10Dq$ should scale as the fifth power of the inverse of the average distance between the paramagnetic ion and its ligands (i.e. the averaged Cr-O distance in ruby). Such laws are often used and hold quite well when one scales the variations of $10Dq$ with the variation of cell parameters in ruby under pressure [15].

Nevertheless, if one wishes to go beyond a pure ionic description of the chemical bond one needs to go for configuration interaction. The ionic picture of the impurity ground state is modified by the charge transfer between the paramagnetic ion and the ligand. In such a description the optical transition around $2\,eV$ in ruby does not anymore correspond to $10Dq$ since the energy of the transition is partly modified by the charge transfer. Then eventual relations between UV–V is spectra and local chromium environment are not expected to be the same as in the ionic model. To check these various points, an initial step is the determination of the local environment of the impurities. In the following section we start from the comparison between red ruby (α-Al_2O_3:Cr^{3+}) and green eskolaite (α-Cr_2O_3) and we stick to the case of Cr^{3+} as coloring agent. The general method could easily be extended to any other paramagnetic coloring ion.

Green *versus* Red

The following results are extracted from the Ph.D. thesis by E. Gaudry where more details can be found. When light goes through ruby (α-Al$_2$O$_3$:Cr^{3+}), the whole yellow-green and violet radiations are absorbed, while red and few blue radiations are transmitted. Therefore, ruby is red with a slight purple overtone. On the contrary, red light is absorbed in eskolaite (α-Cr$_2$O$_3$), so that this mineral looks green. Although it is now well-known that colors in ruby and Eskolaite are due to the chromium ions [16,17], a question remains unexplained. Why does the same chromium chromophore ion, with the same valence state, and with the same kind of distorted octahedral site, yield a red color in ruby and a green one in Eskolaite?

In such oxide minerals, color is generally interpreted within the ligand-field theory. This model is based on an electrostatic interaction between the central cation and the ligands of its coordination sphere. It is based mainly on the geometry and the symmetry around the central cation. The optical data given by Reinen (1969) lead to $10Dq = 2.24\,\text{eV}$ ($18070\ \text{cm}^{-1}$) for ruby and $10Dq = 2.07\,\text{eV}$ ($16700\ \text{cm}^{-1}$) for α-Cr$_2$O$_3$ [18]. Despite the success of the ligand-field theory, some points remained misunderstood. The variations of the $10Dq$ parameters were studied as a function of the temperature T, the pressure P, and the amount x of chromium in the series Cr$_x$Al$_{2-x}$O$_3$ [19,20]. Knowing the relations between T, P or x and the average Cr-O distance, the authors plotted the relations between $10Dq$ and Cr-O. They found three different laws depending on the parameter at the origin of the Cr-O variation. These results tend to prove that either the value of the crystal field as measured by the $^4A_{2g} \rightarrow {}^4T_{2g}$ transition energy is not simply a function of the Cr-O distance or the local Cr-O distance is not well known. When chromium substitutes for aluminum, the chromium site is expected to be different from the aluminum site in α-Al$_2$O$_3$. This is due to the larger ionic radius of chromium ($r_{\text{Cr}^{3+}} = 0.615\,\text{Å}$) compared to that of aluminum ($r_{\text{Al}^{3+}} = 0.535\,\text{Å}$) in an octahedral site [21]. As a consequence, the understanding of the local environment around the chromium atom is absolutely needed to explain the evolution of physical properties with respect to the chromium content x in Cr$_x$Al$_{2-x}$O$_3$.

XAS at the Cr K-edge and $L_{3,3}$ Edges

From the XAFS analysis of five different samples Cr$_x$Al$_{2-x}$O$_3$ with different x, one found that the first peak of the five Fourier transforms were quite similar. Since the first peak contains the Cr-O contribution, one deduces that there should not be significant structural changes in the chromium coordination sphere in the whole range Cr$_2$O$_3$-Al$_2$O$_3$. The local environment of

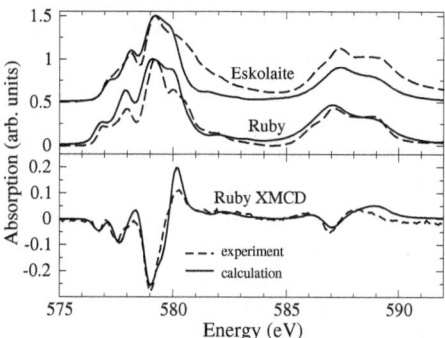

Fig. 2. XAS and XMCD signals at the Cr $L_{2,3}$ edges

chromium in the Al_2O_3-Cr_2O_3 solid solution series is specified by the EX-AFS analysis. One found that the average Cr-O distance is 1.96_5 Å in ruby (x less than 5%) and 1.98 Å in Eskolaite.

Since Cr^{3+} in ruby is an isolated paramagnetic ion, it should carry a net magnetic moment if it is cooled down to low temperature and a large enough magnetic induction is applied. Under such conditions ($T = 10\,K$ and $B = 5\,T$) it is then possible to measure the XMCD signal at the Cr^{3+} $L_{2,3}$ edges (Fig. 2). Such a measurement gives a good measure of $10Dq$ and the Racah parameter B that is a signature of the covalent nature of the Cr-O bond. It was observed that the Cr-O bond in eskolaite is much more covalent than in ruby.

In the point-charge model, the dependence of the crystal field $10Dq$ on the mean metal-ligand distance is given by

$$10Dq = \frac{5}{3}Z_Le^2\langle r^4\rangle R_{Cr-O}^{-5} \tag{1}$$

where R_{Cr-O} is the mean Cr-O distance, Z_Le^2 is the effective charge of the ligands, and $\langle r^4\rangle = \int r^2 dr R_2(r)^2\, r^4$ where $R_2(r)$ is the radial part of the d orbitals [22]. According to Langer (2001), the quantities Z_Le^2 and $\langle r^4\rangle$ would be constant along the Al_2O_3-Cr_2O_3 solid solution series, because chromium lies in the same kind of site, with the same oxygen ligands in the whole range. Equation (1) is then rewritten as

$$\Delta = aR_{Cr-O}^{-5} \tag{2}$$

where a is a constant. As shown from the EXAFS analysis at the Cr K-edge, the variation of the Cr-O distance in the whole composition range is less than $\approx 1\%$, while the variation of Δ crystal field is larger than 8%. According to (2), a variation of the Cr-O distance by 1.6% (more than 0.03 Å) would be necessary to explain the variation of Δ (8% in the Al_2O_3-Cr_2O_3 series). This variation of the Cr-O distance is much larger than the one found from EXAFS analysis (0.01_5 Å). This indicates that the variation of the mean Cr-O distance is not the sole explanation for the variation of Δ in the point charge model. A different metal-ligand charge transfer between α-Cr_2O_3 and ruby probably occurs, so that a from (1) is no longer a constant in the whole solid solution series. The different metal-ligand charge transfer between α-Cr_2O_3 and ruby might be responsible for some percent in the variation of the crystal field. This statement is in agreement with the smaller value for the Racah parameter B in Eskolaite than in ruby, and with the theory of Sugano and Peter (1961), for which the energy of the $^4A_{2g} \rightarrow ^4T_{2g}$ transition is $\Delta + 10\xi B$, where ξ is a covalency parameter, that stands for the different charge transfer between $\sigma(e_g)$ and $\pi(t_{2g})$ Cr-O bonds. The greater Cr-O bonding covalency in α-Cr_2O_3 than in α-Al_2O_3:Cr^{3+} comes from a modification of the electrostatic repulsions and from configuration interaction. This means that the ground state of the chromium atom is described by the $3d^3$ configuration interacting with $3d^4\underline{L}$ where $\underline{L}$ represents a hole on a ligand. A similar theoretical approach has recently been developed by Brik et al. [23].

The local environment of impurities is unknown in most colored materials. It is only from the determination of the actual local environment of the impurities that one can expect to understand the local origin of color.

4 Paleomagnetism

Paleomagnetism is the study of rock magnetization and its relation with geology. Rock magnetism is a branch of magnetism where the magnetic materials are minerals [24]. By definition, a mineral is a well-defined phase that exists naturally on Earth or more generally in the Universe and for which one has found at least one example. Minerals usually receive specific mineralogical names while ancient minerals do not. From the above definition, gold or copper are minerals while aluminum or silicon are not. Since iron is at the center of Earth sciences, ones expects magnetic minerals to be iron bearing minerals. The major magnetic minerals belong to the wide family of magnetite (Fe_3O_4). These are ferrimagnetic or ferromagnetic phases. Another large branch is related to antiferromagnetic phases such as hematite (α-Fe_2O_3) or goethite (α-$FeOOH$). A third group is the ferrihydrite group that contains nanophased iron oxi-hydroxides. These nanophases do not exist as large single crystals (larger than $10\,\mathrm{nm}$) and are precursors of other iron-bearing phases. A last group of iron bearing minerals are iron sulfides. Though iron sulfides exist in a wide class of phases, they have in common the strongly covalent iron-sulfur bond that gives them magnetic properties in between those of iron oxides and those of metallic iron.

On Earth, the main occurrence of magnetism is related to the magnetic dipole field that is the result of magnetohydrodynamics inside the outer core. The intensity of the magnetic dipole field is around 0.5 Gauss with daily variations of several percents as well as secular variations, extinctions or reversals. Since the crust can contain magnetic minerals, the Earth magnetic dipole field creates an induced magnetization of the minerals. For instance, one can observe magnetic anomalies that can be as large as 10^{16} A.m^2 (i.e. 10^{19} emu or $10^{39}\,\mu_B$) in the region of Kursk in Russia. This well known anomaly would be caused by large deposits of hematite. At room temperature, above the Morin transition, hematite is a canted antiferromagnet also called a weak ferromagnet. At room temperature the easy magnetization axis is perpendicular to the trigonal axis. To comply with the antiferromagnetic interactions, the moment on the various subnetworks do not compensate exactly and one observes a magnetization of 0.4 Am^2/kg. Though the average magnetic moment per iron atom can be estimated to one hundredth of what is measured in magnetite, the large amount of hematite (several cubic kilometers) yield the huge magnetic anomaly. One then understands that phases that are most of the time considered as antiferromagnets by physicists can be of primary importance to paleomagneticians.

One of the tasks of paleomagnetism is related to the remanent magnetization that is carried by minerals. Through well controlled conditions, the remanent magnetization can be associated with the history of the Earth magnetic dipole field. Paleomagnetism was the first very strong evidence for plate tectonics and it is daily used as a geologic chronometer. The relation between plate tectonics and paleomagnetism is most clearly exemplified when considering the Medio-Atlantic ridge. From magnetic measurements above the Atlantic ocean, magnetic anomalies have been measured. The anomalies are split into two symmetric groups with respect to the ridge. This has been interpreted as a clear sign of the ocean floor spreading away from the ridge. When basalts from the lithosphere flow, the small grains of magnetite that it contains are cooled down through the Curie temperature and acquire a magnetization. The direction of the of the grain magnetization is imposed by the direction of the Earth magnetic field at the time when they were close to the ridge. When the Earth dipole field is reversed, every 50.000 to 100.000 years, the magnetite magnetization follow the external magnetic field. One then observes magnetic anomalies strips that are polarized in the direction the Earth dipole magnetic field at the time when the strip was close to the ridge. The reversal sequence can be used by simple comparison with the remanent magnetization of other minerals, for instance carrots of sediments, as a chronometer to date some specific levels.

All the major magnetic minerals are iron bearing minerals and the major class belongs to the magnetite structure. Magnetite, Fe_3O_4, is the most studied one. The oxidized form of magnetite is maghemite, γ-Fe_2O_3. At the difference of magnetite where ferrous and ferric iron are present, in maghemite iron is only in the ferric state. In maghemite, the charge balance is made through vacancies on the cationic sublattice. At the bottom of the ocean, one preferentially finds titanomagnetites and titanomaghemites. These are complicated minerals where titanium ions have replaced iron ions. The most abundant titanomagnetite is $Fe_{2.4}Ti_{0.6}O_4$. Titanium is in the tetravalent state and charge compensation is obtained through the ratio of ferrous to ferric iron with some extra cationic vacancies. Titanomaghemites are reduced titanomagnetites whose cation and vacancy distributions are quite complicated [24].

The study of remanent magnetization is strongly related to nanomagnetism. Indeed when magnetite grains are cooled through the Curie temperature the grains acquire some magnetization. Due to the demagnetizing field, the microscopic grains are divided into magnetic domains, yielding a net magnetic moment close to zero for the whole grain. When the grain size is below the monodomain size, i.e. the size of a Bloch wall, the magnetization is blocked by the magnetic anisotropy of the grain and one observes a non-zero net magnetization for the grain. For magnetite, the single-domain size is evaluated to around 60 to 80 nm. If the size of the grains is much less than the single domain size, then thermal excitation can be larger than the magnetic anisotropy. In this case, the magnetization of the grain keeps on flipping between directions of easy magnetization and a superparamagnetic behaviour is observed. One then understands that the grains carrying magnetic information at 300 K are single-domain grains large enough to prevent entering the superparamagnetic regime. Since the single-domain size scales as the inverse of the square of the saturation magnetization, the minerals with small saturation magnetization have large single-domain sizes. For instance the single-domain size for hematite is around a few microns.

Since the grains that interest the paleomagneticians are in the nanometer range, the surface of the grains yields a major contribution. A way to study the surface effects on the nanograins is to work with grains of well controlled sizes that can be varied to change the surface to bulk ratio. In the following we concentrate on one example concerning naomaghemites. The following results are compiled in the Ph.D thesis by Sandrine Brice-Profeta (2004).

4.1 Nanomaghemites

Magnetic properties of nanoparticles of ferro- and ferrimagnetic materials with typical sizes ranging from 3 to 10 nm have been investigated for many years (for a review, see [25] and references therein). Knowledge of the magnetic behavior of fine grains is crucial as regard to their numerous occurrences in technological applications. For instance, iron oxide nanoparticles are components of ferrofluids, biomedical materials, catalysts or magnetic recording media [26]. Magnetic nanoparticles are also research tools in areas of physics, geology, biology and medicine.

Magnetic properties of fine particles depend strongly on surface effects, finite-size effects and interparticles interactions [25,27,28]. Magnetic properties of nanoparticles differ from those of the bulk materials. Below a critical diameter of typically 10–100 nm, nanoparticles contain a single magnetic domain, and can exhibit a superparamagnetic behavior [29]. Oxide nanocrystallites often have larger coercivities than bulk materials, due to a reinforcement of the surface magnetic anisotropy. A reduction of the saturating magnetization generally occurs in ferro- or ferrimagnets such as γ-Fe$_2$O$_3$ (maghemite), NiFe$_2$O$_4$ or CoFe$_2$O$_4$ [27]. On the contrary, nanoparticles of antiferromagnetic materials may have a non-zero magnetic moment. This is the case of hematite (α-Fe$_2$O$_3$) nanoparticles, due to a canted antiferromagnetism [30]. The origin of the magnetization reduction in spinel nanoparticles is still unclear. It is generally accepted that surface spins are canted, due to competing exchange interactions in an incomplete coordination shell for surface ions [31]. In the case of γ-Fe$_2$O$_3$ however, identical degrees of canting have been found using in-field Mössbauer spectroscopy in γ-Fe$_2$O$_3$ nanoparticles, with or without ^{57}Fe enriched surface [32]. This tends to prove that the canting could also be a finite-size effect, and might affect the volume of the particles. Morales and et al. investigated a series of γ-Fe$_2$O$_3$ nanoparticles of $\sim$100 nm with various degrees of vacancy ordering and of chemical disorder ascribed to the presence of hydroxide groups in the structure [33]. Their results show that spin canting can arise from structural disorders in maghemite as well. In general, modifications of the magnetic properties with respect to those of bulk materials are very sensitive to the way the particles are prepared, i.e. to the surface and the local chemical structure of the particles. Spin canting in nanoparticles is probably strongly dependent on chemical irregularities in the whole particle, and more specifically at the surface. A precise knowledge of the chemical state of the surface should be valuable in understanding the magnetic properties of fine particles.

X-ray magnetic circular dichroism (XMCD) at the $L_{2,3}$ edges of transition elements is now widely used as a local probe for the site symmetry and the magnetic moments of transition-metal ions in ferro- and ferrimagnetic materials. The dichroic signal is the difference between the absorption cross-sections for right- and left-circularly polarized x-ray under an external magnetic field applied along the x-ray propagation vector. This section reports about the use of X-ray absorption spectroscopy (XAS) and XMCD at the iron $L_{2,3}$ edges on γ-Fe$_2$O$_3$ nanoparticles to study the effects of the size of maghemite nanoparticles both on their chemical and magnetic structures.

γ-Fe$_2$O$_3$ has a spinel structure, that belongs to the space group symmetry Fd$\bar{3}$m and can be described by the following formula: Fe$_A^{3+}$[Fe$_{5/3}^{3+}\square_{1/3}$]$_B$O$_4$. A represents a tetrahedral site, B an octahedral site and $\square$ a cationic vacancy on the octahedral sublattice. γ-Fe$_2$O$_3$ is the fully oxidized form of magnetite (Fe$_3$O$_4$). The nanoparticles of γ-Fe$_2$O$_3$ investigated here are synthesized by aqueous precipitation of magnetite, followed by an oxidation to maghemite under acidic conditions. Synthesis details can be found in reference [34]. Samples of particles with different average diameters are obtained, depending on the ionic strength of the synthesis medium. They have a mono-modal log-normal size distribution. Due to the elaboration method, there is no ordering of the vacancies. To vary the surface to volume ratio of the particles, powdered samples of small particles (denoted as S), with an average diameter of

2.7 nm, and of medium-sized particles (denoted as M) with an average diameter of 8 nm are studied. In-field Mössbauer spectroscopy and magnetization measurements were carried out to characterize the magnetic properties of those samples. ^{57}Fe Mössbauer spectra in an applied magnetic field of 6 T at 10 K show indications of a canted spin structure over a wide range of temperature [34, 35]. The spin canting increases as the size of the nanoparticles decreases.

Since the surface state is likely to influence strongly the magnetic properties of γ-Fe$_2$O$_3$ nanoparticles, we also work on particles of average diameter 8 nm coated with phosphoric acid (denoted as MP). The binding mode of the ligands at the surface of γ-Fe$_2$O$_3$ has been determined thanks to IR spectra in a previous work [34]. The phosphate groups are bonded to Fe^{3+} in unidentate and bridging bidentate coordination modes. Such a coating modifies the surface state and alters the magnetic order of the surface layer by comparison with an uncoated surface. Indeed, x-ray diffraction patterns exhibit two diffuse bands, that are likely to be due to ferric phosphate complexes with a distorted symmetry around the metal [34]. In addition, zero-field and in-field Mössbauer spectra point out that MP particles consists of a γ-Fe$_2$O$_3$ core surrounded by a magnetically totally disordered phosphated iron shell.

We aim at clarifying the role of the chemical structure of γ-Fe$_2$O$_3$ nanoparticles, especially as far as the surface is concerned, in their magnetic properties. To draw a link between the local environment of atoms and the degree of alignment of the spins along the external field at high magnetic field on the three samples of nanoparticles (S, M and MP), XAS and XMCD signals with an external field of 2 T are recorded at the iron $L_{2,3}$ edges at 4.2 K. In the following, the global lack of alignment of either Fe$_A^{3+}$ or Fe$_B^{3+}$ spins on the external field will be designated as "spin canting". Previous measurements on those samples underline that their static and dynamic magnetic properties are consistent with a core-shell model [35]. At high external magnetic fields (and low temperatures), the magnetization of the surface shell is equal to the core magnetization. At lower external magnetic field, the magnetization due to the surface shell reduces. Hence, XMCD data are also collected on the S, M and MP samples, varying the external magnetic field from 0 T to 2 T at 4,2 K. Thanks to the thin probing depth in the soft x-ray range, surface effects should be emphasized in absorption and dichroism at the iron $L_{2,3}$ edges. According to reference [36], the probing depth at the $L_{2,3}$ edges of iron is typically of 45 Å in the case of iron oxides.

Experimental Details and Theoretical Framework

We recorded the absorption spectra at $L_{2,3}$ edges of iron on three samples of γ-Fe$_2$O$_3$ both on the SU23 beamline of the storage ring Super-ACO at LURE (Orsay, France) and on the BACH beamline (Beamline for Advanced diCHroic experiments) at the ELETTRA Synchrotron Radiation Source (Trieste, Italy) [37]. The experimental end-station is described in [38].

For circularly polarized x-rays, the absorption cross-sections can be labelled as σ^{ab}. a denotes the helicity of the photons. $a = \downarrow$ when the photons are right-hand polarized and $a = \uparrow$ when the photons are left-hand polarized. b denotes the direction of the magnetic field. $b = \uparrow$ (resp. $b = \downarrow$) when the magnetic field is parallel (resp. antiparallel) to the propagation vector $\mathbf{k}$. In the electric dipole approximation, reversing the magnetic field is equivalent to changing the helicity of the beam, i.e.: $\sigma^{\uparrow\uparrow} = \sigma^{\downarrow\downarrow}$ and $\sigma^{\uparrow\downarrow} = \sigma^{\downarrow\uparrow}$ [39].

A dichroic spectrum thus results from the difference of two absorption cross-sections, reversing either the photon helicity or the external magnetic field. By convention, $\sigma_{XMCD} = \sigma^{\uparrow\downarrow} - \sigma^{\uparrow\uparrow}$ for a 100% circular polarization. The experimental dichroic signal has to be normalized to a 100% polarization rate, by taking into account τ, the experimental polarization rate:

$$\sigma_{XMCD} = 1/|\tau| \, (\sigma^{\uparrow\downarrow} - \sigma^{\uparrow\uparrow}) \, .$$

Dichroic signals are obtained by flipping the direction of the magnetic field between two energy scans. In this case,

$$\sigma_{XMCD} = 1/\tau \, [(\sigma^{\downarrow\uparrow} + \sigma^{\uparrow\downarrow}) - (\sigma^{\uparrow\uparrow} + \sigma^{\downarrow\downarrow})]/2 \, .$$

The ability to reverse the photon helicity allows to compensate for systematic errors during the measurement process.

In order to extract quantitative information, the XAS and XMCD spectra at the iron $L_{2,3}$ edges are simulated using ligand field multiplet (LFM) calculations developed by Theo Thole [40]. We expose here briefly the principles of multiplet calculations. More details can be found in [11, 40–45] and in the chapter by G. van der Laan in this book. Iron $L_{2,3}$ edges consist mainly of the dipole allowed $2p^6 3d^5 \rightarrow 2p^5 3d^6$ transitions, as the dipole allowed $2p^6 4s^0 \rightarrow 2p^5 4s^1$ transitions have a negligible contribution to the absorption cross-section. In spherical symmetry, the $2p^6 3d^5$ ground state and the $2p^5 3d^6$ excited state energy levels are split by the interelectronic repulsions and the spin-orbit coupling. The radial integrals for interelectronic repulsions and spin orbit coupling are calculated using Cowan's atomic Hartree-Fock code [43]. One has to take into account for the initial $2p^6 3d^5$ configuration the direct $F^2(d, d)$ and $F^4(d, d)$ Slater integrals and the $3d$ shell spin-orbit coupling parameter ζ_{3d}, and for the excited $2p^5 3d^6$ configuration the direct $F^2(d, d)$, $F^4(d, d)$, $F^2(p, d)$ Slater integrals, the exchange $G^1(p, d)$, $G^3(p, d)$ Slater integrals, the $2p$ shell spin-orbit coupling ζ_{2p} and ζ_{3d}. The Slater integrals are reduced using the κ parameter. This reduction accounts for the electronic delocalization occurring through the chemical bonding in the solid. The effect of the crystal field and of the local magnetic field at the absorbing atom is treated using Butler's group subduction [42]. For octahedral and tetrahedral sites of the spinel structure, the $O_3 - O_h$ and the $O_3 - T_d$ branchings are used. The crystal field is then described by a single parameter, $10Dq$. The local magnetic field appears in the Zeeman hamiltonian: $\mathcal{H}_{Zee} = g \, \mu_B \, H \, S_z$. In ferro- and ferrimagnetic oxides, the $g \, \mu_B \, H$ parameter may typically range from 5 meV to 100 meV [45]. The energy and the intensities of the transitions are calculated in the electric dipole approximation. To simulate the experimental XAS spectra and XMCD signals, the transition lines are broadened by a Lorentzian function to account for the core-hole lifetime, and by a Gaussian function to account for the instrumental resolution.

High resolution transmission electron microscopy (HRTEM) experiments have been carried out on a Jeol 2010F microscope operating at 200 kV, equipped with a high resolution UHR pole piece and a Gatan energy filter GIF 100. Images are taken using elastically scattered electrons. Maghemite particles are deposited on a carbon film.

High Resolution Transmission Electron Microscopy

Determinations of the size of the S, M and MP particles have been carried out previously, using x-ray diffraction and TEM pictures [28]. Since spin disorder in γ-Fe_2O_3 nanoparticles is likely to arise from surface atoms, we are interested in characterizing the surface of the particles. HRTEM pictures of the M particles have been taken. Figure 3 is representative of what has been observed on several tens of particles. From this image, it can be inferred that the nanoparticles are well-crystallized and exhibit well-defined faceted edges. There is no preferential crystallographic orientation of these faceted edges. One or several preferential orientations could have informed on the type of crystallographic sites at the surface of the particles. Indeed, the tetrahedral to octahedral sites ratio, denoted as A/B, can differ at the surface from the ideal sites ratio of $3/5 = 0.6$ in case of existence of preferential surface planes. HRTEM images stand in agreement with former XRD results since no shape anisotropy is observed.

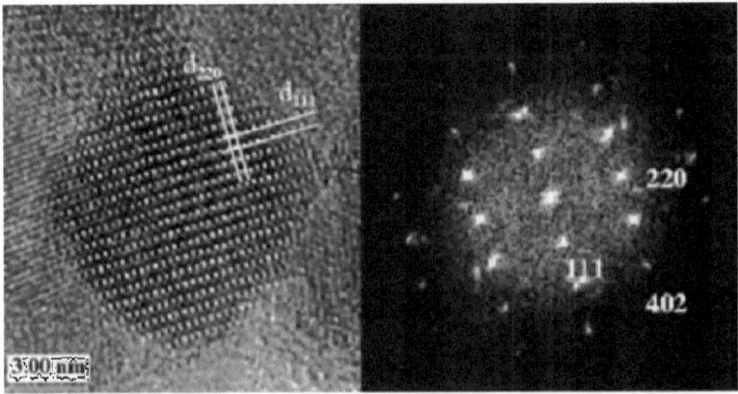

Fig. 3. HRTEM image of the 8 nm γ-Fe_2O_3 nanoparticles (M sample), and the corresponding electronic diffraction pattern

Shape Analysis of XAS and XMCD in the Multiplet Framework

Figure 4 shows the typical aspect of isotropic absorptions (XAS) and XMCD obtained for the three powders. Experimentally, isotropic spectra are obtained by recording the absorption of the sample with no applied field.

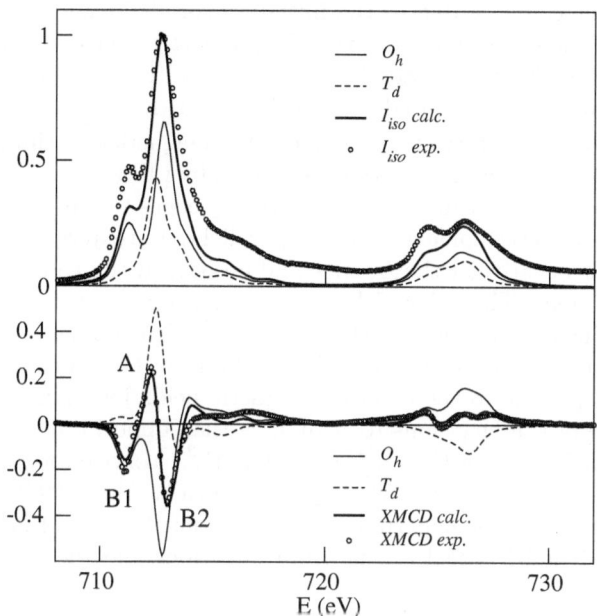

Fig. 4. Multiplet calculation of isotropic absorption and XMCD for Fe_A^{3+} and Fe_B^{3+} ions, along with the weighted averages for $A/B = 3/5$. Comparison to experimental isotropic spectrum and XMCD of 8 nm particles

If the z axis is chosen parallel to $\mathbf{H}$, the right-circular ϵ^-, the left-circular ϵ^+ and the linear $\epsilon^{//}$ polarization vectors are defined as:

$$\epsilon^- = \frac{1}{\sqrt{2}} \begin{pmatrix} 1 \\ -i \\ 0 \end{pmatrix} \qquad \epsilon^+ = -\frac{1}{\sqrt{2}} \begin{pmatrix} 1 \\ i \\ 0 \end{pmatrix} \qquad \epsilon^{//} = \begin{pmatrix} 0 \\ 0 \\ 1 \end{pmatrix}. \qquad (3)$$

In the electric dipole approximation, the three corresponding cross-sections are denoted as: $\sigma^- = \sigma^{\uparrow\downarrow} = \sigma^{\downarrow\uparrow}$, $\sigma^+ = \sigma^{\uparrow\uparrow} = \sigma^{\downarrow\downarrow}$, and $\sigma^{//}$. The isotropic cross-section is given by $\sigma_{iso} = (\sigma^+ + \sigma^- + \sigma^{//})/3$. We know from the HRTEM study that the crystallites are oriented at random with respect to the applied magnetic field. In this case, the magnetic linear dichroism is negligible, and $\sigma^{//} \simeq (\sigma^+ + \sigma^-)/2$ [39]. Consequently $\sigma_{iso} \simeq (\sigma^+ + \sigma^-)/2$, and the experimental quantity $(\sigma^{\uparrow\uparrow} + \sigma^{\downarrow\downarrow} + \sigma^{\uparrow\downarrow} + \sigma^{\downarrow\uparrow})/4$ is a good approximation for σ_{iso}.

For each series of scans, this expression is used to normalize the isotropic absorption to unity at the maximum of the L_3 edge.

XMCD experimental signals are characteristic of the γ-Fe_2O_3 phase [46]. At the L_3 edge, they consist of a positive peak (denoted A) and two negative peaks (denoted B1 and B2). Signals are mostly positive at the L_2 edge. XMCD spectra at the $L_{2,3}$ edges are mainly determined by the strength of the spin-orbit coupling due to the initial $2p$ core-hole, and by the spin polarization of the empty final localized $3d$ states [45]. The core-hole spin-orbit coupling splits the edges into the L_3 edge, resulting from the $2p_{3/2} \to 3d$ transition, and into the L_2 edge, resulting from the $2p_{1/2} \to 3d$ transition. XMCD allows to separate and to quantify the magnetic contributions of Fe_A^{3+} and Fe_B^{3+} ions to the magnetization. As shown by LFM calculations on Fig. 4, XMCD is reversed in sign between L_3 and L_2 edges for Fe_B^{3+} and Fe_A^{3+} ions. For atomic magnetic moments oriented parallel to the external magnetic field, the XMCD is negative at the L_3 edge and positive at the L_2 edge. This is the case of Fe_B^{3+} ions. The spins of Fe_A^{3+} ions are coupled antiferromagnetically with the spins of Fe_B^{3+} ions, as the main magnetic coupling is the antiferromagnetic coupling between A and B sites. Spins of Fe_A^{3+} ions are oriented in a direction opposite to the external magnetic field. The sign of the dichroic contribution of the Fe_A^{3+} ions to XMCD is then opposite to the sign of the dichroic contribution of the Fe_B^{3+} ions. The Fe_A^{3+} contribution is positive at the L_3 edge and negative at the L_2 edge [46, 47]. The XMCD signal of γ-Fe_2O_3 derives from the superposition of the contributions of Fe_A^{3+} and Fe_B^{3+} ions. The B1 and B2 peaks come from Fe_B^{3+} ions and the positive A peak from Fe_A^{3+} ions.

For Fe_A^{3+} and Fe_B^{3+} ions, one needs only the $10Dq$ parameter to describe the symmetry and the strength of the crystal field. Experimental data extracted from reflectance spectra of maghemite gives an octahedral crystal field of $10Dq = 1.9\,eV$ [16, 48, 49]. Concerning the tetrahedral sites, experimental values of $10Dq$ are available for Fe^{3+} ions as impurities in various matrices [50–52]. In our calculations, we use crystal field parameters close to but slightly smaller than the values extracted from optical spectroscopy, due to $2p$ hole in the excited state. For Fe_B^{3+} ions, $10Dq = 1.5\,eV$, and for Fe_A^{3+} ions, $10Dq = 0.7\,eV$. The value of the octahedral crystal field is also consistent with the values found in literature for Fe^{3+} multiplet calculations on iron oxides [53–57]. For the $3d^5$ electronic configuration of Fe^{3+} ions the orbital part is fully symmetric ($^6A_{1g}$ in O_h). The shape of XAS and XMCD signals are then barely affected by slight symmetry distortions around the ion. Such an effect is not detected on the recorded spectra. The above mentioned existence of distortions around Fe^{3+} ions with respect to perfect octahedral or tetrahedral symmetries in the case of MP phosphated particles is thus not detected from the XMCD measurements. However, the values of octahedral and tetrahedral crystal fields are also dependent on the value of κ, that accounts for the covalency of the Fe-O bond. κ is defined by Racah parameters,

$\kappa = B/B_0$, where B_0 is the Racah parameter for an isolated Fe^{3+} ion. Using Slater's integrals, $B_0 = (1/49)F^2(d,d) - (5/441)F^4(d,d)$. For Fe^{3+}, we obtain $B_0 = 160$ meV. We used $\kappa = 55\%$, so the B Racah parameter for Fe^{3+} ions in the solid is of 88 meV. This value is close to the 72 meV value derived from reflectance spectra of γ-Fe_2O_3 [48]. For both the Fe_B^{3+} and the Fe_A^{3+} ions, ζ_{3d} is fixed to its atomic value of 74 meV. To account for the experimental isotropic data, the value of ζ_{2p} is set to 8.4 eV. We performed calculations using an exchange-field parameter of 20 meV, at 4.2 K. $\Gamma = 0.2$ eV for the Lorentzian broadening function at the L_3 edge, $\Gamma = 0.4$ eV at the L_2 edge and $\sigma = 0.25$ eV for the Gaussian instrumental broadening function [41, 58, 59]. Figure 4 presents the calculated Fe_B^{3+} and Fe_A^{3+} contributions to the isotropic absorption (upper panel) and to the XMCD signal (lower panel), and the experimental XAS and XMCD recorded on 8 nm particles at 2 T and 4.2 K. The isotropic cross-section of Fe_B^{3+} ions is much different from the isotropic cross-section of Fe_A^{3+} ions. From the quantitative analysis of Fe_B^{3+} and Fe_A^{3+} isotropic contributions, it is possible to determine the A/B site ratio of the structure. Simulations in Fig. 4 are compatible with the bulk γ-Fe_2O_3 $A/B = 3/5$ sites ratio, with an absolute uncertainty of about 10%. The weighted sum of the isotropic cross-sections using this ratio is thus presented. Nevertheless, from the comparison of the experimental isotropic cross-sections of the different samples, we estimate that the difference in the A/B sites ratio between S, M and MP particles is much smaller than the uncertainty of the simulations, and lies around 1%. In Fig. 4, the calculations are compared to the isotropic XAS of 8-nm particles.

The amplitude of the experimental dichroic signal taken between the positive peak and the second negative peak equals to 59% of the isotropic absorption at the maximum of the L_3 edge. We performed a simulation of XMCD considering a structure with an ideal ferrimagnetic alignment of the spins. The ratio of the magnetic contributions of Fe_B^{3+} and Fe_A^{3+} spins to XMCD, denoted as m_A and m_B, is thus considered to be equal to the A/B site ratio, that is $m_A/m_B = 3/5$. A good agreement is obtained except around 715 eV where features are known to be due to configuration interactions. The B1 peak of the dichroic signal, as well as the corresponding low-energy shoulder of the isotropic absorption at 711.3 eV are not exactly reproduced in width nor in intensity by our calculations. A possible explanation is that the corresponding group of transitions may have a core-hole broadening lower than $\Gamma = 0.2$ eV. Indeed, the intrinsic broadening mechanisms for each transition line are dependent on each final state reached [44, 60]. A similar trend stands for the low-energy peak at 724.5 eV at the L_2 edge. The shoulder at 714 eV in the dichroic signal is overestimated by calculations. For $m_A/m_B = 3/5$, the calculated XMCD signal reproduces nicely the other parts of the dichroic signal of the 8 nm particles. The adequacy between the A/B site ratio and the m_A/m_B magnetic contributions ratio denotes a collinear ferrimagnetic order close to the bulk order in the 8 nm particles at 2 T and 4,2 K.

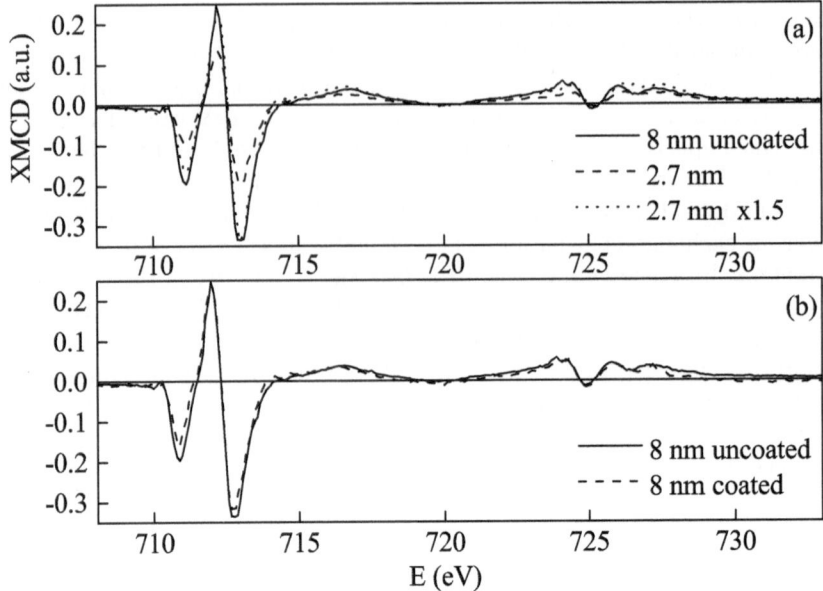

Fig. 5. XMCD of the S (2.7 nm), M (8 nm uncoated) and MP (8 nm phosphate coated) nanoparticles recorded on SU23 (LURE) at 4.2 K under 2 T. (**a**): XMCD of S, M particles, and XMCD of S particles scaled by a 1.5 factor (**b**): XMCD of M and MP particles

A small shoulder on the first negative peak of the L_3 dichroic signal is observed at 710.3 eV. This shoulder is not observed on the dichroic spectra of other bulk spinels, such as $NiFe_2O_4$, $CoFe_2O_4$ or $Li_{0.5}Fe_{2.5}O_{12}$ [47] and is not reproduced by the multiplet calculations either. It does not come from derivative effects and exists in all the recorded dichroic signals that we measured. This shoulder occurs at an energy matching with the chemical shift of Fe^{2+} ions. From multiplet calculations, we estimated this Fe^{2+} dichroic contribution to 4% of the total iron ions that we detect. As we measure XMCD in TEY, we are sensitive to the surface of grains, that may undergo a reduction under UHV. Those traces may also be due to the synthesis method. These Fe^{2+} ions were not seen in Mössbauer because their concentration is below the detection limit.

XMCD of S Particles *versus* M Particles

At the L_3 edge, the positive tetrahedral peak and the two negative octahedral peaks are respectively proportional to the resulting projection of the magnetic moments of Fe^{3+}_A and Fe^{3+}_B ions along the external field. As a result, the amplitude of the dichroic signal measured between A and B_1 peaks is proportional to the magnetization of the sample, as checked below.

A comparison of the dichroic signals of the S, M and MP particles under a 2 T field at 4.2 K is represented in Fig. 5. Variations of the amplitudes of the dichroic signals on the different samples should be similar to the variations of the macroscopic magnetization measured by SQUID at same field and temperature. The upper panel of the Fig. 5 shows the XMCD of the S and M samples. The XMCD amplitude of the S particles is found to be equal only to 66% of the XMCD amplitude of the M particles. From the XMCD amplitude, the ratio of the magnetization of the M particles to the magnetization of S particles is therefore $M_M/M_S = 1.5$ under 2 T and at 4.2 K. This lowering traduces a greater misalignment of the spins along the applied field in small particles. Consequently, a rescaled XMCD of S particles using a 1.5 multiplicative factor is also represented on Fig. 5, to compare the shapes of XMCD of M and S particles. At 2 T and 4.2 K, SQUID measurements give a magnetization of $50 \, A \cdot m^2 \cdot kg^{-1}$ for the S particles, and $81 \, A \cdot m^2 \cdot kg^{-1}$ for the M particles [34]. Magnetization of the S particles thus equals to 62% of the magnetization of the M particles, that is $M_M/M_S = 1.6$, in agreement with XMCD. At last, we noticed above that calculated XMCD considering the bulk structure of maghemite and experimental XMCD on 8 nm particles have close amplitudes. This agrees with the fact that 8 nm particles have a magnetization close to the bulk magnetization of $84 \, A \cdot m^2 \cdot kg^{-1}$.

XMCD of M Particles *versus* MP Particles

The lower panel of the Fig. 5 compares the dichroic signals of the M and of the MP nanoparticles. At the L_3 edge, the corresponding isotropic spectra (not represented) are quite similar in shape and intensity, which allows a direct comparison between the dichroic signals. The XMCD amplitude of the phosphate-coated particles is 93% of the XMCD amplitude of the uncoated particles. At 2 T and 4.2 K, SQUID measurements give magnetization values of $81 \, A \cdot m^2 \cdot kg^{-1}$ and $71 \, A \cdot m^2 \cdot kg^{-1}$ for the M and MP powder samples [34]. The SQUID measurements are given with a 10% experimental uncertainty, due to the H_2O content of the powders. Following these results, the magnetization of phosphate-coated particles equals to 88% of the magnetization of uncoated particles. Magnetization reductions estimated from XMCD and from SQUID are of similar magnitudes, on account of experimental uncertainties.

In-field Mössbauer spectra on the MP powder at 10 K show a strongly perturbed magnetic sub-pattern. It stems from the freezing of a paramagnetic surface layer of Fe^{3+} ions caused by the phosphate ligands, in a totally magnetically disordered shell. The relative area of the perturbed magnetic sub-pattern and the magnetic sextet measured by Mössbauer is of 0.21. 21% of the Fe atoms are thus bounded to phosphate groups, and are magnetically disordered [34]. The comparison of XMCD of M and MP particles also points out this magnetic disorder. There is a 7% reduction of the dichroic amplitude between uncoated and coated particles. As shown on Fig. 5b, the lowering of

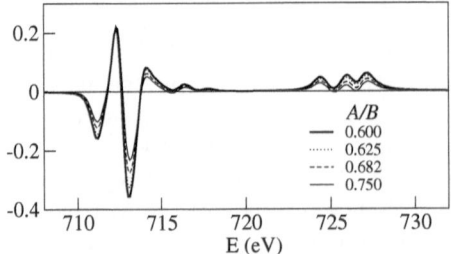

Fig. 6. Multiplet calculations of XMCD for several A/B site ratios in γ-Fe_2O_3

the dichroic amplitude comes from a reduction of the intensities of the two negative octahedral peaks at the L_3 edge, without any modification of the intensity of the positive tetrahedral peak. This peak keeps the same intensity for the M and the MP samples. The intensity of the octahedral peaks is reduced by 10% for MP particles, with respect to M particles. Figure 6 displays results of multiplet calculations with the parameters used above, assuming a progressive reducing of the magnetic contribution of Fe_B^{3+} spins to the dichroic signal. The m_A/m_B magnetic contributions ratio ranges from the ideal bulk ratio ($m_A/m_B = 0.6$) to a $m_A/m_B = 0.75$ ratio. Variations of the magnetic contributions ratio does affect the shape of the dichroic peaks at the L_3 edge. But the ratio of the intensity of the second negative peak to the first negative peak remains unchanged. Magnetic contributions ratio variations only modify the relative intensity of the A tetrahedral peak to the B1 and B2 octahedral peaks. Differences in the intensities of the B1 and B2 peaks between M and MP particles thus happen to be due to a reduction of the magnetic contribution of Fe_B^{3+} ions in phosphate-coated particles. In the case of a phosphate-modified surface, combination of Mössbauer and XMCD measurements indicates that the surface disordered spins could be mainly Fe_B^{3+} spins.

XMCD Field Dependence

Apart from the preceding case of a modified surface, the contrast between core and surface magnetic behaviors shall become perceptible under low magnetic fields, leading us to collect XMCD signals at low applied fields. Hysteresis loops presented in reference [35] for the S and M powders show that the closure field of the loops is 2 T for S particles and 0.5 T for M particles, indicating a surface-related irreversibility. We therefore use two measurement procedures. To follow the hysteresis loops, we raise the magnetic field up to a value of +2 T, before decreasing it to the required positive value for the measurement. We apply an equivalent procedure for negative fields. Figures 7 present the dichroic signals of the S particles at the L_3 edge, varying the external field. On the left panel of the picture, spectra are plotted as-obtained,

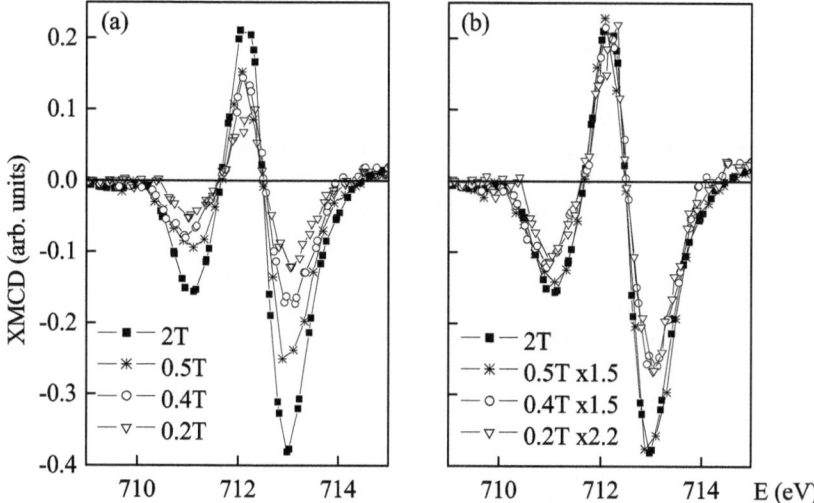

Fig. 7. (**a**): XMCD of the S (2.7 nm) nanoparticles at the L_3 edge recorded on BACH (Elettra) at 4.2 K under 0.2, 0.4, 0.5 and 2 T. (**b**): Rescaled XMCD spectra. The positive tetrahedral peaks all have the same area

the isotropic absorption being normalized to 1 at the L_3 maximum. On the right panel of the picture, they are rescaled so as to get a same area for all the A peaks. Scaling factors are specified on the figure. When the applied magnetic field decreases, the amplitude of XMCD decreases as well, reflecting a loss of the average projection of the magnetic moments of the particles along the applied field. The rescaled spectra show that at 2 T and 0.5 T, the relative intensities of the positive and the negative dichroic peaks remain the same. However, for data collected at 0.4 and 0.2 T, the B1 and B2 peaks become clearly smaller compared to the A peak. The intensity of the B2 peak is reduced by ~30% compared to its intensity at higher fields. In the 2.7 nm particles, this observation evidences a preferential low-field spin disorder involving spins of octahedral Fe^{3+} ions.

To check whether this canting has a surface origin, we performed the same measurements on the bigger M and MP particles. For M particles, the B1 and B2 octahedral peaks are reduced by 4% going down from 2 T to 0.2 T, whereas no reduction is observed for tetrahedral peaks. This reduction is weak (and lies in the uncertainty range) compared to the 30% reduction of the intensity of the B1 and B2 dichroic peaks in the case of S particles. As the relative reduction of the intensity of dichroic octahedral peaks with respect to the tetrahedral peak is connected to a growing magnetic disorder of spins of Fe_B^{3+} ions, the spins misalignment on octahedral sites happens to be weaker in M particles than in S particles, especially at low magnetic fields. As a matter of fact, M particles have a reduced surface to volume ratio, compared to S particles. In S particles, 50% of the Fe^{3+} ions are located at the surface (e.g. in a 0.35 nm thick layer), *versus* only 20% in M particles. Surface effects are therefore expected to be weaker in M particles than in S particles, in agreement with the experimental result. A quantitative comparison between field-induced variations of XMCD in S and M particles must take into account the probing depth of 45 Å [36], larger than the 2.7 nm diameter of S particles and smaller than the 8 nm diameter of M particles. The surface relaxation of M particles might be even smaller than 4%, indicating an even more contrasted behaviour than with S particles.

No relative field-induced variations of the contributions of the Fe_B^{3+} ions *versus* Fe_A^{3+} ions were observed concerning the MP particles. As explained above, at 4.2 K, the surface spins of phosphate-coated particles are frozen in a disordered configuration. As a consequence, the magnitude of the external field has almost no effect on their magnetic order.

The Magnetic Status of Nanomaghemite

The examination of the isotropic cross-sections of 2.7 nm, 8 nm and phosphated 8 nm particles indicates that those three types of particles have similar A/B site occupation ratios. Thanks to XMCD signals recorded with decreasing values of the applied field, we detected at low fields on 2.7 nm and 8 nm particles a greater disorder of Fe_B^{3+} spins with respect to the field direction than for Fe_A^{3+} spins. Furthermore, this disorder is more important for the smaller 2.7 nm particles that have a larger surface to volume ratio than the 8 nm particles. This observation agrees with the comparison of the magnetic contributions of Fe_B^{3+} and Fe_A^{3+} spins in 8 nm and phosphated 8 nm particles, the latter presenting a magnetically disordered surface, as shown previously by Mössbauer. Indeed, we also find a greater spin disorder of Fe_B^{3+} spins in phosphated 8 nm particles than in 8 nm particles, the magnetic contribution of Fe_A^{3+} spins remaining unchanged.

These two results jointly prove the existence of a preferential spin canting of Fe_B^{3+} spins at the surface. This is consistent with the core-shell model of the magnetic structure formerly proposed for the particles. A possible explanation for this structure is that surface spins experience weakened exchange interactions with their neighbors, yielding an increased disorder of Fe_B^{3+} spins at low-magnetic fields, typically below 500 mT for the small particles.

5 Conclusion

The Earth sciences encompass broad fields of Science and most of the questions outlined in the previous pages are expected to be solved through an interdisciplinary approach where physicists and chemists would interact with mineralogists. An excellent exemple of such interdisciplinary approach can be found in the numerous papers of R. Pattrick, D. Vaughan and G. van der Laan, and which are dealing with XMCD or XMLD applied to minerals [61–63]. In the field of magnetism, we have presented some of the questions that would benefit from such an approach. One of the very general features of the questions presented in this paper can be found in the small size of the phases that have to be studied:

- nanophases in meteorites,
- microsamples in diamond anvil cells,
- nanograins of maghemite.

These questions will certainly benefit from the emergence of x-ray microscopes developed on beamlines dedicated to magnetic spectroscopy.

Acknowledgement

We acknowledge the strong instrumental collaboration with Jean-Paul Kappler. Delphine Cabaret and Christian Brouder contributed also to the research on colors in gemstones. Elisabeth Tronc, Jean-Pierre Jolivet and Nicolas Menguy are thanked for their scientific involvement in the Ph.D. thesis of one of us (S.B-P). The staffs from SU23 (Super-ACO), BACH (ELETTRA) and ID8 (ESRF) beamlines contributed to the experimental data acquisition. Philippe Ohresser is thanked for the recording of the chromium $L_{2,3}$ edges in ruby.

References

1. D. Andrault, G. Fiquet, T. Charpin, and T. Le Bihan: Am. Mineralogist **85**, 364 (2000)
2. H.K. Mao, Y. Wu, L.C. Chen, J.F. Shu, and A.P. Jephcoat: J. Geophys. Res. **95**, 21737 (1990)
3. D.G. Rancourt, and R. Scorzelli: J. Magn. Magn. Mat. **150**, 30 (1995)
4. D.G. Rancourt, K. Lagarec, A. Densmore, R. Dunlap, J. Goldstein, R. Reisner, and R. Scorzelli: J. Magn. Magn. Mat. **191**, L255 (1999)
5. S. Odin, F. Baudelet, J.-P. Itié, A. Polian, S. Pizzini, A. Fontaine, Ch. Giorgetti, E. Dartyge, and J.-P. Kappler: J. Applied Phys. **83**, 7291 (1998)
6. O. Mathon, F. Baudelet, J.-P. Itié, S. Pasternak, A. Polian, and S. Pascarelli: J. Synchr. Rad. **11**, 423 (2004)
7. C.D. Auwer, C. Simoni, S. Conradson, and C. Madic: European Journal of Inorganic Chemistry **21**, 3843 (2003)
8. F. Farges. Physics and Chemistry of Minerals **20**, 504 (1994)
9. E. Badro, G. Fiquet, F. Guyot, J.-P. Rueff, W. Struzhkin, G. Vanko, and G. Monaco: Science **300**, 789 (2003)
10. E. Badro, J.-P. Rueff, G. Vanko, G. Monaco, G. Fiquet, and F. Guyot: Science, **305**, 383 (2004)
11. F.M.F. de Groot: Coordin. Chem. Review. **249**, 31 (2005)
12. V. Briois, Ch. Cartier dit Moulin, Ph. Sainctavit, Ch. Brouder, and A.-M. Flank: J. American Chemical Society **117**, 1019 (1995)

13. S. Sugano and Y. Tanabe: J. Phys. Soc. Jpn. **13**, 880 (1958)
14. S. Sugano and M. Peter: Phys. Rev. **122**, 381 (1961)
15. S.J. Duclos, Y.K. Vohra, and A.L. Ruoff: Phys. Rev. B **41**, 5372 (1990)
16. R.G. Burns. Mineralogical Applications of Cristal Field Theory, volume 5 of Cambridge topics in Mineral Physics and Chemistry. Cambridge University Press, 2nd edition (1993)
17. K. Nassau, The Physics and Chemistry of Color. Wiley Interscience (1983)
18. D. Reinen. Structure and Bonding, **6**, 30 (1969)
19. C.P. Poole, and J.F. Itzel: J. Chem. Phys. **39**, 3445 (1963)
20. C.P. Poole. J. Phys. Chem. Solids **25**, 1169 (1964)
21. R.D. Shannon: Acta. Cryst, **A32**, 751 (1976)
22. T. Dunn, D.S. McClure, and R.G. Pearson Some Aspects of Crystal Field Theory. Harper and Row (1965)
23. M.G. Brik, N.M. Avram, and C.N. Avram: Solid State Com. **132**, 831 (2004)
24. D.J. Dunlop and Ö. Özdemir: Rock Magnetism, Cambridge University Press (1997)
25. J.L. Dormann, D. Fiorani, E. Tronc: Magnetic relaxation in fine-particle systems, Advances in Chemical Physics, John Wiley and sons (1997)
26. J.-L. Dormann, D. Fiorani (Eds.), Magnetic Properties of Fine Particles, Delta Series, North-Holland, 1992, *see* Application chapter.
27. R.H. Kodama: J. Magn. Magn. Mat. **200**, 359 (1999)
28. P. Prené: Ph.D. thesis, Université Paris VI, France (1996)
29. H. Zijlstra: Ferromagnetic Materials, Vol. 3, North-Holland, Amsterdam (1982)
30. A.R.B. Castro, R.D. Zysler, M.V. Mansilla, C. Arciperete, M. Dimitrijewits: J. Magn. Magn. Mater. **231**, 287 (2001)
31. J.M.D. Coey: Phys. Rev. Lett. **27**, 1140 (1971)
32. F.T. Parker, M.W. Foster, D.T. Margulies, A.E. Berkowitz: Phys. Rev. B **47**, 7885 (1993)
33. M.P. Morales, C.J. Serna, F. Bødker, S. Mørup: J. Phys: Condens. Matter **9**, 5461 (1997)
34. E. Tronc, A. Ezzir, R. Cherkaoui, C. Chanéac, H.K. M. Noguès, D. Fiorani, A.M. Testa, J.M. Grenèche, J.-P. Jolivet: J. Magn. Magn. Mat. **221**, 63 (2000)
35. E. Tronc, D. Fiorani, M. Noguès, A. Testa, F. Lucari, F. D'Orazio, J. Grenèche, W. Wernsdorfer, N. Galvez, C. Chanéac, D. Mailly, J. Jolivet: J. Magn. Magn. Mat. **262**, 6 (2003)
36. S. Gota, M. Gautier-Soyer, M. Sacchi: Phys. Rev. B **62**, 4187 (2000)
37. M. Zangrando, M. Finazzi, G. Paolucci, G. Comelli, B. Diviacco, R. Walker, D. Cocco, F. Parmigiani: Rev. Sci. Instrum. **72**, 1313 (2001)
38. Ph. Sainctavit, J.-P. Kappler: X-ray magnetic circular dichroïsm at low temperature In: *Magnetism and Synchrotron Radiation*, (Lecture notes in Physics vol 565), ed. by E. Beaurepaire, F. Scheurer, G. Krill, J.-P. Kappler (Springer, Berlin, Heidelberg, New York 2001)
39. Ch. Brouder, J.-P. Kappler: Prolegomena to magnetic circular dichroïsm in x-ray absorption spectroscopy In: *Magnetism and Synchrotron Radiation* (Les éditions de physique 1997)
40. B.T. Thole, G. van der Laan, J.C. Fuggle, G. Sawatzky, R.C. Karnatak, J.-M. Esteva: Phys. Rev. B **32**, 5107 (1985)
41. M.-A. Arrio, Ph. Sainctavit, Ch. Cartier dit Moulin, T. Mallah, M. Verdaguer, E. Pellegrin, C.T. Chen: J. Am. Chem. Soc. **118**, 6422 (1996)

42. P.H. Butler: Point Group Symmetry Applications: Methods and Tables (Plenum Press 1981)
43. R.D. Cowan: The theory of atomic structure and spectra, Los Alamos series in basic and applied sciences (University of California Press 1981)
44. F.M.F. de Groot, J.C. Fuggle, B.T. Thole, G.A. Sawatzky: Phys. Rev. B **42**, 5459 (1990)
45. G. van der Laan, B. Thole, Phys. Rev. B 43 (1991) 13401–13411.
46. E. Pellegrin, M. Hagelstein, S. Doyle, H.O. Moser, J. Fuchs, D. Vollath, S. Schuppler, M.A. James, S. Saxena, L. Niesen, O. Rogojanu, G.A. Sawatsky, C. Ferrero, M. Borowski, O. Tjernberg, N.B. Brookes: Phys. Stat. Sol. (b) **215**, 797 (1999)
47. F. Sette, C.T. Chen, Y. Ma, S. Modesti, N. Smith: X-ray absorption fine structure (Ellis Horwood 1991)
48. D.M. Sherman, T.D. Waite: Am. Min. **70**, 1262 (1985)
49. D.M. Sherman: Phys. Chem. Minerals **12**, 161 (1985)
50. N.T. Melamed, F. de S. Barros, P.J. Viccaro, J.O. Artman: Phys. Rev. B **5**, 3377 (1972)
51. G.A. Waychunas, G.R. Rossman: Phys. Chem. Minerals **9**, 212 (1983)
52. Z. Zheng-Wu, W. Ping-Feng, Y. Jian-Hua, Z. Kang-Wei: Phys. Rev. B **48**, 16407 (1993)
53. G. Cressey, C. Henderson, G. van der Laan: Phys. Chem. Minerals **20**, 111 (1993)
54. J. Crocombette, M. Pollak, F. Jollet, N. Thromat, M. Gautier-Soyer: Phys. Rev. B **52**, 3143 (1995)
55. P. Kuiper, B.G. Searle, P. Rudolf, L.H. Tjeng, C.T. Chen: Phys. Rev. Lett. **70**, 1549 (1993)
56. P. Kuiper, B. Searle, L.-C. Duda, R. Wolf, P. van der Zaag: J. Electron. Spectrosc. Relat. Phenom. **86**, 107 (1997)
57. G. van der Laan, I. Kirkman: J. Phys: Condens. Matter **4**, 4189 (1992)
58. J.C. Fuggle, J. Inglesfield (Eds.), Unoccupied Electronic States, Vol. 69 of Topics in Applied Physics (Springer-Verlag 1992)
59. M.O. Krause, J.H. Olivier: J. Phys. Chem. Ref. Data **8**, 329 (1979)
60. W.A. Caliebe, C.C. Kao, J.B. Hastings, M. Taguchi, A. Kotani, T. Uozumi, F.M.F. de Groot: Phys. Rev. B **58**, 13452 (1998)
61. G. van der Laan, C.M.B. Henderson, R.A.D. Pattrick, and D. Vaughan: Phys. Rev. B **59**, 4314 (1999)
62. R.A.D. Pattrick, G. van der Laan, J.M. Charnock, and D. Vaughan: American Mineralogist **89**, 541 (2004)
63. R.A.D. Pattrick, G. van der Laan, C.M.B. Henderson, J.M. Charnock, and D. Vaughan: European Journal of Mineralogy **14**, 1095 (2002)
64. D.G. Rancourt: Review in Mineralogy and Geochemistry **44**, 217 (2004)

Materials for Spintronics

Agnès Barthélémy and Richard Mattana

Unité Mixte de Physique CNRS-Thalès, Domaine de Corbeville, 91404 Orsay, and
Université Paris Paris-Sud 91405 Orsay Cedex, France
agnes.barthelemy@thalesgroup.com

Abstract. The spin dependent conduction of ferromagnetic materials in heterostructures gives rise to new phenomena such as giant magnetoresistance (GMR) in magnetic multilayers or tunnel magnetoresistance (TMR) in magnetic tunnel junctions. We review the mechanism of GMR and TMR and experimental results which emphasize the influence of different parameters on these phenomena, taking as an example the case of the widely used transition metals. The integration of ferromagnetic materials and semiconductors is also discussed. The potential of newly emerging materials for spin-electronics or spintronics such as half-metals for fully spin-polarized electrodes, ferromagnetic insulators as spin-dependent tunnel barriers, and ferromagnetic semiconductors is described.

1 Introduction

The birth of spintronics dates back to the observation in 1988 of large changes of resistance by application of a magnetic field to magnetic multilayers, composed of a stack of ferromagnetic metals (such as Fe) and non-magnetic metallic spacer (Cr) [1,2]. This phenomenon, called giant magnetoresistance (GMR), is illustrated in Fig. 1. The GMR ratio is defined as the variation of resistance between the antiparallel and the parallel states of the magnetizations of the ferromagnetic layers $(R_{AP} - R_P)$ over the resistance in the parallel state (R_P). A 80% variation was obtained in these first experiments but GMR as high as 220% were obtained later [3]. To measure this effect, the current can be either applied parallel to the plane of the layers, i.e. in the so-called CIP (current in-plane) geometry, or perpendicular to the interfaces, i.e. CPP (current perpendicular to the plane) geometry [4]. Extensive reviews can be found in [5,6].

Another well-studied heterostructure in the field of spintronics is the ferromagnetic tunnel junction. It is composed of two ferromagnetic metallic electrodes separated by an ultrathin insulating barrier. As for the GMR phenomena, the tunnel magnetoresistance $(TMR = \frac{R_{AP}-R_P}{R_P})$ is related to the variation of conductance between the parallel (P) and the antiparallel (AP) state of the magnetizations of the two magnetic electrodes but, in that case, the origin arises from the tunnelling of the electrons through the insulating barrier. First results have been obtained by Jullière in 1975 [7], but it is only after the discovery in 1995 of a 16% effect at room temperature by Moodera's

A. Barthélémy and R. Mattana: *Materials for Spintronics*, Lect. Notes Phys. **697**, 429–462 (2006)

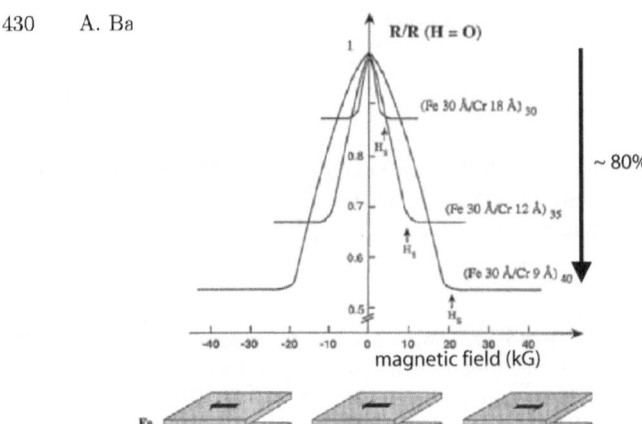

Fig. 1. Resistance versus field curve of Fe/Cr multilayers. The GMR ratio is defined as $(R_{AP} - R_P)/(R_P)$. From [1]

group [8] that a tremendous research has been performed in this field. An example of Tunnel Magnetoresistance (TMR) is shown in Fig. 2. An extensive description of TMR can be found in [9].

Applications for GMR and TMR structures are expanding. GMR read heads, now in all computers, have allowed a large increase in the hard drive storage densities (60–100% by year). TMR-MRAMs (Magnetic Random Access Memories) are currently under development.

Combining magnetic materials with semiconductors is the new challenge of spintronics. This will allow spin injection and transport in semiconductor heterostructures and represent a promising avenue to combine new spin dependent functionalities to the usual properties of "traditional" electronics [10].

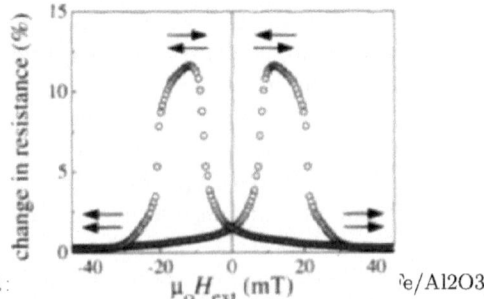

Fig. 2. First TMR   e/Al2O3/Co. From [8]

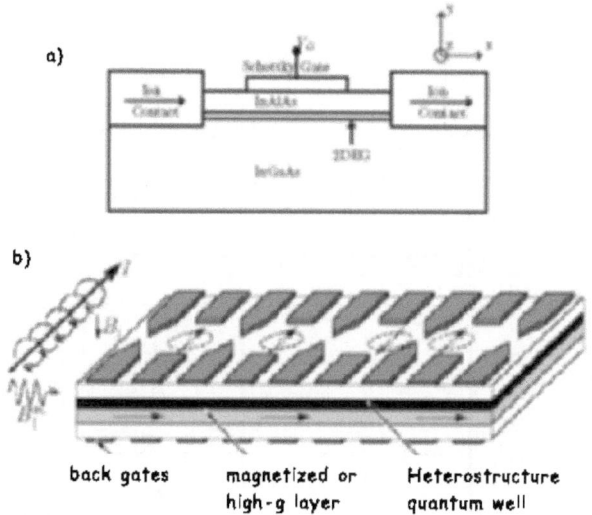

Fig. 3. Spin field effect Transistor (**a**) proposed by Datta and Das [11]. Quantum dot array (**b**) where electrons are confined and can be individually manipulated [12]

Figure 3 presents two designs of devices combining ferromagnets and semi-conductors. The Spin-FET of Datta and Das [11] (Fig. 3a) is a traditional FET (field effect transistor) in which the source and the drain are replaced by magnetic contacts that should allow the injection of spins in the 2D electron gas. The spin of an electron in a quantum dot is a potential way to achieve qubits and/or quantum gates (Fig. 3b). Qubits and quantum gates consti-tute the basic requirement for quantum information (in particular quantum computing and quantum cryptography). The integration of the two types of properties encountered for a long time the problem of efficient spin injection. Large spin injection effects have been observed only recently [13].

In the following chapter, we describe the mechanism of GMR, TMR and experimental results, which emphasize the influence of different parameters on these phenomena. We will mainly discuss the case of transition metals that have been widely used. The problem of spin injection in semiconductors is also described. We try to define the properties of the ideal materials necessary to obtain a large GMR or TMR and efficient spin injection in semiconductors.

2 Mechanism of GMR, TMR and the Problem of Spin Injection in Semiconductors

Spintronics is based on the spin dependence of the electric conduction. The calculated band structures of ferromagnetic transition metals (Fe, Co, Ni)

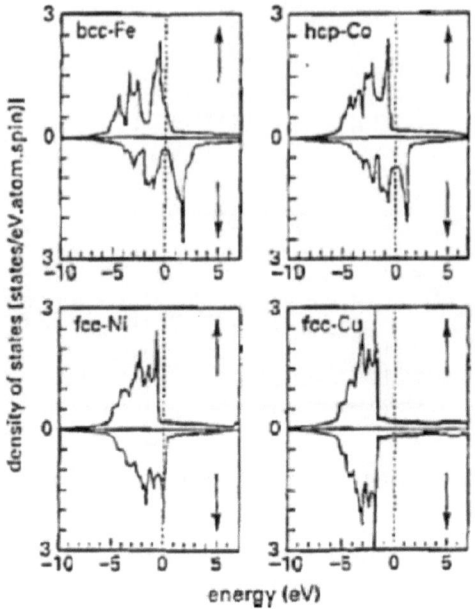

Fig. 4. Band structure of Fe, Co, Ni and Cu. From [15]

are presented in Fig. 4. They are composed of broad sp bands similar to free electron bands, mainly responsible for the conductivity, superposed with narrow localized $3d$ bands responsible for ferromagnetism. All these materials, Fe, Co, Ni and their alloys, fulfil the Stoner criterion

$$N(\varepsilon_F) \cdot U > 1$$

where $N(\varepsilon_F)$ is the density of state (DOS) at the Fermi level and U the exchange interaction (around $1\,\mathrm{eV}$).

Ferromagnetism results from the $3d$ exchange-split bands and different number of occupied states for majority-spin and minority-spin electrons (see chapter by C. Lacroix in this book). In the case of strong ferromagnets like Co or Ni the majority $d_\uparrow$ band is fully occupied whereas the minority $d_\downarrow$ band is only partly occupied and crosses the Fermi level. The conduction of these materials is well described in terms of two parallel conduction channels, one for spin-$\uparrow$ and the other for spin-$\downarrow$ s-electrons [14].

The shift of the two $3d$ bands is also responsible for the asymmetry in the density of state (DOS) at the Fermi level (see Fig. 4). This has two important consequences: The first one is the existence of a finite spin polarization, defined as the normalized difference in the DOS at the Fermi level for spin $\uparrow$ and $\downarrow$

$$SP = \frac{N\uparrow(E_F) - N\downarrow(E_F)}{N\uparrow(E_F) + N\downarrow(E_F)}.$$

This spin polarization associated by to the barrier effect is exploited in magnetic tunnel junction to give the TMR effect. Another consequence is the different resistivities of spin $\uparrow$ and spin $\downarrow$ conduction electrons due to their different probabilities of $s \rightarrow d$ transitions. In this two-current model the conductivity is the sum of the conductivities of spin $\uparrow$ and spin $\downarrow$ electrons which carry the current in two different channels with resistivities

$$\rho_\sigma = \frac{m_\sigma}{n_\sigma e^2 \tau_\sigma} = \frac{1}{n_\sigma e \mu_\sigma} .$$

where n_σ, τ_σ, m_σ and μ_σ are respectively the number of electrons, the relaxation time, the effective mass and the mobility of spin σ conduction electrons. In the Born approximation the relaxation time is related to the scattering potential V_σ and to the density of states at the Fermi level $N(E_F)$ ($\tau_\sigma^{-1} \propto |V_\sigma|^2 N_\sigma(E_F)$).

The intrinsic origin for the spin dependence of the conductivity is related to the spin dependence of n_σ, m_σ and $N_\sigma(E_F)$, due to the splitting of the band by the exchange interaction. Extrinsic origins are related to the presence of defects or impurities within the layers. The asymmetry in the conductance of the two channels is characterized by a spin asymmetry coefficient [16]

$$\alpha = \frac{\rho_\downarrow}{\rho_\uparrow} \text{ or } \beta = \frac{\rho_\downarrow - \rho_\uparrow}{\rho_\downarrow + \rho_\uparrow}$$

For example, in the case of Ni, $s \rightarrow d$ transition are not allowed for spin $\uparrow$ conduction electrons due to the absence of available $d_\uparrow$ states at the Fermi level, whereas $s \rightarrow d$ transitions can occur for spin $\downarrow$ s-electrons. It results a small resistivity, $\rho_\uparrow$, for the spin $\uparrow$ channel, a larger one, $\rho_\downarrow$, for the spin $\downarrow$ channel and an α coefficient generally larger than one due to the intrinsic origin. However when 1% Cr impurities are added into Ni an α coefficient smaller than one is measured because of scattering by impurities with opposite spin asymmetry, counterbalancing the matrix spin asymmetry. This spin asymmetry of the conduction is exploited in magnetic metallic multilayers and is responsible for the GMR effect described below.

2.1 Giant Magnetoresistance

GMR can be interpreted using an extrapolation of the Mott model [14] of two parallel current channels, one for majority-spin and one for minority-spin electrons. In magnetic multilayers the contribution to the intrinsic potential arises from the band structure of each layer but the effect of band mismatch at the interface between the different materials constitutive of the multilayer must also be taken into account. In the case of a Co/Cu interface for example, a good matching is observed only for spin-$\uparrow$ s-type electrons at the Fermi level (see DOS of Co and Cu in Fig. 4). Conversely, for spin-$\downarrow$ electrons, the wave functions are of s and d type at the Fermi level in the Co, whereas they

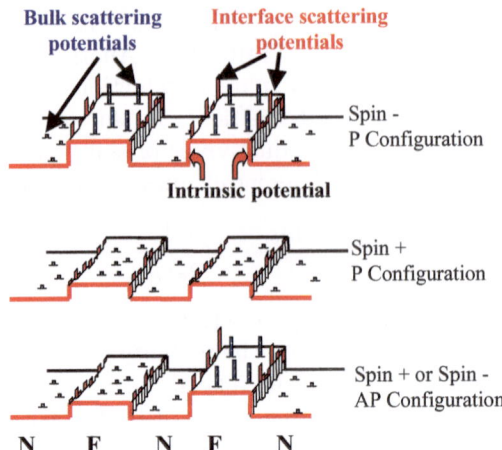

Fig. 5. Potential landscape for spin ↑ and spin ↓ conduction electrons in the parallel and antiparallel configuration of the magnetizations. Spin ↑ and spin ↓ refer to the orientation of the spin ($S_Z = 1/2$ and $S_Z = -1/2$ respectively) in an absolute referential. In the parallel configuration of the magnetizations, electrons with spin ↑ (↓) are majority (minority) spin electrons in all the layers and are weakly (strongly) scattered in the whole structure. A short circuit in the parallel configuration is obtained for the fast electron channel, resulting in a low resistance for this configuration. In the antiparallel configuration spin ↑ and spin ↓ are alternatively the majority and the minority spin directions. The electrons of the two spin directions are alternatively weakly and strongly scattered in each channel and the resistance is higher than in the parallel configuration. This picture holds in CIP and CPP geometry when the thickness of the layers is smaller than the scaling length of the problem, i.e. the mean free path in CIP and the spin diffusion length in CPP. The mean free path is defined as the mean distance between two collisions, and the spin diffusion length as the mean distance between two spin reversal. From [23]

are of s type in Cu. A bad matching at the interface is then obtained for the spin-↓ channel. The extrinsic potential in multilayers takes into account the presence of impurities or defects within the layer as well as the roughness at the interface, which is also spin dependent. The whole microscopic origin of GMR is well illustrated by the potential landscape experienced by conduction electrons in the heterostructures represented in Fig. 5 for spin ↑ and ↓ electrons in the parallel and antiparallel configurations of the magnetizations. In the potential landscape of Fig. 5 (corresponding to $\alpha > 1$, i.e. weakly scattered majority spin electrons), for a parallel configuration of the magnetizations, electrons with spin ↑ (↓) are majority (minority) spin electrons in all the layers and are weakly (strongly) scattered in the whole structure. A short circuit in the parallel configuration is obtained for the spin ↑ fast electron channel, leading to a low resistance. Conversely, in the antiparallel configuration, spin ↑ and spin ↓ are alternatively the majority

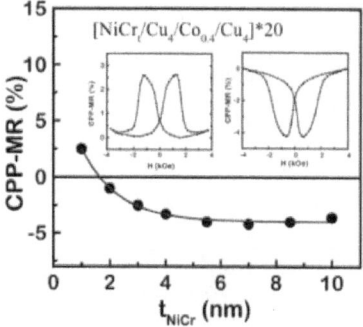

Fig. 6. Variation of the CPP-GMR as a function of the thickness of the $Ni_{0.95}Cr_{0.05}$ layers in $\underline{Ni}Cr/Cu/Co/Cu$ multilayers. The competition between the bulk scattering (characterized by $\beta_{NiCr} < 0$) and interface scattering (with $\gamma_{NiCr/cu} > 0$) gives rise to a compensation thickness t^*_{NiCr} at which they cancel each other (t^*_{NiCr} is determined by $\beta_{\underline{Ni}Cr} \times \rho^*_{\underline{Ni}Cr} \times t^*_{\underline{Ni}Cr} + 2\gamma_{\underline{Ni}Cr/Cu} \times r_{\underline{Ni}Cr/Cu} = 0$). Below this compensation thickness $t^*_{\underline{Ni}Cr}$, the interface scattering is predominant, the global spin asymmetry of the $\underline{Ni}Cr$ layer is positive as that of the Co layer (characterized by $\beta_{Co} > 0$ and $\gamma_{Co/Cu} > 0$) and the effect is normal (see insert a). Above t^*, the bulk scattering in $\underline{Ni}Cr$ becomes predominant so that the global asymmetry of the NiCr layers is negative. Consequently, with alternating spin asymmetries for the NiCr and Co layers, the effect is inverse (see insert b). From [23]

and the minority spin directions and the electrons of the two spin directions are alternatively weakly and strongly scattered in each channel; the resistance is higher than in the parallel configuration.

The influence of the different types of potentials has been corroborated experimentally. For example, the role played by the scattering at the interface has been underlined by dusting the interface with another magnetic element, which leads to an increase or decrease of the GMR [17,18]. In the same manner, the role played by the scattering within the layers has been evidenced by introducing bulk scattering centers [19–22]. The most spectacular examples of impurity effects on GMR are obtained when impurities are added to induce spin asymmetries either with opposite signs in consecutive layers or with opposite and competing signs at interfaces and in the bulk [21, 22]. A typical example is presented in Fig. 6 for $Ni_{0.95}Cr_{0.05}/Cu/Co/Cu$ multilayers. Doping Ni by 5% Cr impurities is enough to reverse the asymmetry coefficient of NiCr layers which, in that case, compete with the interface asymmetry coefficient. At small thickness of the NiCr layers, the global asymmetry of the NiCr/Cu layer is positive like that of the Co/Cu layer, which results in a normal GMR effect. At larger NiCr thicknesses, the global asymmetry of the NiCr/Cu layers becomes negative, i.e., opposite to that of Co/Cu. An inverse effect, with a resistance in the antiparallel state smaller than in the parallel configuration of the magnetizations, is observed. An extensive quantitative study has been performed in the CPP geometry and has allowed the

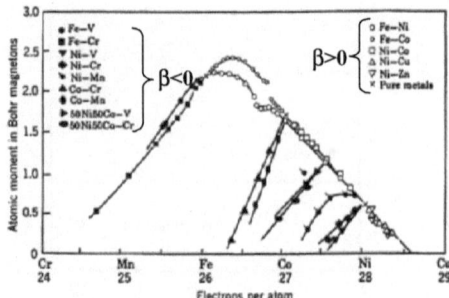

Fig. 7. Slater-Pauling curve representing the variation of the atomic moment of alloys as a function of the atomic number. For all the alloys on the branch with a negative slope a coefficient larger than one is found, whereas it is smaller than 1 for the alloys on the branch with a positive slope, due to the formation of a virtual bound state leading to a high spin ↑ DOS at the Fermi level on the impurity site. From [5]

determination of the spin asymmetry coefficient [23]. The sign of the β coefficients is related to the position of the alloy on the Slater-Pauling curve. As shown in Fig. 7, β is positive for all the alloys corresponding to a negative slope in the Slater-Pauling curve, whereas it is negative for the alloys corresponding to a positive slope. The positive values of β are associated with a higher density of states (DOS) at the Fermi level for the minority spin direction in pure metals or on the impurity sites. In contrast, for Cr or V impurities in Fe, or Cr impurities in Co or Ni, the d levels of the impurities are well above the host d bands for the majority spin direction. Therefore they cannot be hybridized with the d band states. The resulting resonant scattering of spin ↑ sp electrons with empty d states of the impurity just above the Fermi level accounts for the large spin ↑ resistivity and for the negative sign of β. Alternatively, one can say that the formation of a virtual bound state leads to a high spin ↑ DOS at the Fermi level on the impurity site.

In conclusion, the GMR phenomena are now well understood. The different contributions of scattering at the interface or within the bulk of the layers can be identified by GMR experiments and analysis in the CPP geometry. To obtain a large GMR very different mobilities for spin-up and spin-down electrons are required.

2.2 Tunnelling Magnetoresistance

TMR of magnetic tunnel junctions is related to spin-dependent tunnelling, as shown by Meservey and Tedrow in the seventies in superconductor/Al_2O_3/ferromagnet junctions [24, 25]. This type of experiments uses the Zeeman splitting of the quasiparticle density of state of the superconductor (with a

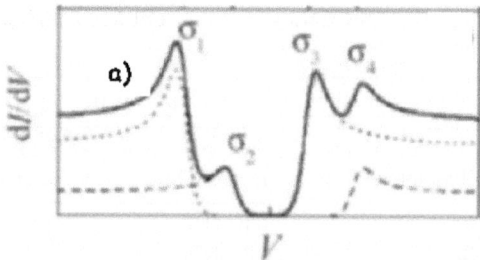

Fig. 8. Typical conductance curve as a function of the bias applied to FM/Insulator /superconductor tunnel junctions. From [26]

sharp peak at the gap band edge) in spin-up and spin-down densities as a detector of the spin-polarization of various magnetic electrodes (FM). The separation in spin-up and spin-down densities by application of a magnetic field on the superconductor allows a determination of the spin-up and spin-down contributions to the tunnelling current of FM/Insulator/superconductor tunnel junctions. The variation of the conductance with the bias applied to the junction is presented in Fig. 8. From the area of the conductance peaks, the spin-polarization of the tunnelling current can be deduced:

$$SP = \frac{(\sigma_4 - \sigma_2) - (\sigma_1 - \sigma_3)}{(\sigma_4 - \sigma_2) + (\sigma_1 - \sigma_3)} .$$

An extensive review of this work is presented in [26]. Values of the spin polarization obtained by Meservey-Tedrow type experiments are reported in Table 1.

The surprising conclusion of these first experiments was the always positive sign of the spin-polarization in the case of the Al_2O_3 barrier (few attempts have been performed later to measure the spin-polarization with other types of barriers). In fact, from the DOS of Ni and Co presented in Fig. 4, a negative spin-polarization reflecting the larger density of states for the minority spin band would be expected. This positive spin-polarization is attributed to the fact that only s-like itinerant electrons are responsible for tunnelling [33, 34]. Due to the $s - d$ hybridization in the minority band, the density of state of the s band is smaller for the minority spin electrons than for the majority ones. A positive spin-polarization results. The amplitude of the TMR effect is related to the spin-polarization via Jullière's model [7]. This model assumes that the spin is conserved in the tunnelling process. It also considers that the conductance of each spin channel is proportional to the product of the density of states of the emitting electrode (number of electrons available for tunnelling) by the density of states of the collecting electrode (number of empty states available for electrons to tunnel into). In the parallel configuration of the magnetizations, the conductance is expressed as

Table 1. Tunnelling spin polarization SP (%) obtained in FM/Insulator/ superconductor tunnel junctions

Meservey-Tedrow Type Experiments			
Material	SP for Al_2O_3 barrier [26]	SP new values for Al_2O_3 barrier	SP for other barriers
Ni	23	33 [15]	
Co	35	42 [15]	
Fe	40	45 [27]	74 (MgO) [28]
$Ni_{80}Fe_{20}$	32	48 [15]	
$Ni_{40}Fe_{60}$	37	55 [29]	
$Co_{50}Fe_{50}$		55 [15]	
$Co_{70}Fe_{30}$			85 (MgO) [28]
$Co_{84}Fe_{16}$		55 [15]	
CrO_2		>90 [30]	
$La_{0.7}Sr_{0.3}MnO_3$			72 ($SrTiO_3$) [31]
$SrRuO_3$			$-9.5(SrTiO_3)$ [32]

$$G_P \propto N_{1\uparrow}N_{2\uparrow} + N_{1\downarrow}N_{2\downarrow}$$

In the antiparallel configuration, the majority spin in one electrode becomes a minority spin in the other, which corresponds to a conductance

$$G_{AP} \propto N_{1\uparrow}N_{2\downarrow} + N_{1\downarrow}N_{2\uparrow}$$

One deduces the TMR ratio:

$$TMR = \frac{R_{AP} - R_P}{R_P} = \frac{G_{AP} - G_P}{G_{AP}} = \frac{2SP_1SP_2}{1 - SP_1SP_2}$$

with SP_i the spin-polarization of electrode i defined as

$$SP_i = \frac{N_{i\uparrow} - N_{i\downarrow}}{N_{i\uparrow} + N_{i\downarrow}}$$

This model has been used to correlate the amplitude of the TMR ratio with the spin-polarization for transition metals and alloys deduced from Meservey-Tedrow experiments with a quite good agreement. TMR ratio around 70% at room temperature are currently measured. From the expression of TMR as a function of the spin-polarization, it is clear that a fully spin-polarized electrode could lead to very high TMR effects (theoretically infinite). Such materials called half-metals are presented in paragraph 3. The physics of TMR is more complicated than the simple model of Jullière.

De Teresa et al. demonstrated the role played by the barrier in tunneling in $La_{0.7}Sr_{0.3}MnO_3$/I/Co tunnel junctions, taking advantage of the positive nearly full polarization of the half-metal $La_{0.7}Sr_{0.3}MnO_3$ [35, 36]. Results are

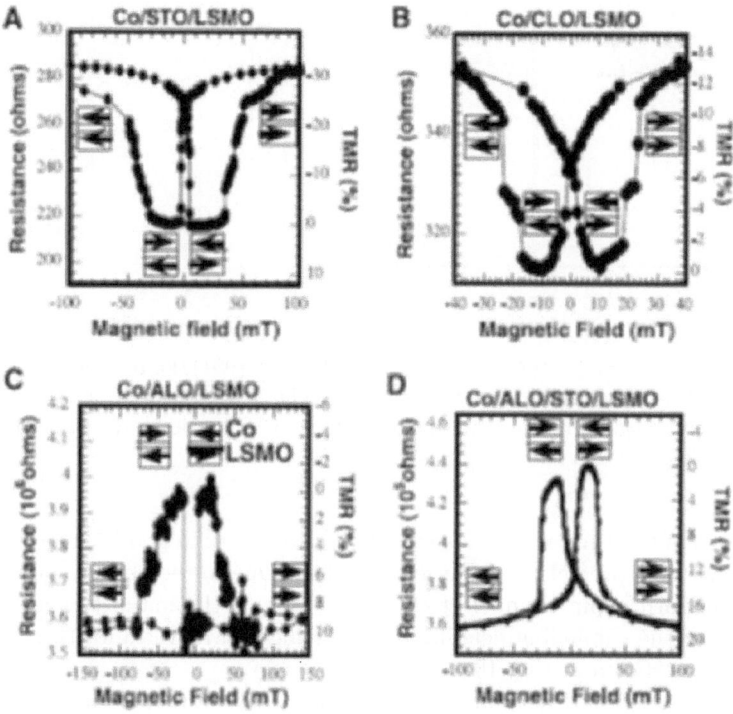

Fig. 9. TMR curves obtained at small bias ($-10\,$mV) on $La_{0.7}Sr_{0.3}MnO_3/I/Co$ tunnel junctions; $I = SrTiO_3$ (**A**), $(Ce_{0.69}La_{0.31})O_{1.85}$ (**B**), Al_2O_3 (**C**) and $SrTiO_3/Al_2O_3$ (**D**). From [35]

reported in Fig. 9. In the case of an Al_2O_3 barrier, a normal effect corresponding to a positive spin-polarization for the counter electrode of Co is found as always observed with this kind of barrier. On the contrary, when $SrTiO_3$ or $(Ce_{0.69}La_{0.31})O_{1.85}$ is used for the barrier, an inverse TMR effect (with a resistance smaller in the antiparallel state) is observed. Because of the positive spin-polarization of $La_{0.7}Sr_{0.3}MnO_3$, this means a negative spin-polarization for the Co in that case. Furthermore, when a thin Al_2O_3 layer is inserted between $SrTiO_3$ and Co (D) the normal TMR is recovered, indicating the crucial role played by the bonding at the interface in the determination of the effective spin-polarization.

Confirmation of the effect of bounding at the interface to fix the effective spin-polarization has been theoretically provided by Oleinik and coworkers [37,38]. Considering the $Co/SrTiO_3$ and Co/Al_2O_3 interface, they have shown that metal induced gap states (MIGS) propagate through the insulator. These MIGS states are spin-polarized and responsible for tunnelling.

More recently, very large TMR ratios were observed in $Fe(001)/\,MgO(001)/Fe(001)$ tunnel junctions [28, 39] thanks to the coherence of electron wave

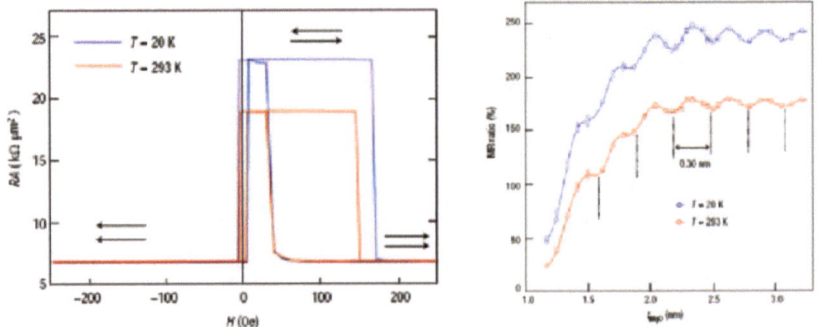

Fig. 10. Tunnel magnetoresistance of Fe(001)/MgO(001)/Fe(001) junctions. *Left*: Magnetoresistance curves at 20 K and 293 K. The TMR ratio are 247% and 180% respectively. *Right*: thickness dependence of the TMR ratio as the function of the MgO thickness. From [39]

functions in the tunnelling process. These results are in good agreement with the theoretical prediction of Butler et al. [40]. Due to the absence of Δ_1 band in the minority spin channel at the Fermi energy, the majority spin current dominates the tunnelling, leading to a very high TMR ratio for thick enough barriers, as shown in Fig. 10.

The TMR is not as well understood as GMR perhaps because most of the experiments have been performed on magnetic tunnel junctions with an amorphous Al_2O_3 barrier. Recent experiments on crystalline junctions allow a better understanding of TMR, underlying its complexity but also new possibilities to engineer the magnetic tunnel junctions with larger effects.

2.3 Electrical Spin Injection and Detection in Semiconductors

Spin-polarized transport in semiconductors is the key to future development of spintronics. In the nineties, first attempts used transition metals to electrically inject spins in semiconductors. The use of transition metals is hampered by two problems: the chemical reactivity between metals and semiconductors and the so-called "conductivity mismatch". The chemical reactivity at the ferromagnetic metal/semiconductor (FM/SC) interface (Fe/GaAs, Fe/Si for example) generally induces the formation of a non-magnetic layer at this interface. Therefore the spin-polarization at the interface tends towards zero and the spin injection is not efficient. In 2000, Schmidt et al. [41] have highlighted a fundamental obstacle linked to the large difference of conductivity between ferromagnetic metals and semiconductors. In a diffusive regime, when spins are injected from a ferromagnetic metal into a non-magnetic semiconductor, it creates a spin accumulation both in the metal and in the semiconductor on a length characterized by the spin diffusion length l_{sf} (see Fig. 11). As the number of spin-flips is inversely proportional to the resistivity and the spin

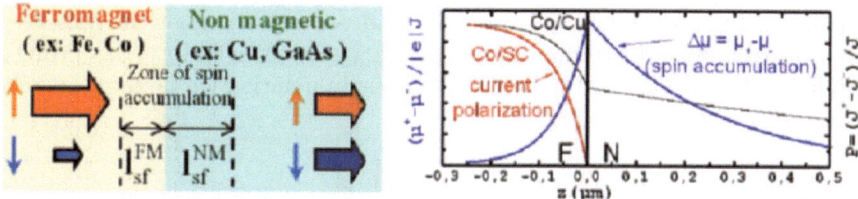

Fig. 11. Scheme illustrating a ferromagnet/non-magnetic interface. A spin accumulation over a distance equal to the spin diffusion length l_{sf} is created at the interface. The number of spin-flips, inversely proportional to the resistivity is larger in the ferromagnetic metal than in the semiconductor leading to a complete depolarization of the injected current at the interface and a non-efficient spin injection

diffusion length, all the spin accumulation relaxes in the metal and therefore the net spin polarization is zero at the interface.

One solution proposed by Rashba [42] and by Fert et al. [43], consists of inserting a resistance at the FM/SC interface (Schottky barrier [45], tunnel barrier or a sufficiently resistive ohmic contact). The insertion of a spin-dependent resistance has the effect of restoring the spin-polarization at the interface through a large spin accumulation in the semiconductor[1].

Another solution is to use ferromagnetic semiconductors since non-magnetic and magnetic semiconductors have better compatible growth and closer density of states and conductivities.

3 Alternative Ferromagnetic Conducting Materials

$3d$ metals and even strong ferromagnets lead to incomplete spin-polarization of the conduction electrons due to the presence of the non-polarized $4s/4p$ bands at the Fermi level. A better situation is expected from half-metallic materials, as called initially by De Groot et al. to characterize the band structure of Heusler alloys [46]. In half-metals, one of the spin band is metallic and the other presents a semiconducting behaviour with a Fermi level within the gap, or a localization of the carriers (the compound is then called a "transport half-metal" [47]). The expected 100% spin-polarization explains the recent interest in these materials. An increasing number of compounds have been predicted to be half-metallic; some Heusler alloys such as NiMnSb, mixed valence manganites [48], Fe_3O_4 [49], CrO_2 [50], and some double perovskites [51] but also $Tl_2Mn_2O_7$ [52], sulfurospinels [53]. Attempts have been made to confirm experimentally the half-metallic nature of some of these compounds using Spin

---

[1] Conditions linked to electrical spin injection exist of course for an electrical spin detection. The condition is now related to the finite carrier spin lifetime in the semiconductor. This interface resistance should not be too large and its value depends on the device geometry [43, 44].

Polarized Photoemission Spectroscopy, Andreev reflection[2] or the Meservey Tedrow technique. Until now, only CrO_2, $La_{0.7}Sr_{0.3}MnO_3$, $La_{0.7}Ba_{0.3}MnO_3$, Sr_2FeMoO_6 and $Ga_{1-x}Mn_xAs$ have shown spin polarization larger than 85% from the point of view of transport.

3.1 CrO$_2$

Chromium dioxide is the simplest half metal in the sense that it contains only Cr^{4+} and oxygen ions. It crystallizes in the tetragonal rutile structure in which each Cr^{4+} is surrounded by a distorted oxygen octahedron with two short bonds and four longer in-plane bonds. The Cr d orbitals are then split by the crystal field into an e_g doublet and a t_{2g} triplet, which is further split due to the distortion of the octahedron. Ferromagnetism arises from the exchange interaction of the $3d^2$ ions in the rutile structure and is enhanced by delocalization of the $t_{2g\downarrow}$ electrons in a d band. The expected saturation magnetization is $2\,\mu_B$ by formula unit, in good agreement with experimental results, and the curie temperature Tc is around 400 K. Many band structure calculations indicate the half-metallic nature of chromium dioxide [50, 55], with a finite DOS at the Fermi level in the majority spin subband and a gap of at least 1 eV in the minority one (i.e. a positive spin-polarization). An example of such band structure is presented in Fig. 12 [50].

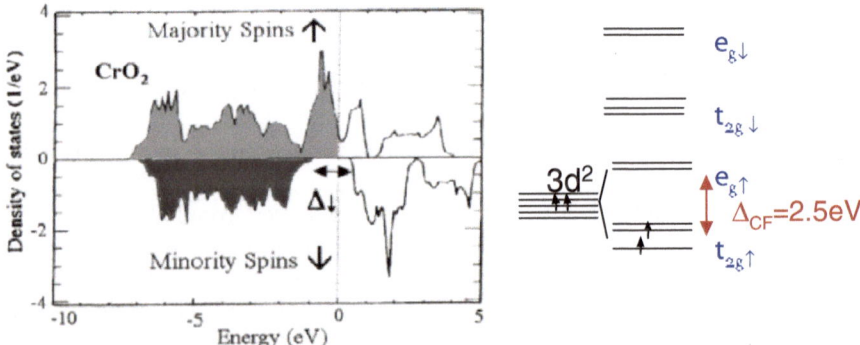

Fig. 12. Band structure calculation and ionic picture of CrO_2. From Lewis et al. [50]

Spin-polarized photoemission [56] and point contact measurements have evidenced the half-metallic nature of CrO_2 [57–59] with a spin-polarization of 98% at 1.6 K (see Fig. 13a).

---

[2] Andreev reflection weights the contribution of the different electronic states differently than in the tunnelling process. An agreement between spin-polarized tunneling and Andreev reflections is then not expected [54].

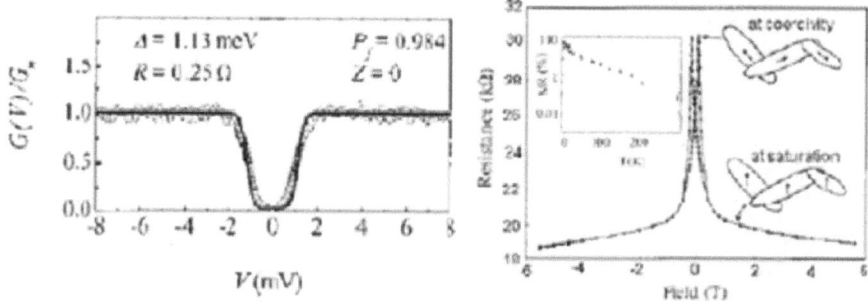

Fig. 13. (a) Point contact measurement of the spin-polarization of CrO_2. From [59]. (b) Magnetoresistance of CrO_2-Cr_2O_3 pressed powders [60]. The large low-field magnetoresistance, due to tunneling across insulating Cr_2O_3 grain boundaries, corresponds to a spin-polarization of 67%. The inset represents the temperature variation of the low-field magnetoresistance

CrO_2 powders exhibit a large low-field magnetoresistance as presented in Fig. 13b [60]. This low-field magnetoresistance, due to tunneling across insulating grain boundaries, a mechanism close to the one in magnetic tunnel junctions, corresponds to a spin-polarization of 67% at 4 K. This magnetoresistance decreases by two orders of magnitude between low and room temperature as shown in the inset of Fig. 13b. CrO_2/native oxide of Cr_2O_3/Co tunnel junctions have been studied but exhibit disappointingly a very small magnetoresistance even at low temperature [61]. Furthermore, the observed inverse TMR, corresponding to a negative spin-polarization, is in contradiction to what is expected from band structure calculations.

3.2 Mixed Valence Manganites

Other materials, widely studied in the field of spintronics, are manganites. Manganites are well known to exhibit the Colossal Magnetoresistance effect associated to the metal-insulator transition at their Curie temperature [62, 63]. But colossal magnetoresistance, which arises from the realignment of the spins by an applied magnetic field restoring the conduction, necessitates very large fields (several teslas), impeding its application. In this chapter we concentrate on spin-dependent tunnelling, i.e., TMR experiment on tunnel junctions or tunnelling experiments across grains boundaries. $La_{1-x}A_xMnO_3$ (with A = Ca, Sr, or Ba), crystallizes in the perovskite structure ABO_3 Fig. 14a), in which each Mn atom is surrounded by an oxygen octahedron. Due to the corresponding crystalline field, the d bands of Mn are split in t_{2g} and e_g states (see Fig. 14b). Replacing La^{3+} by a divalent ion such as Ca, Sr or Ba results in mixed valence compounds with alternating Mn^{3+} and Mn^{4+} ions on the B sites. The mechanism at the origin of the ferromagnetism and the conduction in these compounds is the double exchange

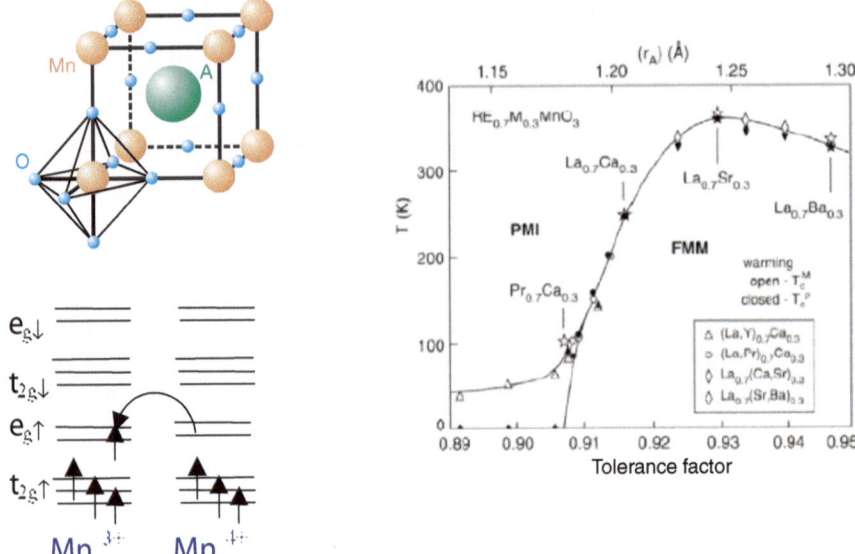

Fig. 14. (a) Crystalline structure of manganite. (b) Ionic picture of the conduction mechanism. (c) variation of the Curie temperature of manganites as a function of the tolerance factor, which is defined as $f = \frac{1}{\sqrt{2}} \cdot \frac{RB-RA}{RA+RO}$, where R is the ionic size of each element in the ABO_3 perovskite structure (A = Rare earth, B = Mn, O = oxygen). T_c is the highest for the compound $La_{0.7}Ba_{0.3}MnO_3$ for which the distortion of the oxygen octahedra is the smallest leading to a better delocalization of the carrier at the origin of ferromagnetism as well as conduction. From [54]

mechanism [64]. A maximum of the Curie temperature and metallicity is observed for the compound $La_{0.7}Sr_{0.3}MnO_3$ (see Fig. 14). From band structure calculations, a half-metallic character with only majority spins at the Fermi level is expected for the $x = 1/3$ compounds [48,65]. Spin-polarized [66] and inverse spin-polarized [67] photoemission experiments also evidenced the half-metallic character of these compounds with only majority spin at the Fermi level as shown in Fig. 15.

This half-metallic nature is supported by transport measurements on polycrystalline films [68] presented in Fig. 16. On single crystals only the colossal magnetoresistance varying linearly is observed, due to the reduction of the spin-canting by the applied field. In polycrystalline films, both low-field and high-field magnetoresistance are present. This low-field magnetoresistance arises from tunneling across grain. It amounts to 23% and corresponds to a spin-polarization of 55% considering the random orientation of the magnetizations of different grains. The half-metallic nature is also confirmed by spin-dependent tunnelling experiments on $LaSrMnO_3/SrTiO_3/$ superconducting Al junctions [69], which allow the determination of a spin-polarization at low temperature of +72% for the LSMO/$SrTiO_3$ interface.

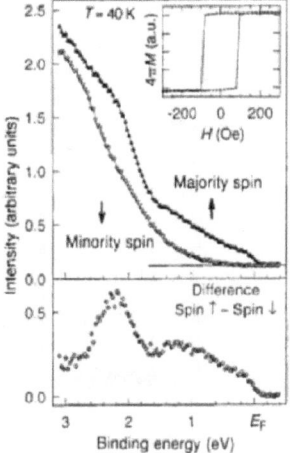

Fig. 15. Spin-resolved photoemission spectra of thin film of $La_{0.7}Sr_{0.3}MnO_3$ near the Fermi energy; Measurements are taken at 40 K. *Bottom panel*: difference spectra between majority and minority spin spectra. From [66]

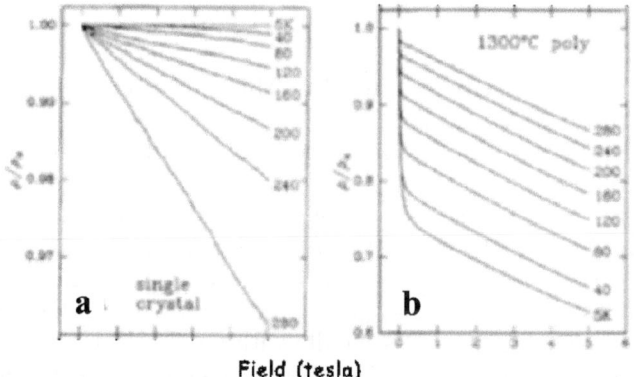

Fig. 16. Resistivity versus magnetic field curves of $LaSrMnO_3$ single crystal and polycrystalline films. A large low field magnetoresistance due to tunnelling across grain boundaries is observed for polycrystalline films in good agreement with the expected half metallic nature of the compound. From [68]

Tunnel magnetoresistance experiments on manganite based magnetic tunnel junctions lead to very large TMR ratios at low temperature [70]. Moon Ho Jo et al. [71] obtained a large TMR effect with a very well defined antiparallel arrangement, corresponding to a spin-polarization of 86% at 77 K using a $NdGaO_3$ tunnel barrier (see Fig. 17). However, the TMR of these junctions rapidly decreases and vanishes at 150 K, well below the T_c of the material. This rapid decrease of the TMR and corresponding spin-polarization at the interface has been attributed to the phase separation mechanism, which arises

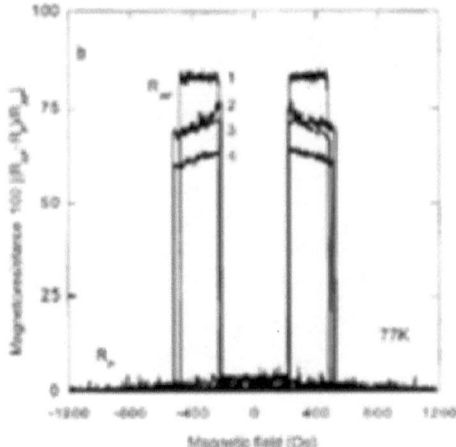

Fig. 17. Resistance versus applied magnetic field curve of $La_{2/3}Ca_{1/3}MnO_3/$ $NdGaO_3/$ $La_{2/3}Ca_{1/3}MnO_3$ tunnel junctions at 77 K. The R(H) curve shows very well defined switching filed and a large magnetoresistance corresponding to a spin-polarization of 86% at 77 K; (1), (2), (3), (4) correspond to junctions with in creasing aeras. From [71]

at the interface in the case of $La_{2/3}Ca_{1/3}MnO_3$. This has been confirmed by NMR studies of a $La_{2/3}Ca_{1/3}MnO_3/SrTiO_3$ interface [72], underlining the localization of charges at the interface leading to the presence of a more insulating phase.

Magnetic tunnel junctions with Sr doped manganites, $La_{2/3}Sr_{1/3}MnO_3$, have shown TMR as large as 1900% [73] corresponding to a nearly total spin-polarization of 95% as expected from the half-metallic nature of this compound. As in the case of $La_{2/3}Ca_{1/3}MnO_3$, the TMR of $La_{2/3}Sr_{1/3}MnO_3$ based tunnel junction decreases rapidly with temperature but vanishes only around room temperature [74, 75]. The temperature dependence of both TMR and spin-polarization of $La_{2/3}Sr_{1/3}MnO_3/I/La_{2/3}Sr_{1/3}MnO_3$ tunnel junctions (where $I = SrTiO_3$, $LaAlO_3$, TiO_2) is presented in Fig. 18 [75]. By comparison with the temperature dependence found by Park and coworkers by photoemission experiments for a free surface, Garcia and coworkers concluded that the polarization is more preserved at the interface due to the continuity of the oxygen octaedra. Nevertheless, spin-polarization at the interface possesses a T_c around 300 K, slightly smaller than the bulk value of 350 K, which hampers the use of such materials for application at room temperature.

3.3 Magnetite

The best known predicted half-metal is certainly magnetite. Fe_3O_4 is a ferrimagnetic oxide crystallizing in the inverse spinel structure (see Fig. 19a).

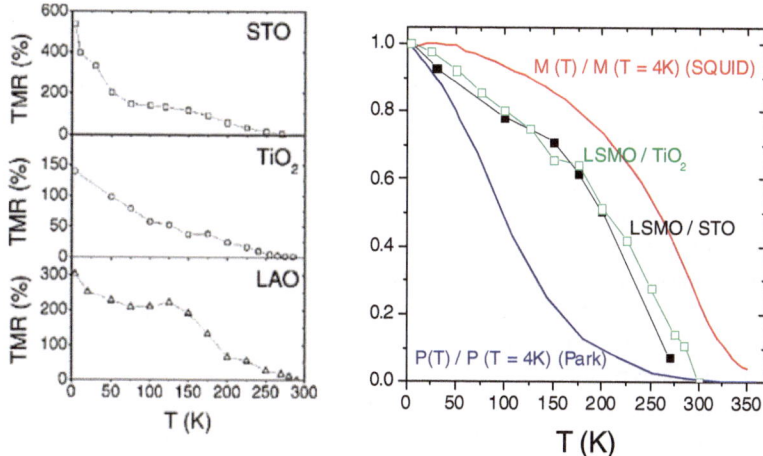

Fig. 18. (a) Temperature dependence of $La_{2/3}Sr_{1/3}MnO_3/I/$ $La_{2/3}Sr_{1/3}MnO_3$ tunnel junctions, where $I = SrTiO_3$, $LaAlO_3$, TiO_2. (b) The variation as a function of temperature of the spin-polarization, extracted from the TMR, is compared to the reduced magnetization, $M(T)/M(T = 4\,K$, and the surface spin polarization at E_F, $P(T)/P(T = 4\,K)$. From [75]

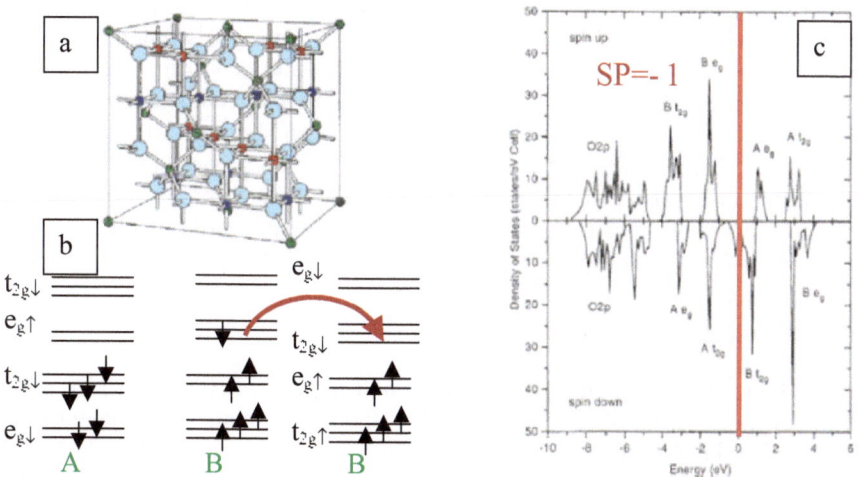

Fig. 19. (a) Crystalline structure of Fe_3O_4. (b) Ionic model for the conduction and ferrimagnetism of the compound. (c) Band structure calculation of Fe_3O_4

It has the highest Curie temperature, $(T_c = 860\,K)$ and, in that sense, can be considered as the ideal candidate for room temperature applications. In this structure, tetrahedrally coordinated sites A are occupied by Fe^{3+} ions whereas octahedrally coordinated sites B are occupied by both Fe^{2+} and Fe^{3+} ions. The magnetic moments on A and B sites are antiparallel due to a super-exchange interaction whereas a double exchange interaction takes place on the B site between Fe^{3+} and Fe^{2+}. Fe_3O_4 is ferrimagnetic with a net magnetic moment of $4\,\mu_B$ per formula unit. The temperature dependence of the resistivity is quite complex, changing from a semiconducting to a metallic behaviour slightly above room temperature, and back to semiconducting above T_c [76, 77]. The conductivity at room temperature is due to an electron transfer between Fe^{2+} and Fe^{3+} ions on the octahedral sites. Below the Verwey transition, which takes place around 120 K, carrier transport occurs via electron hopping. Band structure calculations [78–80], (Fig. 19c), indicate the half-metallic nature of this compound with only minority spin density crossing the Fermi level. The expected spin-polarization of -100% is however in disagreement with the observed semiconducting behaviour up to room temperature. Photoemission experiments on Fe_3O_4 single crystals [81] and thin films grown on Pt(111) [82] corroborate the half-metallic nature of the compound.

Transport measurements on magnetic tunnel junctions with magnetite electrodes and a MgO barrier give disappointingly a very small TMR ratio of 0.5% at room temperature which increases to 1.5% at 150 K [83]. In $Fe_3O_4/Al_2O_3/Co$ tunnel junctions a positive TMR of 43% at room temperature has been observed [84]. Since the spin-polarization at the Co/Al_2O_3 interface is positive, such positive TMR corresponds to a positive spin-polarization at the Fe_3O_4/Al_2O_3 interface, in contrast to what is expected from the band structure of magnetite. Recent results of Parkin and coworkers show, on the contrary, a negative TMR of -22% at room temperature in $Fe_3O_4/Al_2O_3/CoFe$ magnetic tunel junctions when the crystallographic orientation of the Fe_3O_4 is (100) [85]. A positive TMR is always observed if the crystalline orientation of the layer is (111). This negative spin-polarization is also in agreement with the results of Hu and Suzuki on $Fe_3O_4/CoCr_2O_4/La_{2/3}Sr_{1/3}MnO_3$ tunnel junctions presented in Fig. 20 [86].

3.4 Double Perovskites

Sr_2FeMoO_6 has also been predicted to be half-metallic. It belongs to the class of double perovskites of formula $A_2BB'O_6$ where A is an alkaline element (Ca, Ba, Sr), B = Fe, Cr and B' = Mo, Re… The Curie temperature of Sr_2FeMoO_6 is 420 K. Sr_2FeMoO_6 has been the most studied material of this family but the highest T_c has been obtained for the Sr_2FeReO_6 and Sr_2CrReO_6 compounds. Electron doping in the $Sr_{2-x}La_xFeMoO_6$ raises the Curie temperature to 490 K [87]. Sr_2FeMoO_6 is a ferrimagnet with a localized $S = 5/2$ spin on the Fe site (see Fig. 21) and a delocalized spin down electron [88–90]. In order to explain the strong antiferromagnetic coupling between the localized Fe^{3+} and the delocalised electron extended double exchange has been invoked [91,92]. The expected magnetic moment is $4\,\mu_B$ per formula unit but magnetization measurements generally lead to a reduced magnetic moment around 3 or $3.5\,\mu_B$ due to cation disorder on the B sites [93]. Band structure calculations

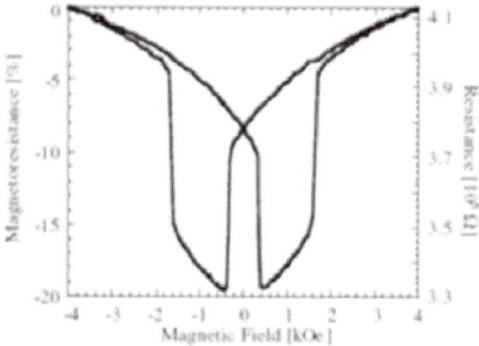

Fig. 20. Magnetoresistance curve of a $Fe_3O_4/CoCr_2O_4/$ $La_{2/3}Sr_{1/3}MnO_3$ tunnel junction [86]. The inverse TMR observed is in agreement with the expected negative spin polarization of the material

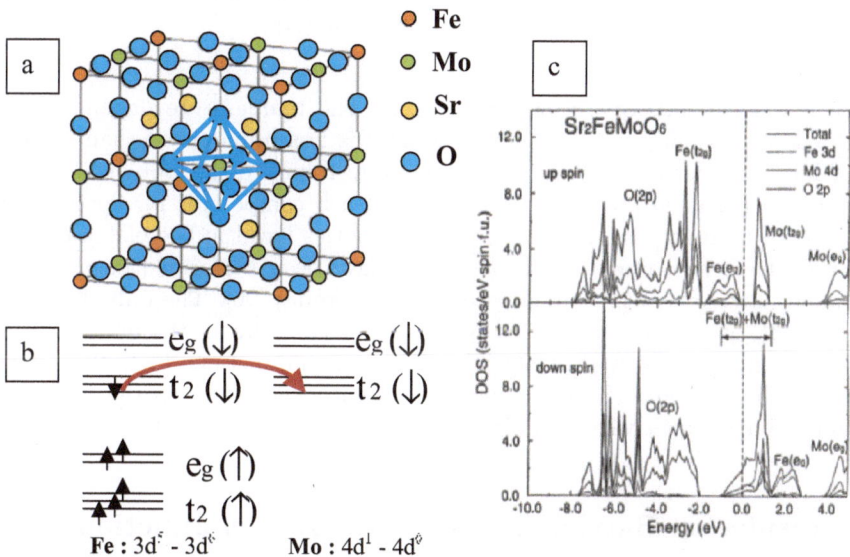

Fig. 21. (a) Double perovskite structure of Sr_2FeMoO_6. (b) Ionic model for the conduction and ferrimagnetism of the compound. (c) Band structure calculation of Sr_2FeMoO_6 [51]

of Sr_2FeMoO_6 and Sr_2FeReO_6 predicted the half-metallic nature of these compounds [51, 94]. In the case of Sr_2FeMoO_6 (see Fig. 21), a gap of 1 eV is present at the Fermi level in the majority spin band between the occupied Fe e_g states and the unoccupied Mo t_{2g} level in the majority band. The minority band crosses the Fermi level and is composed of Fe and Mo hybridized states.

Magnetoresistance measurements on polycrystalline samples, presented in Fig. 22a, evidence a large low-field magnetoresistance, which amount to

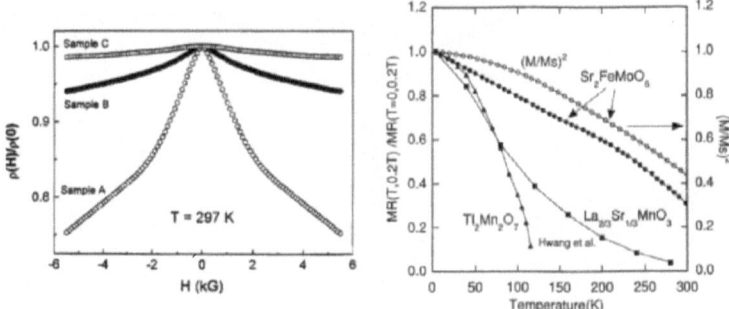

Fig. 22. (a) Resistance versus field curve of Sr_2FeMoO_6 powders. These curves present a low field magnetoresistance behaviour associated to tunnelling across grain boundaries which amplitude depends on the grain size. From [95]. The temperature dependence of the low field magnetoresistance is less pronounced in the case of Sr_2FeMoO_6 than in the case of $La_{2/3}Sr_{1/3}MnO_3$ which evidences a more robust behaviour of this compound in temperature [51]

30% and depends on the grain size [95]. The temperature dependence of this low-field magnetoresistance (Fig. 22b) is less pronounced than the one of $La_{2/3}Sr_{1/3}MnO_3$ powders [51]. The more robust behaviour of SFMO in temperature is encouraging for the use of such compounds for applications at room temperature. Due to growth difficulties faced in the elaboration of Sr_2FeMoO_6 thin films, few experiments have been performed on magnetic tunnel junctions with double perovskite electrodes [96]. The only TMR experiment on magnetic tunnel junctions reported so far in the literature has been performed on $Sr_2FeMoO_6/SrTiO_3/Co$ nanojunctions [97]. The -50% effect observed at $4\,K$ corresponds to a negative spin-polarization of -90% in good agreement with band structure calculations.

4 Insulating Barrier for Magnetic Tunnel Junctions

Materials used for the barrier in magnetic tunnel junctions have been mostly amorphous Al_2O_3. Attempts have been made to use other barriers, AlO_xN_y [98, 99], Ga_2O_3 [100], ZrO_2 [101] and $ZrAlO_x$ [102], in order to decrease the resistance area product for applications[3]. The influence of the barrier appears already in the introduction, with the possibility, depending on the barrier, to obtain positive or negative spin-polarization [35], or depending on the crystallinity to tunnel to selected bands [28, 39]. In this part, we concentrate on a new type of barriers composed of ferromagnetic insulators, which allows the filtering of the electrons depending on their spins. EuS

---

[3] Typical values which are required are $100\,\Omega\mu m^2$ for MRAMs and $0.5\,\Omega\mu m^2$ for sensors with at least 10% of TMR.

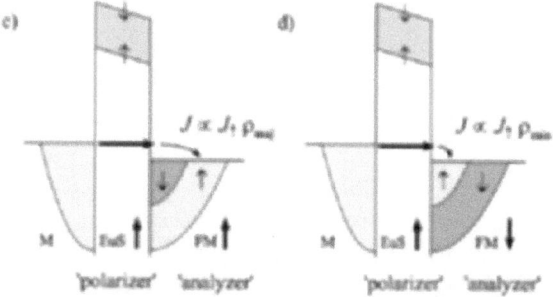

Fig. 23. Schematic representation of the spin filter composed of a nonmagnetic electrode, a ferromagnetic insulating barrier and a ferromagnetic counter-electrode. In the parallel configuration of the magnetization of the barrier and the analyser, a large current is expected due to both the large majority spin current emerging of the barrier and the large density of state at the Fermi level for the majority spins in the counter-electrode. In the antiparallel configuration, the current is smaller. From [105]

[103–105], EuSe [106], EuO [107], $CoCr_2O_4$ [86], $NiFe_2O_4$ [108], $BiMnO_3$ [109] and $La_{0.1}Bi_{0.9}MnO_3$ [110] have been used as ferromagnetic insulators in such Spin Filters. In a ferromagnetic insulator, the conduction bands are spin split by exchange leading to two different barrier heights for spin ↑ and spin ↓ electrons. The combination of a non-magnetic electrode with a ferromagnetic barrier leads to two very different currents for spin ↑ and spin ↓ electrons due to the exponential dependence of the transmission with the barrier height. This bilayer constitutes an artificial half-metal with nearly full spin-polarization. This spin polarization can be checked by a metallic ferromagnetic counter-electrode, which acts as an analyser as represented on Fig. 23. Depending on the magnetic configuration of the magnetization of the barrier and the counter-electrode, the resistance will be small (for parallel) or high (antiparallel) and should result in very large TMR effect. Such an artificial half-metal can also be used to inject spins in semiconductors. In the case of double spin filters, with two ferromagnetic insulating barriers and nonmagnetic electrodes, effects as large as 10^5 % are expected [111].

In order to check this spin filtering concept, Moodera and coworkers performed tunnel experiments with a superconducting counter-electrode and Eu-based tunnel barriers [103, 106, 107]. By combining such a EuS barrier with a Gd ferromagnetic counter-electrode, Leclair et al. obtained a TMR effect larger than 100% at 2 K (Fig. 24) which has disappeared at 30 K as expected from the 16 K T_c of the barrier, since above T_c, spin ↑ and spin ↓ electrons experience the same barrier height and the current emerging from the barrier is no more spin polarized. Additional evidence of the spin filtering effect is obtained from the temperature dependence of the junction resistance (see Fig. 24b). If spin filtering is present, one of the barrier heights (in that case

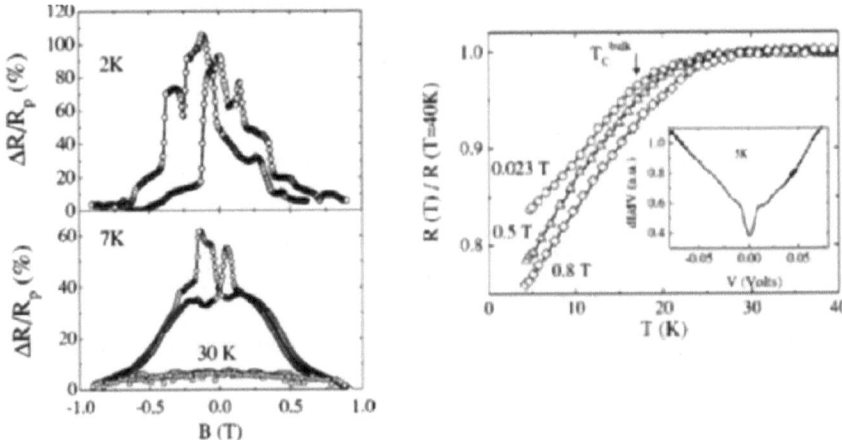

Fig. 24. *Left*: Resistance versus field curves at 2 K, 7 K and 30 K measured on Al/EuS/Gd spin filter (well above the T_c of the EuS barrier). *Right*: temperature dependence of the resistance of a Al/EuS/Gd spin filter. From [105]

spin ↑) should be smaller below T_c than the barrier of either spin ↑ or spin ↓ above T_c. This results in a decrease of the resistance of the junction below T_c as well as an increase of the spin polarization of the current when temperature decreases. A similar spin filtering effect has been obtained in $LaSrMnO_3/La_{0.1}Bi_{0.9}MnO_3/Au$ tunnel junctions, with a TMR ratio of 190% at 4 K [110].

5 Diluted Magnetic Semiconductors

In order to inject spins in semiconductors it is highly desirable to have materials that present ferromagnetic and semiconducting properties both together. Diluted Magnetic Semiconductors combine these two properties. The general idea is to take a semiconductor and to replace some of its atoms by magnetic atoms (for example Mn atoms) in order to induce ferromagnetism. A large number of recent reviews concern these materials [112–117]. In II–VI semiconductors, replacing the II element by a transition element with equivalent valence leads to paramagnetic or spinglass magnetic compounds, because of the lack of carriers to couple the isolated magnetic atoms. Ferromagnetism can be induced by co-doping with Nitrogen (see for instance nitrogen doped $Zn_{1-x}Mn_xTe$ [118]) but the Curie temperature is low (in the order of 1 K). Such paramagnetic semiconductors have been used under large applied magnetic field to inject spin in Spin-LED devices [119]. The spin-LED shown schematically in Fig. 25 enables to determine the efficiency of the injection process. Spin-polarized carriers injected in the AlGaAs/GaAs quantum well recombine in the quantum well. The spin-polarization of the carrier injected

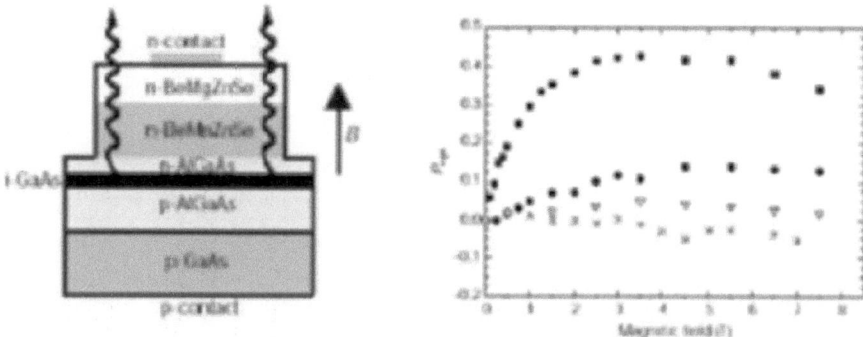

Fig. 25. Schematics of a spin-LED device (*left*). Polarization of the circular light emitted by the diode as a function of the applied magnetic field (*right*). From [119]

from the paramagnetic BeMnZnSe under magnetic field is determined from the degree of polarization of the circular light emitted by the diode. Increasing the magnetic field leads, in that case, to an increase of the spin-polarization of the carrier up to 40%.

Replacing an element of column III in III–V semiconductors by a magnetic ion such as Mn induces ferromagnetism. This results from the 2 valence of the doping transition element. The insertion of a Mn ion (i.e. a localized magnetic moment) corresponds also to the insertion of a hole. This hole mediates the coupling between the localized magnetic moments, producing ferromagnetism. In 1999, Ohno et al. [120] reported on the effective spin injection of carriers from GaMnAs in a InGaAs quantum well of a spin-LED. The temperature dependence of the polarization follows the one of the magnetization up to the Curie temperature ($T_c = 16\,\mathrm{K}$) of the material.

Since the magnetic properties depend on itinerant carriers concentration (holes), they can be modified by playing with carrier density. Several studies (in Mn doped InAs [121], Ge [122] or CdTe quantum wells [123]) have shown that the Curie temperature can be tuned by an electric field or by light. Figure 26 presents the results of Ohno et al. [121] in which a field effect transistor is used to control the hole concentration in the $In_{1-x}Mn_xAs$ ferromagnetic semiconductor channel. In Fig. 26 right, the hall resistance used to measure the small magnetization of the channel is plotted as a function of the magnetic field for different values of the applied gate voltage. For negative values of the gate voltage, holes accumulate in the channel. A clear ferromagnetic response results. Conversely, for positive values of the gate voltage, holes are partially depleted and a paramagnetic response is observed.

Ferromagnetic semiconductors, $Ga_{1-x}Mn_xAs$ in particular, have been inserted in semiconductor-based heterostructures. For instance, one can note experiments on spin dependent tunnelling in magnetic tunnel junctions [124, 125], electrical detection of spin accumulation in quantum wells [126, 127], current induced magnetic switching in magnetic tunnel junctions

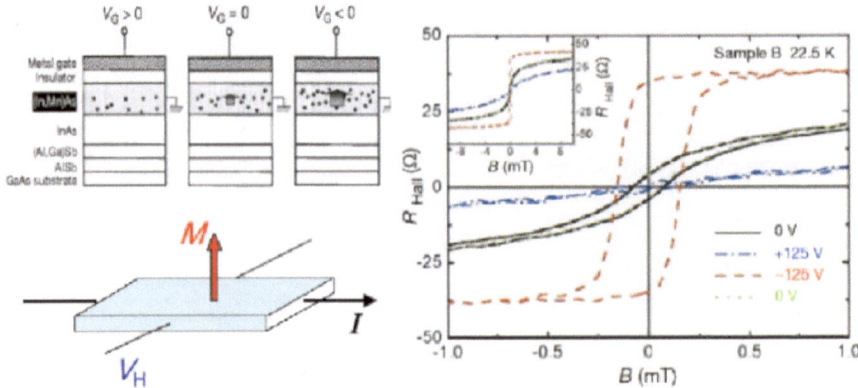

Fig. 26. Schematics of the electrical field control of hole-induced ferromagnetism in $In_{1-x}Mn_xAs$. A gated voltage modifies the hole concentration and thus changes the Curie temperature. Holes are depleted (accumulated) in the channel for a positive (negative) gate voltage resulting in a decrease (increase) of the Curie temperature. In these experiments ferromagnetism is detected by Hall measurements. From [121]

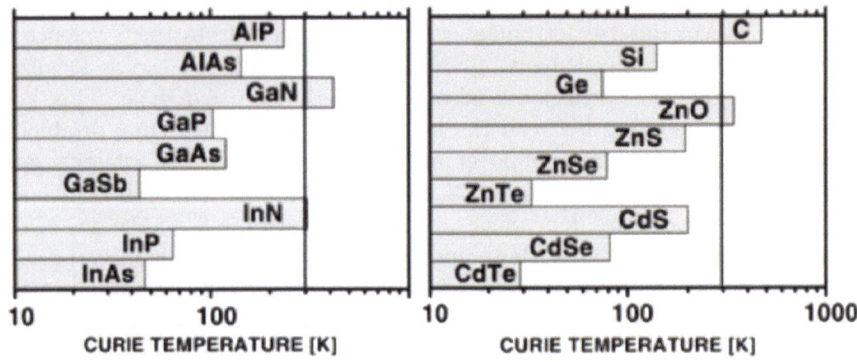

Fig. 27. Curie temperature calculated for III–V, IV and II–VI semiconductors containing 5% of Mn^{2+} and a hole density of 3.5×10^{20} cm^{-3}. From Dietl et al. [130]

[128] or electrical spin injection in light-emitting diodes [129]. The challenge is now to find ferromagnetic semiconductors with Curie temperatures higher than room temperature. In order to define the most attractive materials, T. Dietl et al. have developed a mean field model which considers that ferromagnetism is driven by the exchange interaction between carriers and the localized moment on the Mn sites [130]. The calculated Curie temperature of ferromagnetic semiconductors is proportional to $T_c \propto p^{1/3} N_0 \beta^2 \rho_s$, where p is the hole density, N_0 the cation site concentration, β the $p-d$ exchange integral and ρ_s the density of states. Figure 27 represents the Curie temperature calculated for III–V and II–VI semiconductors with 5% of Mn and 2.5% for IV semiconductors and for a hole density of 3.5×10^{20} cm^{-3}.

The Exchange energy $N_0\beta$ is inversely proportional to the cubic lattice parameter, thus T_c is higher for semiconductors with light anions where the lattice parameter is small. Among III–V semiconductors, nitride and phosphoride are expected to have a high T_c. The second parameter is the density of states $\rho(E_F) = \frac{m^* k_F}{\pi^2 \hbar^2}$, where m^* is the effective mass. Therefore, the Curie temperature is high for a heavy effective mass, i.e. for flat bands. High Curie temperature (near or above room temperature) have been observed in wide band gap semiconductors [131] such as GaN [132] or ZnO [133, 134]. Curie temperatures above room temperature have been also observed in "exotic" semiconductors like chalcopyrites (Mn doped $ZnGeP_2$) or oxides (Co or Fe doped TiO_2, SnO_2 for instance). All these compounds have been widely studied. Up to now, the growth technique and growth conditions have a strong influence on the properties of the films and the exact origin (intrinsic or due to clusters) has to be clarified.

6 New Materials Emerging in the Field of Spintronics

Materials that have coupled electric, magnetic and/or structural order parameters, leading to ferroelectricity, ferromagnetism and ferroelasticity are called multiferroics. This opens the way to control the electric polarization by a magnetic field or the magnetization by an electric field, and to create, for example, ferroelectric memories, which can be magnetically or electrically recorded. The coupling between magnetic and electric orders has been evidenced in $BiMnO_3$ [135] by an anomaly in the dielectric constant at the Curie temperature (105 K) (Fig. 28).

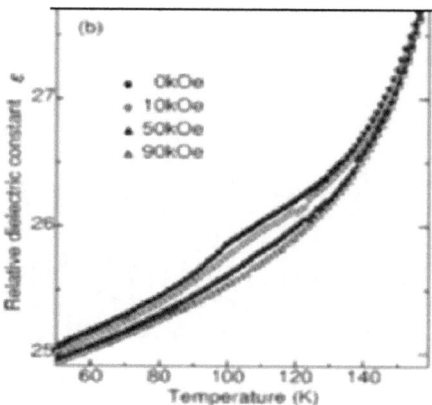

Fig. 28. Temperature variation of the dielectric constant of $BiMnO_3$. An anomaly in the dielectric constant appears at the Curie temperature which evidenced the presence of a magnetoelectric coupling between the magnetic and electric orders [135]

Another manifestation of the coupling between magnetic and electric orders has been obtained on YMnO$_3$ thin films [136]. Spatial maps have been obtained by imaging with optical second harmonic generation. In Fig. 29a, the ferroelectric contribution appears, with $+P$ and $-P$ domains corresponding to white and black regions. In Fig. 29b, the antiferromagnetic domains are evidenced. The interaction between magnetic and electric domain walls leads to a magnetic configuration which is dominated by the ferroelectromagnetic product of the order parameters (see Fig. 29c). Hur et al. [137] report on the electric polarization reversal with an applied magnetic field in TbMn$_2$O$_5$. Results are presented in Fig. 30, evidencing a highly reproducible reversal

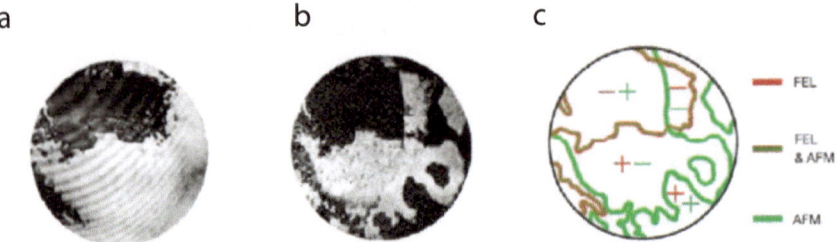

Fig. 29. Coexisting electric and magnetic domains in YMNO3 imaged with second harmonic generation. (**a**) ferroelectric domains. (**b**) Antiferromagnetic domains. (**c**) To a ferroelectric domain wall corresponds always an antiferromagnetic domain wall [136]

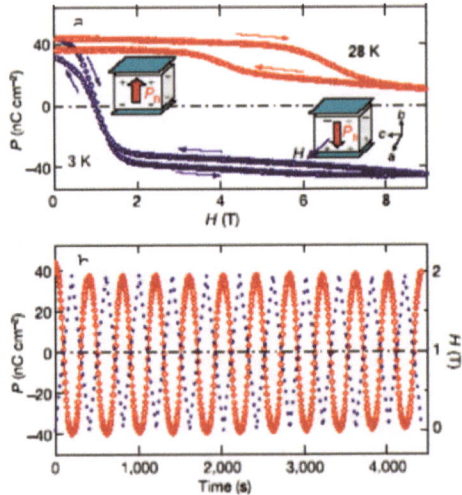

Fig. 30. (**a**) Change of the total polarization of TbMn$_2$O$_5$ with a magnetic field at 3 K and 28 K. (**b**) Polarization flipping at 3 K by linearly varying the magnetic field from 0 to 2 T. A highly reproducible polarization switching is observed [137]

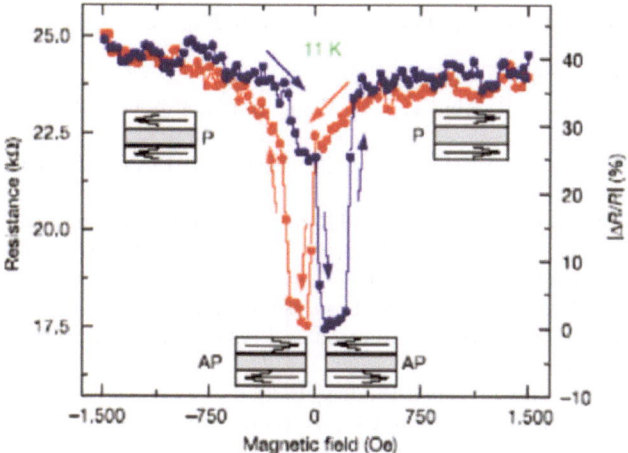

Fig. 31. Magnetoresistance curve of a $La_{0.7}Sr_{0.3}MnO_3/Alq_3/Co$ tunnel junction. From [139]

with no decrease of the amplitude of the electrical polarization after several reversals (Fig. 30b).

Organic materials are also emerging in spintronics. In 2004, two papers report on TMR experiments using such barriers [138,139]. Xiong et al. have performed tunnel experiments on $La_{0.7}Sr_{0.3}MnO_3/Alq_3/Co$ tunnel junctions [139]. An inverse TMR effect, with well-defined antiparallel and parallel states (see Fig. 31), is found. Results are closed to the one found in $La_{0.7}Sr_{0.3}MnO_3/SrTiO_3/Co$ tunnel junctions [35]. Petta et al. report on tunnel experiments in Ni/octanethiol/Ni samples fabricated in nanopores [138]. The dispersion in the results and the bias and temperature dependence suggest that the spin polarized transport is degraded by localized states in the molecular barrier.

References

1. M. Baibich and J.M. Broto and A. Fert and F. Nguyen van Dau and F. Petroff and P. Etienne and G. Creuzet and A. Friedrich and J. Chazeles: Phys. Rev. Lett. **61**, 2472 (1988)
2. G. Binash, P. Grünberg, F. Saurenbach, and W. Zinn: Phys. Rev. B **39**, 4828 (1989)
3. R. Schad, C.D. Potter, P. Beliën, G. Verbanck, V.V. Moshchalkov, and Y. Bruynseraede: Appl. Phys. Lett. **64**, 3500 (1994)
4. W.P. Pratt, S.F. Lee, J.M. Slaughter, R. Loloee, P.A. Schroeder, and J. Bass: Phys. Rev. Lett. **66**, 3060 (1991)
5. A. Barthélémy, A. Fert and F. Petroff: *Handbook of Magnetic Materials*, vol. 12, edited by K.H.J. Bushow (1999), Elsevier p. 1

6. E.Y. Tsymbal and D.G. Pettifor: *Solid State Physics* Vol. 56, (Academic Press 2001)
7. M. Jullière: Phys. Lett. A **54**, 225 (1975)
8. J.S. Moodera, L.R. Kinder, T.M. Wong and R. Mesevey: Phys. Rev. Lett. **74**, 3273 (1995)
9. E.Y. Tsymbal, O.N. Mryasov and P. Leclair: J. Phys.: Condens. Matter **15**, R109 (2003)
10. "Semiconductor Spintronics and Quantum Computation", edited by D.D. Awschalom, D. Loss and N. Samarth (Springer, Berlin 2002)
11. S. Datta and B. Das: Appl. Phys. Lett. **56**, 665, (1990)
12. Vitaly N Golovach and Daniel Loss: Semicond. Sci. Technol. **17**, 355 (2002)
13. B.T. Jonker, S.C. Erwin, A. Petrou and A.G. Petukhov: MRS Bulletin, **740**, October 2003
14. N. Mott: Proc. Roy. Soc. **156**, 368 (1936)
15. J.S. Moodera, J. Nassar and G. Mathon: Annu. Rev. Mater. Sci. **29**, 381 (1999)
16. I.A. Cambell and A. Fert: Ferromagnetic materials p 769, ed. E.P. Wohlfarth; North Holland (Amsterdam)
17. B. Dieny, V.S. Speriosu, J.P. Nozières, B.A. Gurney, A. Vedyayev, N. Ryzhanova: Proc. Nato ARW on *Structure and magnetism in systems of reduced dimensions*, vol. 309, NATO ASI series B: Physics, (Plenum Press, NY 1993), p. 279
18. S.S.P. Parkin: Phys. Rev. Lett. **71**, 1641 (1993)
19. L.H. Chen, T.H. Tiefel, S. Jin, R.B. van Dover, E.M. Gyorgy, R.M. Fleming: Appl. Phys. Lett. **63**, 1279 (1993)
20. H. Kubota, M. Sato, T. Miyazaki: J. Magn. Magn. Mater. **167**, 12 (1997)
21. J.P. Renard, P. Bruno, R. Mégy, B. Bartenlian, P. Beauvillain, C. Chappert, C. Dupas, E. Kolb, M. Mulloy, P. Veillet and E. Velu: Phys. Rev. B **51**, 12821 (1995)
22. S.Y. Hsu, A. Barthélémy, P. Holody, R. Loloee, P.A. Schroeder, A. Fert: Phys. Rev. Lett. **78**, 2652 (1997)
23. C. Vouille, A. Barthélémy, F. Elokan, A. Fert, P.A. Schroeder, S.Y. Hsu, A. Reilly, R. Loloee, W.P. Pratt: Phys. Rev. B **60**, 6710 (1999)
24. P.M. Tedrow and R. Meservey and P. Fulde: Phys. Rev. Lett. **25**, 1270 (1970)
25. P.M. Tedrow and R. Meservey: Phys. Rev. Lett. **26**, 192 (1971)
26. R. Meservey and P.M. Tedrow: Physics Reports **238**, 173 (1994)
27. S.S.P. Parkin et al.: J. Appl. Phys. **85**, 5828 (1999)
28. S.S.P. Parkin, C. Kaiser, A. Panchula, P.M. Rice, B. Hughes, M. Samant and S.H. Yang: Nature Materials **3**, 862 (2004)
29. D.J. Monsma and S.S.P. Parkin: Appl. Phys. Lett. **77**, 720 (2000)
30. J.S. Parker, S.M. Watts, P.G. Ivanov, and P. Xiong: Phys. Rev. Lett. **88**, 196601 (2002)
31. D.C. Worledge and T.H. Geballe: Phys. Rev. B **62**, 447–451 (2000)
32. D.C. Worledge and T.H. Geballe: Phys. Rev. Lett. **85**, 5182 (2000)
33. B.A. Politzer and P.H. Cutler: Phys. Rev. Lett. **28**, 1330 (1972)
34. M.B. Stearns: J. Magn. Magn. Mater. **5**, 167 (1977)
35. J.M. De Teresa, A. Barthélémy, A. Fert, J.P. Contour, F. Montaigne, A. Vaurès: Science **286**, 507 (1999)
36. J.M. de Teresa, A. Barthélémy, A. Fert, J.P. Contour, R. Lyonnet, F. Montaigne, P. Seneor, A. Vaurès: Phys. Rev. Lett. **82**, 4288 (1999)

37. I.I. Oleinik, E.Y. Tsymbal and D.G. Pettifor: Phys. Rev. B **62**, 3952 (2000)
38. I.I. Oleinik, E.Y. Tsymbal and D.G. Pettifor: Phys. Rev. B **65**, 020401 (2002)
39. S. Yuasa, T. Nagahama, A. Fukushima, Y. Suzuki and K. Ando: Nature Materials **3**, 868 (2004)
40. W.H. Butler, X.G. Zhang, T.C. Schulthess, J.M. Mac Laren: Phys. Rev. B **63**, 054416 (2001)
41. G. Schmidt, D. Ferrand, L.W. Molenkamp, A.T. Filip, and B.J. van Wees: Phys. Rev. B **62**, R4790 (2000)
42. E.I. Rashba: Phys. Rev. B. **62** R16267, (2000)
43. A. Fert, and H. Jaffrès: Phys. Rev. B. **64**, 184420 (2001)
44. H. Jaffrès, and A. Fert: J. Appl. Phys. **91**, 8111 (2002)
45. A.T. Hanbicki, B.T. Jonker, G. Itskos, G. Kioseoglou and A. Petrou: Appl. Phys. Lett. **80**, 1240 (2002)
46. R.A. De Groot, F.M. Mueller, P.G. Van Engen and K.H.J. Buschow: Phys. Rev. Lett. **50**, 2024 (1983)
47. J.M.D. Coey and M. Venkatesan: J. Appl. Phys. **91**, 8345 (2002)
48. W.E. Pickett, D.J. Singh: Phys. Rev. B **53**, 1146 (1996)
49. M. Penicaud, B. Silberchiot, C.B. Sommers and J. Kubler: J. Magn. Magn. Mater. **103**, 212 (1992)
50. S.P. Lewis, P.B. Allen and T. Sasaki: Phys. Rev. B **55**, 10253 (1997)
51. K.I. Kobayashi, T. Kimura, H. Sawada, K. Terakura and Y. Tokura: Nature **395**, 667 (1998)
52. D. Singh: Phys. Rev. B **55**, 313 (1997)
53. M.S. Park, S.K. Kwon, S.J. Youn and B.I. Min: Phys. Rev. B **59**, 10018 (1999)
54. I.I. Mazin: Phys. Rev. Lett. **83**, 1427 (1999)
55. M.A. Korotin, V.I. Anisimov, D.I. Komskii and G.A. Sawatsky: Phys. Rev. Lett. **80**, 4305 (1998)
56. K.T. Kamper, W. Schmitt, G. Güntherodt, R.J. Gambino and R. Ruf: Phys. Rev. Lett. **59**, 2788 (1987)
57. R.J. Soulen, J.M. Byers, M.S. Osofsky, B. Nadgorny, T. Ambrose, S.F. Cheng, P.R. Broussard, C.T. Tanaka, J. Nowak, J. Moodera, A. Berry and J.M.D. Coey: Science **282**, 85 (1998)
58. J.S. Parkar, S.M. Watts, P.G. Ivanov and P. Xiong: Phys. Rev. Lett. **88**, 196601 (2002)
59. A. Anguelouch, A. Gupta, Gang Xiao, D.W. Abraham, Y. Ji, S. Ingvarsson and C.L. Chien: Phys. Rev. B **64**, 180408 (2002)
60. J.M. Coey: J. Appl. Phys. **85**, 5576 (1999)
61. Gupta, X.W. Li and Gang Xiao: Appl. Phys. Lett. **78**, 1894 (2001)
62. J.M.D. Coey, M. Viret and S. Von Molnar: J. Phys.: Condens. Matter **12**, L173 (2000)
63. M. Ziese: Rep. Prog. Phys. **65**, pp. 143–249 (2002)
64. C. Zener: Phys. Rev. **82**, 403 (1951)
65. E.A. Livesay et al.: J. Phys.: Condens. Matter **11**, L279 (1999)
66. Park et al.: Nature **392**, 794 (1998)
67. R. Bertacco, M. Portalupi, M. Marcon, L. Duo, F. Ciccacci, M. Bowen, J.P. Contour, A. Barthélémy: J. Magn. Magn. Mater. **242**, 710 (2002)
68. Hwang et al.: Phys. Rev. Lett. **77**, 2041 (1996)
69. Worledge and Geballe: Appl. Phys. Lett. **76**, 900 (2000)
70. M. Viret, M. Drouet, J. Nassar, J.P. Contour, C. Fermon and A. Fert: Europhys. Lett. **39**, 545 (1997)

71. Moon Ho Jo et al.: Phys. Rev. B **61**, R14905 (2000)
72. M. Bibes et al.: Phys. Rev. Lett. **87**, 067210 (2001)
73. M. Bowen, M. Bibes, A. Barthélémy, J.P. Contour, A. Anane, Y. Lemaître, A. Fert: Appl. Phys. Lett. **82**, 233 (2003)
74. T. Obata, T. Manako, Y. Shimakawa and Y. Kubo: Appl. Phys. Lett. **74**, 290 (1999)
75. V. Garcia, M. Bibes, A. Barthélémy, M. Bowen, E. Jacquet, J.P. Contour, A. Fert: Phys. Rev. B **69**, 052403 (2004)
76. V.A.M. Brabers: Handbook of Magnetic Materials, vol.8 (1995), edited by K.H.J. Buschow (North-Holland Publishing Company, Amsterdam)
77. M. Imada, A. Fujimori and Y. Tokura: Rev. Mod. Phys. **70**, 1039 (1998)
78. R.A. de Groot and K.H.J. Buschow: J. Magn. Magn. Mater. **54–57**, 1377 (1986)
79. M. Pénicaud, B. Siberchicot, C.B. Sommers and J. Kübler: J. Magn. Magn. Mat. **103**, 212 (1992)
80. A. Yanase and N. Hamada: J. Phys. Soc. Japan **68**, 1607 (1999)
81. A. Chainanai, T. Yokoya, T. Moromoto, T. Takahashi and S. Todo: Phys. Rev. B **51**, 17976 (1995)
82. Y.Q. Cai, M. Ritter, W. Weiss and A.M. Bradshaw: Phys. Rev. B **58**, 5043 (1998)
83. X. Li, A. Gupta, G. Xiao, W. Qian and V.P. Dravid: Appl. Phys. Lett. **73**, 3282 (1998)
84. P. Seneor, A. Fert, J.L. Maurice, F. Montaigne, F. Petroff and A. Vaurès: Appl. Phys. Lett. **74**, 4017 (1999)
85. S.S.P. Parkin et al., communication at ICF9, San Franscisco, August 2004.
86. G. Hu and Y. Suzuki: Phys. Rev. Lett. **89**, 276601 (2002)
87. J. Navarro, C. Frontera, L.L. Balcells, B. Martiez and J. Fontcuberta: Phys. Rev. B **64**, 092411 (2001)
88. J. Linden et al.: Appl. Phys. Lett. **76**, 2925 (2000)
89. L.L. Balcells et al.: Appl. Phys. Lett. **78**, 781 (2001)
90. M. Besse, V. Cros, A. Barthélémy, H. Jaffrès, J. Vogel, F. Petroff, A. Mirone, A. Tagliaferri, P. Bencok, P. Decorse, P. Berthet, Z. Szotek, W.M. Temmerman, S.S. Dhesi, N.B. Brookes, A. Rogalev, A. Fert: Eur. Phys. Lett. **60**, 608 (2002)
91. Z. Fang, K. Terakura and J. Kanamori: archiv/cond-Mat/0103189 (2001)
92. D.D. Sarma et al.: Phys. Rev. Lett. **85**, 2549 (2000)
93. L.L. Balcells, J. Navarro, M. Bibes, A. Roig, B. Martinez and J. Fontcuberta: Appl. Phys. Lett. **74**, 4014 (2001)
94. K.I. Kobayashi, T. Kimura, Y. Tomioka, H. Sawada, K. Terakura and Y. Tokura: Phys. Rev. B **59**, 11159 (1999)
95. Yuan et al.: Appl. Phys. Lett. **75**, 3853 (1999)
96. M. Besse, F. Pailloux, A. Barthélémy, K. Bouzehouane, A. Fert, J. Olivier, O. Durand, F. Wyczisk, R. Bisaro and J.P. Contour: J. of Crystal Growth **241**, 448 (2002)
97. M. Bibes, K. Bouzehouane, A. Barthélémy, M. Besse, S. Fusil, M. Bowen, P. Seneor, J. Carrey, V. Cros, A. Vaures, J.P. Contour and A. Fert: Appl. Phys. Lett. **83**, 2629 (2003)
98. M. Sharma, J.H. Nickel, T.C. Anthony and S.X. Wang: Appl. Phys. Lett. **77**, 2219 (2000)

99. J. Wang, P.P. Freitas, P. Wei, N.P. Barradas, J.C. Soares: J. Appl. Phys. **89**, 6868 (2001)

100. Z. Li, C. de Groot, J.S. Moodera: Appl. Phys. Lett. **77**, 3630 (2000)

101. J. Wanng, P.P. Freitas, E. Snoeck, P. Wei and J.C. Soares: Appl. Phys. Lett. **79**, 4387 (2001)

102. J. Wang, P.P. Freitas and E. Snoeck: Appl. Phys. Lett. **79**, 4553 (2001)

103. J.S. Moodera, X. Hao, G.A. Gibson and R. Meservey: Phys. Rev. Lett. **61**, 637 (1988)

104. X. Hao, J.S. Moodera and R. Meservey: Phys. Rev. B **42**, 8235 (1990)

105. P. Leclair, J.K. Ha, J.M. Swagten, J.T. Kohlhepp, C.H. Van de Vin and W.J.M. de Jonge: Appl. Phys. Lett. **80**, 625 (2002)

106. J.S. Moodera, R. Meservey and X. Hao: Phys. Rev. Lett. **70**, 853 (1993)

107. T.S. Santos and J.S. Moodera: Phys. Rev. B **69**, R241203 (2004)

108. U. Lüders, M. Bibes, A. Barthélémy, K. Bouzehouane, S. Fusil, J. Bobo and A. Fert: preprint

109. M. Gajek, M. Bibes, A. Barthélémy, K. Bouzehouane, S. Fusil, M. Varela, J. Fontcuberta and A. Fert: preprint

110. M. Gajek, M. Bibes, A. Barthélémy, K. Bouzehouane, S. Fusil, M. Varela, J. Fontcuberta and A. Fert: preprint

111. D.C. Worledge and T.H. Geballe: J. Appl. Phys. **88**, 5277 (2000)

112. F. Matsukura, H. Ohno, and T. Dietl: *Handbook of Magnetic Materials*, Vol. 14, ed by K.H.J. Buschow, (North-Holland, Amsterdam, 2002), pp. 1–87

113. J. König, J. Schliemann, T. Jungwirth, and A.H. MacDonald: *Electronic Structure and Magnetism of Complex Materials*, ed D.J. Singh and D.A. Papaconstantopoulos (Springer Verlag, Berlin, 2003)

114. MRS Bulletin October 2003

115. I. Zutic, J. Fabian, S. Das Sarma: Rev. Mod. Phys. **76**, 323 (2004)

116. S.J. Pearton, W.H. Heo, M. Ivill, D.P. Norton, T. Steiner: Semicond. Sci. Technol. **19**, R59 (2004)

117. W. Prellier, A. Fouchet and B. Mercey: J. Phys.: Condens. Matter **15**, R1583 (2003)

118. D. Ferrand, J. Cibert, A. Wasiela, C. Bourgognon, S. Tatarenko, G. Fishman, T. Andrearczyk, J. Jaroszynski, S. Kolesnik, T. Dietl, B. Barbara and D. Dufeu: Phys. Rev. B **63**, 085201 (2001)

119. R. Fiederling, M. Keim, G. Reuscher, W. Ossau, G. Schmidt, A. Waag, and L.W. Molenkamp: Nature 402, 787 (1999)

120. H. Ohno et al.: Nature **402**, 790 (1999)

121. H. Ohno, D. Chiba, F. Matsukura, T. Omiya, E. Abe, T. Dietl, Y. Ohno, and K. Ohtani: Nature **408**, 944 (2000)

122. Y.D. Park, A.T. Hanbicki, S.C. Erwin, C.S. Hellberg, J.M. Sullivan, J.E. Mattson, T.F. Ambrose, A. Wilson, G. Spanos, and B.T. Jonker: Science **295**, 651 (2002)

123. H. Boukari, P. Kossacki, M. Bertolini, D. Ferrand, J. Cibert, S. Tatarenko, A. Wasiela, J.A. Gaj, and T. Dietl: Phys. Rev. Lett. **88**, 207204 (2002)

124. M. Tanaka, and Y. Higo: Phys. Rev. Lett. **87**, 026602 (2001)

125. R. Mattana, M. Elsen, J.-M. George, H. Jaffrès, F. Nguyen Van Dau, A. Fert, M.F. Wyczisk, J. Olivier, P. Galtier, B. Lépine, A. Guivarc'h, and G. Jézéquel: Phys. Rev. B **71**, 075206 (2005)

126. R. Mattana, J.-M. George, H. Jaffrès, F. NGuyen Van Dau, A. Fert, B. Lépine, A. Guivarc'h, and G. Jézéquel: Phys. Rev. Lett. **90**, 166601 (2003)

127. J.-M. George, H. Jaffrès, R. Mattana, M. Elsen, F. Nguyen Van Dau, A. Fert, B. Lépine, A. Guivarc'h, and G. Jézéquel: Molecular physic reports **40**, 23 (2004)
128. D. Chiba, Y. Sato, T. Kita, F. Matsukura, and H. Ohno: Phys. Rev. Lett. **93**, 216602 (2004)
129. P. Van Dorpe, Z. Liu, W. Van Roy, V.F. Motsnyi, M. Sawicki, G. Borghs, and J. De Boeck: Appl. Phys. Lett. **84**, 3495 (2004)
130. T. Dietl, H. Ohno, and F. Matsukura: Phys. Rev. B **63**, 195205 (2001)
131. S.J. Pearton, C.R. Abernathy, M.E. Overberg, G.T. Thaler, D.P. Norton, N. Theodoropoulou, A.F. Hebard, Y.D. Park, F. Ren, J. Kim, and L.A. Boatner: J. Appl. Phys. **93**, 1 (2003)
132. S. Sonoda, S. Shimizu, T. Sasaki, Y. Yamamoto, and H. Hori: Journal of Crystal Growth **237**, 1358 (2002)
133. K. Ueda, H. Tabata, and T. Kawai, Appl. Phys. Lett. **79**, 988 (2001)
134. K. Rode, A. Anane, R. Mattana, J.-P. Contour, O. Durand and R. LeBourgeois: J. Appl. Phys. **93**, 7676 (2003)
135. T. Kimura, S. Kawamoto, I. Yamada, M. Azuma, M. Takano and Y. Tokura: Phys. Rev. B **67**, 180401 (2003)
136. M. Fiebig, Th. Lottermoser, D. Fröhlich, A.V. Goltsev and R.V. Pisarev: Nature **419**, 818 (2002)
137. N. Hur, S. Park, P.A. Sharma, J.S. Ahn, S. Guha and S.W. Cheong: Nature **429**, 393 (2004)
138. J.R. Petta, S.K. Slater and D.C. Ralph: Phys. Rev. Lett. **93**, 136601 (2004)
139. Z.H. Xiong, Di Wu, Z. Valy Vardeny and Jing Shi: Nature **427**, 821 (2004)

Index

Lecture Notes in Physics

For information about earlier volumes
please contact your bookseller or Springer
LNP Online archive: springerlink.com